AF182232

LAVERN J. WEBER
PHARMACOLOGY

WENNER-GREN CENTER
INTERNATIONAL SYMPOSIUM SERIES

VOLUME 5

MECHANISMS OF RELEASE OF BIOGENIC AMINES

Already published in this series:

OLFACTION AND TASTE *Edited by* Y. Zotterman, 1962.

LIGHTING PROBLEMS IN HIGHWAY TRAFFIC *Edited by* E. Ingelstam, 1963.

THE STRUCTURE AND METABOLISM OF THE PANCREATIC ISLETS *Edited by* S. E. Brolin, B. Hellman and H. Knutson, 1964.

TOBACCO ALKALOIDS AND RELATED COMPOUNDS *Edited by* U. S. von Euler, 1965.

MECHANISMS OF RELEASE OF BIOGENIC AMINES

Proceedings of an International Wenner-Gren Center Symposium held in Stockholm, February 1965

EDITED BY

U. S. VON EULER

S. ROSELL and B. UVNÄS

SYMPOSIUM PUBLICATIONS DIVISION

PERGAMON PRESS

OXFORD · LONDON · EDINBURGH · NEW YORK

TORONTO · PARIS · BRAUNSCHWEIG

Pergamon Press Ltd., Headington Hill Hall, Oxford
4 & 5 Fitzroy Square, London W.1

Pergamon Press (Scotland) Ltd., 2 & 3 Teviot Place, Edinburgh 1

Pergamon Press Inc., 44-01 21st Street, Long Island City, New York 11101

Pergamon of Canada Ltd., 6 Adelaide Street East, Toronto, Ontario

Pergamon Press S.A.R.L., 24 rue des Ecoles, Paris 5e

Friedr. Vieweg & Sohn Verlag, Postfach, 185, 33 Braunschweig, W. Germany

Copyright © 1966
Pergamon Press Ltd

First edition 1966

Library of Congress Catalog Card No. 65-28545

PRINTED IN GREAT BRITAIN BY BLACKIE & SON LTD., BISHOPBRIGGS, GLASGOW

(2684/66)

CONTENTS

INTRODUCTORY REMARKS

B. Uvnäs

Ladies and Gentlemen.

On behalf of the Organizing Committee I would like to extend a most cordial welcome to all participants in this Symposium. We are very pleased that so many have accepted our invitation. We are very happy, too, to see so many of those whose contributions may be regarded as milestones on the road leading to our present knowledge of the biogenic amines.

It is especially gratifying to see in the audience my predecessor, the Grand Old Man of Swedish pharmacology, well known for his scientific achievements as well as for his eloquence, Professor Göran Liljestrand. We regret that illness has prevented two members of the Organizing Committee, Professor Hugo Theorell and Professor Nils-Åke Hillarp, from attending this Symposium. May I suggest that we send them some flowers as a token of our sympathy and esteem.

The rapid expansion of our knowledge covering the biochemistry, physiology and pharmacology of the biogenic amines stems from the brilliant observations of a few. It suffices for me to mention two names only: Henry Dale and Otto Loewi, whose discoveries of the chemical mediation of nerve effects gained them Nobel Prize awards in 1936. Since their original observations on acetylcholine, adrenaline and histamine, the family of the biogenic amines has considerably increased. One of the new family members, noradrenaline, warrants special mention. The recognition by Ulf von Euler of noradrenaline as the adrenergic transmitter is one of the fundamental observations which have guided research on the biogenic amines for two decades.

I could name many other researchers—present and absent—whose work has contributed essentially to our present knowledge. However, I will confine myself to only *one* more. He is not present personally, but he is well known to us by virtue of his classical monograph, *Die Biogene Amine*. The numerous editions of this work have long been an indispensable source of knowledge for many of us. Markus Guggenheim, Emeritus Professor, celebrates his 80th birthday during our Symposium, on Wednesday 24 February to be exact. May I suggest that we cable our congratulations to him.

You may wonder: Why arrange a Symposium around the biogenic amines? These substances have already been the principal topics at a number of Symposia and other scientific gatherings. That may be so, but there is one aspect which has been relatively little discussed. Not only do these amines play an important role as chemical mediators in the peripheral and central nervous system; more and more drugs are found to exert a similar action via the release of amines or by interfering with the release. Release of amines is implicated in functional disturbances characteristic of various diseases. Despite the accumulated evidence which demonstrates the significance of the release of amines in physiology, pharmacology and pathology, we still possess but a fragmentary knowledge of the mechanism of that release.

It is the hope of the Organizing Committee that discussions among specialists—morphologists, biochemists, physiologists, pharmacologists, etc. —will result—if not this week, at least later on—in a better understanding of this fundamental process.

With these introductory remarks I have the pleasure of opening our Symposium. I hope we shall have some rewarding days with a spirited but friendly scientific debate during our working hours and an equally enjoyable though not strenuous, social life in our free time.

CERTAIN CONCEPTS OF CELL STRUCTURE RELEVANT TO RELEASE OF NEUROHUMORAL AGENTS

J. DAVID ROBERTSON
Harvard Medical School, Research Laboratory,
McLean Hospital, Belmont, Massachusetts

I SHOULD like to begin by outlining the development of the unit membrane concept and then go on to tell you how this is applicable to the formulation of a concept of cell structure relevant to the release of agents secreted by cells in general, including, of course, neurohumoral agents of various kinds. Figure 1 is an electron micrograph of a human red blood cell that shows the pattern of structure that I refer to by the term "unit membrane". This is a triple-layered structure measuring somewhere around 75 Å in overall thickness. It consists of a pair of roughly parallel dense lines each about 20 Å thick, separated by a light central zone on the order of 35 Å in thickness. This configuration is now well recognized to be a constant feature of the surfaces of cells, as well as of cell organelles, though about ten years ago when it was first seen it posed a controversial problem. This arose mainly because the structure is not constantly seen, and indeed for a long time was only found in cells that had been fixed with $KMnO_4$. Even in these cells it was not always demonstrable. Sometimes the cell surface appeared as a thin dense line $\sim$50–60 Å thick, of rather fuzzy quality, which was very difficult to define. At intercellular boundaries two thin dense lines were found separated by an interspace $\sim$100–150 Å wide. The thin dense lines were regarded as protein. It was speculated that the light interspace between cells might be filled by lipid and that at the free surfaces of cells outside the thin dense line there might be a variable thickness of lipid.

The problem of deciding whether or not the particular structure defined as the unit membrane was a meaningful one was settled mainly by studies of nerve myelin and its development. I should like now to trace the evolution of these ideas, filling in for you the most significant steps. In general, I shall not illustrate my points with electron micrographs because I have published these elsewhere.[1–5] Diagrams seem sufficient for the purposes of this presentation. Only a few electron micrographs will be used to give a few concrete examples of the type of information being dealt with.

It was postulated about ten years ago that the nerve myelin sheath might in fact be formed as a spiral evolution of a mesaxon like the ones described

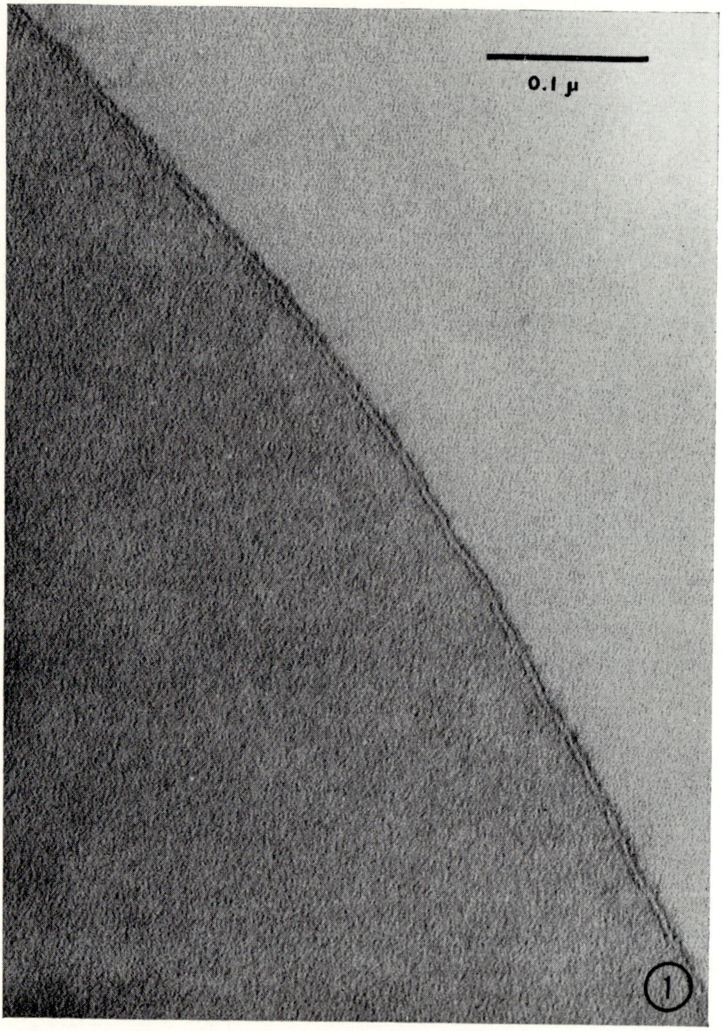

FIG. 1. Electron micrograph of a thin section of a human red blood cell, showing a portion of the unit membrane at its surface. $KMnO_4$ fixed, $\times 280{,}000$.

by Gasser[6–8] in non-myelinated nerve fibers. He had shown that the Schwann cell of unmyelinated axons was not a syncytial cell completely enclosing the axons as previously supposed, but that instead the axons were simply pushed into deep troughs in its surface, with the Schwann cell surface membrane running down around the axons as illustrated in Fig. 2. Dr. Betty Geren Uzman[9] observed certain non-myelinated fibers in developing chick peripheral nerves in which the mesaxon was elongated in a spiral around the axon. She postulated that myelin might be formed by a further elaboration of such a spiral although the evidence at that time was not sufficient to make this entirely clear. The next step came when the first micrograph of an adult, fully matured, myelinated peripheral nerve fiber was obtained[5] in which the spiral nature of the myelin sheath was made evident by the demonstration of both the outer and the inner ends of the mesaxon running off the compact myelin sheath. The original micrograph published in 1955 that showed these mesaxons is reproduced in Fig. 2. The quality of the reproduction demonstrates the deficiencies of the techniques at that time, since the membrane of the Schwann cell appears only as a very fuzzy dense line and it is not at all clear exactly how the surface membrane is related to the compact myelin sheath. This micrograph made it clear that no non-membranous components were laid down between the cytoplasmic surfaces of the mesaxon loops, but it was not at all certain that additional non-membranous constituents were not laid down between the outside surfaces of the spiral loops of the mesaxon as the compact myelin sheath evolved. Somewhat later, in 1957,[4] with the introduction of permanganate fixation,[10] we were able to demonstrate conclusively that compact myelin indeed consisted of nothing but closely packed Schwann cell membrane material with no additional components laid down during its evolution. This was shown by demonstration of the triple layered unit membrane structure at the surface of the Schwann cell, and by tracing it directly into the compact myelin sheath (see ref. 11). Figure 3 is an illustrative micrograph. Not only did the major dense line originate by intimate apposition of the cytoplasmic surfaces of the mesaxon loop; but also the intraperiod line was demonstrated to originate by a similar apposition, or possibly partial fusion, of the outside surfaces of the Schwann cell membranes making up the mesaxon. At this point it became quite reasonable to conclude that the myelin sheath was in fact nothing but cell membrane material. Figure 4 summarizes diagrammatically the information available at that point. With these facts established, it became possible to extrapolate the known information about myelin directly to the Schwann cell surface and make a deduction about its molecular organization. Previous chemical studies had indicated that the myelin sheath consisted in large part of phospholipids and the protein neurokeratin. Polarized light studies by W. J. Schmidt[12] and by Chinn and Schmitt[13] had indicated that the lipid molecules were probably arranged with their long axes in the radial direction,

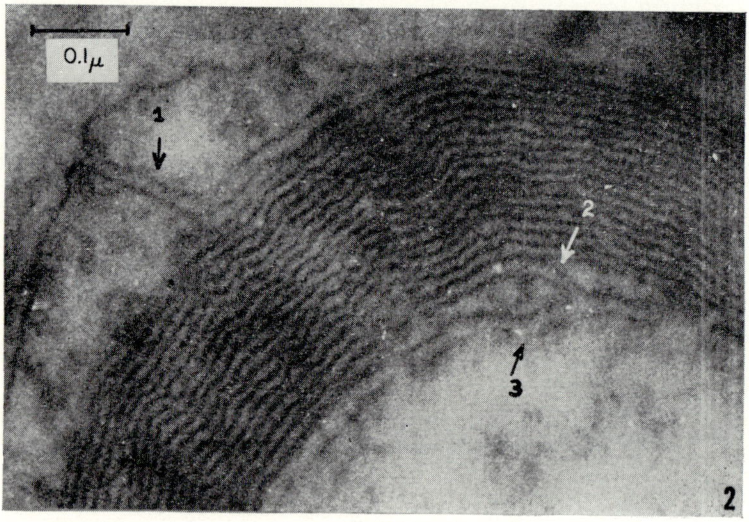

②

FIG. 2. Electron micrograph of a portion of lizard sciatic myelinated nerve fiber, taken in 1954 (1, outer mesaxon; 2, inner mesaxon; 3, axon-Schwann-membrane). OsO_4 fixed, methacrylate embedded. ×160,000.

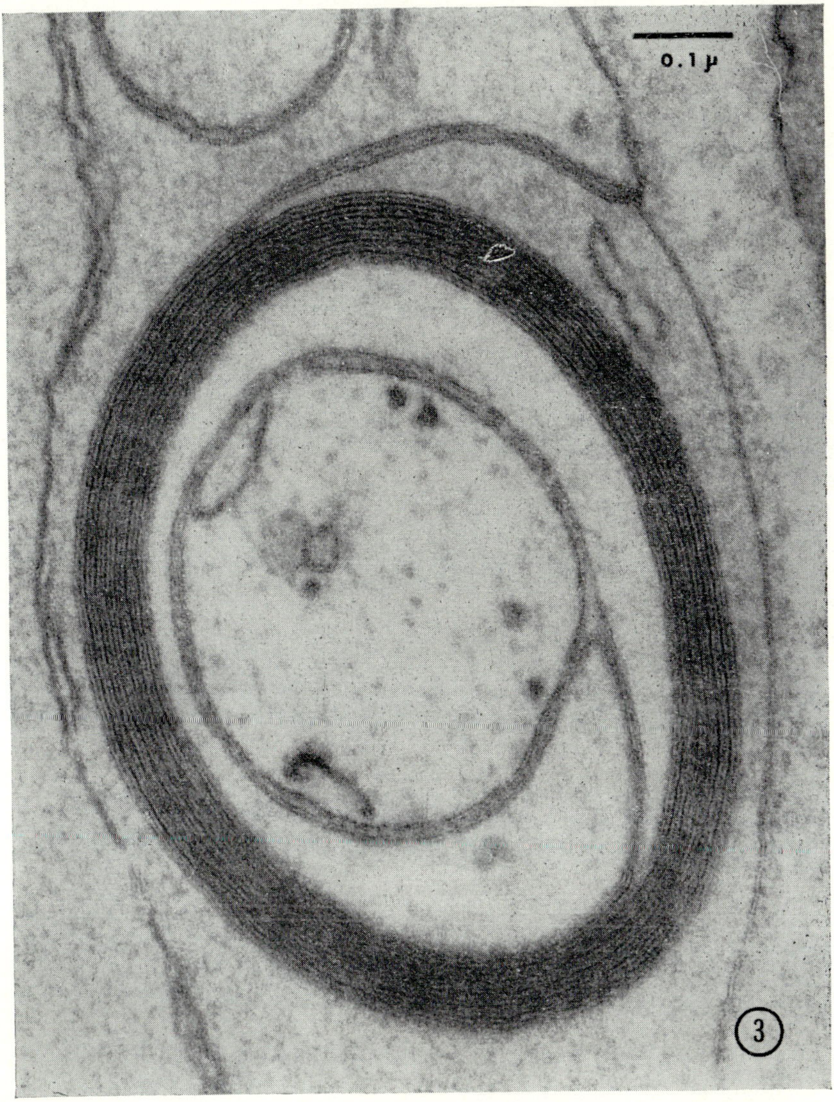

FIG. 3. Portion of a developing mouse sciatic nerve fiber, showing myelin in early stage of development of the mesaxon with the relation to the myelin sheath clearly displayed. Note the two unit membranes making up the mesaxon coming together to make the intraperiod line, and the apposition of the mesaxon with the compact myelin to produce the major dense line. $KMnO_4$ fixed, Araldite embedded. ×88,000.

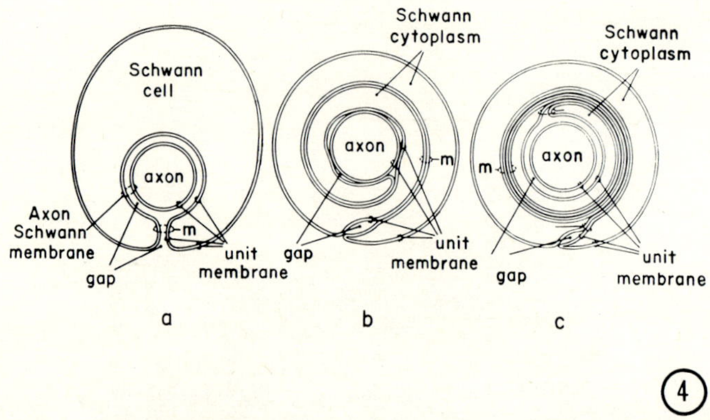

FIG. 4. Diagram showing stages in the evolution of nerve myelin. m, mesaxon.

and the protein molecules tangentially, around the circumference of the nerve fiber. Of course, it was impossible from the polarized light data to decide how thick any of these layers might be. W. J. Schmidt's diagram is reproduced in Fig. 5. In 1935, the first small angle X-ray diffraction patterns were obtained from nerve myelin by Schmitt, Bear and Clark.[14] This established the presence of a radially repeating unit measuring between 171–186 Å in thickness, depending on the species. One of our recent patterns from frog sciatic nerve showing this is illustrated in Fig. 6.

Studies of monolayers of lipids by Langmuir[15] had established the thickness of monomolecular films of certain lipids and X-ray diffraction studies of

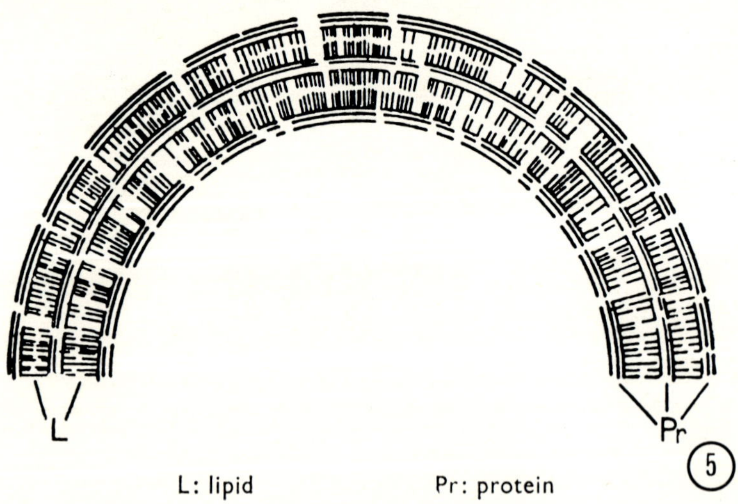

FIG. 5. Diagram from W. J. Schmidt, showing his conception of the organization of the nerve myelin sheath based on his polarized light analyses.

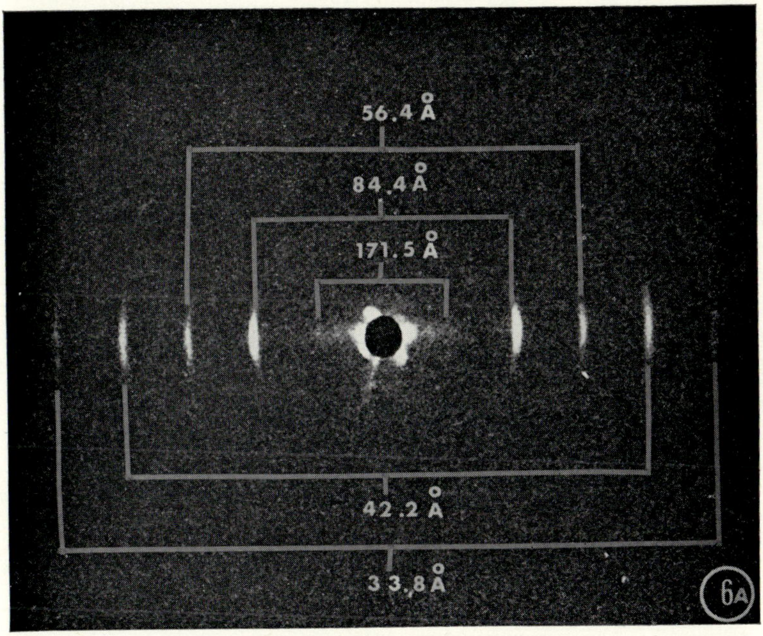

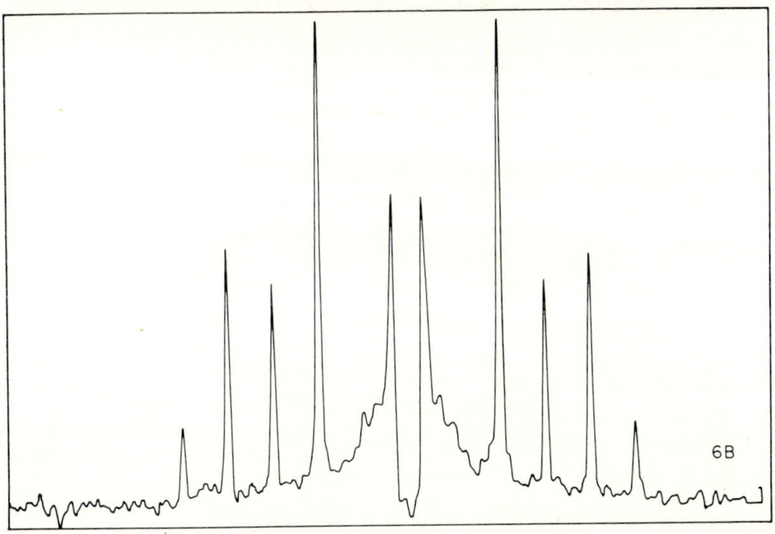

FIG. 6. (a) X-ray diffraction pattern of frog sciatic nerve myelin in Ringer solution at room temperature. This was made with a Jarrell–Ash microfocus X-ray unit and a Franks optical focusing X-ray camera. The scatter around the backstop is artifact. The first order is faintly visible. The much more intense second order is clearer, and the third through fifth orders are visible as indicated. The spacings are as indicated. (b) Microdensitometer tracing across the pattern at a gain of 1:20.

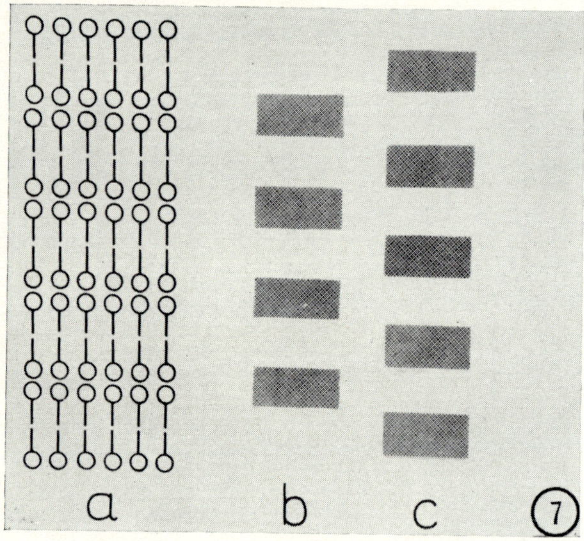

FIG. 7. (a) Diagram of lipid in smectic state. (b) Appearance of electron micrographs of material of this type with the dense layer lines shown produced by the polar heads of the lipid molecules. (c) Alternative possibility with the non-polar carbon chains producing the dense layer lines.

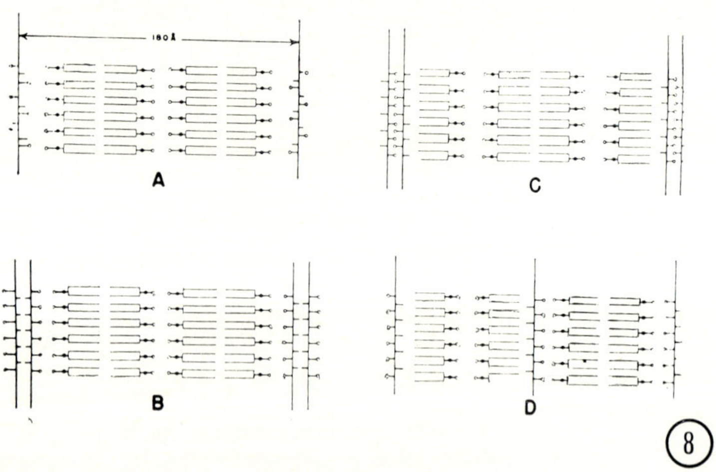

FIG. 8. Four possible molecular organizations of the repeating unit in the radial direction in the myelin sheath, as drawn by Schmitt, Bear, and Palmer in 1941 (18) from a consideration of the X-ray diffraction and polarized light data.

isolated lipids by Finean[16] and others[17] had shown that lipids under certain conditions tend to arrange themselves in layers in the so-called smectic state illustrated in Fig. 7. Here the lipid molecules are each represented by a circle for the polar end of the molecule and a bar for the non-polar carbon chain. Monolayer studies had also established roughly the thickness of monomolecular films of protein at an air–water interface. After consideration of all such data, Schmitt, Bear and Palmer[18] postulated that the organization of the myelin repeating unit might be like one or the other of the diagrams

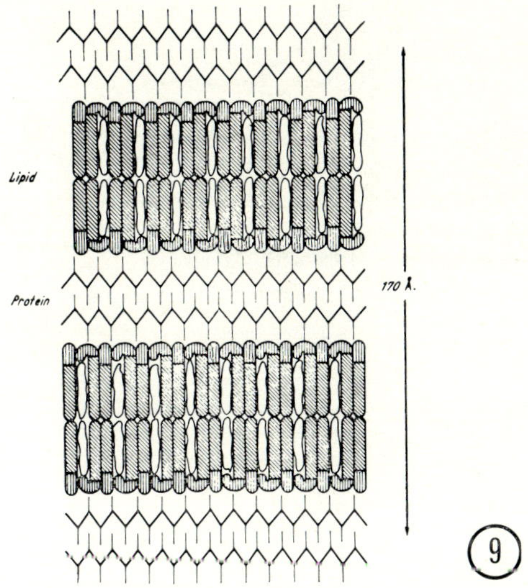

FIG. 9. Finean's conception of the molecular organization of the radially repeating unit in the myelin sheath based on polarized light, X-ray diffraction, and electron microscopical evidence.

illustrated in Fig. 8. In the late 1950's, after the electron microscopical evidence was available, Finean[19] suggested that the most probable arrangement of the repeating unit was as in Fig. 9. However, there was nothing in the evidence which made this pattern in any way a unique one. The polar ends of the lipid molecules could just as well have been drawn the other way around, with the non-polar carbon chains being associated with the protein monolayers, and, indeed, there was nothing to prove that the non-lipid monolayers consisted entirely of protein. The data did, however, show clearly that the non-lipid monolayers must be chemically different from one another in some important way, and furthermore that there must be no more than two monolayers present for each bilayer of lipid. It remained for electron microscope studies of purified lipids and proteins to prove that the pattern of organization in Finean's diagram was indeed essentially correct. I shall return to this point later on.

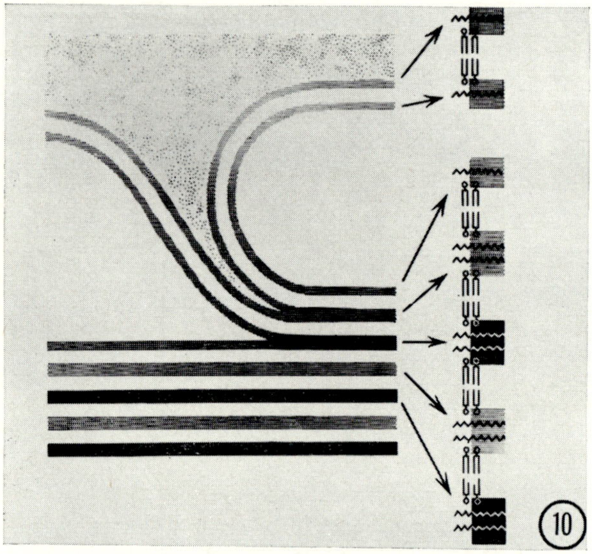

FIG. 10. Diagram in which the X-ray repeating unit is superimposed upon the repeating unit detected by electron microscopy, with extrapolation to the Schwann cell surface to show the general molecular organization of the Schwann cell unit membrane.

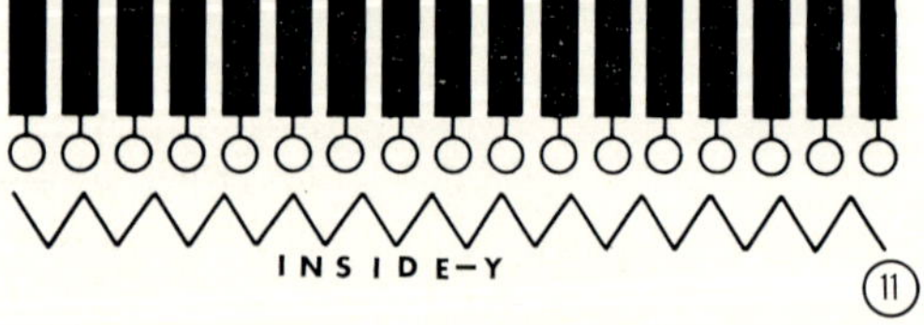

FIG. 11. Highly schematic diagram of the molecular organization of the unit membrane. The circles represent the polar heads of lipid molecules; the bars, non-polar carbon chains. The zig-zag lines indicate non-lipid monolayers with the outer one blocked in to indicate that there is a chemical difference between the outside (X) and inside (Y) surfaces of the membrane.

Let us assume the Finean diagram to be correct. Now it is clear on symmetry grounds alone in the electron micrographs that the mesaxon is the repeating unit. Therefore, we may superimpose the molecular diagram on a diagram of the myelin sheath, as indicated in Fig. 10. One sees then that the Schwann cell surface membrane must consist of a bimolecular layer of lipid with the polar ends directed outward and covered by monolayers of non-lipid as indicated in Fig. 11. It was further evident from the electron microscope data as well as the X-ray diffraction data that the outside surface is chemically different from the inside surface, and there was some reason to suppose that the difference might be due to a high content of polysaccharides in the outer monolayer. One reason for this is the fact that $KMnO_4$ fixes this layer much

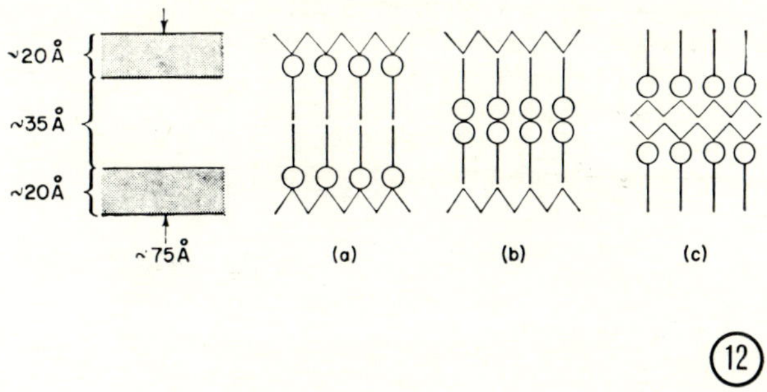

FIG. 12. Diagram showing the observed electron microscope image to the left, with three possible molecular interpretations to the right.

better than OsO_4 and that $KMnO_4$ also tends to fix polysaccharides better than OsO_4. Another reason was the relationship of this layer to basement membrane material and its hydration properties.[20] Nevertheless, this was only a guess and there remained the very serious problem of the lack of uniqueness in the X-ray evidence. I should now like to indicate how this problem was solved.

Figure 12 shows three possible configurations of the unit membrane which could satisfy the X-ray evidence. The decision between these three possible models came from studies of purified lipids and proteins fixed, embedded and sectioned in the same way as the tissues. It was shown that crystalline proteins such as chymotrypsinogen[1] displayed a uniform density in electron micrographs after fixation with either $KMnO_4$ or OsO_4. This confirmed the impression already accepted, that proteins gave a uniform density because it was known that collagen produced a uniformly dense appearance in micrographs as in Fig. 13. In addition, the myosin filaments in muscle appeared as uniformly dense structures. The picture obtained from a fixed, embedded and sectioned specimen of a phospholipid, however, was very different, as indicated in Fig. 14. This shows a specimen of egg cephalin, fixed with OsO_4, embedded, and sectioned. Here one sees a system of regular dense lines alternating with light bands, repeating at a period of about 40 Å. Similar patterns were also obtained with $KMnO_4$-fixed lipids. This band pattern corresponds quite closely to the spacings detected by X-ray diffraction studies of such lipids.[16] One basic problem was to decide whether the dense bands represented the polar ends of the lipid molecules or the non-polar ends, as indicated in the diagram (Fig. 7). This problem was solved in our laboratory in 1958[21] by the demonstration that lipid materials giving the layered pattern

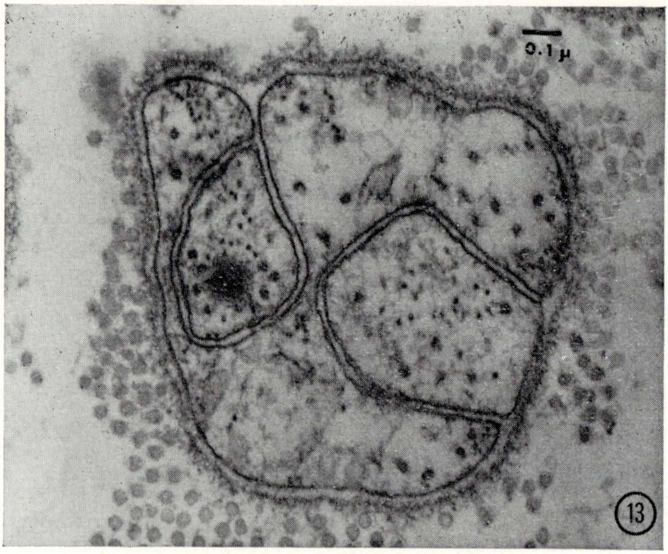

FIG. 13. Cross-section of OsO_4 fixed unmyelinated frog sciatic nerve fiber showing two axons surrounded by one Schwann cell. The axons and Schwann cell contain microtubules as well as filaments. Note the dense collagen fibrils and the heavy basement membrane. ×77,000.

like that illustrated in Fig. 7, when reduced to their minimal structure, i.e. to one layer, showed a pair of dense lines separated by a light zone. This provided good evidence that it was the polar ends of the lipid molecules that were responsible for the densities in the larger aggregates. If the non-polar ends were responsible, one would expect to see for each bimolecular leaflet a single dense line instead of a pair. This point is made by Fig. 15, which shows a specimen of egg cephalin in which the individual bimolecular leaflets separated out during the fixation procedure into separate layers. Here one sees a pair of dense lines for each bilayer. These observations made it possible to choose Fig. 12a as the correct model, for the other two did not fit the observed facts. This agreed with the general arrangement in the X-ray pattern drawn by

FIG. 14. Section of egg cephalin fixed with OsO_4, embedded and sectioned. The dense layers repeated at a period of about 40 Å. ×1,040,000.

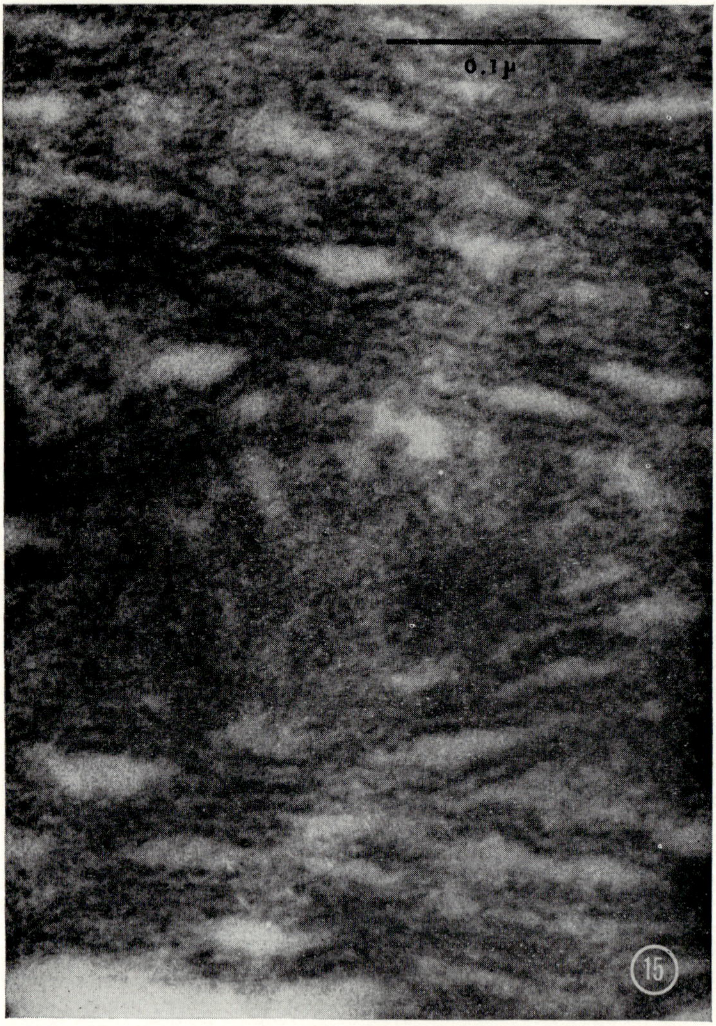

FIG. 15. Section of egg cephalin with individual bimolecular leaflets separated out. These each appear as pairs of dense lines with a light intervening zone. ×36,000.

Finean and it was concluded that the model of the unit membrane structure that was chosen was the correct one.

It should be pointed out that the unit membrane concept only deals with a pattern of organization of the lipid and non-lipid in the membrane. It says nothing about the particular molecular species that are present. The important point is made, however, that the outside and inside surfaces are chemically different in some important way. The membrane is chemically asymmetric. There are definite outside and inside surfaces that can be referred to as the "X" and "Y" surfaces.

After the demonstration of the unit membrane at the surface of the Schwann cell and its characterization in terms of a particular molecular organization, it remained necessary to show that this was a general structure, not just one confined to Schwann cells. This was done by means of an extensive organ survey conducted in our laboratory in the late 1950's.[22] It was found that this pattern could be demonstrated at the surfaces of a wide variety of cells. To be sure, the demonstration of the pattern was by no means constant, when fixatives other than $KMnO_4$ were used. However, it was always possible by the use of proper techniques to demonstrate this particular structure, and it was the only structure to which a particular molecular meaning could be ascribed. Therefore, I concluded that this was the most meaningful structure demonstrable at cell surfaces, and that it was of general significance. This same structure was also demonstrated in all membranous cell organelles, including mitochondria, the endoplasmic reticulum, and nuclear membranes.[22,23] The endoplasmic reticulum was, however, quite erratic in displaying the pattern and often failed to show it even after $KMnO_4$ fixation. However, the important point, I believe, is the fact that it could sometimes be demonstrated. I attribute this erratic fixation feature to differences in the chemical composition of the endoplasmic reticulum membranes in terms of molecular species, but not in the fundamental arrangement. The pattern was particularly well shown in synaptic vesicles, which, after all, are a part of the endoplasmic reticulum. Here one may perhaps say that a closer similarity in chemical composition to cell surface membranes is indicated.

The generality of the unit membrane pattern since its original proposal has been substantiated by extensive studies by a number of investigators who have been able to demonstrate the structure with increasing frequency, using a variety of different fixation methods, and it is now generally accepted that this structure is a basic feature of living membranes.

It should be pointed out that the unit membrane concept of the organization of cell membranes, while resembling the earlier Danielli–Davson model, is based on quite different evidence, having had its origin mainly in direct electron microscope studies. The original Danielli–Davson model[24,25] was deduced on indirect grounds and there was nothing in their original model which would allow any firm statements to be made about the total thickness of the membrane or the total number of molecules present. However, it was guessed that there might be as few as 2–4 lipid molecules, mainly because of the electrical capacitance of some membranes. However, membrane capacitance was found to vary in certain cells, and it was thought that the thickness of the membrane might accordingly be quite variable in terms of the number of molecules present. Perhaps the most important contribution of the electron microscope studies is setting the limit of the total number of molecules present in the membrane to a single bilayer of lipid and two monolayers of non-lipid. Of course, there are other constituents added to membranes in many locations, as, for example, ribosomes in the endoplasmic reticulum and layers of protein as in desmosomes, but the basic structure that is universally present is that defined by the unit membrane. This structure thus appears to be a biological unit of unusual constancy in all living things.

It is a well-known fact that membranes of the endoplasmic reticulum at some time become continuous with the surface membranes of cells and that the nuclear membrane is simply made up of a group of flattened sacs of the endoplasmic reticulum, with which it maintains continuity. Certain components of the sarcoplasmic reticulum of skeletal muscle fibers are now believed to be continuous with the surface membrane. Other instances of continuity between the endoplasmic reticulum and the surface membranes are well known, as, for example, the continuity that exists between the surface membrane of pancreatic acinar cells at the time the zymogen granule is expelled at the surface into the lumen of the pancreatic acinus.

We have also observed continuity between the outer mitochondrial membrane and tubular components of the endoplasmic reticulum as in Fig. 16. This sort of micrograph has led me to postulate that mitochondria may be formed as indicated in the diagram.[26] They might originate by folding of the endoplasmic reticulum as indicated diagramatically in Fig. 17. Of course, this is speculation and by no means proven as yet. The extent to which the various membranes of cells regardless of their origin are maintained in a state of continuity during life is another controversial point on which there is no conclusive evidence at the present time. It may be that connections are continually formed and broken down between the endoplasmic reticulum components such as vesicles and the surface membrane. Be this as it may, there is one point that stands out. Nobody has yet produced a convincing electron micrograph that shows a unit membrane in a cell bounded on both sides by cytoplasm. They always occur in closed loops. I interpret this as evidence of either actual or potential continuity and an expression of the X–Y asymmetry of the membranes referred to above. Another important general feature is the fact that vesicles or tubules of the endoplasmic reticulum are never smaller than ~300 Å. I think there may be a steric reason for this in that below a radius of 150 Å the outer lipid polar heads of the membrane would tend to be pulled apart (see ref. 1). This would be opposed by strong forces.

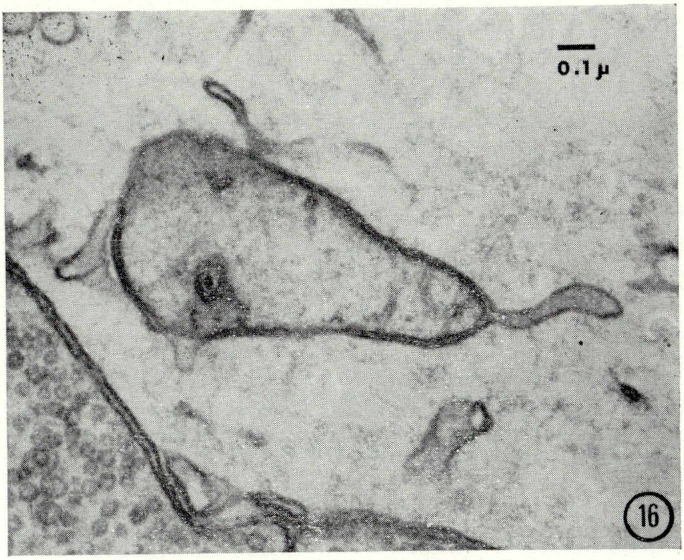

FIG. 16. Micrograph of a mitochondrion showing possible continuity of the outer membrane with a component of the endoplasmic reticulum. ×60,000.

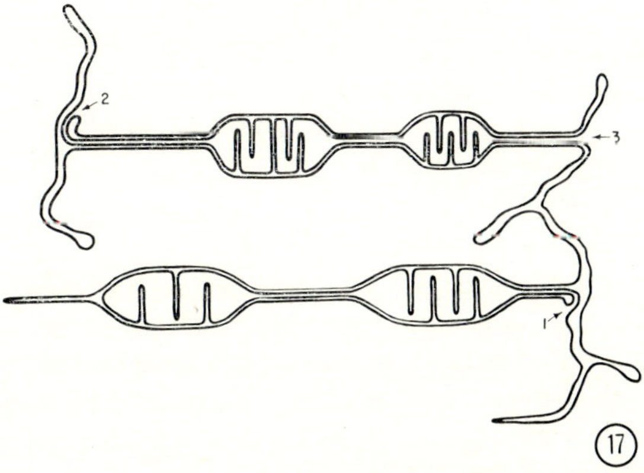

FIG. 17. Diagram of possible origin of mitochondria from endoplasmic reticulum by a folding mechanism.

Perhaps whole vesicles of this size suddenly crystallize out in cytoplasm when new membranes are formed, but I think it more likely that the synthetic machinery is built into the membrane and that new membranes come from membranes with new molecules being added singly or in small groups rather than as large micellar aggregates.

All these considerations, facts, and speculations have led me to advance a unified concept of cell structure as indicated in the diagram in Fig. 18. This diagram says, in essence, that the cell is a three-phase system. First, there is a

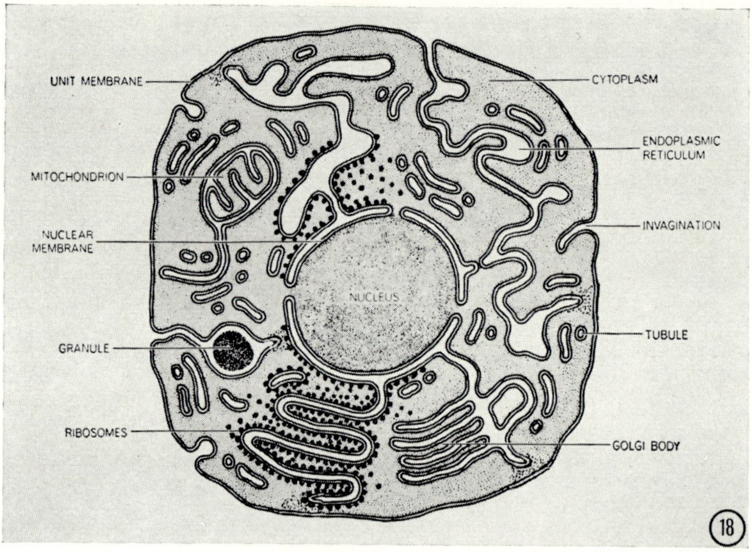

FIG. 18. Theoretical diagram of a cell, showing the three-phase relationship discussed in the text.

nucleo-cytoplasmic matrix phase, which is in continuity throughout and bounded everywhere by an asymmetric membrane that makes up a second phase. This is in turn bounded by a third phase that is at some time in continuity directly or indirectly with outside stuff. The third phase can then be called outside stuff, even though it may lie deep in the cell as, for instance, in the nuclear membrane. The diagram is, of course, an oversimplification and the degree of continuity shown may never exist in a given cell. However, it is conceivable that it does and that during the process of fixation a continuous system breaks down into vesicles and isolated sacs to give the usual picture. We are unable at present to decide this issue, but there are reasons to believe that this may happen. There is a good illustration of this in the sarcoplasmic reticulum of muscle fibers. In 1955[27] I observed an extremely regular array of sarcoplasmic reticulum tubules in lizard muscle, located regularly at the A.I. junction between the myofibrils.[28] I believed that this system might in fact be continuous with the surface membrane of the muscle cell, and that the action potential occurring at the surface of the cell might spread throughout the muscle fibers so as to cause synchronous contraction of the myofibrils. This, I believed, would resolve the difficulty that Hill[29, 30] pointed to when he showed that it was impossible for any substance released at the muscle surface to diffuse fast enough to cause contraction of the deeper myofibrils in synchrony with the more peripheral ones. I, therefore, postulated that there was in fact a continuity that was not demonstrable at that time. Huxley and Taylor had shown in 1955[31] that a micro-electrode placed over an "I" band in frog muscle could cause a local contraction involving both "I" bands of the nearest surface myofibril. Following up my observations, Huxley and Straub[23] tested lizard muscle and found that a micro-electrode placed over an A–I junction caused local contractions of a single "I" band. They attributed the local contractions in frog muscle to the triads described by Porter and Palade in 1957[33] at the "Z" line in frog and in lizard muscle to the similar endoplasmic reticulum components at the A.I. junctions described in my earlier paper.[28] Continuity of the endoplasmic reticulum membranes with the surface membrane was implied by all this, but it was not until rather recently that such continuity has been demonstrated by the use of improved fixation methods.[34–36] It is now generally accepted that at least some components of the sarcoplasmic reticulum are continuous with the surface of the muscle fiber in life and that the electrical potential changes occurring at the surface probably spread into the sarcoplasmic reticulum. An important point here is the implication that failure to see such interconnections of membrane in cells is not very good evidence that they do not exist in life, for it seems that membranes tend to break up into vesicles spontaneously—perhaps during fixation. Thus, for instance, we should say that we simply do not know whether synaptic vesicles are derived by breakdown of a continuous tubular system during fixation, with continuity being maintained with the surface of the cell during life, or whether the vesicles exist as we see them in nerve endings as in Fig. 19, as isolated vesicles. Nevertheless, whether the continuity is maintained during life or is something that is set up actively at the time of release of a secretory product as in a pancreatic acinar cell or at the time of ingestion of a phagocytic granule, is really not of crucial importance to the unit membrane concept or the unified cell theory, because the main point is that the asymmetry of the membrane is maintained and that all membranes of the cell are potentially continuous with one another. The crucial point is that membranes are never formed as isolated structures in the cytoplasm bounded on both sides by cytoplasm. The essential polarity of the membrane is always maintained.

Pinocytosis, secretion, and the release of biogenic amines all become understandable in the following terms. In pinocytosis, material may be taken into the cell by the formation of an invagination of the surface membrane, transported into a particular region of the endoplasmic reticulum where certain enzymes are located which can break the material down into small molecules which can pass across the membrane into the cytoplasmic matrix. How the material gets from the surface of the cell to the region at which it is acted upon is not yet clear. It may occur by some kind of microperistalsis occurring in continuous channels of the endoplasmic reticulum, or it might

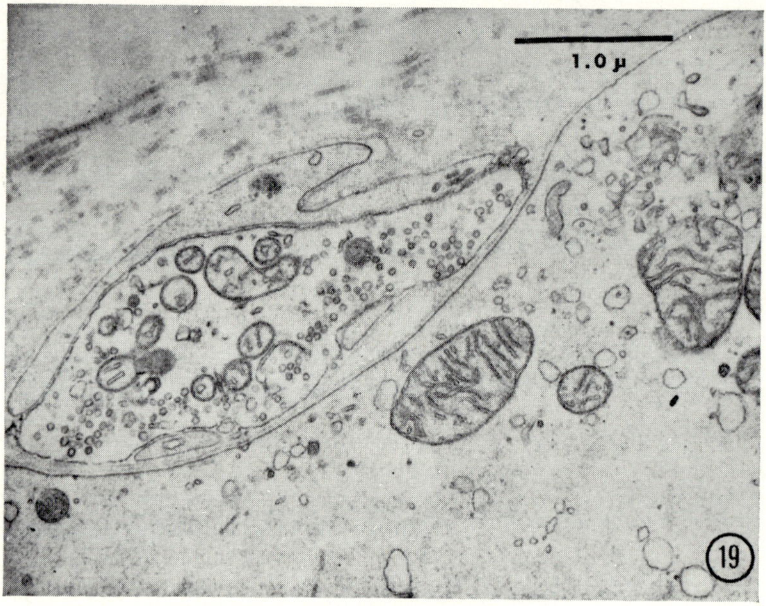

FIG. 19. Electron micrograph of a frog motor end plate on skeletal muscle. Note the numerous vesicles in the ending with moderate numbers of mitochondria. ×27,000.

occur by movement of isolated vesicles acting in a sense as carriers. However, it seems well established that the membrane always separates the material from the cytoplasmic matrix and that the material gets into the cytoplasmic matrix only by passage through the membrane. Conversely, secretion of whatever kind involves the formation of material in a particular part of the endoplasmic reticulum where the correct synthetic mechanisms are built into the membrane. The material is probably formed on the "Y" surface of the membrane, passes through to the "X" surface and is segregated in a cavity of the endoplasmic reticulum from which it then moves to the surface of the cell either through a continuous system by microperistalsis or by means of isolated vesicles which find their way by some unknown mechanism to the correct portion of the cell surface. The release of the material may occur as a result of depolarization of the surface membrane or some influence exerted on the surface membrane by some effector agent. The theory thus offers a unifying concept of cell structure relating to these various problems of cell function.

Before closing, I should say that the continuous smectic bilayer concept contained in the original theory might have to be modified on the basis of current studies which have suggested that at some times and in some locations there may be a micro-spherical phase transformation of the lipid core of a kind found in certain purified lipids by Luzzati and Husson[37] by X-ray diffraction methods and Stoeckenius by electron microscopy.[38] We have observed appearances in membranes, as in the synaptic disc of Mauthner cell club endings, illustrated in Figs. 20 and 21, which are interpretable in this way. Figure 21 shows a diagram in which the appearances of the synaptic discs are illustrated. The vague transverse densities traversing the light central zones of the membranes could be due to the conversion of the continuous smectic bilayer into a system of closely packed lipid microspheres with the lipid carbon chains radially oriented. We, as well as Sjöstrand and others [39,40] have observed such transverse densities in a variety of different membrane systems, but I believe that we are dealing with some kind of artifact. I should like to indicate my reasons for this belief.

The kind of globular substructure in question is very prominent in retinal rod outer segments. One would expect that if any such structure exists in the living membrane of the rod outer segment, it should be demonstrable by a combination of polarized light and X-ray diffraction studies. We are now doing such studies and have obtained what we believe to be the first X-ray diffraction patterns from fresh, unfixed retinal rod outer segments (Fig. 22). Here we have found two orders of a 300 to 320 Å lamellar repeat in the direction of the rod axis.[41, 42] However, our patterns show no evidence of any microspherical organization of the lipid in the planes of the membranes. If this were present, we should have detected it and I am forced, therefore, to conclude that on present evidence the globular appearance of the membranes represents some kind of artifact. However, this is a matter that requires further analysis and study before a definite conclusion can be reached. If such phase transformations in the membrane lipids actually occur in life, then the implications for the transport of materials through the membrane are obviously of great importance. However, if such a change of state occurs, it must be of fleeting nature and restricted location to account for the negative X-ray evidence. At the present time, I believe that the evidence is not in favor of such an organization as a feature of living membranes. I believe that we are dealing with some form of fixation or image artifact in our electron microscope preparations. Therefore, I believe that the original smectic bilayer arrangement of the lipid core of the unit membrane remains the only one which is substantiated by evidence from electron microscopy, polarized

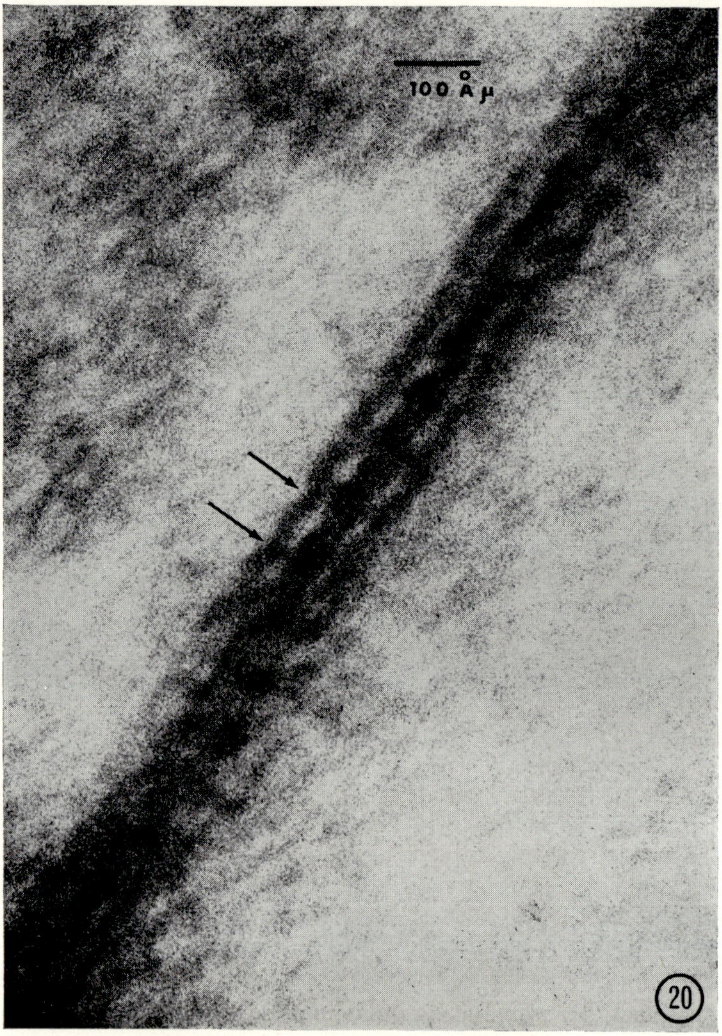

FIG. 20. High magnification micrograph of a synaptic disc in a Mauthner cell club ending of goldfish medulla fixed with $KMnO_4$ and stained with lead. Note the scalloping of the cytoplasmic surfaces as indicated by the arrows and the beading of the central dense line opposite the scalloped indentations. There is a hazy density running from the indentations to the beads.

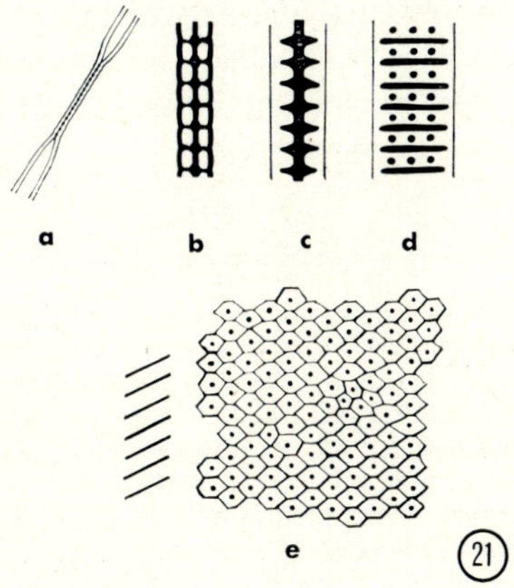

FIG. 21. Diagram of structures observed in synaptic discs in Mauthner cell club endings (a) shows the appearance of the synaptic discs at low power; (b) shows a high-power view of the central region. Compare with Fig. 20. (c) Appearance of the synaptic disc slightly tilted. (d) Appearance of the disc at a greater degree of tilt. (e) Frontal view of the disc showing an hexagonal array of subunits repeating at a period of about 80 Å as indicated by the lines to the left. Each facet consists of a dense border and central dense spot. The lattice is hexagonal, but there appear to be a few pentagonal facets interspersed. The facets are believed produced mainly by the apposed outside surfaces of the two unit membranes. This conclusion is reached mainly because of the characteristics of the image during tilt. The facet's structure is referred to as granulo-fibrillar substructure. The scalloping and vague transverse densities in (b) might indicate a micellarphase transformation in the membrane lipids, or could equally well represent some kind of optical image superimposition artifact. It is referred to as globular fine structure.

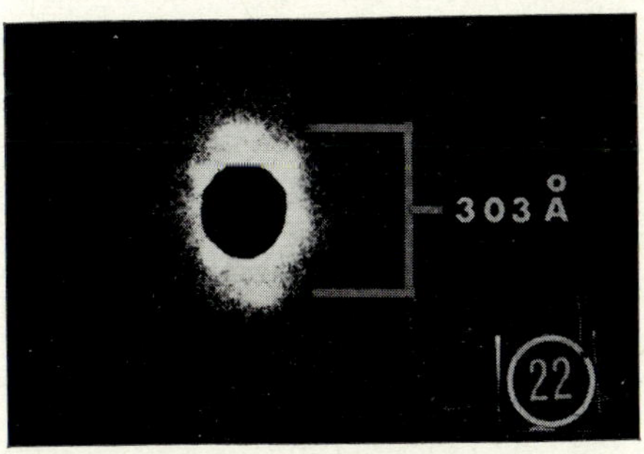

FIG. 22. X-ray diffraction pattern of fresh retinal rods *in situ* in Ringer solution. The rods were oriented vertically and the two spots marked in the print are due to the first order meridional reflections. These occur at 303 Å in this case and are clearly due to the spacing of the lamellae. It is important that no equatorial diffractions such as would be expected from any such globular lipid micellar phase transformation as discussed in the text are detected. These would be expected because of the reciprocal Bragg relationship if there were any such order in the system. The central scatter around the backstop is artifact. ×18.

light, and X-ray diffraction analysis, as well as by chemical information. Until some positive evidence from some source other than electron microscopy is forthcoming, I am not willing to accept such an arrangement of the lipid as a feature of living membranes.

REFERENCES

1. J. D. Robertson, Unit membranes: a review with recent new studies of experimental alterations and a new subunit structure in synaptic membranes, in *Cellular Membranes in Development, Proceedings of the XXII Symposium of the Society for the Study of Development and Growth* (Edited by Dr. M. Locke), Academic Press.
2. J. D. Robertson, Electron microscopy of the motor end plate and the neuromuscular spindle, *American Journal of Physical Medicine* **39**, 1–43 (1960).
3. J. D. Robertson, The molecular structure and contact relationships of cell membranes, in *Progress in Biophysics* (Edited by B. Katz and J. A. V. Butler), Pergamon Press, pp. 343–418 (1960).
4. J. D. Robertson, New observations on the ultrastructure of the membranes of frog peripheral nerve fibers, *J. Biophys. Biochem. Cytol.* **3**, 1043–8 (1957).
5. J. D. Robertson, The ultrastructure of adult vertebrate peripheral myelinated fibers in relation to myelinogenesis, *J. Biophys. Biochem. Cytol.* **1**, 271–8 (1955).
6. H. S. Gasser, Discussion in *The Hypothesis of Saltatory Conduction* (Edited by B. Frakenhaeuser), *Cold Spr. Harb. Symp. Quant. Biol.* **17**, 32–6 (1952).
7. H. S. Gasser, Properties of dorsal root unmedullated fibers on the two sides of the ganglion, *J. Gen. Physiol.* **38**, 709–28 (1955).
8. H. S. Gasser, Olfactory nerve fibers, *J. Gen. Physiol.* **39**, 473–96 (1956).
9. B. Geren Uzman, The formation from the Schwann cell surface of myelin in the peripheral nerves of chick embryos, *Expt. Cell Res.* **7**, 558–62 (1954).
10. J. H. Luft, Permanganate—a new fixative for electron microscopy, *J. Biophys. Biochem. Cytol.* **2**, 799–801 (1956).
11. J. D. Robertson, The unit membrane, in *Electron Microscopy in Anatomy, Proceedings of a Symposium held by the Anatomical Society of Great Britain on the Ultrastructure of Cells* (Edited by J. D. Boyd, F. R. Johnson and J. D. Lever), Edward Arnold Ltd. London, pp. 74–99 (1961).
12. W. J. Schmidt, Doppelbrechung und Feinbau der Markscheide der Nervenfasern, *Z. Zellforsch u. Mikr. Anat.* **23**, 657–76 (1936).
13. P. Chinn and F. O. Schmitt, On the birefringence of nerve sheaths as studied in cross-sections, *J. Cell Comp. Physiol.* **9**, 289–96 (1937).
14. F. O. Schmitt, R. S. Bear and G. L. Clark, X-ray diffraction studies on nerve, *Radiology* **25**, 131–51 (1935).
15. I. Langmuir, The constitution and fundamental properties of solids and liquids, II. Liquids, *J. Amer. Chem. Soc.* **37**, 1848–1905 (1917).
16. J. B. Finean, X-ray diffraction studies on the polymorphism of phospholipids, *Biochim. Biophys. Acta* **12**, 371–84 (1953).
17. J. W. McBain, *Colloid Science*, Heath and Company, Boston, 1950.
18. F. O. Schmitt, R. S. Bear and K. J. Palmer, X-ray diffraction studies on the structure of the nerve myelin sheath, *J. Cell Comp. Physiol.* **18**, 31–41 (1941).
19. J. B. Finean, Recent ideas on the structure of myelin, in *Biochemical Problems of Lipids*, Proc. 2nd Int. Conf., Univ. Ghent, 27–30, July 1955, pp. 127–31, Butterworths, London (1956).
20. J. D. Robertson, Structural alterations in nerve fibers produced by hypotonic and hypertonic solutions, *J. Biophys. Biochem. Cytol.* **4**, 349–64 (1958).
21. J. D. Robertson, A molecular theory of cell membrane structure, in *Proc. Fourth Internat. Conf. on Electron Microscopy*, Berlin, 1958, pp. 159–71, Springer (1960).
22. J. D. Robertson, The ultrastructure of cell membranes and their derivatives, *Biochem. Soc. Symposia* **16**, 3–43 (1959).

23. J. D. Robertson, A new unit membrane organelle of Schwann cells, in *Biophysics of Physiological and Pharmacological Action* (Edited by A. M. Shanes), Pub. No. 69 of the American Association for the Advancement of Science, pp. 63–96, Washington (1961).
24. J. F. Danielli and H. A. Davson, A contribution to the theory of permeability of thin films, *J. Cell Comp. Physiol.* **5**, 495–508 (1935).
25. H. Davson and J. F. Danielli, *The Permeability of Natural Membranes*, Cambridge University Press (1943).
26. J. D. Robertson, Cell membranes and the origin of mitochondria, in *Regional Neurochemistry, Proceedings of the Fourth Neurochemical Symposium* (Edited by S. S. Kety and J. Elkes), Pergamon Press, pp. 497–530, Oxford (1961).
27. J. D. Robertson, The ultrastructure of skeletal muscle, Abstract of paper presented at the Scientific Session of the 44th Ann. Meeting of the Int. Assoc. of Medical Museums, Hotel Shamrock, Houston, Texas, 5 April, 1955.
28. J. D. Robertson, Some features of the ultrastructure of reptilian skeletal muscle, *J. Biophys. Biochem. Cytol.* **2**, 369–80 (1956).
29. A. V. Hill, On the time required for diffusion and its relation to processes in muscle, *Proc. Roy. Soc. London*, Series B, **135**, 446–53 (1948).
30. A. V. Hill, The abrupt transition from rest to activity in muscle, *Proc. Roy. Soc. London*, Series B, **136**, 399–420 (1949).
31. A. F. Huxley and R. E. Taylor, Function of Krause's membrane, *Nature* **176**, 1068 (1955).
32. A. F. Huxley and R. W. Straub, Local activation and interfibrillar structures in striated muscle, *J. Physiol.* **143**, 40–1P (1958).
33. K. R. Porter and G. E. Palade, Studies on the endoplasmic reticulum, III. Its form and distribution in striated muscle cells, *J. Biophys. Biochem. Cytol.* **3**, 269–300 (1957).
34. L. D. Peachey and K. R. Porter, Intracellular impulse conduction in muscle cells, *Science* **129**, 721–2 (1959).
35. D. S. Smith, The structure of insect fibrillar flight muscle. A study made with special reference to the membrane systems of the fiber, *J. Biophys. Biochem. Cytol.* **10**, No. 4, Suppl., 123–58 (1961).
36. Clara Franzini-Armstrong and K. R. Porter, Sarcolemmal invaginations constituting the T system in fish muscle fibers, *J. Cell Biol.* **22**, 675–96 (1964).
37. V. Luzatti and F. Husson, The structure of the liquid crystalline phases of lipid–water systems, *J. Cell Biol.* **12**, 207–20 (1962).
38. W. Stoeckenius, Some electron microscopical observations on liquid crystalline phases in lipid–water systems, *J. Cell Biol.* **12**, 221–9 (1962).
39. F. S. Sjöstrand, A new repeat structural element of mitochondrial and certain cytoplasmic membranes, *Nature* **199**, 1262–4 (1963).
40. S. E. G. Nilsson, Receptor cell outer segment development and ultrastructure of the disk membranes in the retina of the tadpole, *J. Ultrastruc. Res.* **11**, 581–620 (1964).
41. J. D. Robertson, Current problems of unit membrane structure and substructure, In *Intracellular Membrane Structure*, Eds. Seno, S. and Cowdry, E. V., *Okayama Jap. Soc. Cell Biol.* pp. 379–433 (1965).
42. J. D. Robertson, In *Nerve as a Tissue*, Ed. K. Rodahl, Hoeber Med. Div. Harper & Row, Chap. II, in press.

DISCUSSION

Pletscher: Do you have an idea on the junction of sympathetic nerve endings and smooth muscle fibers?

Robertson: I cannot say anything about sympathetic nerves in relation to smooth muscle. However, I might mention the fact that adjacent smooth muscle cells show some regions of close contact where the gap between the membranes is closed as in an electrical synapse. Dewey and Barr call such contact a "nexus" and believe it serves to effect electrical transmission.

UVNÄS: What function do you ascribe to the vesicles in an electrical synapse?

ROBERTSON: There is one thing that seems to apply to all nerve terminals. They all seem to show vesicles. Perhaps they are there because of some such general process as axoplasmic flow. However, in some endings they are very much more prominent. While they do occur in electrical endings they are not so abundant. In fact there is a relative paucity of vesicles.

BROWN: Do you find vesicles in nerve endings which are not in contact with their effector cell? I am thinking particularly of regenerating nerves.

ROBERTSON: Vesicles occur at nodes of Ranvier, but I have no knowledge about their occurrence in regenerating fibres. However, Weis has shown that if you put a constricting band round a nerve, vesicles accumulate on the central side. It is also interesting that Dr. Bischoff has found in my laboratory that chicken nerves show marked accumulations of vesicles near nodes of Ranvier after poisoning with tri-ortho-cresyl-phosphate.

HOLMSTEDT: In how many places do you see synaptic vesicles postsynaptically?

ROBERTSON: They are seen in the crayfish median-giant-to-motor synapse and the lateral giant septal synapses. Sometimes one sees them in fair numbers postsynaptically here, but they are not really a prominent feature. I should add that in both these synapses one sometimes sees a row of rather regularly spaced vesicles lined up along both the pre- and postsynaptic membranes. Strictly speaking I might add that one can find the odd vesicles in almost any cell and so one can find a few vesicles in almost any postsynaptic region.

PEKKARINEN: Have you observed any changes in the structure of the synaptic membrane after neurogenic or electrical stimulation or after central nervous system acting drugs, e.g., neuroleptics, psychosedatives or antiolytic drugs?

ROBERTSON: I studied the crayfish median-giant-to-motor synapse after electrical stimulation in 1951. However, at that time techniques were so poor that I could draw no definite conclusions. It is timely now to restudy this problem and include the effects of drugs.

CEDERGREN: 1. I wonder Dr. Robertson if you in your studies of the structure of central synapses have observed other components than the so-called synaptic vesicles, which might be of importance for the localization of the bioamines, for example structures similar to or identical with features observed in peripheral sensory nerve endings, such as small (about 200 Å or less) heavily stained particles, heavily stained large particles of different appearance, a characteristic network organization of small tubular elements, etc.?

2. What was the significant structure in the synapse you showed us as an example of an "electrical synapse"? The synaptic membranes were partly in close apposition, partly separated by a small space (200 Å). Material having cell membranes organized close to each other do generally show, if preserved by permanganate or aldehyde, either close apposition of the cell membranes or cell membranes separated by a small space.

3. You said you don't believe in electron micrographs. In your electron micrographs of membrane structure you showed us structures that you identified as globular structures· You do not accept the interpretation given by certain other investigation, but you admit that globular structures are present. What is your opinion about the structure? Do you consider it as an artifact introduced by permanganate or aldehyde fixation or by some staining treatment? The globular structure has been observed in frozen-dried material.

ROBERTSON: 1. There are other components that might be important. In the crayfish giant synapse system I pointed out in my presentation that one sometimes sees granules that resemble glycogen in the boutons on the motor fibre synaptic processes. I have also observed similar granules pre- and postsynaptically in frog motor end plates. One also sees in the crayfish median-giant-to-motor electrical synapse presynaptically flattened endoplasmic reticulum cisternae resembling the subsurface cisternae described in vertebrate neurons by Rosenblueth. Sometimes these sacs are collapsed with intimate adhesion of the limiting membranes making external-compound membranes.

Vesicles are a dominant feature in many presynaptic endings, but mitochondria are also very prominent and tubular smooth endoplasmic reticulum components sometimes appear amongst mitochondria. In such endings neurofilaments and neurocanaliculae or "microtubules" are absent. The latter are, however, found in the presumably electrical goldfish endings and seem to end at presynaptic membranes.

2. If I understand you correctly, you are wondering whether I regard the closed cleft as significant and I presume you have in mind here the fact that Schultz and Karlsson have been

able to produce micrographs in brain in which almost all intermembrane gaps are closed. I think the point of this is simply that the gap is labile and can be closed or opened by milieu variations. We already knew this from our experiments in which gap occlusions were experimentally produced in frog non-myelinated fibers by treatment with strong salt solutions. Therefore, we may reasonably ask which state is significant. In general, I believe the most generally prevalent open gap is significant, partly because of independent physiological evidence from Frankenhaeuser and Hodgkin, and Ritchie and Straub suggesting its presence in living nerves. But perhaps the most important point is that we can demonstrate an open cleft produced in nerve myelin by hypotonic solutions and detected independently by X-ray diffraction or polarization optics on unfixed material. When these independent methods tell us the gap is open, we see it open in our fixed material; and when they tell us the gap is closed, we see it closed. This produces some faith in the reality of open or closed gaps seen in tissues fixed by similar methods. The other important point here is the fact that in our material closed synaptic clefts are very regularly seen in certain consistent and specific places, whereas in adjacent regions of a different character but only a few hundred Ångstrom units away we see the gap open. This leads me to believe we are seeing something meaningful. If it is all one way or the other, questions are reasonably raised; but when we find it both ways at distances $\sim 0{\cdot}1\ \mu$ or so apart in the same tissue, one cannot escape the conclusion that some significant differences exist in the membrane contacts concerned.

3. I certainly do mean to say that I do not believe everything I see in my or anyone else's electron micrographs. This is the problem that we must guard against most strongly—the tendency to jump to immediate conclusions upon seeing something in an electron micrograph that is compatible with some idea we have. The pitfalls are particularly dangerous when we are dealing with molecular organization. There is nothing more treacherous in my opinion than looking at an electron micrograph and drawing a conclusion about molecular organization. Indeed, I have spent over ten years dwelling on the problem of drawing the conclusions I have drawn about the meaning of the unit membrane pattern in terms of arrangements of molecules. These conclusions have only been reached after a most careful consideration of the convergence of evidence from polarized light and X-ray diffraction studies in fresh material as well as evidence from electron micrographs. Hence, I feel very cautious about interpreting the globular appearances of unit membranes in the way some people have proceeded. I think there needs to be much more consideration of what kinds of artifact might be involved from the standpoint of fixation and the standpoint of strictly electron optical and specimen preparation aspects. I have attempted this kind of analysis by taking membranes that show the globular substructure most clearly and putting them to the test of seeing if polarized light or X-ray diffraction will support the electron microscope observations. So far, no support has been forthcoming. Admittedly, more efforts should be made and we are making them. Meanwhile, I do not believe that what appears in the electron micrographs is meaningful in terms of the kinds of molecular organizations we are discussing. The relevance of saying that the globular structure has been seen in frozen-dried material escapes me, for such material is known to be generally less well fixed than chemically fixed preparations.

DOUGLAS: As you have drawn it, the membrane is organized with a "Y" surface oriented to the cell sap and an "X" surface oriented to the extracellular environment, while the vesicle membrane has its "X" surface directed towards its own interior—although again the "Y" surface is presented to the cell sap. The two membranes, cell and vesicle, are thus, in your scheme, oppositely oriented, the inner surface of the first facing the outer surface of the second. What is your evidence for this?

ROBERTSON: You have correctly stated my position except possibly for the last sentence. I would say that the "Y" surface of the vesicle faces the "Y" surface of the cell membrane across the cell sap or "matrix" as I prefer to say. They are oppositely oriented in a sense and I believe isolated synaptic vesicles if they come out intact present the "Y" surface to the external milieu.

The evidence for this viewpoint is of several kinds. First, there is the general question of whether the outside and inside surfaces of any membrane are chemically different in terms of anything other than particular molecular species. The only clear-cut evidence for this is the electron microscopical and X-ray diffraction evidence about the membrane relationships in myelin dealt with in my paper. The main point is the fact that the major dense line and the intraperiod line differ considerably from one another, proving the asymmetric nature of the Schwann cell membrane. Second, there are some observations of so-called cavealae at cell surfaces and even at synaptic membranes, in which one sees continuity with a vesicle or tubule extending into the cell. One simply follows the outside surface in and sees that it borders the lumen of the tubule or vesicle. This is all part of my three-phase cell concept illustrated in my paper.

CENTRAL MONOAMINE NEURONS*

N.-Å. Hillarp, K. Fuxe and A. Dahlström

Department of Histology, Karolinska Institutet,
Stockholm, Sweden

To date, over the course of many years, a great deal of work has been carried out with the purpose of identifying the synaptic transmitters in the central nervous system. Gaddum (1963) states that "no good evidence exists that any substance except acetylcholine is liberated as transmitter in the CNS". However, strong evidence for the view that acetylcholine is a central transmitter has been given only in the case of the Renshaw cells (Eccles, Fatt and Koketsu, 1954), on which acetylcholine exerts an excitatory action. Studies on the possible role of acetylcholine as a neurotransmitter in the CNS have been greatly facilitated by the existence of a specific histochemical technique, developed by Koelle *et al.* for the demonstration of acetylcholinesterases (see review by Koelle, 1963).

In the case of the monoamines DA, NA and 5-HT, which for several years have been known to occur in significant amounts and to have important functions in the brain, no histochemical methods of corresponding importance have existed. Until three years ago, not even the cellular localization of the monoamines was known and one of the greatest difficulties in this field has, in fact, been our lack of knowledge as to the structural basis for the formation, storage, release and inactivation of the amines.

In co-operation with Prof. A. Carlsson (Department of Pharmacology, Göteborg), experiments were started several years ago with a view to developing histochemical methods for the demonstration of neuronal monoamines. In 1961 it was discovered that certain CA can be converted into highly fluorescent trihydroxyindoles in tissues, and it was found possible to use this highly sensitive reaction in a histochemical procedure (Carlsson, Falck, Hillarp, Thieme and Torp, 1961), the sensitivity of which was of an entirely different order of magnitude from that of previous methods. Owing, however, to certain technical difficulties this procedure could not be used for the demonstration of CA in the CNS. The results obtained were even so of the utmost importance, since they showed that highly sensitive and specific histochemical fluorescence methods could be developed for the study of

* *Abbreviations used:* A, adrenaline; CA, catecholamine(s); CNS, central nervous system; COMT, catechol-*o*-methyltransferase; DA, dopamine; DOPA, 3,4-dihydroxyphenylalanine; FRA, flexor reflex afferents; 5-HT, 5-hydroxytryptamine; 5-HTP, 5-hydroxytryptophan; MAO, monoamine oxidase; NA, noradrenaline.

neuronal monoamines and in fact it was this that led to the development of the present method along similar lines. The chemical basis of the present method has been clarified mainly by the work of Corrodi and Hillarp (1963, 1964).

The *present* fluorescence technique is quite outstanding in specificity and above all in sensitivity, as compared to the methods previously available for the histochemical demonstration of CA and 5-HT. These previous methods have been markedly inadequate with respect to the localization of monoamines in the CNS and the peripheral nervous system, although they have been of great help in studies on cells containing high concentrations of CA or 5-HT, such as the entero-chromaffin cells and adrenal medullary cells. The most important of these methods have been the chromaffin reaction, the potassium–iodate reaction of Hillarp and Hökfelt for the selective demonstration of NA (cf. 1955), and the fluorescence methods of Erös (1932), Hamperl (1932) and Eränkö (1955).

With the help of the present histochemical fluorescence technique, it has now been shown that DA, NA and 5-HT in the mammalian CNS (including the area postrema) are localized to DA, NA and 5-HT neurons respectively (Dahlström and Fuxe, 1964a, 1965a; Fuxe, 1965a and b). This is in all probability true also for the CA and 5-HT present in the CNS of birds and invertebrates (Dahl, Falck, Lindqvist and Mecklenburg, 1963; Fuxe and Ljunggren, 1965).

THE PRESENT METHOD

Chemistry, Histochemistry, Specificity and Sensitivity

The idea underlying this method originated from the procedure of Hess and Udenfriend (1959) for the spectrophotofluorimetric determination of tryptamine (in which the amine is condensed with formaldehyde to tetrahydronorharmane, which is oxidized to the strongly fluorescent norharmane). While trying to utilize this and similar principles for histochemical purposes, there was made a very unexpected observation. Although—as expected—no or at the most a very weak fluorescence could be obtained when CA, such as DA and NA, were treated—in solution or on paper—with aqueous formaldehyde or its vapors, the amines were found to be very readily converted into products with a remarkably intense fluorescence when treated with formaldehyde gas under very mild conditions, *provided that the reaction took place in a dried protein.* It was found also that the compounds formed become enclosed in reactive groups in the protein—and maybe also partly bound through methylene bridges to such groups—in such a way as not to be extracted by organic solvents or even by the lower alcohols. This peculiar fluorescence reaction was the primary discovery, which together with observations as to the essential role of proteins and the nature and properties of the fluorescent products, made it possible to work out a practical histochemical procedure.

The present fluorescence method has been thoroughly studied, above all with the use of model systems, and its chemical and histochemical background and specificity are by now well understood (Falck, Hillarp, Thieme and Torp, 1962; Corrodi and Hillarp, 1963, 1964; Corrodi, Hillarp and Jonsson, 1964). The first step in the fluorescence reaction in the case of CA is a condensation, resulting in the formation of a 6,7-dihydroxy-1,2,3,4-tetrahydroisoquinoline. Under the mild conditions used, the great activation of the 3-OH group is required for this condensation to be achieved and the catecholamines must obviously be either primary or secondary amines. Their 3-O-methylated and acid metabolites therefore do not react. The next step is a quite unexpected protein-promoted dehydrogenation, which gives the final fluorescent product, a 6,7-dihydroxy-3,4-dihydroisoquinoline (main peaks of activation and emission 410 and 480 mμ, respectively). In the case of a secondary amine, such as A, this dehydrogenation gives rise to a quaternary 3,4-dihydroisoquinoline and requires more severe reaction conditions. Primary and secondary catecholamines can thus be differentiated from each other histochemically in a simple manner. Not only the CA (and DOPA) but also 5-HT (and 5-HTP) react readily and in principle in the same manner as described above (Corrodi and Hillarp, 1963; Corrodi and Jonsson, 1965a). The fluorescent product, 6-hydroxy-3,4-dihydro-β-carboline, has an emission spectrum with a peak about 50 mμ higher than that of the fluorescent product formed from CA. CA and 5-HT can consequently be distinguished by the differing colour of their fluorescence (green to yellow-green and yellow, respectively, in the fluorescence microscope), provided that certain precautions are taken (see Dahlström and Fuxe, 1964a; Norberg and Hamberger, 1964; Caspersson, Hillarp and Ritzén, 1965; Corrodi, Hillarp and Jonsson, 1964). The studies made on the chemistry of the fluorescence reaction and other results (Hillarp and Malmfors, 1964; Malmfors, 1965) disproved incidentally the view of Csillik (1964) that the method suffers from a serious limitation in that only the amines bound in the storage granules can be demonstrated.

Extensive studies on models and tissues have shown that the method has a high specificity for CA and 5-HT. With the help of a number of histochemical criteria—based on the characteristics of the fluorescence reaction and the properties of the fluorescent products—almost conclusive evidence can usually be obtained that the fluorescence observed in a particular case is due in fact to the presence of such amines. It is possible also to differentiate between primary and secondary amines, and between CA and 5-HT. Analyses of activation and emission spectra by means of a new microspectrophotometric technique developed by Caspersson have proved to be of great value both for the evaluation of specificity and for differentiation between CA and 5-HT (Caspersson, Hillarp and Ritzén, 1965). The histochemical criteria currently used and found to be of value have now been compiled and discussed in detail (Dahlström and Fuxe, 1964a; Norberg and Hamberger, 1964). Recent model experiments show that a direct histochemical differentiation between NA and DA is possible. It has been found in model experiments that the 4,6,7-trihydroxy-3,4-dihydroisoquinoline formed from NA has an unstable 4-hydroxy group which can be easily split off. The fully aromatic 6,7-dihydroxyisoquinoline thus formed shows other fluorescence characteristics, than does the 3,4-dihydroisoquinolines and it is not reduced by sodium borohydride (Corrodi and Jonsson, 1965b).

Certain drugs and analogues of the amines and their amino acid precursors (e.g. reserpine, tetrabenazine, guanethidine, aramine, *m*-tyrosine, α-MMT) deplete or reduce the monoamine stores in the peripheral and/or central nervous system. Other drugs (e.g. MAO inhibitors) can increase the amine content. On the basis of these findings a pharmacological approach was introduced—in close co-operation with Prof. A. Carlsson (Dept. of Pharmacology, Göteborg)—in studies made on the nervous monoamines with the fluorescence method, with a view to ensuring specificity and differentiating between DA, NA and 5-HT (Carlsson, Falck and Hillarp, 1962; Falck, 1962; Carlsson, Falck, Fuxe and Hillarp, 1964). This approach has proved to be of great value in studies on both the peripheral tissues (Norberg and Hamberger, 1964) and, above all, the central nervous system (Dahlström and Fuxe, 1964a; Fuxe, 1965a). Of special interest is the fact that there now exist reliable means of distinguishing between the DA and NA terminals in the central nervous system, at least in some species (Carlsson, Dahlström, Fuxe and Hillarp, 1965; Fuxe, 1965a).

The present method has an exceptionally high sensitivity (see Norberg and Hamberger, 1964). Thus, for instance, less than one picogram (10^{-6} μg) of NA, present in an adrenergic ganglion cell with a volume of 10,000 to 20,000 μ^3 can be readily demonstrated. The limit of detection seems to be about 0·001 per cent NA (w/v) in such cells (10 μ sections). Small varicosities of adrenergic nerve terminals (see below) show a fairly strong fluorescence due to the presence of NA in amounts probably in the order of one-thousandth of a picogram (10^{-3} pg; 10^{-9} μg).

TECHNICAL IMPROVEMENTS

Studies on the most critical step in the histochemical procedure, the formaldehyde treatment, have greatly improved the possibilities of standardizing the method (Hamberger, Malmfors and Sachs, 1965).

Techniques permitting the demonstration of both monoamines and enzymes, such as acetylcholinesterase, in cryostat sections have been developed (Jacobowitz and Koelle, 1963; Hamberger and Norberg, 1964; Hamberger, Norberg and Sjöqvist, 1964).

At least on peripheral tissues, excellent results can be obtained with the use of a modified Pearse apparatus for high speed freeze-drying (Thieme, 1965). This has made the histochemical procedure both easier to handle and much less time-consuming.

Some Experimental Methods for the Direct Demonstration of Monoamine Neurons and Pathways

The axons belonging to monoamine neurons can be specifically demonstrated by raising their amine levels, and their presence, distribution and direction in peripheral nerves and the CNS can be readily studied in this way. Simple procedures have been introduced on the basis of the findings that the amines (in fact, the amine storage granules) accumulate rapidly in the proximal part of transected axons (Dahlström and Fuxe, 1964a; Dahlström, 1965a) and that the adrenergic axons readily accumulate exogenous NA along their entire lengths due to a specific uptake-concentration mechanism (Hamberger, Malmfors, Norberg and Sachs, 1964; Malmfors, 1965). Peripheral adrenergic nerve cells and fibres can thus be readily visualized and distinguished from those belonging to cholinergic or other types of neurons (Norberg and Hamberger, 1964). The central 5-HT neurons can be readily demonstrated in their entire extension following drug-induced increases in their amine levels (Dahlström and Fuxe, 1964a; Fuxe, 1965a). Although it has so far proved impossible to trace the central CA nerve fibres back to their cell bodies, the central CA nerve cell bodies of transected axons can still be identified, thanks mainly to a marked accumulation of CA in the cell bodies shortly after the transection (Dahlström and Fuxe, 1965a). This phenomenon makes it possible to map out the CA neurons in the central nervous system.

A very important achievement, however, is the finding that monoamine nerve terminals rapidly lose their stored transmitter following activation in combination with an inhibition of amine resynthesis (Fuxe and Gunne, 1964; Malmfors, 1964; Dahlström, Fuxe, Kernell and Sedvall, 1965). Quite unique opportunities now exist for direct studies on the state of activity of different neuron systems under physiological or experimental conditions.

Since there is an efficient blood–brain barrier against CA (Weil-Malherbe, Whitby and Axelrod, 1961) it is very difficult to study the uptake of CA in the intact brain. In brain slices, however, this barrier is no longer present and after incubation the CA have been shown to accumulate in the entire CA neuron due to the presence in the cell membrane of a specific uptake-concentration mechanism, which is thus present not only in peripheral (see below) but also in central CA neurons (Hamberger and Masuoka, 1965). This *in vitro* approach with brain slices in combination with the histochemical fluorescence technique will be a useful tool when studying the effects of drugs on the uptake-concentration mechanism in the central CA neurons. Furthermore, the CA axons of the brain slice can be easily visualized after incubation with CA if MAO inhibition has been performed.

RESULTS AND DISCUSSION

1. *Evidence for the Existence of Monoamine Neurons in the Central Nervous System*

The detection of fine, varicose fibres of the same appearance as the adrenergic terminals and with extremely high concentrations of CA or 5-HT in their enlargements or "varicosities" (Carlsson, Falck, Hillarp, 1962; Carlsson, Falck, Fuxe and Hillarp, 1964; Fuxe 1965a and b), and the discovery of a large number of specific nerve cell groups in the lower brain stem with low concentrations of CA or 5-HT (Dahlström and Fuxe, 1964a) previously gave strong support to the existence of central monoamine neurons. Final evidence for this view has now been obtained with the help of the changes in intraneuronal amine levels occurring after axotomy (see Andén, Carlsson, Dahlström, Fuxe, Hillarp and Larsson, 1964; Carlsson, Falck, Fuxe and Hillarp, 1964; Andén, Dahlström, Fuxe and Larsson, 1965a; Dahlström and Fuxe, 1965a). After total transection (at C8) of the spinal cord, for example, the fine, varicose CA and 5-HT fibres below the lesion lost their fluorescence after 6 to 8 days. Biochemical determinations of the part below the lesion likewise revealed an almost complete disappearance of both NA and 5-HT after this time-period (Andén, Häggendal, Magnusson and Rosengren, 1964). Experiments of this kind show clearly that the fine, varicose fibres that contain NA and 5-HT, and which have the same characteristic appearance in all parts of the central nervous system, are true terminal arborizations of neurons. Those in the spinal cord must derive from descending axons in the white matter of the spinal cord; these axons originate in all probability from supraspinal levels, since no CA or 5-HT nerve cells have been found in the spinal cord.

With the help of the above-mentioned methods for mapping out monoamine neuron systems, there have been obtained almost complete characterizations of a large nigro-neostriatal DA neuron system and several different bulbo-spinal NA and 5-HT neuron systems (Andén, Carlsson, Dahlström, Fuxe, Hillarp and Larsson, 1964; Andén, Dahlström, Fuxe and Larsson, 1965a; Dahlström and Fuxe, 1965a). Large ascending CA (mainly NA) and 5-HT neuron systems from the lower brain stem to the hypothalamus, the preoptic area, the limbic forebrain structures and the neocortex have also been discovered (Dahlström, Fuxe, Olson and Ungerstedt, 1964; Andén, Dahlström, Fuxe and Larsson, 1965b). The ascending CA and 5-HT neuron systems are at least to a large extent uncrossed. Furthermore, the existence of a tubero-infundibular DA neuron system has been demonstrated (Fuxe, 1964).

Since the essential characteristics of the 5-HT and CA nerve cells in the lower brain stem are in principle similar, *all the specific nerve cells discovered give rise in all probability to synaptic terminals, which have an extremely high concentration of monoamine in their varicosities.* Practically the entire amine content in the brain is localized to these varicosities. The adrenergic vasomotor terminals in the brain are of no importance from this quantitative point of view, since they are found almost exclusively around the arteries on the basal surface of the brain.

Central CA and 5-HT neurons have been found in all the mammals studied, i.e. mouse, rat, guinea pig, dog, cat and monkey (Dahlström and Fuxe, 1964a; Fuxe 1965b), and also in birds (Fuxe and Ljunggren, 1965). Although the content of A in the mammalian brain is very low (Bertler and Rosengren, 1959a; Gunne, 1962; Häggendal, 1963) the existence of a small number of special neurons forming and storing A cannot be excluded and/or some of the CA neurons may in addition to their primary CA also contain small amounts of A. In the frog brain—where metanephrine and A but not normetanephrine and NA are found in large amounts (Bertler and Rosengren, 1959b; De Caro and Roseghini, 1964)—there is good evidence for the existence of A neurons (Dahlström and Fuxe, unpublished data). It should be pointed out that, in bird brain, the monoamines are localized not only to specific monoamine neurons but also to the neuroglandular cells of the "modified ependyma", which contains two types of cell forming and storing CA and 5-HT respectively (Fuxe and Ljunggren, 1965). The same is true of the "modified ependyma" in fish (Bertler, Falck and Mecklenburg, 1963).

2. *The Central Monoamine Neuron*

The intraneuronal distribution of NA, DA and 5-HT is highly characteristic. All these monoamines are very strongly accumulated in the terminal parts of their respective neurons. In fact, practically the entire monoamine content of the brain is stored in these parts. The cell bodies and the non-terminal axons show only very low amine concentrations.

A. *The CA neuron*

(a) *CA nerve cell bodies: General characteristics.* The perinuclear cytoplasm of the nerve cells exhibits a weak to strong, diffuse, green to yellow-green fluorescence, which is usually evenly distributed throughout the cytoplasm, except in certain cell groups in the pons (A5 to A7 according to the terminology of Dahlström and Fuxe, 1964a) where the cells show an accumulation of fluorescence around the nucleus (Fig. 1). This fluorescent zone has a dusty appearance. The cell nucleus is non-fluorescent unless diffusion has occurred. The cells are usually small to medium-sized and round to oval and sometimes spindle-shaped. Many of the cells appear to be multipolar. Since the first step in the monoamine synthesis is in all probability rate-limiting (Axelrod, 1963; Carlsson, Corrodi and Waldeck, 1963; Udenfriend and Zaltzman-Nirenberg, 1963; Carlsson and Corrodi, 1964), the precursors DOPA and—in NA cell bodies—DA exist at the most in very low concentrations and cannot be expected to accumulate under normal conditions.

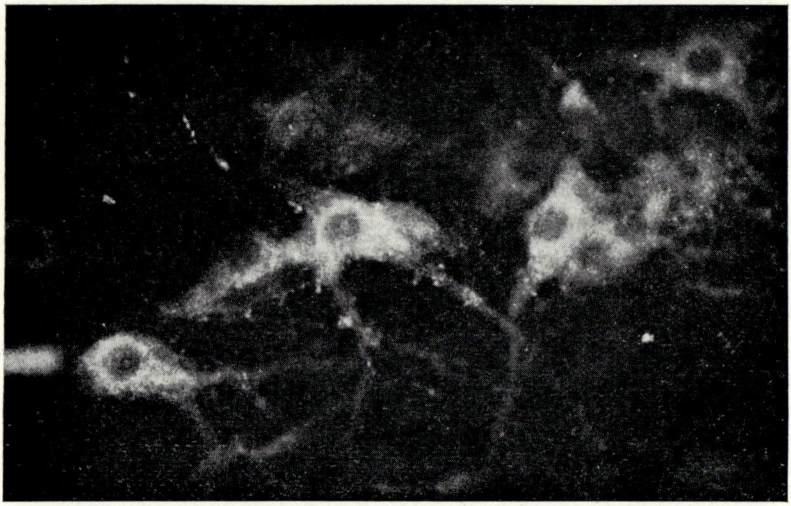

FIG. 1. CA nerve cells of group A6 (see Dahlström and Fuxe, 1964a). Normal rat. Medium-sized nerve cells with an accumulation of green fluorescence around the nucleus. The processes are clearly seen due to their weak green fluorescence. ×420.

It therefore seems safe to conclude that of the reactive monoamines which give a green fluorescence only DA or NA—depending whether β-hydroxylase activity is present or not—exist in the CA neurons in histochemically significant amounts. The content of DA or NA is probably in the order of one-half of a picogram, since an adrenergic ganglion cell probably contains at maximum $0{\cdot}7 \times 10^{-6}$ μg (see Norberg and Hamberger, 1964). The average concentration would then be 0·003 per cent, which is extremely low in comparison with the average concentration (0·5 per cent) of another important cytoplasmic constituent, ribonucleic acid (cf. Hydén, 1960). The concentrations in the cells with the lowest and highest fluorescence intensities are probably about 0·001 and 0·01 per cent respectively.

A simple explanation for the peculiar heterogeneity as regards fluorescence intensity may exist in the formation and transport of the amine granules. There are certain indications that the cells showing low and high concentrations of CA are neurons with short and long axons respectively. It may well be, however, that the decisive factor is not the distance between the cell bodies and their terminals but the total volume of the terminals belonging to a given neuron. An extensive terminal system could require a high production and down transport of granules, which might leave the cell bodies more or less depleted, whereas a piling up could occur in neurons with short axons and/or short terminals.

It should also be pointed out that this heterogeneity persists after MAO inhibition, which causes no obvious increase, or only a small increase, in the

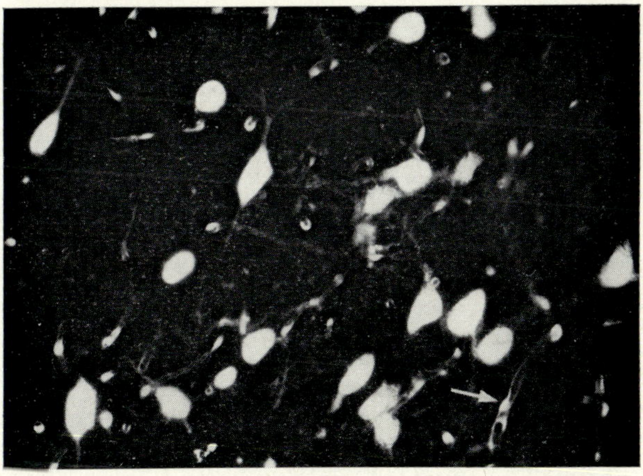

FIG. 2. CA nerve cells of group A8 (see Dahlström and Fuxe, 1964a). Nialamide-L-DOPA treated rat. A very marked increase in the fluorescence of the nerve cell bodies is observed. In the capillary walls the pericytes are observed (→). ×200.

fluorescence intensity of the CA nerve cells in rat. After a reserpine-induced depletion, the same pattern develops again within 24 hr. This heterogeneity is even more marked in the ganglia, where the adrenergic ganglion cells form a very heterogeneous population in respect of NA content (cf. Norberg and Hamberger, 1964).

In animals given a moderate dose of nialamide (100 mg/kg), treatment with L-DOPA (50 mg/kg) produced a marked to very marked increase in the CA contents of almost all the CA nerve cells (Fig. 2) in the CNS (Fuxe, 1965a). This makes it easier to differentiate between the true CA nerve cells and non-monoamine-containing nerve cells, since the latter do not take up DOPA. CA nerve cells that have previously been missed owing to their very weak green fluorescence become more intense and can easily be detected. The adrenergic ganglion cells have the same properties (Norberg and Hamberger, 1964). The fact that MAO-inhibition was necessary for an accumulation of fluorescence clearly suggests the presence of MAO-activity in the CA nerve cell bodies. In the case of the CA nerve cells in the locus coeruleus, it has actually been possible to demonstrate a strong MAO activity histochemically (Fuxe, Hamberger and Norberg, to be published). Electron microscopical investigations have also been made on the CA nerve cell bodies of the locus coeruleus, which showed no unique ultrastructural characteristics, and appeared as ordinary nerve cells (Fuxe, Hökfelt, Nilsson and Reinius, 1966). The Golgi apparatus occupied a large part of the perinuclear zone, where the CA's are highly accumulated. Small numbers of vesicles of an appearance similar to that of the "synaptic vesicles" were found throughout the cytoplasm, above all in relation to the Golgi apparatus.

Localization of the CA nerve cells. These cells have a widespread distribution within the lower brain stem and small groups of CA nerve cells are found also in the hypothalamus. No CA nerve cells have been discovered in the telencephalon, diencephalon (with the exception mentioned above), cerebellum, tectum or spinal cord. The cells are mainly present within the mesencephalon. The CA nerve cells have been divided into 13 groups (A1 to A13), which have been described in detail by Dahlström and Fuxe (1964a). The division was made not only on purely topographical grounds but also with consideration of the degree to which the cells seemed to form more or less continuous systems. In view, however, of our scanty knowledge as to what functional or anatomical systems these cells belong to, it is possible that the division is partly artificial. For details see Dahlström and Fuxe, 1964a and 1965a.

It may be mentioned that by compression of the ventral spinal roots it has been shown that some CA fibres descending in the anterior and lateral funiculi leave the spinal cord via these roots without any relays on motoneurons (Dahlström and Fuxe, 1965b). Inhibitory fibre systems of this kind have recently been postulated by Beck (1964).

CA nerve cells that probably give rise to intraretinal synaptic terminals exist between the inner plexiform layer and the inner granular layer of the retina (Malmfors, 1963). These neurons have been shown to contain DA (Häggendal and Malmfors, 1963). A varying number of CA nerve cells has been found in the area postrema of different mammals. In view of these findings, the possibility arises that these neurons may be involved in the reflex path of central emesis (Fuxe and Owman, 1965).

(b) *CA non-terminal axons.* The thin axons emanating from the CA nerve cells usually exhibit a distinct fluorescence only when running together in bundles, and seem to have about the same low monoamine concentrations as the cell bodies. This agrees with biochemical data on the peripheral nervous system (Euler, 1956; Muscholl and Vogt, 1958). Their presence, distribution and course can be readily studied with the help of the methods developed for raising their amine contents. Taking into consideration the thinness of these axons, they can be judged to have CA concentrations of around 100–500 μg.

(c) *CA terminals: Characteristics.* As the axons approach the nucleus that is to be innervated, there occur drastic and highly characteristic changes. They become extremely fine and abundant, and round to fusiform enlargements (varicosities) appear more or less regularly along the entire length of

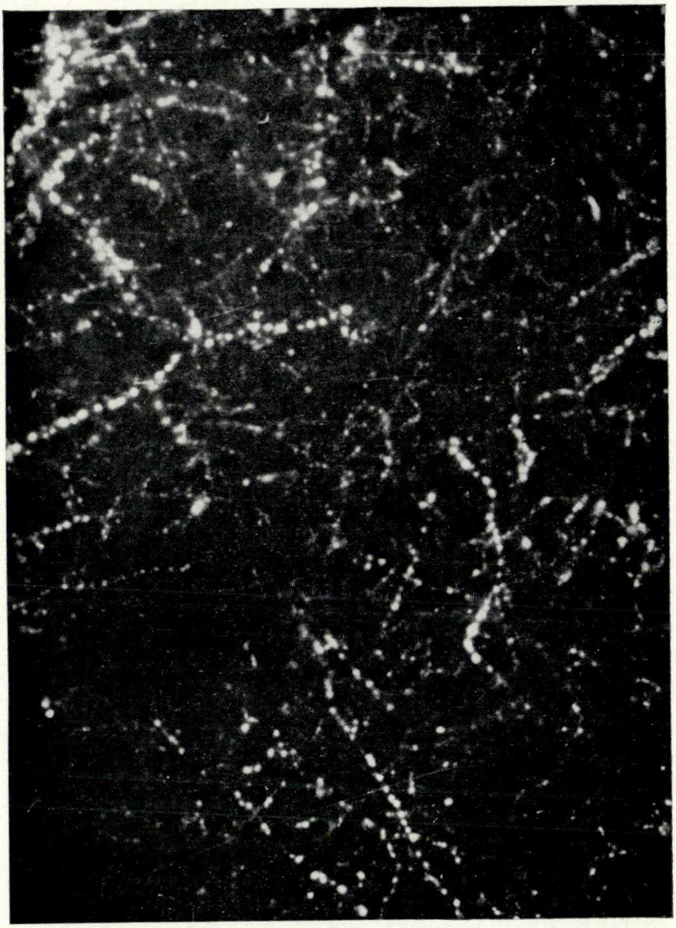

FIG. 3. Dorsomedial part of the retrochiasmatic area of normal rat. Fine to thick, varicose CA nerve terminals are observed. ×420.

the fibre (Fig. 3). These varicosities have a strong fluorescence intensity and contain very high concentrations of CA (in the order of 10,000 μg/g) (see Carlsson, Falck, Fuxe and Hillarp, 1964; Fuxe, 1965a). They thus have the same characteristics as the peripheral adrenergic terminals (Norberg and Hamberger, 1964). The fibre segments between the varicosities are very thin (down to 0·2 μ) and barely visible (Fig. 3). Due to the highly irregular course of the terminals, they can be traced in 10 μ sections only for about 50–100 μ, except in parts of the retrochiasmatic area and the substantia gelatinosa, where they are oriented in a certain direction and have been traced up to 400 μ. The varicosities are of a thickness varying mainly between 0·3 and 2 μ, and 15 to 30 varicosities per 100 μ of fibre is a normal figure (Fuxe, 1965a). The diameter of the varicosities of the NA terminals varies greatly as between different regions of the brain. In some areas, e.g. in many hypothalamic nuclei, such as the ventral part of the nuc. interstitialis striae terminalis, a large proportion of the CA terminals are *fairly thick* (diameter ranging between 1 to 2·5 μ; Fig. 3), while in many other nuclei, e.g. the nuc. motorius n. trigemini, most of them are *fine* (diameter around 1 μ), and in other regions, e.g. the neocortex, they are almost exclusively *very fine* (diameter around 0·5 μ). In certain regions there are observed *thick* terminals, with diameters between 2·5 μ and 4 μ. (For further details, see Fuxe, 1965b.) The biological significance of these variations in size is unknown. It may be that the terminals with small varicosities innervate neuron systems requiring a somewhat finer regulation. The DA terminals, on the other hand, are almost exclusively very fine.

The finely dotted fluorescence observed in, for example, the median eminence and the nuc. amygdaloideus centralis, and the diffuse green fluorescence observed in, for example, the nuc. caudatus putamen, the nuc. accumbens and the tuberculum olfactorium (Fuxe, 1965a and b) are due to the presence of abundant numbers of densely packed very fine—partly sublight-microscopical—CA terminals, the vast majority being of the DA type. It is thus possible under optimal reaction conditions to reveal the presence of abundant numbers of very fine, varicose terminals in, for example, the neostriatum. Final evidence for this view has been obtained with the help of large lesions destroying the crus cerebri, the medial forebrain bundle and the subthalamus, for after these lesions the diffuse green fluorescence will disappear (Andén, Carlsson, Dahlström, Fuxe, Hillarp and Larsson, 1964; Dahlström, Fuxe, Olson and Ungerstedt, 1964; Andén, Dahlström, Fuxe and Larsson, 1965a and b). Furthermore, electron microscopical studies on the caudate nucleus revealed the presence of a densely packed plexus built up by *inter alia* abundant synaptic nerve terminals, many of which had a diameter below 0·4 μ (Fuxe, Hökfelt and Nilsson, 1964).

Distribution of the CA terminals. Strong support has been obtained—*inter alia* from experiments with α-methyl-DOPA (Carlsson, Dahlström, Fuxe and Hillarp, 1965)—for the view that the CA terminals showing rapid and slow recovery after α-MMT are DA and NA terminals respectively (see below). On the basis of these pharmacological experiments and from the biochemical data available, the distribution patterns of the DA and NA terminals have been determined. It was found that the DA terminals were concentrated mainly to certain circumscribed areas in the forebrain. The NA terminals, on the other hand, had a widespread distribution in the central nervous system, from the cortex to the spinal cord. It should be pointed out that the CA terminals are not confined in any way to the sympathetic nuclei, but are just as frequent in the parasympathetic nuclei.

The DA terminals are accumulated mainly in the neostriatum, the tuberculum olfactorium, the median eminence (Fig. 4a) and in all probability also in the nuc. accumbens. The DA terminals in these areas are very fine and react to treatment with reserpine, α-MMT, α-methyl-DOPA-reserpine (Fig. 4b) and nialamide-DOPA in the same way. For further details, see Fuxe, 1965a and b and Carlsson, Dahlström, Fuxe and Hillarp, 1965.

The NA terminals are found to form dense plexuses in *inter alia* the vegetative hypothalamus, various parts of the limbic lobe, many of the visceral efferent and afferent nuclei of the cranial nerves, and the ventral horn, sympathetic lateral column and substantia gelatinosa of the spinal cord. For further details, see Fuxe, 1965b.

CA non-terminal axons have also been shown in certain mammals to descend in the infundibular stalk, giving rise to a rich plexus of terminals in the pars intermedia (Fuxe, 1964). These CA terminals may possibly inhibit the release of melanocyte stimulating hormone (see Iturriza, 1964). The anterior lobe, on the other hand, contains no CA nerve terminals. It contains instead PAS-positive cells, which probably store primary CA and form primary CA from L-DOPA. (Dahlström and Fuxe, 1965c.)

Contacts made by the CA terminals: Axosomatic contacts. CA terminals have been observed in various regions of the brain to make intimate contacts with, above all, non-fluorescent cell bodies and their processes, but also with CA cell bodies and 5-HT cell bodies (above all, groups A1 and B1). In some nuclei, such as the nuc. hypoglossus, 5-HT and CA terminals were observed to surround the same nerve cell. For further details, see Fuxe, 1965a.

These observations do not, however, exclude the possibility that the contact is made with a non-fluorescent terminal, which in its turn forms a synaptic contact with the cell body. Electronmicroscopic investigations on certain brain areas rich in monoamine terminals have failed, on the other hand, to reveal any axo-axonic contacts (Fuxe, Hökfelt and Nilsson, 1965).

Axodendritic contacts. These are more difficult to demonstrate. It seems very probable, however, that the large numbers of CA terminals present in the apical dendritic layer of the hippocampus (CA 3) for the most part form contacts of this kind (Fuxe, 1965a), especially since there are to be found in this area fine, varicose fibres which form axodendritic contacts via "boutons en passage" (cf. Blackstad, 1963). These boutons are furthermore in the same size range as the varicosities. A large proportion at least of the CA terminals in the geniculate probably also form such contacts (see Fuxe, 1965b). In the substantia gelatinosa in the spinal cord and the retrochiasmatic area, parallel

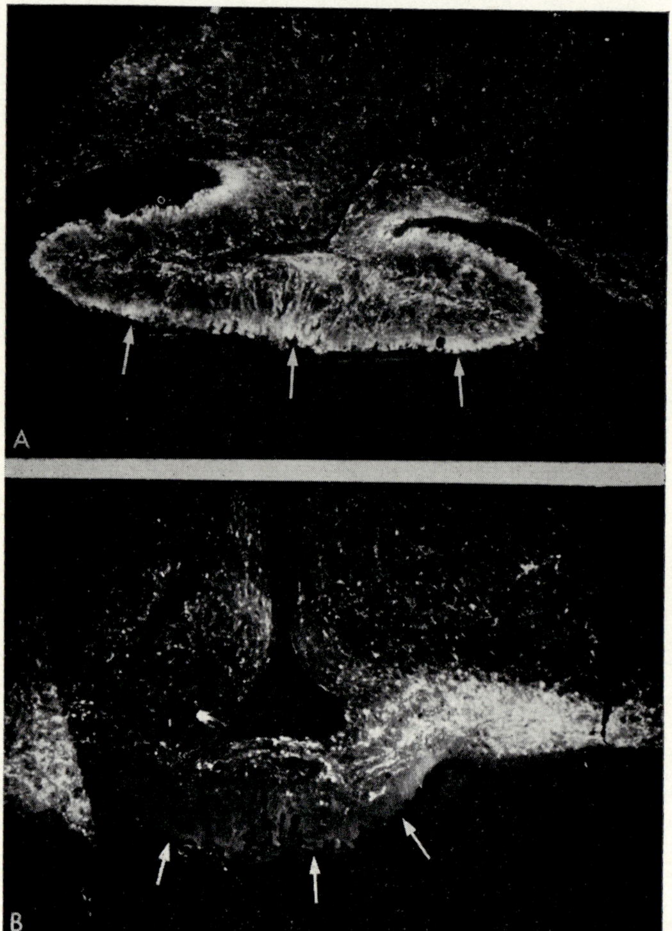

FIG. 4a. Median eminence of normal animal. A fluorescent zone of very fine DA nerve terminals is present in the external layer (→). ×80.

FIG. 4b. Median eminence of an α-methyl-DOPA-reserpine treated rat. The fluorescent zone containing DA nerve terminals has disappeared (→), while the NA nerve terminals in the hypothalamus remain. ×100.

axodendritic contacts are probably made by most of the CA terminals (Dahlström and Fuxe, 1965a; Fuxe, 1965b).

B. *5-HT neurons*

(a) *5-HT nerve cell bodies: Characteristics.* The perinuclear cytoplasm shows a very weak to weak yellow fluorescence, which in normal animals is evenly distributed throughout the cytoplasm. After MAO-inhibition it shows—in contrast to the CA nerve cells—a marked to very marked increase in fluorescence intensity and some of the larger processes also become clearly

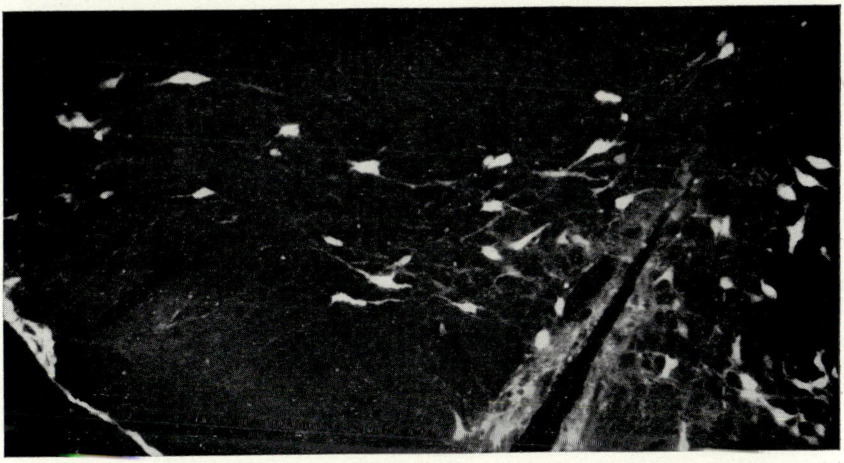

FIG. 5. 5-HT nerve cells of mainly group B3 (see Dahlström and Fuxe, 1964a). Reserpine-nialamide treated rat. Medium-sized multipolar 5-HT nerve cells with a strong to very strong yellow fluorescence are present in the area surrounding the pyramidal tract at the level of the nuc. n. facialis. ×120.

visible. The cell nucleus is non-fluorescent. On account of this, the following description is based mainly on findings from animals treated with a potent MAO inhibitor, such as nialamide. The cells are usually small to medium-sized, and round to oval in shape. The medium-sized cells are localized mainly to a group of 5-HT cells (group B3 according to Dahlström and Fuxe, 1964a) present in the part of the nuc. reticularis gigantocellularis that immediately surrounds the tractus pyramidalis. These cells are clearly multipolar (Fig. 5).

Since the first step is in all probability rate-limiting (for references see the chapter on CA nerve cells), the precursor 5-HTP is present in negligible concentrations in normal or nialamide-treated animals. The monoamine giving the yellow fluorescence is therefore 5-HT. Further evidence for this is furnished by the strong increase in fluorescence intensity obtained after treatment with MAO inhibitors.

In contrast to the CA nerve cells, the 5-HT nerve cells showed no marked differences with regard to fluorescence intensity. The 5-HT cells reacted as a fairly homogeneous population also with respect to the effects of different drugs. The CA nerve cells belonging to one and the same cell group were also pharmacologically homogeneous. They differed, however, between groups (Dahlström and Fuxe, 1964a). The high intraneuronal 5-HT levels obtained after nialamide treatment make it easy to differentiate between true 5-HT nerve cells and non-monoamine nerve cells, and between 5-HT and CA nerve cells.

Localization of the 5-HT nerve cells. These nerve cells are present only in the lower brain stem, where they have an entirely different localization from the CA nerve cells. The 5-HT nerve cells are localized predominantly to the raphe nuclei, the remainder being present in the most ventral and medial part of the reticular formation. The 5-HT cells have been divided into 9 groups (B1–B9), which have been described in detail by Dahlström and Fuxe (1964a), see also Dahlström and Fuxe (1965a).

(*b*) *5-HT non-terminal axons.* The 5-HT non-terminal axons are not observed at all in normal animals, but show a strong yellow fluorescence after MAO inhibition. They are then observed to have the same appearance as the CA axons.

(c) *5-HT terminals.* These terminals, like the rest of the 5-HT neuron, need better reaction conditions for optimal reaction than the CA neurons. This is in agreement with the results obtained in model experiments (Corrodi, Hillarp and Jonsson, unpublished data). They are best observed, in fact, under conditions giving a slight diffusion of the CA terminals. There is no problem, however, in a section to observe green and yellow fluorescent terminals in the microscope simultaneously.

Characteristics. The 5-HT axons, when they reach the area to be innervated, undergo the same dramatic changes as the CA axons. The terminals have the same general characteristics as the CA terminals. They are very fine to fine and many are probably sublight-microscopical, especially in the telencephalon and diencephalon. In normal animals, the varicosities exhibit a weak to strong intensity of fluorescence. After MAO inhibition they practically all show a strong yellow fluorescence.

Distribution. Fairly large numbers of 5-HT terminals have been detected in the lower brain stem and especially in the spinal cord, while in the diencephalon and telencephalon only certain nuclei show any considerable numbers of terminals, e.g. the nuc. suprachiasmaticus and the nuc. corporis geniculati lateralis. In the lower brain stem, numerous visceral efferent and afferent nuclei have a dense innervation. In the spinal cord, very densely packed 5-HT terminals are present in the sympathetic lateral column, and quite abundant numbers are found also in the ventral horn, especially in the lumbosacral part. For further details, see Fuxe, 1965b.

Contacts made by the 5-HT terminals. 5-HT terminals have so far been observed to make intimate contacts only with non-fluorescent cell bodies and their processes, e.g. the α-moto-neurons in the spinal cord. At least some of the 5-HT terminals in the nuc. corporis geniculati lateralis probably form axodendritic contacts (cf. Fuxe, 1965b).

3. *Effects of Certain Drugs Interfering with Monoamine Metabolism*

Certain drugs—especially reserpine, tetrabenazine, MAO inhibitors, and certain DOPA-analogues (*m*-tyrosine, α-MMT, α-methyl-DOPA)—have afforded a great deal of information on the monoamine metabolism in the central nervous system (cf. Carlsson, 1964, 1965). Their effects on the monoamine neurons have therefore been studied (see Carlsson, Falck and Hillarp, 1962; Carlsson, Falck, Fuxe and Hillarp, 1964; Dahlström and Fuxe, 1964a; Fuxe, 1965a, b and c). Unfortunately, space does not permit a discussion of these important experiments in the present paper.

4. *Evidence for the Existence of Monoamine Neurotransmission in the Central Nervous System*

The evidence is summarized in Table 1. This evidence has been arrived at mainly by comparing the results obtained from studies on the peripheral adrenergic neuron with those obtained from studies on the central CA neuron. Especially due to the work of Euler and his school, NA has been generally accepted to be the adrenergic transmitter in the peripheral adrenergic nervous system (see Euler, 1959, 1961).

TABLE 1

Evidence for Monoamine Neurotransmission in CNS

A. The central CA neurons[1] and peripheral adrenergic neurons[2] have the same basic properties and characteristics in all essential respects.

1. Morphology: highly characteristic synaptic terminals.
2. Intraneuronal distribution of the amines[3]: highly characteristic. Practically the entire amine content is stored in the terminals in very high concentrations, while the other parts of the neurons have low concentrations.
3. Effects of axotomy[4]: (*a*) proximally, (*b*) distally, (*c*) cell body reactions.
4. Biochemistry: contain enzyme systems for synthesis and inactivation of the amines.
   (*a*) Dopa-decarboxylase.[5]
   (*b*) Dopamine β-oxidase (the NA neurons).[6]
   (*c*) MAO: important for inactivation of "free" intraneuronal amines.[7]

5. Contain specific mechanisms for uptake and storage of the amines.
   (*a*) An uptake-concentration mechanism in the cell membrane of the entire neuron.[8]

   Characteristics:
   (α) Highly efficient.
   (β) Insensitive to reserpine but blocked by drugs like cocaine and desmethyl-imipramine.

   (*b*) An uptake-storage mechanism in specific granules.[9]

   Characteristics:
   (α) Specifically blocked by reserpine.[9]
   (β) Formation in the cell bodies and transport down to the terminals.[10]
6. Reactions to drugs interfering with the neuronal monoamine metabolism[11] (e.g. reserpine, tetrabenazine, *m*-tyrosine, α-MMT, α-methyl-DOPA, MAO-inhibitors).
7. Release of the amines from and increased synthesis in the synaptic terminals by nerve impulses.
   (*a*) Biochemical data.[12]
   (*b*) Histochemical data[13]: *the varicosities are the presynaptic structures specialized for the synthesis, storage and release of the amines.*
8. High specificity for uptake (or possibly decarboxylation) of L-DOPA and 5-HTP, respectively.[14]

B. *Pharmacological and Neurophysiological Data*

Brief review made on the basis of the discussion in Dr. Andén's dissertation.[15]

1. Inhibition by 5-HTP of insulin-induced CA secretion from the adrenal medulla.
2. Effect of stimulation of descending bulbospinal 5-HT tracts and 5-HTP respectively on moto-neurons.
3. FRA: effect of L-DOPA, 5-HTP, adrenergic α- and β-receptor blockers, and reserpine. Effect of DOPA-decarboxylase inhibitors nialamide, etc.

References to Table 1

(An asterisk * indicates a reference to the peripheral adrenergic neuron.)

[1] Reviews in Dahlström and Fuxe, 1964a, 1965a; Fuxe, 1965a and b.
[2] Reviews in Norberg and Hamberger, 1964*; Malmfors, 1965*; Norberg, 1965*.
[3] See Carlsson, Falck, Fuxe and Hillarp, 1964; Norberg and Hamberger, 1964*; Fuxe, 1965a.
[4] Dahlström and Fuxe, 1964b*, c, 1965; Dahlström, 1965.
[5] Andén, Magnusson and Rosengren, 1965; Fuxe, 1965b; Norberg, 1965*.
[6] Carlsson and Lindqvist, 1962; Carlsson and Waldeck, 1963; Andén, 1964a*; Carlsson, Dahlström, Fuxe and Hillarp, 1965; Fuxe, 1965a.

[7] Carlsson, Rosengren, Bertler and Nilsson, 1957; Carlsson, 1960; Carlsson, 1964; Hamberger, Malmfors, Norberg and Sachs, 1964*; Hillarp and Malmfors, 1964*; Kopin, 1964; Dahlström, Fuxe and Hillarp, 1965; Hamberger and Masuoka, 1965; Malmfors, 1965*.

[8] Fuxe and Hillarp, 1964; Hamberger, Malmfors, Norberg and Sachs, 1964*; Hillarp and Malmfors, 1964*; Norberg and Hamberger, 1964*; Carlsson, Dahlström, Fuxe and Lindqvist, 1965; Hamberger and Masuoka, 1965; Malmfors, 1965*.

[9] See Carlsson, 1965; Dahlström, Fuxe and Hillarp, 1965.

[10] Dahlström and Fuxe, 1964b*, c; Dahlström, 1965*; Dahlström, Fuxe and Hillarp, 1965; Dahlström and Häggendal, 1965b*.

[11] Dahlström and Fuxe, 1964a; Norberg and Hamberger, 1964*; Carlsson, Dahlström, Fuxe and Hillarp, 1965; Fuxe, 1965a, b; Norberg, 1965*.

[12] Andén, Carlsson, Hillarp and Magnusson, 1964, 1965.

[13] Fuxe and Gunne, 1964; Malmfors, 1964*; Dahlström, Fuxe, Kernell and Sedvall, 1965.

[14] Fuxe, 1965b; Norberg, 1965*.

[15] Andén, 1964b.

Comments to Table 1

A1 and A2. The amine concentrations in the varicosities are extremely high, namely in the order of 10,000μg/g wet weight. It may be mentioned that ten years ago Euler and co-workers had already presented strong evidence for the view that the adrenergic transmitter is highly concentrated somewhere in ther terminal part of the postganglionic fibres (cf. Euler, 1956, 1961).

In brain homogenates, Masuoka (unpublished data) has been able to show that the varicosities containing CA are localized to the B fraction of Whittaker (1959) which contains pinched-off nerve endings.

Fine, varicose fibres in the central nervous system have previously been described on the basis of conventional neurohistological studies (for references see Fuxe, 1965a) and electron microscopic studies have shown that these fibres make synaptic contacts via the varicosities or "boutons en passage" (Blackstad, 1963). In the superior cervical ganglion, Elfvin (1963) has shown that the preganglionic neurons make contact only via the varicosities. It has been shown furthermore that several of the varicosities in the iris are partly denuded of their Schwann cells (Nilsson, 1964; Hökfelt and Nilsson, 1965), which should facilitate chemical transmission. These circumstances alone and the fact that the central CA terminals have the same appearance as the peripheral adrenergic terminals give very strong reason to believe that monoamine neurotransmission takes place in the central nervous system.

A3. The terminals have been found to react uniformly throughout their entire length after axotomy. In other words, they react to degeneration as a unit. The adrenergic terminal has no specialized "nerve endings", as has been bemonstrated by Malmfors and Sachs (1956) in a study on the termination of a single adrenergic neuron.

A4a. As regards the central nervous system, DOPA-decarboxylase activity has been found to have the same distribution pattern as the monoamines and a very marked reduction in DOPA-decarboxylase activity was obtained after transection of the bulbospinal monoamine neuron systems (Andén, 1964b; Andén, Magnusson and Rosengren, 1965).

A4b. In biochemical and histochemical studies it has been possible to show that, following the administration of α-MMT and α-methyl-DOPA, metaraminol and α-methyl-NA respectively probably are formed in the NA neurons, revealing the presence of dopamine-β-oxidase activity.

A4c. "Free" means not bound to specific storage granules. Studies on the formation of amines and their O-methylated and deaminated metabolites in the brain following MAO inhibition and the administration of amino acid precursors (Carlsson, 1960; Carlsson and Hillarp, 1962; Andén, Roos and Werdinius, 1963; Carlsson, 1964) have furnished evidence that MAO is located intraneuronally and plays an important role in preventing the accumulation and overflow of free amines. Studies on the effects of different MAO inhibitors on brain amines give further support to such a view (Brodie, Spector and Shore, 1959; Shore, 1962; Spector, 1963). The present finding that the administration of chemically different

MAO inhibitors in doses that effectively block MAO leads to a very rapid accumulation of 5-HT in the entire 5-HT neuron (Dahlström and Fuxe, 1964a; Bartoniček, Dahlström and Fuxe, 1964) also strongly supports this view. The same rapid accumulation occurs also in the peripheral part of transected axons up to 5 days after transection, which further emphasizes the importance of MAO for the amine metabolism in these neurons (Dahlström and Fuxe, unpublished data). More direct evidence for the intraneuronal localization of MAO has been obtained in studies on the uptake of CA (Fuxe and Hillarp, 1964; Hamberger and Masuoka, 1965).

A5b. The fact that reserpine and α-MMT depleted the NA terminals of their amines for at least several days speaks strongly against the view that the varicosities, which are full of amine storage granules, are the site where the granules are manufactured (cf. de Robertis, 1964). The observation after depletion induced by reserpine and α-MMT of the recovery first of a fluorescent zone around the nucleus, where CA accumulated in high concentrations before filling up the cell body and—finally—the processes (Dahlström, Fuxe and Hillarp, 1965), is in complete agreement with the axoplasma flow theory of Weiss (cf. 1961). Many independent data strongly indicate that enzymes and cytoplasmic particles are formed in the cell bodies and transported down to the terminals (for references, see Dahlström, 1965). Furthermore, the axonal transport of acetylcholinesterase has recently been convincingly demonstrated (Lubinska, Niemierko, Oderfelt and Szwarc, 1964). The protein synthesis of neurons seems to occur exclusively in the cell bodies (and probably in the dendrites containing Nissl-substance), i.e. not in the axons (for references, see Hydén, 1960; Droz and Leblond, 1963), and complex protein structures such as the zymogen granules in the pancreatic exocrine cells are formed in the Golgi-region (Caro and Palade, 1964). It is thus quite unlikely that the amine granules, which are complex structures containing, e.g.enzymes (Hillarp, 1958; Kirshner, 1962), can be manufactured in any other part than the cell bodies. Furthermore, the electron microscopical data on the CA nerve cell bodies (Fuxe, Hökfelt, Nilsson and Reinius, unpublished data) suggest that they may be formed in the Golgi Zone. Studies on the formation and transport of the amine granules after the axotomy of monoamine fibres have also furnished strong support for the view that the amine granules are formed in the cell bodies (see, Dahlström and Fuxe, 1964b and c; Dahlström, 1965; Dahlström and Häggendal, 1965). They provide almost conclusive evidence that the amine storage granules are manufactured in the cell bodies at a high rate—which varies greatly, however, from neuron to neuron—and transported down to the terminals. It has further been found that amine granules accumulate in large numbers in the peripheral adrenergic axons above a constriction made simultaneously with the administration of a very large dose of reserpine. Such accumulations can occur within 6–12 hr, at which time no or a very small recovery can be detected in the cell bodies and which is several days before the terminals show the first sign of recovery (Dahlström, unpublished data). This furnishes independent evidence that reserpine does not act primarily on the cell membrane. From the results obtained in this and similar experiments it can be concluded also that the drug can act directly on the granules and thus block their storage mechanism.

A7a. Electrical stimulation of the descending bulbospinal NA and 5-HT neuron systems *in vitro*, followed by biochemical determinations of the NA and 5-HT content in the medium and in the spinal cord, has revealed an increased synthesis of NA and 5-HT, and a release of NA and 5-HT into the incubation medium.

A7b. Electrical stimulation of various peripheral and central monoamine neuron systems *in vivo*, in combination with the inhibition of the rate-limiting first step in the amine synthesis by the new inhibitors of Carlsson, Corrodi *et al.* (Carlsson, Corrodi and Hillarp, 1963; Carlsson and Corrodi, 1964), has shown that a depletion is obtained mainly in the varicosities of the terminal. The non-terminal axons are not affected (Malmfors, 1964). Together with the results obtained from electron microscopy (cf. Blackstad, 1963; Elfvin, 1963), these findings give very strong—not to say conclusive—evidence that the *varicosities are the presynaptic structures* (see Norberg and Hamberger, 1964; Fuxe, 1965a) *specialized for the storage, formation and release of the amines.* There is no evidence of any other general and direct mechanism for the nervous release of momoamines.

There is thus very strong evidence that DA, NA and 5-HT act as neurotransmitters in the central nervous system.

5. *The Function of the Monoaminergic Neuron Systems*

The readers are referred to the following articles: Andén, 1964b; Dahlström and Fuxe, 1964a, 1965a and c; Fuxe, 1965b and c.

SUMMARY

With the help of a highly sensitive and specific histochemical fluorescence method for the direct demonstration of CA and 5-HT, a large number of previously unknown central neuron systems have been discovered which store and form DA, NA and 5-HT respectively. The amines exist in low concentrations in the cell bodies and the non-terminal axons and in very high concentrations in the varicosities of the synaptic terminals, which are the presynaptic structures specialized for the storage, formation and release of the amines. Very strong evidence has been obtained for the view that monoamine neuro-transmission exists in the CNS. The evidence is mainly based on the fact that the central monoamine neurons have in all essential respects, the same basic properties and characteristics as the peripheral adrenergic neurons.

ACKNOWLEDGMENTS

For generous supplies of drugs we are indebted to the following companies: Swedish Ciba, Sweden (reserpine); Swedish Pfizer (nialamide); Hoffman–La Roche, Basle, Switzerland (iproniazide); Abbott Research Laboratories, Chicago, Ill., U.S.A. (MO-911); and Firma E. Lindblom & Co., Stockholm as agents of Merck, Sharp & Dohme (α-methyl-DOPA).

The skilful technical assistance of Miss Inger Elm, Mrs. Mirta Baidins, Mrs. Gunilla Carlsson and Miss Ingalill Nordgren is gratefully acknowledged.

This investigation has been supported by research grants (Y 247 and 482) from the Swedish Medical Research Council, by a Public Health Service Research Grant (NB 05236-01) from the National Institute of Neurological Diseases and Blindness and by grants from Stiftelsen Th. and J. Anderssons Minne and Gustaf and Tyra Svenssons Minne.

REFERENCES

ANDÉN, N.-E. (1964a) On the mechanisms of noradrenaline depletion by alpha-methyl meta-tyrosine and metaraminol, *Acta Pharmacol.* (*Kbh.*) **21**, 260–71.

ANDÉN, N.-E. (1964b) Studier över monoaminer och deras roll som transmittorsubstanser. M.D. Thesis, Göteborg.

ANDÉN, N.-E., A. CARLSSON, A. DAHLSTRÖM, K. FUXE, N.-Å. HILLARP and K. LARSSON (1964) Demonstration and mapping out of nigro-neostriatal dopamine neurons, *Life Sci.* **3**, 523–30.

ANDÉN, N.-E., A. CARLSSON, N.-Å. HILLARP and T. MAGNUSSON (1964) 5-hydroxytryptamine release by nerve stimulation of the spinal cord, *Life Sci.* **3**, 473–8.

ANDÉN, N.-E., A. CARLSSON, N.-Å. HILLARP and T. MAGNUSSON (1965) Noradrenaline release by nerve stimulation of the spinal cord, *Life Sci.* **4**, 129–32.

ANDÉN, N.-E., A. DAHLSTRÖM, K. FUXE and K. LARSSON (1965a) Further evidence for the presence of nigro-neostriatal dopamine neurons in the rat, *Amer. J. Anat.* **116**, 329–33.

ANDÉN, N.-E., A. DAHLSTRÖM, K. FUXE and K. LARSSON (1965b) Mapping out of CA and 5-HT neuron systems innervating the telencephalon and diencephalon, *Life Sci.* **4**, 1275–9.

ANDÉN, N.-E., J. HÄGGENDAL, T. MAGNUSSON and E. ROSENGREN (1964) The time course of the disappearance of noradrenaline and 5-hydroxytryptamine in the spinal cord after transection, *Acta Physiol. Scand.* **62**, 115–18.

ANDÉN, N.-E., T. MAGNUSSON and E. ROSENGREN (1965) Occurrence of dihydroxyphenylalanine decarboxylase in nerves of the spinal cord and sympathetically innervated organs, *Acta Physiol. Scand.* (in press).

ANDÉN, N.-E., B.-E. ROOS and B. WERDINIUS (1963) 3,4-dihydroxyphenylacetic acid in rabbit corpus striatum normally and after reserpine treatment, *Life Sci.* **2**, 319–25.

AXELROD, J. (1963) The function, metabolism, uptake and release of noradrenaline and adrenaline. In: *The Clinical Chemistry of Monoamines*. Ed.: Varley-Govenlock Amsterdam, Elsevier Publ. Comp., pp. 5–18.

BARTONICEK, V., A. DAHLSTRÖM and K. FUXE (1964) Some effects of certain psychopharmaca on the intraneuronal levels of 5-HT and catecholamines in the specific monoamine neurons of the rat brain, *Experientia* (*Basel*) **20**, 690–3.

BECK, L. (1964) A new concept of autonomic interactions in the peripheral sympathetic nervous system, *Tex. Rep. Biol. Med.* **22**, 375–409.

BERTLER, Å., B. FALCK and C. v. MECKLENBURG (1963) Monoaminergic mechanisms in special ependymic areas. In saline irideus L. *Gen. Comp. Endocr.* 3: 6.

BERTLER, Å and E. ROSENGREN (1959a) Occurrence and distribution of catecholamines in brain, *Acta Physiol. Scand.* **47**, 350–61.

BERTLER, Å. and E. ROSENGREN (1959b) Occurrence and distribution of dopamine in brain and other tissues, *Experientia* (*Basel*) **15**, 10.

BLACKSTAD, T. W. (1963) Ultrastructural studies on the hippocampal region, *Prog. Brain Res.* **3**, 122–48.

BRODIE, B. B., S. SPECTOR and P. A. SHORE (1959) Interaction of drugs with norepinephrine in the brain, *Pharmacol. Rev.* **11**, 548–64.

CARLSSON, A. (1960) Discussion in: *Ciba Found. Symp. on Adrenergic Mechanisms*. Eds: J. R. Vane, G. E. W. Wolstenholme and M. O'Connor London, J. and A. Churchill Ltd, pp. 558–9.

CARLSSON, A. (1964) Functional significance of drug-induced changes in brain monoamine levels, *Prog. Brain Res.* **8**, 67–71.

CARLSSON, A. (1965) Drugs which block the storage of 5-hydroxytryptamine and related amines. In: *Handbuch der exp. Pharmak. Ergänzungswerk*. Ed.: V. Erspamer. Berlin-Göttingen-Heidelberg, Springer-Verlag (in press).

CARLSSON, A. and H. CORRODI (1964) In den Catecholamin-Metabolismus eingreifende Substanzen. 3. 2,3-Dihydroxyphenylacetamide und verwandte Verbindungen, *Helv. Chim. Acta* **47**, 1340–9.

CARLSSON, A., H. CORRODI and B. WALDECK (1963) α-substituierte Dopacetamide als Hemmer der Catechol-O-methyltransferas unter der enzymatischen-hydroxylierung aromatischer Aminosäueren in den Catecholamin-Metabolismus eingreifende Substanzen, *Helv. Chim. Acta* **46**, 2271–85.

CARLSSON, A., A. DAHLSTRÖM, K. FUXE and N.-Å. HILLARP (1965) Failure of reserpine to deplete noradrenaline neurons of α-methyl-noradrenaline formed from α-methyl-DOPA, *Acta Pharmacol.* (*Kbh.*) **22**, 270–6.

CARLSSON, A., A. DAHLSTRÖM, K. FUXE and M. LINDQVIST (1965) Histochemical and biochemical detection of monoamine release from brain neurons, *Life Sci.* **4**, 809–16.

CARLSSON, A., B. FALCK, K. FUXE and N.-Å. HILLARP (1964) Cellular localization of monoamines in the spinal cord, *Acta Physiol. Scand.* **60**, 112–19.

CARLSSON, A., B. FALCK and N.-Å. HILLARP (1962) Cellular localization of brain monoamines, *Acta Physiol. Scand.* **56**, Suppl. 196.

CARLSSON, A., B. FALCK, N.-Å. HILLARP, G. THIEME and A. TORP (1961) A new histochemical method for visualization of tissue catecholamines, *Med. Exp.* (*Basel*) **4**, 123–5.

CARLSSON, A. and N.-Å. HILLARP (1962) Formation of phenolic acids in brain after administration of 3,4-dihydroxy-phenyl-alanine, *Acta Physiol. Scand.* **55**, 95–105.

CARLSSON, A. and M. LINDQVIST (1962) DOPA analogues as tools for the study of dopamine and noradrenaline in brain. In: *Monoamines et Système Nerveux Central.* Ed.: J. de Ajuriaguerra, Symp. Bel-Air, Geneva 1961. Genève, George Co., pp. 89–92.

CARLSSON, A., E. ROSENGREN, Å. BERTLER and J. NILSSON (1957) Effect of reserpine on the metabolism of catecholamines. In: *Psychotropic Drugs.* Eds: S. Garattini and V. Ghetti. Amsterdam, Elsevier Publ. Co., pp. 363–72.

CARLSSON, A. and B. WALDECK (1963) *β*-hydroxylation of tyramine *in vivo*, *Acta Pharmacol.* (*Kbh.*) **20**, 371–4.

CARO, L. G. and G. E. PALADE (1964) Protein synthesis, storage and discharge in the pancreatic exocrine gland, *J. Cell Biol.* **20**, 473–95.

CASPERSSON, T., N.-Å. HILLARP and M. RITZÉN (1965) Fluorescence microspectrophotometry of cellular catecholamines and 5-hydroxytryptamine (in manuscript).

CORRODI, H. and N.-Å. HILLARP (1963) Fluoreszenzmethoden zur histochemischen Sichtbarmachung von Monoaminen. 1. Identifizierung der fluoreszierenden Produkte aus Modellversuchen mit 6,7-Dimethoxy-isochinolinderivaten und Formaldehyd, *Helv. Chim. Acta* **46**, 2425–30.

CORRODI, H. and N.-Å. HILLARP (1964) Fluoreszenzmethoden zur histochemischen Sichtbarmachung von Monoaminen. 2. Identifizierung der fluoreszierenden Produkte aus Dopamin und Formaldehyd, *Helv. Chim. Acta* **47**, 911–18.

CORRODI, H., N.-Å. HILLARP and G. JONSSON (1964) Fluorescence methods for the histochemical demonstration of monoamines. 3. Sodium borohydride reduction of the fluorescent compounds as a specificity test, *J. Histochem. Cytochem.* **12**, 582–6.

CORRODI, H. and G. JONSSON (1965a) Fluoreszenzmethoden zur histochemischen Sichtbarmachung von Monoamiene. 5. Identifizierung des fluoreszierenden Produktes aus Modellversuchen mit 5-Methoxy-Tryptamin und Formaldehyd, *Acta Histochem.* (in press).

CORRODI, H. and G. JONSSON (1965b) Fluorescence methods for the histochemical demonstration of monoamines. 4. Histochemical differentiation between dopamine and noradrenaline in models, *J. Histochem. Cytochem.* **13**, 484–7.

CSILLIK, B. (1964) Histochemical model experiments on the effect of various drugs on the catecholamine content of adrenergic nerve terminals, *J. Neurochem.* **11**, 351–5.

DAHL, E., B. FALCK, M. LINDQVIST and C. V. MECKLENBURG (1963) Monoamines in mollusc neurons, *Kgl. Fysiogr. Sällsk. Lund Förh.* **32**, 89–92.

DAHLSTRÖM, A. (1965) Observations on the accumulation of noradrenaline in the proximal and distal parts of peripheral adrenergic nerves after compression, *J. Anat.* (*Lond.*) (in press).

DAHLSTRÖM, A. and K. FUXE (1964a) Evidence for the existence of monoamine-containing neurons in the central nervous system: I. Demonstration of monoamines in the cell bodies of brain stem neurons, *Acta Physiol. Scand.* **62**, Suppl. 232.

DAHLSTRÖM, A. and K. FUXE (1964b) A method for the demonstration of adrenergic nerve fibres in peripheral nerves, *Z. Zellforsch.* **62**, 602–7.

DAHLSTRÖM, A. and K. FUXE (1964c) A method for the demonstration of adrenergic nerve fibres in the central nervous system, *Acta Physiol. Scand.* **60**, 293–5.

DAHLSTRÖM, A. and K. FUXE (1965a) Evidence for the existence of monoamine neurons in the central nervous system: II. Experimentally induced changes in the intraneuronal amine levels of bulbospinal neuron systems, *Acta Physiol. Scand.* **64**, Suppl. 247.

DAHLSTRÖM, A. and K. FUXE (1965b) Evidence for the existence of an outflow of noradrenaline nerve fibres in the ventral roots of the rat spinal corp, *Experientia* (*Basel*) **21**, 409–11.

DAHLSTRÖM, A. and K. FUXE (1965c) Monoamines and the pituitary gland, *Acta Endocrinol.* (in press).

DAHLSTRÖM, A., K. FUXE and N.-Å. HILLARP (1965) Site of action of reserpine, *Acta Pharmacol.* (*Kbh.*) **22**, 277–92.

DAHLSTRÖM, A., K. FUXE, D. KERNELL and G. SEDVALL (1965) Reduction of the monoamine stores in the terminals of bulbospinal neurons following stimulation in the medulla oblongata, *Life Sci.* **4**, 1275–9.

DAHLSTRÖM, A., K. FUXE, L. OLSON and U. UNGERSTEDT (1964) Ascending systems of catecholamine neurons from the lower brain stem, *Acta Physiol. Scand.* **62**, 485–6.

DAHLSTRÖM, A. and J. HÄGGENDAL (1965a) Some quantitative data on the noradrenaline content in the terminals of sympathetic adrenergic neurons (in manuscript).

DAHLSTRÖM, A. and J. HÄGGENDAL (1965b) Biochemical studies on the transport and turnover of amine storage granules in a peripheral adrenergic neuron system. *Acta Physiol. Scand.* (in press).

DE CARO, G. and M. ROSEGHINI (1964) 5-hydroxytryptamine, metanephrine and histamine in the brain of Rana esculenta, *J. Neurochem.* **11**, 825.

DROZ, B. and C. P. LEBLOND (1963) Axonal migration of proteins in the central nervous system and peripheral nerves as shown by radioautography, *J. Comp. Neurol.* **121**, 325–46.

ECCLES, J. C., P. FATT and K. KOKETSU (1954) Cholinergic and inhibitory synapses in a pathway from motor-axon collaterals to motoneurons, *J. Physiol.* **126**, 524–62.

ELFVIN, L.-G. (1963) The ultrastructure of the superior cervical sympathetic ganglion of the cat. II. The structure of the preganglionic end fibres and the synapses as studied by serial sections, *J. Ultrastruct. Res.* **8**, 441–76.

ERÄNKÖ, O. (1955) Histochemistry of noradrenaline in the adrenal medulla of rats and mice, *Endocrinology* **57**, 363–8.

ERÖS, G. (1932) Eine neue Darstellungsmethode der sogenannten "gelben" argentaffinen Zellen des Magendarmtraktes, *Zbl. allg. Path. path. Anat.* **54**, 385–91.

EULER, U. S. VON (1956) *Noradrenaline.* Ch. C. Thomas, Springfield, Ill.

EULER, U. S. VON (1959) Autonomic neuroeffector transmission. In: *Handbook of Physiology*, Ed.: J. Field. Sect. 1. **1**, 215–37. Washington Am. Physiol. Soc.

EULER, U. S. VON (1961) Neurotransmission in the adrenergic nervous system. In: *Harvey Lect.* **55**, 43–65.

FALCK, B. (1962) Observations on the possibilities of the cellular localization of monoamines by a fluorescence method, *Acta Physiol. Scand.* **56**, Suppl. 197.

FALCK, B., N.-Å. HILLARP, G. THIEME and A. TORP (1962) Fluorescence of catecholamines and related compounds condensed with formaldehyde, *J. Histochem. Cytochem.* **10**, 348–54.

FUXE, K. (1964) Cellular localization of monoamines in the median eminence and the infundibular stem of some mammals, *Z. Zellforsch.* **61**, 710–24.

FUXE, K. (1965a) Evidence for the existence of monoamine neurons in the central nervous system: III. The monoamine nerve terminal, *Z. Zellforsch.* **65**, 573–96.

FUXE, K. (1965b) Evidence for the existence of monoamine neurons in the central nervous system: IV. Distribution of monoamine nerve terminals in the central nervous system, *Acta Physiol. Scand.* **64**, Suppl. 247.

FUXE, K. (1965c) Evidence for the existence of monoamine-containing neurons in the central nervous system. M.D. Thesis, Stockholm.

FUXE, K., A. DAHLSTRÖM and N.-Å. HILLARP (1965) Central monoamine neurons and monoamine neurotransmission. In: *Proc. of the 23rd International Congress of Physiological Sciences, Tokyo.* Exerpta Medica Foundation, Amsterdam 419–34.

FUXE, K. and L.-M. GUNNE (1964) Depletion of the amine stores in brain catecholamine terminals on amygdaloid stimulation, *Acta Physiol. Scand.* **62**, 493–4.

FUXE, K. and N.-Å. HILLARP (1964) Uptake of L-DOPA and noradrenaline by central catecholamine neurons, *Life Sci.* **3**, 1403–6.

FUXE, K., T. HÖKFELT and O. NILSSON (1964) Observations on the cellular localization of dopamine in the caudate nucleus of the rat, *Z. Zellforsch.* **63**, 701–6.

FUXE, K., T. HÖKFELT and O. NILSSON (1965) A fluorescence and electron microscopic study on certain brain regions rich in monoamine terminals, *Amer. J. Anat.* **117**, 33–46.

FUXE, K., T. HÖKFELT, O. NILSSON and S. REINIUS (1966) A fluorescence and electron-microscopic study on central monoamine nerve cells, *Anat. Rec.* (in press).

FUXE, K. and L. LJUNGGREN (1965) Cellular localization of monoamines in the upper brain stem of the pigeon, *J. Comp. Neurol.* (in press).

FUXE, K. and CH. OWMAN (1965) Cellular localization of monoamines in the area postrema of some mammals, *J. Comp. Neurol.* (in press).

GADDUM, J. H. (1963) Chemical transmission in the central nervous system, *Experientia (Basel)* **197**, 741–3.

GUNNE, L. M. (1962) Relative adrenaline content in brain tissue, *Acta Physiol. Scand.* **56**, 324–33.

HAMBERGER, B., T. MALMFORS, K.A. NORBERG and CH. SACHS (1964) Uptake and accumulation of catecholamines in peripheral adrenergic neurons of reserpinized animals, studied with a histochemical method, *Biochem. Pharmacol.* **13**, 841–4.

HAMBERGER, B., T. MALMFORS and CH. SACHS (1965) The standardization of paraformaldehyde and of certain procedures for the histochemical demonstration of monoamines, *J. Histochem. Cytochem.* (in press).

HAMBERGER, B. and D. MASUOKA (1965) Localization of catecholamine uptake in rat brain slices, *Acta Pharmacol. (Kbh.)* **22**, 363–8.

HAMBERGER, B. and K.-A. NORBERG (1964) Histochemical demonstration of monoamines in fresh-frozen sections, *J. Histochem. Cytochem.* **12**, 48–9.

HAMBERGER, B., K.-A. NORBERG and F. SJÖQVIST (1964) Correlated studies of monoamines and acetylcholinesterase in sympathetic ganglia, illustrating the relative distribution of adrenergic and cholinergic neurons. In: *Symposium on cholinergic and adrenergic transmission. Proc. 2nd Int. Pharmacol. Meeting*, Prague 1963. Oxford, Pergamon Press.

HAMPERL, H. (1932) Was sind argentaffine Zellen? *Virschows Arch. path. Anat.* **286**, 811–33.

HESS, S. and S. UDENFRIEND (1959) A fluorimetric procedure for the measurement of tryptamine in tissues, *J. Pharmacol. Exp. Ther.* **127**, 175–7.

HILLARP, N.-Å. (1958) Enzymatic systems involving adenosinephosphates in the adrenaline- and noradrenaline-containing granules of the adrenal medulla, *Acta Physiol. Scand.* **42**, 144–65.

HILLARP, N.-Å. and B. HÖKFELT (1955) Histochemical demonstration of noradrenaline and adrenaline in the adrenal medulla, *J. Histochem. Cytochem.* **3**, 1–5.

HILLARP, N.-Å. and T. MALMFORS (1964) Reserpine and cocaine blocking of uptake and storage mechanisms in adrenergic nerves, *Life Sci.* **3**, 703–8.

HYDÉN, H. (1960) The neuron. In: *The Cell*, IV, Eds: J. Brachet and A. Mirsky, Academic Press, New York, London, pp. 215–323.

HÄGGENDAL, J. (1963) Studier över adrenalin och noradrenalin i vävnader och kroppsvätskor. M.D. Thesis, Göteborg.

HÄGGENDAL, J. and T. MALMFORS (1963) Evidence of dopamine-containing neurons in the retina of rabbits, *Acta Physiol. Scand.* **59**, 295–6.

HÖKFELT, T. and O. NILSSON (1965) The relationship between nerves and smooth muscle cells in the rat iris. II. The sphincter muscle, *Z. Zellforsch.* **66**, 848–56.

ITURRIZA, F. C. (1964) Electron-microscopic study of the pars intermedia of the pituitary of the toad *Bupo arenarum*, *J. Gen. Comp. Endocrinol.* **4**, 492–502.

JACOBOWITZ, D. and G. B. KOELLE (1963) Demonstration of both acetylcholinesterase and catecholamines in the same trunk, *Pharmacologist* **5**, 270.

KIRSHNER, N. (1962) Uptake of catecholamines by a particulate fraction of the adrenal medulla, *J. Biol. Chem.* **237**, 2311–17.

KOELLE, G. B. (1963) Cytological distributions and physiological functions of cholinesterases. In: Cholinesterases and anticholinesterase agents. *Handbuch des exp. Pharmakol.* Suppl. 15, 187–298.

KOPIN, I. J. (1964) Storage and metabolism of catecholamines: The role of monoamine oxidase, *Pharmacol. Rev.* **16**, 179–91.

LUBIŃSKA, L., S. NIEMIERKO, B. ODERFELT and B. L. SZWARC (1964) Behaviour of acetylcholinesterase in isolated nerve segments, *J. Neurochem.* **11**, 493–503.

MALMFORS, T. (1963) Evidence of adrenergic neurons with synaptic terminals in the retina of rats demonstrated with fluorescence and electron microscopy, *Acta Physiol. Scand.* **58**, 99–100.

MALMFORS, T. (1964) Release and depletion of the transmitter in adrenergic terminals produced by nerve impulses after the inhibition of noradrenaline synthesis or reabsorption, *Life Sci.* **3**, 1397–1402.

MALMFORS, T. (1965) Studies on adrenergic nerves. The use of rat and mouse iris for direct observations on their physiology and pharmacology at cellular and subcellular levels, *Acta Physiol. Scand.* **64**, Suppl. 248.

MALMFORS, T. and CH. SACHS (1965) Direct demonstration of the systems of terminals belonging to an individual adrenergic neuron and their distribution in rat iris, *Acta Physiol. Scand.* (in press).

MUSCHOLL, E. and M. VOGT (1958) The action of reserpine on the peripheral sympathetic system, *J. Physiol. (Lond.)* **141**, 132–55.

NILSSON, O. (1964) The relationship between nerves and smooth muscle cells in the rat iris. I. Dilator muscle, *Z. Zellforsch.* **64**, 166–71.

NORBERG, K.-A. (1965) Drug-induced changes in catecholamine levels in the sympathetic adrenergic cell bodies and terminals. A histochemical study, *Acta Physiol. Scand.* (in press).

NORBERG, K.-A. and B. HAMBERGER (1964) The sympathetic adrenergic neuron. Some characteristics revealed by histochemical studies on the intraneuronal distribution of the transmitter, *Acta Physiol. Scand.* **63**, Suppl. 238.

DE ROBERTIS, E. (1964) *Histophysiology of Synapses and Neurosecretion.* Pergamon Press, Oxford, London, New York, Paris.

SHORE, P. A. (1962) Release of serotonin and catecholamines by drugs, *Pharmacol. Rev.* **14**, 531–50.

SPECTOR, S. (1963) Monoamine oxidase in control of brain serotonin and norepinephrine content, *Ann. N. Y. Acad. Sci.* **107**, 856–64.

THIEME, G. (1965) Small tissue dryers with high capacity for rapid freeze-drying, *J. Histochem. Cytochem.* **13**, 386–9.

UDENFRIEND, S. and P. ZALTZMAN-NIRENBERG (1963) Norepinephrine and 3,4-dihydroxyphenetylamine turnover in guinea-pig brain *in vivo*, *Science* **142**, 394–6.

WEIL-MALHERBE, H., L. G. WHITBY and J. AXELROD (1961) The uptake of circulating (H^3) norepinephrine by the pituitary gland and various areas of the brain, *J. Neurochem.* **8**, 55–64.

WEISS, P. (1961) The concept of perpetual neuronal growth and proximo-distal substance convection. In: *Regional Neurochemistry*. Eds: S. S. Kety and J. Elkes. Pergamon Press, New York, London, pp. 220–42.

WHITTAKER, V. P. (1959) The isolation and characterization of acetylcholine-containing particles from brain, *Biochem. J.* **72**, 694–706.

DISCUSSION

MITCHELL: The suggestion of momoamine neurotransmission in the CNS should be supported by clear evidence of cells which are sensitive to these substances. Do you feel that iontophoretic application experiments support your views?

FUXE: It is very difficult to say if the results obtained from microelectrophoretic studies agree with the results obtained with the histochemical fluorescence method. In some cases the neurophysiologists obtain effects of NA or 5-HT in areas where very few or no NA or 5-HT terminals are present. Thus it seems that the depression or excitation obtained in nerve cells by microelectrophoretic application of NA and 5-HT does not necessarily indicate the existence of true noradrenergic and 5-hydroxytryptaminergic mechanisms.

COSTA: What concentrations of NE were used to study the blockade of NE uptake by reserpine? I should like to make a comment concerning some studies which were carried out to investigate the physiological role of dopamine in corpus *striatum*. Dopamine injected by microelectrophoresis slows the rate of discharge of caudate cells; in contrast acetylcholine increases it. Unfortunately the effect of dopamine is not specific because also NE acts like dopamine: it decreases the spontaneous rate of discharge of caudate neurons and blocks the effects of acetylcholine (Bloom, Costa and Salmoiraghi, in press).

HAMBERGER: We have definite evidence for the view that an uptake in nerve terminals in brain slices from reserpine-pretreated rats is highly efficient and not due to simple diffusion (Hamberger and Masuoka, *Acta Pharmacol.* 1965, in press). Thus under the experimental conditions used the concentration in the incubation medium (1 μg/ml) cannot give sufficient fluorescence to be detected in the microscope. On the contrary the intense fluorescence found in the terminals suggests an efficient uptake-concentration mechanism which is capable of concentrating at least 1000 times. In addition the neuronal uptake was strongly inhibited by cocaine.

KOPIN: Iversen (1965) has shown that there are two independent mechanisms for active norepinephrine uptake in the isolated perfused rat heart, the second beginning to operate only at high concentrations of norepinephrine—above those of apparent saturation of the first uptake mechanism.

AXELROD: There are differences in the use of brain slices and in the intact brain on the uptake of noradrenaline. In brain slices, both imipramine and chlorpromazine will block the uptake of radioactive noradrenaline while in the intact brain imipramine will block the uptake of noradrenaline but chlorpromazine will not.

FUXE: After intraventricular injections of NA, it is not possible to say which areas NA will reach and the concentrations will be unknown. On account of these facts it is very difficult to compare the results from such uptake studies with those from brain slices.

BRODIE: With so many discordant results in this field, it is important that a working hypothesis is not called a conclusion. For example the author on the basis of his beautiful techniques, concludes that reserpine does not prevent the uptake of NA into nerve endings. This conclusion is based in part on *in vitro* experiments using tissue slices. Of course reserpine will not block uptake if the concentrations of NE in outer fluid are so high that the active transport system is saturated and amine is now taken up mainly by passive diffusion. However, if tracer amounts of noradrenaline are used it is not possible to show active uptake after reserpine. Again, the conclusion is made that imipramine and cocaine act on the nerve membrane. The results could also be explained by assuming changes in permeability of granules for which evidence is available. Finally, much of our current thinking is based on experiments in which reserpine prevents uptake of NA into granules *in vitro*. It must be kept in mind that reserpine in this case does not release NA. In other words conclusions cannot be made without a sophisticated application of kinetic principles.

FUXE: This will be commented later on by Dr. Carlsson and Dr. Malmfors.

HISTOCHEMISTRY OF MONOAMINERGIC MECHANISMS IN PERIPHERAL NEURONS

B. Falck and Ch. Owman

Departments of Histology and Anatomy,
University of Lund, Lund, Sweden

During the twenty years since von Euler found that noradrenaline is the sympathetic transmitter in mammals, the views on chemical transmission have developed rapidly, and many additional substances have been brought into discussion as possible neurohumours. One of the great difficulties in this field has always been the lack of adequate histochemical methods for the direct study, at the cellular level, of the formation, storage, release and inactivation of such substances. A most important development was made by Koelle and his co-workers (see Koelle, 1962, 1963a and b) who introduced specific methods for the cytological study of acetylcholinesterase. These methods have contributed much to our knowledge of cholinergic mechanisms. Many basic problems concerning nervous transmission, however, demand for their solution methods which allow direct cellular demonstration of the transmitters themselves. This was made possible with respect to the monoaminergic system by the development of a fluorescence microscopical method (Falck and Torp, 1961, 1962; Falck, 1962) based upon the principle that the catecholamines and 5-hydroxytryptamine in a dry milieu and in the presence of protein condense with formaldehyde to intensely fluorescent derivatives (Falck *et al.*, 1962; Corrodi and Hillarp, 1963, 1964). Within a short time the method was found to possess unusual sensitivity, specificity and applicability for demonstrating tissue monoamines, and thus represents an additional valuable tool for the study of monoaminergic mechanisms. A detailed account of the technique and its methodological problems has recently been given by Falck and Owman (1965).

Only results obtained with this fluorescence microscopical method will be considered in the present paper. This is justified by the abundance of material now at our disposal and also by the fact that alternative approaches to the problem of monoaminergic mechanisms will be considered by other contributors to this Symposium.

The fluorescent material obtained upon formaldehyde treatment has—as might be expected—a quite diffuse distribution in the cytoplasm of the monoaminergic neuron. The existence of an additional cytoplasmic, granular fluorescence in the perikarya of neurons in the rat cervical ganglion, suggesting that the monoamine is present in both a "soluble" and "bound" form, was reported by Eränkö and Härkönen (1963) and Härkönen (1964), but analysis of extensive material including different types of monoaminergic neurons from coelenterates to primates has not confirmed this observation. Apart from unspecific autofluorescence, any granular fluorescence related to monoaminergic ganglion cells is probably due to artifacts or to fluorescent varicosities belonging to superimposed adrenergic terminals. Moreover, there are strong reasons for believing that the whole transmitter content must be stored in the submicroscopic amine granules (Hillarp and Malmfors, 1964; Norberg and Hamberger, 1964). Further, it seems that "free" monoamines (or L-DOPA and 5-HTP) are not retained in the cytoplasm but also appear in the nucleus. This can be seen in cells which can take up DOPA and 5-HTP and form monoamines, but lack a specific cytoplasmic storage mechanism. The brain capillaries, for example, form an enzymatic blood-brain barrier in that they readily take up L-DOPA or 5-HTP (but not the corresponding amines and D-DOPA) which are decarboxylated to dopamine and 5-hydroxytryptamine, respectively, within the endothelium (Bertler *et al.*, 1965). This fluorescence is located also in the nuclei of the endothelial cells and pericytes, and is especially clearly demonstrable if the break-down of the amines formed in this way is arrested by inhibition of monoamine oxidase, which gives rise to a very high fluorescence. The endothelial "trap" for these amino acids is not always effective since a fluorescence in the contiguous pericytes can also be seen, and if the blood concentration of the amino acids is sufficiently high, they will diffuse into the brain parenchyma. Where there are leakages of the barrier, the capillaries behave like peripheral capillaries, i.e. the monoamines and their precursors pass freely through the wall. Thus, in the median eminence they are then taken up and concentrated in adrenergic neurons. If the specific storage mechanism in these is not functional, as after reserpine treatment, the increased fluorescence due to the "free" amine appears also in the nucleus.

There are, however, a few examples in which the monoaminergic transmitter is not revealed as a diffuse cytoplasmic fluorescence. In the ventral nerve cord of the leech *Hirudo medicinalis*, ganglion cells occur with the fluorescence characteristics of 5-hydroxytryptamine, in which the fluorescence is confined to numerous small ring-shaped structures, and in some Annelida the specific fluorescence of certain adrenergic neurons appear as a net-work (Myhrberg, 1965).

The transmitter is present in the whole monoaminergic neuron but shows characteristic differences in concentration in the different parts of the neuron (see below). Great variations in transmitter content also seem to occur between different neurons. Thus, in mammalian sympathetic and prevertebral ganglia, the fluorescence intensity of the adrenergic nerve cells varies from low to relatively strong—a phenomenon which cannot be completely explained by differences in the levels at which the cells are sectioned. The intensity differences seem to be unaffected by preganglionic denervation (Härkönen, 1964; Norberg and Hamberger, 1964) but may reflect variations in the rates of granule formation (see Norberg and Hamberger, 1964). The length of the neurons also may have a bearing on these variations. The perikarya of short monoaminergic neurons (see below), e.g. in the internal male accessory genital organs of many mammalian species and in the cod swimbladder often exhibit only slight individual variations in fluorescence intensity, which is quite high. On the other hand, ganglia having a wide spectrum of fibre length, e.g. the superior cervical ganglion, are typical examples of great heterogeneity in cell body fluorescence. The fluorescence intensity of monoaminergic perikarya, however, cannot always be used as a criterion of neuron length. In the swimbladder of the trout *Salmo irideus* (Fahlén, Falck and Rosengren, 1965) the noradrenergic nerve cells are close to the musculature which they innervate and the axons are therefore very short—whole neurons can in fact be observed in a single section—but in spite of this, they display a rather weak perikaryon fluorescence. Further, the short noradrenergic neurons to the internal male genital organs of the rabbit have a remarkably low perikaryon fluorescence, barely visible under normal conditions (Owman and Sjöstrand, 1965). It should also be borne in mind that cell bodies may contain an equal total amount of monoamine, but exhibit great variations in intensity due to corresponding variations in perikaryon volume. The very small ganglion cells related to the internal male genital organs of the monkey emit an unusually high and uniform fluorescence intensity (Owman and Sjöstrand, 1965).

When applying a combination of formaldehyde gas treatment and cholinesterase staining to the cat "sweat secretory" ganglia it has been found that acetylcholinesterase activity is only present in cells lacking specific fluorescence (Hamberger, Norberg and Sjöqvist, 1963a). In similar studies of the rat superior cervical ganglion, however, Eränkö and Härkönen (1964) and Härkönen (1964) found acetylcholinesterase activity in adrenergic cells without any apparent correlation to fluorescence intensity. The discrepancy in these results may reflect a species variation in distribution of acetylcholinesterase. For the present, it cannot be excluded, however, that the last-mentioned authors may also have demonstrated propionylcholinesterase which may interfere despite the use of appropriate inhibitors in the histochemical analysis. This unspecific cholinesterase is known to be present in rat but not in cat tissues (Ord and Thompson, 1951).

When the monoaminergic axon is divided a short distance from the perikaryon, distinct retrograde changes can be observed (Härkönen, 1964.) These appear after 24 hr, when a decrease in fluorescence intensity is seen, concomitant with a swelling of the cell body and disappearance of Nissl granules in stained preparations. The chromatolytic cell exhibits hardly any fluorescence during the next two to three weeks. Many of the cells recover during 50 to 150 days after the operation, and then show an even more intense fluorescence than normally.

The reaction in the cell body is quite different if the axon has been cut or crushed at a greater distance from the cell body and no decrease in fluorescence intensity is seen during a period of 12 hr to one month (Dahlström and Fuxe, 1964). Instead, there is a slight but clearly visible increase in the fluorescence intensity 5 to 55 days after division of the nerve, and this is especially apparent in those ganglion cells normally emitting a low or medium intensity (Norberg and Hamberger, 1964; Owman, 1964a; Owman and Sjöstrand, 1965).

The dendrites and axons from the perikaryon normally exhibit a specific fluorescence of moderate to low intensity, sometimes barely visible, although the short axons from, for example, ganglia related to the male internal genital organs show a fairly considerable fluorescence intensity.

In spite of this low fluorescence in the monoaminergic preterminals they can be specifically demonstrated by raising their amine levels, and therefore their distribution and pathways can easily be studied. The whole sympathetic neuron readily accumulates exogenous noradrenaline, or its precursors L-DOPA and dopamine (Dahlström and Fuxe, 1964; Norberg and Hamberger, 1964; Owman and Sjöstrand, 1965), due to an efficient uptake-concentration mechanism (Hamberger *et al.*, 1964; Hillarp and Malmfors, 1964). The effect is more pronounced after pretreatment with a monoamine oxidase inhibitor. No fluorescence, except the generally increased intensity in the whole tissue, is seen in cholinergic neurons, and therefore these experiments provide a useful tool for distinguishing monoaminergic neurons in doubtful cases (see Owman and Sjöstrand, 1965).

The preterminal axon can also be demonstrated along its whole length by cutting or crushing the nerve, so causing an accumulation of the monoamine storage granules in a proximal direction from the point of lesion (Dahlström and Fuxe, 1964; Härkönen, 1964; Owman, 1964; Owman and Sjöstrand, 1965). A slight accumulation of fluorescent material is observed near the swollen and distorted cut end even after 21 hr. The fluorescence is markedly enhanced after 24 to 36 hr, and during the following days the accumulation spreads towards the perikaryon. The whole proximal stump exhibits a very intense fluorescence after 4 days and this accumulation has been found to be essentially unchanged even after 55 days.

The smooth, weakly fluorescent preterminals, as they approach the effector cells, are transformed into varicose fibres emitting an intense light especially from their varicosities (Falck, 1962). The transition is sharply demarked and can be easily observed in individual axons, e.g. from the whole iris mounted as a stretch-preparation (see Falck and Owman, 1965). The varicosities have a diameter of about 0·4 to 2μ and are richly distributed along the thin (about 0·2μ) terminal fibre. It can be calculated that the amine concentration in these varicosities is in the order of 10·000 μg/g. For comparison, the concentration in the cell body and its preterminal is probably 10–100 μg/g (Norberg and Hamberger, 1964). The sympathetic terminals are conspicuously thin in the pineal gland of rat (Bertler, Falck and Owman, 1963, 1964) and guinea-pig (Owman, 1965), as well as in collateral sprouts growing into transplanted tissue (Owman, 1964a). In these thin terminals the varicosities are distributed very irregularly along the fibre.

With a combination of methods for demonstrating monoaminergic nerves and cholinesterases, cholinesterase-positive fibres have been localized in the dilatator zone of the rat iris in a plexus of the same density and topography as the monoaminergic fibres. To a large extent the enzyme is an unspecific cholinesterase, as demonstrated with various inhibitors. It is probably located in the Schwann cells belonging to the adrenergic plexus (Ehinger, 1965).

The fluorescence and morphology of the axons are not altered after sectioning the preganglionic nerves (Falck, 1962; Falck and Torp, 1962; Norberg and Hamberger, 1964; Owman, 1964a; Falck, Owman and Sjöstrand, 1965; Owman and Sjöstrand, 1965). After excision of the superior cervical chain in the rat, a fluorescence decrease can be seen in the iris nerves even after 15 hr. During the following few hours the fluorescence is considerably reduced, and 48 hr post-operatively it has disappeared from all nerves (Falck, 1962). This rapid disappearance of the monoamine thus precedes the structural disintegration of the sympathetic fibres, as studied in the rat pineal gland (Håkanson and Owman, 1965).

The accumulated evidence that the varicose part of the neuron has a synaptic function has been discussed at length by Norberg and Hamberger (1964) and there are also good reasons for believing that the swellings, i.e. the varicosities, form the synaptic contact with the effector cells (Elfvin, 1963). However, it is not immediately evident that all varicosities have a transmitter releasing function. In many tissues, e.g. the iris, pineal tissue and the male internal genital organs, nerve bundles contain, in addition to smooth, weakly fluorescent axons, intensely fluorescent varicose fibres which often can be followed into an autonomic ground-plexus of adrenergic terminals (Falck, 1964; Owman, 1964a, 1965; Ehinger, 1965; Owman and Sjöstrand, 1965). In the central nervous system of various species, beaded fibres have repeatedly been observed travelling for a comparatively long distance before reaching and synaptically enclosing nerve cells. This is especially evident in the chicken brain (Ljunggren, 1965) in the tract homologous with the dopaminergic nigro-neostriatal pathway in mammals (Bertler *et al.*, 1964). The fact that some monoaminergic axons transform into varicose fibres at a certain distance from the effector sites does not exclude a transmitter release in these parts although it may also imply a transmitter deposit that is not immediately available for transmission. Finally it should be mentioned that in central and peripheral tissues of many submammalian species (e.g. teleosts, cyclostomes, annelids, coelenterates) (Dahl *et al.*, 1963) there are monoaminergic cells which issue varicose fibres directly.

In many ganglia and nerve bundles in both mammalian and submammalian species there occurs a peculiar type of small, polymorph cell with coarse processes of varying length and exhibiting a catecholamine fluorescence of an extreme intensity comparable to that of adrenal medullary cells (Hamberger, Norberg and Sjöqvist, 1963a; Falck, Owman and Sjöstrand, 1965; Owman and Sjöstrand, 1965). Some of these cells store a primary catecholamine while others contain adrenaline (Owman and Sjöstrand, 1965).

Soon after the development of the histochemical fluorescence method (Falck and Torp, 1961, 1962; Falck, 1962; Falck *et al.*, 1962) it appeared that not only noradrenaline but also dopamine and 5-hydroxytryptamine (Dahl *et al.*, 1962) and adrenaline (Falck, Häggendal and Owman, 1963) appear in peripheral neurons possessing the same characteristics as noradrenergic sympathetic neurons with respect to morphology and intraneuronal distribution of the amines (see also Falck, 1964). It was also demonstrated (Owman, 1964a) that a primary catecholamine and 5-hydroxytryptamine occur simultaneously in certain sympathetic terminals (see below). It should be noted that the presence of peripheral dopamine in certain bovine nervous tissue may well be explained by the presence of mast cells which carry high amounts of dopamine in ruminants (Falck *et al.*, 1964; Owman and Falck, 1965).

By use of the fluorescence method important findings have been obtained by Hamberger *et al.* (1964) and Hillarp and Malmfors (1964) regarding the uptake and storage mechanisms in adrenergic nerves. The whole adrenergic neuron possesses a highly efficient uptake mechanism in the cell membrane for noradrenaline and related amines. Thus, noradrenaline is readily taken up and accumulated against a considerable gradient even after suppressing the storage mechanism with reserpine. The amine is retained intra-axonally only for a short time if the monoamine oxidase has not previously been inhibited. This and other findings (see Norberg and Hamberger, 1964; Hillarp, personal communication) provides further strong evidence that monoamine oxidase is located within the whole neuron and seems definitely to demonstrate that the monoaminergic transmitter cannot exist outside the storage granules to any great extent. The findings disprove the view of Csillik (1964) that the fluorescence method has the serious limitation that only those amines bound in the storage granules can be demonstrated.

A remarkable uptake-storage mechanism has been disclosed in the pineal nerves of rat and guinea-pig. The pineal sympathetic nerves, which arise in the superior cervical ganglion, display in their preterminal parts the characteristics of noradrenergic fibres (Owman, 1964a and b). However, when running within the pineal gland their terminals take up considerable amounts of 5-hydroxytryptamine (originating from the pineal parenchyma) with a displacement of a certain amount of the noradrenaline. On the other hand, in the rat pineal all this nervous 5-hydroxytryptamine can be displaced by exogenous noradrenaline (Bertler, Falck and Owman, 1963, 1964; Owman, 1964a and b, 1965). That this uptake mechanism is not unique for the pineal nerves is shown by the fact that adrenergic sympathetic nerves growing into the transplanted pineal gland also take up and store 5-hydroxytryptamine together with noradrenaline (Owman, 1964a). Moreover, it has been found that in, or in the vicinity of human and simian pineals, but not in other tissues, even the mast cells exhibit the yellow fluorescence characteristic of 5-hydroxytryptamine.

Transmitter synthesis has been found to initiate surprisingly early in the developing sympathetic system (Enemar, Falck and Håkanson, 1965). Thus, noradrenergic cells in the primary sympathetic chain of the chick embryo can be demonstrated even during the fourth day of incubation, slightly preceded by the appearance of DOPA-decarboxylase activity. The number of fluorescent cells increases rapidly and an intensive migration of adrenergic cells forming structures such as the aortic plexus and adrenal medulla can be followed. At the same stage outgrowing, smooth outlined sympathetic fibres appear showing a considerable fluorescence along their whole length. The adrenergic innervation of different organs appears at various stages. For example there is a rich invasion of nerves (mostly related to the vessels) in the pancreas already by the 10th day, whereas the adrenergic terminals to the submucous and myenteric plexuses are first observed in the 20-day embryo. In a current investigation of human embryos a final adrenergic innervation pattern has been found in 20-week fetuses at many sites, e.g. different skin areas, the cholinergic ganglia of the intestines, and some genital organs. On the other hand, only a sparse distribution of adrenergic nerves is present in the heart at that stage.

The fluorescence method has been extensively applied to studies of the peripheral adrenergic innervation apparatus by many investigators. The accumulated data (see discussion in Norberg and Hamberger, 1964) completely conform with Hillarp's fundamental concept on the construction of the autonomic neuro-effector system (Hillarp, 1946, 1959).

The knowledge of the topographic distribution of the adrenergic terminals in different tissues is of profound importance for the understanding of the response to sympathetic activation. In this connection only some basic examples of general interest will be given. A conspicuous feature of the vascular system is that an adrenergic ground-plexus which does not penetrate into the smooth muscle cells is superimposed on the media of the arterial vessels. Irrespective of the thickness of the media, only few fibres can be seen in its superficial layer (Falck, 1962; Norberg and Hamberger, 1964). This plexus exhibits great variations in the density of fibres, being very dense around mesenterial arteries but much less developed in the arteries of the skeletal musculature (Fuxe and Sedvall, 1964). The veins are as a rule provided with far fewer fibres or, as can be seen in many tissues, even none at all. The vessels of the central nervous system usually have no or a remarkably poor adrenergic innervation (Carlsson *et al.*, 1964; Falck, Mchedlishvili and Owman, 1965). Arteries with a special type of innervation implying a more or less complete penetration of adrenergic nerves into the media have been found to occur simultaneously with arteries of the ordinary type in a variety of tissues: pineal gland (Owman, 1965), iris of various mammals (Ehinger, 1964, 1965), gingiva (dog), finger-tips and peni (cynomolgous monkey).

The vas deferens is frequently used as a model structure for physiological and pharmacological studies of monoaminergic mechanisms and so its sympathetic innervation has aroused new interest in recent years. The vas deferens contains high levels of noradrenaline (Sjöstrand, 1962b) in accordance with an extensive innervation by way of adrenergic terminals (Falck, 1962; Norberg and Hamberger, 1964; Falck, Owman and Sjöstrand, 1965; Owman and Sjöstrand, 1965). In both chemical (Sjöstrand, 1962b) and histochemical (Falck, Owman and Sjöstrand, 1965; Owman and Sjöstrand, 1965) studies, the vas deferens noradrenaline has turned out to be resistant to hypogastric denervation. This, and the finding that ganglionic-blocking substances abolish the motor response of the guinea-pig vas deferens to hypogastric nerve stimulation (Sjöstrand, 1962a), suggested the presence of a peripheral synapse in close relation to the internal male genital organs. The presence of peripheral sympathetic ganglia was accordingly demonstrated by Falck, Owman and Sjöstrand (1965) using the fluorescence microscopic technique, and has been confirmed in studies by Norberg and Hamberger (1964). Extensive comparative studies on the innervation apparatus of the internal accessory male genital organs of several species are now on record and a similar arrangement has been found to occur (Owman and Sjöstrand, 1965; Sjöstrand, 1965). This pattern is remarkable since here adrenergic neurons arise in ganglia located in the vicinity of, or within, the innervated organs, thus resembling that seen in the parasympathetic nervous system. However, in lower vertebrates similar short sympathetic neurons exist in other organ systems (Fahlén, Falck and Rosengren, 1965). To judge from the data given by Risley and Skrepetos (1964) there seems to be no correlation between cholinesterase-positive structures in the vas deferens and the noradrenergic nerve fibres. Instead, the cholinesterase activity probably outlines a system of cholinergic neurons which run in the smooth muscle coats of the vas deferens to the secretory epithelium.

In the intestinal wall the muscle coats contain only very few adrenergic terminals, whereas a rich supply of such fibres enclose synaptically the ganglion cells of the submucous and myenteric plexuses (Falck, 1964; Norberg, 1964; Enemar, Falck and Håkanson, 1965). It might be concluded from this distributional pattern that the adrenergic inhibitory function on the intestinal wall is exerted in the ganglia and not directly on the muscle cells.

Adrenergic terminals superimposed on cells in various ganglia may have different functions. There is good evidence for the presence of intraganglionic adrenergic neurons in various ganglia (Hamberger, Norberg and Sjöqvist, 1963a and b; Owman and Sjöstrand, 1965), and these may exert some modulatory influence in the ganglia. This system also includes a rich occurrence of adrenergic fibres synaptically terminating upon adrenergic cell bodies. It may well be that both these neurons contain the same transmitter since it has been proved that noradrenergic neurons arising in the coeliac ganglion of the cod terminate synaptically upon the peripheral noradrenergic neurons supplying the musculature of the swimbladder (Fahlén, Falck and Rosengren, 1965).

The fact that corresponding structures in different species may have a quite divergent adrenergic innervation pattern is of great importance for the interpretation of tissue response to sympathetic activation. Among numerous examples the following may be mentioned. The mammalian heart possesses a typical autonomic ground-plexus whereas the teleost heart is completely devoid of monoaminergic innervation, except for a small number of terminals in the bulbus arteriosus. Nor does the cyclostome heart receive any monoaminergic fibres (Dahl *et al.*, 1965) but contains a peculiar system of branching cells storing huge amounts of catecholamines, allowing the demonstration of them by the chromaffin reaction (Bloom *et al.*, 1961). In the pineal gland the types of monoamines present in nerves (and also the parenchyma) as well as the distribution of monoaminergic nerves between pineal vessels and the parenchyma differ widely between mammalian species (Owman, 1964b, 1965). The pupillary sphincter of the cat has—in contrast to that of many other mammals—a remarkably rich supply of adrenergic terminals. The circular muscle layer of the dog vas deferens differs from the innervation pattern described above in that it contains a rich distribution of almost exclusively non-monoaminergic fibres (Owman and Sjöstrand, 1965). Finally, the lacrimal gland possesses an adrenergic innervation in cat and monkey but not in rat, rabbit and guinea-pig (Ehinger, 1965).

SUMMARY

Basic features of the peripheral monoaminergic neuron are discussed in terms of findings obtained by a recently devised fluorescence microscopic method for the cellular localization of dopamine, noradrenaline, adrenaline, and 5-hydroxytryptamine.

ACKNOWLEDGMENTS

The investigations presented in this paper have been supported by grants from Air Force Office of Scientific Research under Grant AF EOAR 64–5 through the European Office of Aerospace Research (OAR), United States Air Force; United States Public Health Service (NB 05236–01); Association for the Aid of Crippled Children, New York, USA; and the Swedish Medical Research Council.

REFERENCES

BERTLER, Å., B. FALCK and CH. OWMAN (1963) Cellular localization of 5-hydroxytryptamine in the rat pineal gland, *Kgl. Fysiogr. Sällsk. Lund Förh.* **33**, 13–16.

BERTLER, Å., B. FALCK and CH. OWMAN (1964) Studies on 5-hydroxytryptamine stores in pineal gland of rat, *Acta Physiol. Scand.* **63**, Suppl. 239, 1–18.

BERTLER, Å., B. FALCK, CH. OWMAN and E. ROSENGREN (1965) Monoaminergic barrier mechanisms in the brain vessels of the mouse, *Pharmacol. Rev.* (in press).

BERTLER, Å., C. G. GOTTFRIES, L. LJUNGGREN and E. ROSENGREN (1964) Some observations on adrenergic connections between mesencephalon and cerebral hemispheres, *Acta Pharmacol. (Kbh.)* **21**, 283–9.

BLOOM, G., E. ÖSTLUND, U. S. VON EULER, F. LISHAJKO, M. RITZÉN and J. ADAMS-RAY (1961) Studies on catecholamine-containing granules of specific cells in cyclostome hearts, *Acta Physiol. Scand.* **53**, Suppl. 185, 1–35.

CARLSSON, A., B. FALCK, K. FUXE and N.-Å. HILLARP (1964) Cellular localization of monoamines in the spinal cord, *Acta Physiol. Scand.* **60**, 112–19.

CORRODI, H. and N.-Å. HILLARP (1963) Fluoreszenzmethoden zur histochemischen Sichtbarmachung von Monoaminen. 1. Identifizierung der fluoreszierenden Produkte aus Modellversuchen mit 6,7-Dimethoxyisochinolinderivaten und Formaldehyd, *Helv. chim. Acta* **46**, 2425–30.

CORRODI, H. and N.-Å. HILLARP (1964) Fluoreszenzmethoden zur histochemischen Sichtbarmachung von Monoaminen. 2. Identifizierung des fluoreszierenden Produktes aus Dopamin und Formaldehyd, *Helv. chim. Acta* **47**, 911–18.

CSILLIK, B. (1964) Histochemical model experiments on the effect of various drugs on the catecholamine content of adrenergic nerve terminals, *J. Neurochem.* **11**, 351–5.

DAHL, E., B. FALCK, M. LINDQVIST and C. VON MECKLENBURG (1962) Monoamines in mollusc neurons, *Kgl. Fysiogr. Sällsk. Lund Förh.* **32**, 89–92.

DAHL, E., B. FALCK, C. VON MECKLENBURG and H. MYHRBERG (1963) An adrenergic nervous system in sea anemones, *Quart. J. Micr. Sci.* **104**, 531–4.

DAHL, E., B. FALCK, C. VON MECKLENBURG, H. MYHRBERG and E. ROSENGREN (1965) On the monoamine storing cells in the heart of *Lampetra fluviatilis* and *L. planeri* (Cyclostomata). To be published.

DAHLSTRÖM, A. and K. FUXE (1964) A method for the demonstration of adrenergic nerve fibres in peripheral nerves, *Z. Zellforsch.* **62**, 602–7.

EHINGER, B. (1964) Adrenergic nerves to the eye and its adnexa in rabbit and guinea pig, *Acta Univ. Lund II* **20**, 1–23.

EHINGER, B. (1965) Unpublished observations.

ELFVIN, L.-G. (1963) The ultrastructure of the superior cervical sympathetic ganglion of the cat. II. The structure of the preganglionic end fibres and the synapses as studied by serial sections, *J. Ultrastruct. Res.* **8**, 441–76.

ENEMAR, A. B. FALCK, and R. HÅKANSON (1965) Observations on the appearance of norepinephrine in the sympathetic nervous system of the chick embryo, *Develop. Biol.* **11**, 268–83.

ERÄNKÖ, O. and M. HÄRKÖNEN (1963) Histochemical demonstration of fluorogenic amines in the cytoplasm of sympathetic ganglion cells of the rat, *Acta Physiol. Scand.* **58**, 285–6.

ERÄNKÖ, O. and M. HÄRKÖNEN (1964) Noradrenaline and acetylcholinesterase in sympathetic ganglion cells of the rat, *Acta Physiol. Scand.* **61**, 299–300.

FAHLÉN, G., B. FALCK and E. ROSENGREN (1965) Monoamines in the swimbladder of *Gadus callarias* and *Salmo irideus*, *Acta Physiol. Scand.* **64**, 119–26.

FALCK, B. (1962) Observations on the possibilities of the cellular localization of monoamines by a fluorescence method, *Acta Physiol. Scand.* **56**, Suppl. 197, 1–25.

FALCK, B. (1964) Cellular localization of monoamines. In: *Progress in Brain Research*, H. E. Himwich and W. A. Himwich (Eds), Vol. 8 (Biogenic Amines). Amsterdam, Elsevier, pp. 28–44.

FALCK, B., N.-Å. HILLARP, G. THIEME and A. TORP (1962) Fluorescence of catechol amines and related compounds condensed with formaldehyde, *J. Histochem. Cytochem.* **10**, 348–54.

FALCK, B., J. HÄGGENDAL and CH. OWMAN (1963) The localization of adrenaline in adrenergic nerves in the frog, *Quart. J. Exp. Physiol.* **48**, 253–7.

FALCK, B., G. I. MCHEDLISHVILI and CH. OWMAN (1965) Histochemical demonstration of adrenergic nerves in cortex-pia of rabbit, *Acta Pharmacol.* (*Kbh.*) **23** (in press).

FALCK, B., T. NYSTEDT, E. ROSENGREN and J. STENFLO (1964) Dopamine and mast cells in ruminants, *Acta Pharmacol.* (*Kbh.*) **21**, 51–8.

FALCK, B. and CH. OWMAN (1965) A detailed methodological description of the fluorescence method for the cellular demonstration of biogenic monoamines, *Acta Univ. Lund II* **7**, 1–23.

FALCK, B., CH. OWMAN and N. O. SJÖSTRAND (1965) Peripherally located adrenergic neurons innervating the vas deferens and the seminal vesicle of the guinea pig, *Experientia* (*Basel*) **21**, 98–100.

FALCK, B. and A. TORP (1961) A fluorescence method for histochemical demonstration of noradrenaline in the adrenal medulla, *Med. Exp.* (*Basel*) **5**, 429–32.

FALCK, B. and A. TORP (1962) New evidence for the localization of noradrenaline in the adrenergic nerve terminals, *Med. Exp.* (*Basel*) **6**, 169–72.

FUXE, K. and G. SEDVALL (1964) Histochemical and biochemical observations on the effect of reserpine on noradrenaline storage in vasoconstrictor nerves, *Acta Physiol. Scand.* **61**, 121–9.

HAMBERGER, B., T. MALMFORS, K.-A. NORBERG and CH. SACHS (1964) Uptake and accumulation of catecholamines in peripheral adrenergic neurons of reserpinized animals, studied with a histochemical method, *Biochem. Pharmacol.* **13**, 841–4.

HAMBERGER, B., K.-A. NORBERG and F. SJÖQVIST (1963a) Cellular localization of monoamines in sympathetic ganglia of the cat. A preliminary report, *Life Sci.* **9**, 659–61.

HAMBERGER, B., K.-A. NORBERG and F. SJÖQVIST (1963b) Evidence for adrenergic nerve terminals and synapses in sympathetic ganglia, *Int. J. Neuropharmacol.* **2**, 279–82.

HILLARP, N.-Å. (1946) Structure of the synapse and the peripheral innervation apparatus of the autonomic nervous system, *Acta Anat.* (*Basel*) **2**, Suppl. 4, 1–153.

HILLARP, N.-Å. (1959) The construction and functional organization of the autonomic innervation apparatus, *Acta Physiol. Scand.* **46**, Suppl. 157, 1–38.

HILLARP, N.-Å. and T. MALMFORS (1964) Reserpine and cocaine blocking of the uptake and storage mechanisms in adrenergic nerves, *Life Sci.* **3**, 703–8.

HÅKANSON, R. and CH. OWMAN (1965) Effect of denervation and enzyme inhibition on DOPA decarboxylase and monoamine oxidase activities of rat pineal gland, *J. Neurochem.* **12**, 417–29.

HÄRKÖNEN, M. (1964) Carboxylic esterases, oxidative enzymes and catecholamines in the superior cervical ganglion of the rat and the effect of pre- and postganglionic nerve division, *Acta Physiol. Scand.* **63**, Suppl. 237, 1–94.

KOELLE, G. B. (1962) A new general concept of the neurohumoral functions of acetylcholine and acetylcholinesterase, *J. Pharm. Pharmacol.* **14**, 65–90.

KOELLE, G. B. (1963a) The use of histochemistry in pharmacological studies, *Proceedings of the First International Pharmacological Meeting*, Vol. 5. Oxford, Pergamon Press, pp 5–14.

KOELLE, G. B. (1963b) Cytological distributions and physiological functions of cholinesterases, *Handbuch der exp. Pharmakol.* **15**, 187–98.

LJUNGGREN, L. (1965) Unpublished observations.

MYHRBERG, H. (1965) Unpublished observations.

NORBERG, K.-A. (1964) Adrenergic innervation of the intestinal wall studied by fluorescence microscopy, *Int. J. Neuropharmacol.* **3**, 379–82.

NORBERG, K.-A. and B. HAMBERGER (1964) The sympathetic adrenergic neuron. Some characteristics revealed by histochemical studies on the intraneuronal distribution of the transmitter, *Acta Physiol. Scand.* **63**, Suppl. 238, 1–42.

ORD, M. G. and R. H. S. THOMPSON (1951) The preparation of soluble cholinesterases from mammalian heart and brain, *Biochem. J.* **49**, 191–9.

OWMAN, CH. (1964a) Sympathetic nerves probably storing two types of monoamines in the rat pineal gland, *Int. J. Neuropharmacol.* **3**, 106–12.

OWMAN, CH. (1964b) New aspects of the mammalian pineal gland, *Acta Physiol. Scand.* **63**, Suppl. 239, 1–40.

OWMAN, CH. (1965) Localization of neuronal and parenchymal monoamines under normal and experimental conditions in the mammalian pineal gland. In: *Progress in Brain Research*, Vol. 10 (Structure and Function of the Epiphysis Cerebri). Ariëns Kappers and J. P. Schadé (Eds), Amsterdam, Elsevier.

OWMAN, CH. and B. FALCK (1965) Localization of neuronal monoamines at the cellular level, *Proceedings of the Second International Pharmacological Meeting*, Vol. 3, pp. 165–83.

OWMAN, CH. and N. O. SJÖSTRAND (1965) Short adrenergic neurons and catecholamine-containing cells in vas deferens and accessory male genital glands of different mammals, *Z. Zellforsch.* **66**, 300–20.

RISLEY, P. L. and C. N. SKREPETOS (1964) Histochemical distribution of cholinesterases in the testis, epididymis and vas deferens of the rat, *Anat. Rec.* **148**, 231–49.

SJÖSTRAND, N. O. (1962a) Inhibition by ganglionic blocking agents of the motor response of the isolated guinea-pig vas deferens to hypogastric nerve stimulation, *Acta Physiol. Scand.* **54**, 306–15.

SJÖSTRAND, N. O. (1962b) Effect of reserpine and hypogastric denervation on the noradrenaline content of the vas deferens of the guinea-pig, *Acta Physiol. Scand.* **56**, 376–80.

SJÖSTRAND, N. O. (1965) The adrenergic innervation of the vas deferens and the accessory male genital glands. An experimental and comparative study, *Acta Physiol. Scand.* **65**, Suppl. 257.

DISCUSSION

PLETSCHER: Are there sympathetic nerve terminals in organs like iris and vas deferens or do the sympathetic fibres form a syncytium?

FALCK: The fluorescence microscopic studies on these as well as many other organs have demonstrated that the adrenergic fibres do not form a syncytium but a ground-plexus in the sense of Hillarp (Hillarp, N.-Å., *Acta Physiol. Scand.*, 1959, suppl. 157).

MALMFORS: Do the sympathetic nerve terminals form a syncytial network?

FALCK: As it is possible to demonstrate systems of branching terminals of individual adrenergic neurons after partial denervation it is obvious that the adrenergic ground-plexus does not have any syncytial organization.

VON EULER: Dr. Falck, would you suggest that the noradrenaline demonstrated in the preterminal parts between the varicosities of the adrenergic nerves is in bound or free form?

FALCK: Probably mainly in the bound form since catecholamines taken up by adrenergic nerves in reserpinized animals are retained only for a short time unless MAO has been inhibited.

LINDOR BROWN: You would expect decentralization to produce an increase in the amine content of postganglionic nerves but the amount would probably not be more than a doubling. I do not suppose that the fluorescent method is sufficiently discriminating to show a difference of that sort.

FALCK: No, it is not.

ERÄNKÖ: Together with HÄRKÖNEN, we have observed the small branched cells in the sympathetic ganglion of the rat to exhibit a yellow fluorescence, electronmicroscopically these cells appear to look like adrenal medullary cells. Have you any comment on this?

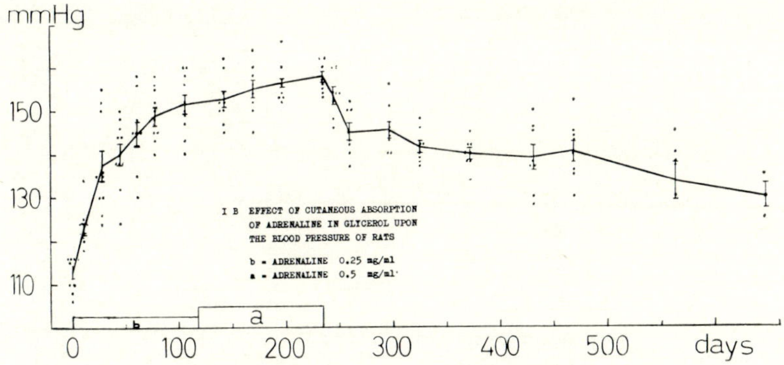

FIG. 1. Effect of cutaneous absorption of adrenaline and noradrenaline in glycerol through the tail of rat upon the blood pressure in acute tests. Noradrenaline bitartr. 0·5 mg/ml. Adrenaline 0·25 mg/ml.

RITZÉN: By fluorescence microspectrographic methods (Caspersson, Hillarp, Ritzén, 1965) it is possible to define the fluorescence emission from parts of cells in sections. Emission (and excitation) spectra from the small "yellowish" cells in sympathetic ganglia are identical with those of the ordinary more "greenish" cells (maximal at 470 mμ). This is easily differentiated from the emission of 5-HT (maximal at 520 mμ).

PEKKARINEN: I would like to ask your opinion on the location and storage of adrenaline and noradrenaline in varicosities or peripheral postganglionic sympathetic neurons, especially in the walls of blood vessels after chronic administration of adrenaline and

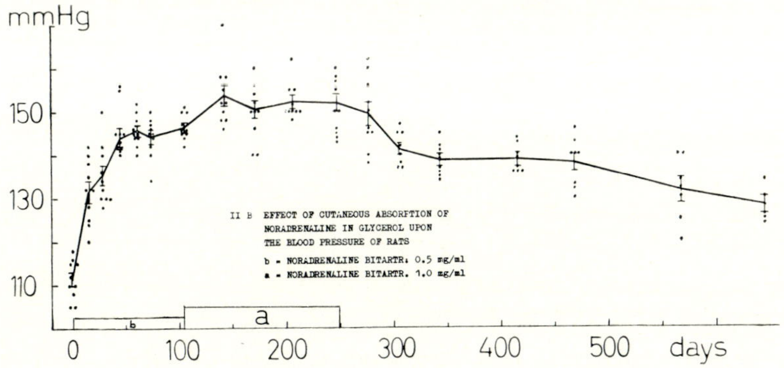

FIG. 2. Experimental production of hypertension by prolonged administration of noradrenaline in rats. Effect of cutaneous absorption of noradrenaline in glycerol through the tail of rat upon the blood pressure. Noradrenaline bitartrate 0·5 and 1·0 mg/ml. in glycerol administered daily.

noradrenaline. We have produced experimental hypertension in rats by prolonged administration of adrenaline or noradrenaline through percutaneous adsorption in glycerol through rat tails (Konu, Pekkarinen, Loajoki). Adrenaline and noradrenaline are administered in 8 months' time (Figs. 1 and 2). The blood pressure increases slightly also after single (Fig. 3) administration of adrenaline and noradrenaline in acute experiments. During the prolonged treatment the increase of blood pressure has been 40–50 mm Hg above the control level.

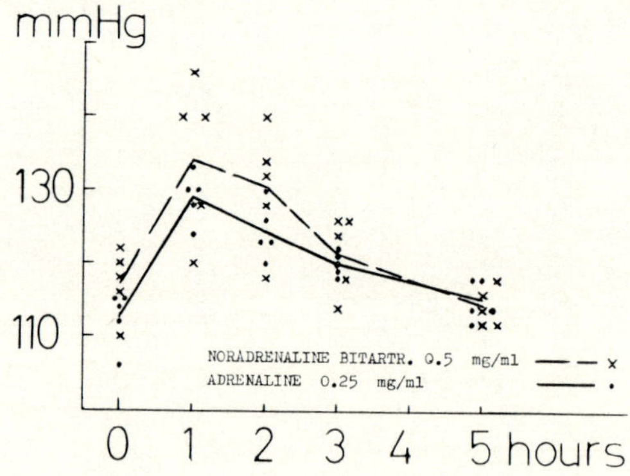

FIG. 3. Experimental production of hypertension by prolonged administration of adrenaline in rats. Effect of cutaneous absorption of adrenaline in glycerol through the tail of rat upon the blood pressure. Adrenaline 0·25 and 0·5 mg/ml. in glycerol, administered daily.

This increase continued then also without treatment by adrenaline and noradrenaline several months afterwards, showing continuous and slow tendency to normalize. After one year, the mean blood pressure is still slightly above the control level. We believe, that the maintenance of continuous vasoconstriction for 8 months causes the peripheral vasoconstriction which continued then also without further treatment. I suppose that it is interesting to know the localization and storage of adrenaline and noradrenaline after this prolonged treatment with adrenaline and noradrenaline, which produces the experimental hypertension.

FALCK: Unfortunately, we have no data which can explain your findings.

IN VITRO RELEASE AND UPTAKE OF NORADRENALINE IN THE RAT IRIS

OLAVI ERÄNKÖ and LIISA RÄISÄNEN
Department of Anatomy, University of Helsinki,
Siltavuorenpenger, Helsinki

NERVE fibres whose intrinsic stores of noradrenaline (NA) have been depleted with reserpine rapidly restore the normal NA content after exogenous administration of this amine (Hillarp and Malmfors, 1964). Moreover, Angelakos (1964) has recently reported that the NA content of a normal iris increases 600 per cent after incubation *in vitro* in Tyrode's solution containing 50 μg/ml of NA and 1 μg/ml of adenosine triphosphate (ATP).

The present report deals with the histochemically demonstrable NA content of the iris of the rat after *in vitro* incubation in solutions containing NA, dihydroxyphenylalanine (DOPA), tyramine (TA), dopamine (DA), adrenaline (A) or 5-hydroxytryptamine (5-HT).

MATERIAL AND METHODS

Albino rats of the Sprague–Dawley strain were used. The animals were killed by decapitation after a short ether anaesthesia. The iris was removed as a whole and transferred immediately into the incubation medium.

The basic medium consisted of Krebs' salt solution supplemented with calcium chloride (1 ml of 1·22 per cent $CaCl_2$ into 100 ml). This medium was used as such or after addition of 1 mg/ml of glucose and/or 0·001–1·0 mg/ml of ATP.

After dissolving in the above media–100 μg/ml of DOPA, TA, DA, NA, A or 5-HT, the pH was adjusted at 7·0. The incubation was carried out at 37°C without agitation. After incubation the iris was washed in three changes of the medium in which no amine was dissolved, for 10 min each. It was then stretched on a slide and allowed to dry for 10 min. The slide was subsequently exposed to dry formaldehyde vapour for 15 min–16 hr at 50–80°C to render the amines into fluorescent compounds, which are formed at different rates from different amines (Eränkö, 1964). The specimens were then mounted in Entellan and examined in the fluorescence microscope.

RESULTS

Incubation in the Krebs' salt solution, with or without glucose or ATP, was followed by a slow weakening in the intensity of the catecholamine fluorescence in the sympathetic nerve net of the iris. No essential difference was observed between specimens exposed directly to formaldehyde and those

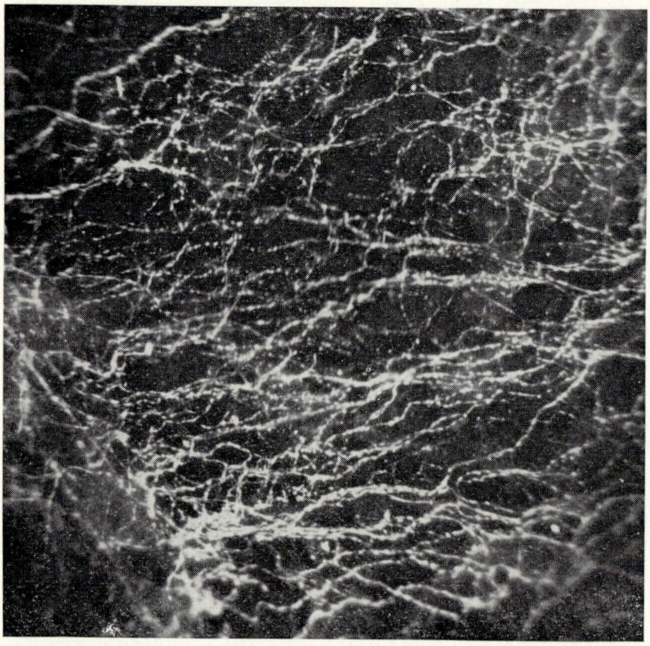

FIG. 1. Formaldehyde-induced fluorescence in an iris incubated for 2 hr in Krebs-glucose solution. The sphincter is in the lower left corner. ×100.

first incubated for 1–2 hr in the Krebs' solution. After 4–5 hr incubation, a slight loss of fluorescence was observed. Specimens incubated overnight in Krebs-glucose still exhibited, though more weakly, the specific fluorescence in the nerve fibres. If glucose was not present, a total loss of fluorescence occurred.

Incubation of the iris with up to 100 μg/ml of NA for 2–4 hr caused but a slight or no increase in the fluorescence intensity of the adrenergic nerve net, whether or not 1 mg/ml of glucose and up to 1 mg/ml of ATP were present.

Incubation in solutions containing 100 μg/ml of DOPA, TA, DA, A or 5-HT caused in 1·5–2 hr a marked decrease in the fluorescence intensity of the nerve fibres (Table 1). While formaldehyde does not render TA fluorescent, the other compounds employed are demonstrable with the present

TABLE 1. INTENSITY OF CATECHOLAMINE FLUORESCENCE IN THE NERVE FIBRES OF THE RAT IRIS INCUBATED FOR 2 HOURS IN KREBS-GLUCOSE WITH DIFFERENT COMPOUNDS (100 μg/ml)

Incubation medium	*Fluorescence intensity*
None (fresh control)	++++
Krebs-glucose (incubation control)	++++
Noradrenaline	+++++
Dihydroxyphenylalanine	+
Tyramine	+
Dopamine	+
Adrenaline	+
5-Hydroxytryptamine	+

technique, although a longer exposure is necessary to convert A into a fluorescent compound. An increase in the fluorescence intensity of the tissues between the nerve fibres was indeed observed, but only after incubation in DOPA or DA, while the nerve fibres had become almost indistinguishable.

The effect of incubation in A is illustrated in Figs. 1 and 2 showing,

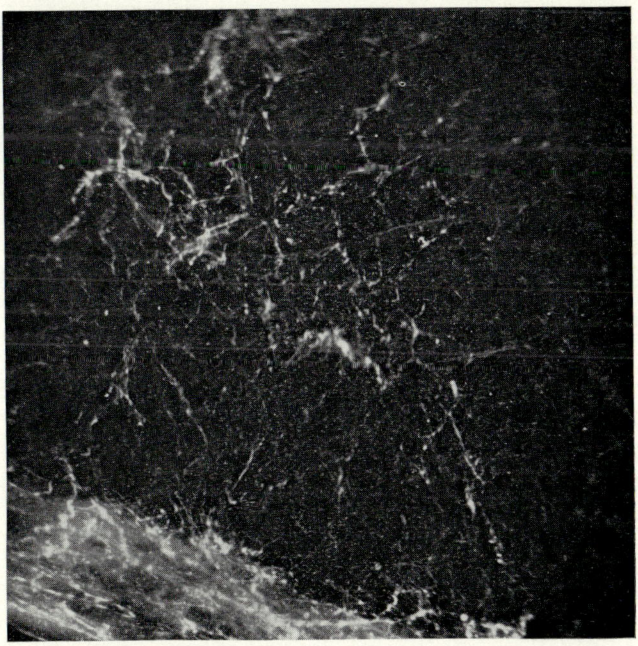

FIG. 2. Formaldehyde-induced fluorescence in an iris incubated for 2 hr in Krebs-glucose solution containing 100 μg/ml. of adrenaline. The reaction is less intense in the dilatator fibres but it is somewhat diffusely increased in the sphincter (lower left corner). $\times 100$.

respectively, an iris incubated for 2 hr in Krebs-glucose and another incubated in the same solution in which 100 μg/ml of A was dissolved. There was often an almost total loss of fluorescence. If the iris was incubated in a mixture containing besides 100 μg/ml of A also 10 μg/ml of NA, no loss of fluorescence occurred.

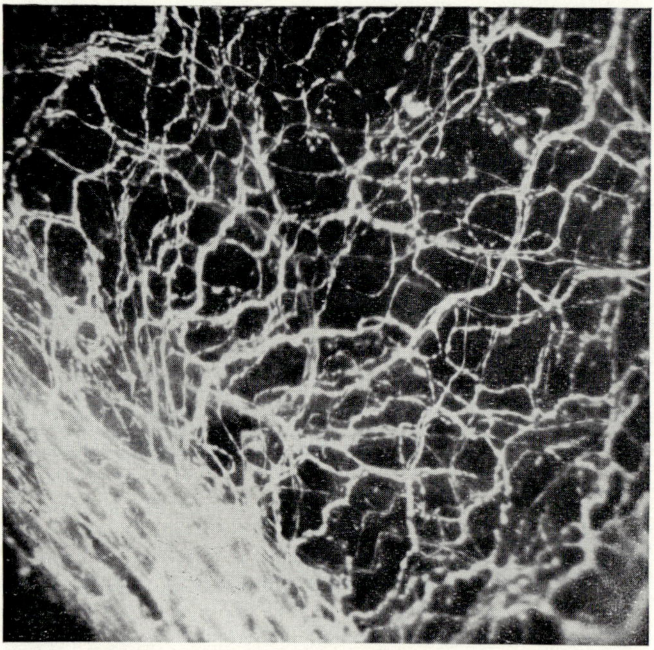

FIG. 3. Formaldehyde-induced fluorescence in an iris first incubated for 2 hr with 100 μg/ml. of adrenaline, washed in Krebs-glucose, and incubated for 2 hr with 100 μg/ml. of noradrenaline. Note the intense fluorescence of all fibres including the preterminal ones and those in the sphincter (lower left corner). ×100.

If an iris whose NA stores were thus depleted with A was subsequently incubated with NA for 2 hr, the adrenergic nerve net became fluorescent again (Fig. 3, Table 2). The intensity of the fluorescence often somewhat exceeded that in a control iris exposed to formaldehyde without incubation. A concentration of 1 μg/ml of NA was sufficient to restore the fluorescence. However, the presence of glucose in the incubation medium was necessary to obtain such restitution. Addition of ATP in a concentration of 1 μg–1 mg/ml had no effect.

Besides the terminal fibres, which normally exhibit an intense fluorescence after exposure to formaldehyde, incubation in NA-containing solution rendered strongly fluorescent also the smooth preterminal fibres of the iris, which normally contain little NA. This occurred equally, whether the iris

TABLE 2. INTENSITY OF CATECHOLAMINE FLUORESCENCE IN THE NERVE FIBRES OF THE RAT IRIS INCUBATED WITH ADRENALINE AND NORADRENALINE (100 μg/ml)

Incubation medium and time	*Fluorescence intensity*
Krebs-glucose 4 hr (control)	++++
Adrenaline 2 hr	+
Adrenaline 4 hr	±
Adrenaline 2 hr, Noradrenaline 2 hr	+++++
Noradrenaline 4 hr	+++++

was first exposed to A or some other amine and thereafter to NA, or whether it was directly incubated with NA.

DISCUSSION

Our results supplement the *in vivo* experiments carried out by Hillarp and Malmfors (1964), who observed that exogenous administration restores the NA content of the nerve fibres of the iris, if these were first depleted with reserpine. On the other hand, our observations are at variance with those made by Angelakos (1964), who found a sevenfold increase in the NA concentration of the normal iris after incubation *in vitro* with NA, ATP being necessary to obtain this effect. This discrepancy is difficult to explain but it is possible that Angelakos employed longer incubation times (not given in his paper).

Observations of the present study indicate that DOPA, TA, DA, A and 5-HT are capable of releasing NA from the adrenergic fibres of the iris, apparently without occupying the binding sites of NA permanently. Since the loss of fluorescence was marked also in the normally intensely fluorescent varicosities of the nerve net, in which NA is probably bound in granules, these compounds must affect both the axonal and the granular pool of NA.

Why DOPA, TA, DA, A or 5-HT did not become attached to the nerve fibres is an interesting problem. One explanation is that NA is chemically bound in many different ways (Stjärne, 1964). Musacchio *et al.* (1965) proposed that the amino group, the catechol hydroxyl group and the β-hydroxyl group all play a part in binding amines in sympathetic nerves. Of the compounds employed in the present study, only NA possesses all of these three groups, except for A, whose amino group is methylated. Whatever the mode of binding may be, the experiments described demonstrate its high specificity to NA.

Finally, it seems to be of considerable physiological interest that amines such as DA, A and 5-HT are capable of releasing NA largely without being bound in its binding sites in the sympathetic nerve terminals.

SUMMARY

Incubation *in vitro* of rat's iris in Krebs' salt solution containing dihydroxyphenylalanine, tyramine, dopamine, adrenaline or 5-hydroxytryptamine caused a dramatic loss of histochemically demonstrable noradrenaline from the sympathetic nerve net of the rat iris. Incubation of normal iris with noradrenaline had little effect on the noradrenaline content of the fibres but restored normal histochemical reaction in iris whose noradrenaline was first depleted with one of the above compounds. Glucose was necessary for such restoration but adenosine triphosphate had no effect on it.

REFERENCES

ANGELAKOS, E. T. (1964) Histochemical demonstration of uptake of exogenous norepinephrine by adrenergic fibres *in vitro*, *Science* **145**, 503–4.

ERÄNKÖ, O. (1964) Histochemical demonstration of catecholamines by fluorescence induced by formaldehyde vapour, *J. Histochem. Cytochem.* **12**, 487–8.

HILLARP, N.-Å. and T. MALMFORS (1964) Reserpine and cocaine blocking of the uptake and storage mechanisms in adrenergic nerves, *Life Sci.* **3**, 703–8.

MUSACCHIO, J. M., V. K. WEISE and I. J. KOPIN (1965) Mechanism of norepinephrine binding, *Nature* (*Lond.*) **205**, 606–7.

STJÄRNE, L. (1964) Studies of catecholamine uptake, storage and release mechanisms, *Acta Physiol. Scand.* **62**, Suppl. 228.

DISCUSSION

HAMBERGER: I would like to comment on the depletion of noradrenaline by dopamine in rat iris incubated *in vitro*. In extensive *in vitro* and *in vivo* experiments with dopamine, other catecholamines and DOPA we have in Hillarp's laboratory never found any signs of lowering of the catecholamine fluorescence. On the contrary, we have found that the whole adrenergic neuron has a very efficient uptake-concentration mechanism demonstrable at low concentrations (*in vitro* down to 0·01 ug/ml), located to the level of the cell membrane and insensitive to reserpine.

WEINER: Angelakos (1964) has shown, both by fluorimetric and histochemical studies, that norepinephrine is taken up and concentrated in the iris *in vitro*, presumably in nerve terminals, when ATP is added to the incubation medium. Perhaps you are not obtaining increased fluorescence because of an absence of energy for this uptake mechanism. Have you correlated your studies with biochemical values?

ERÄNKÖ: I am glad to hear Dr. Hamberger's comment, which shows that our observations are new. The loss of catecholamine fluorescence due to incubation in the amines which we employed and to subsequent washing in Krebs-glucose to remove any loosely bound amines is readily reproducible. An iris depleted in such a mild way of itw noradrenaline stores seems a useful preparation for the study of the uptake of catecholamines by sympathetic nerves.

To Dr. Weiner I should like to say that we also observed, like Angelakos, an increase in the fluorescence intensity after incubation with noradrenaline, although a smaller one and not dependent on the presence or absence of ATP. We did not make quantitative determinations with biochemical methods.

STRUCTURAL ASPECTS OF HISTAMINE BINDING: THE MAST CELL GRANULE*

D. LAGUNOFF

Department of Pathology, University of Washington,
Seattle, Washington, U.S.A.

THERE is evidence that not all cells with the capacity for making histamine are competent to bind and on appropriate stimulus release this amine (Kahlson, 1960; Telford and West, 1961; Schayer, 1963). It seems likely that the binding of histamine requires some form of subcellular structural specialization that is restricted to certain cell types. The most extensively studied cell which stores histamine is the tissue mast cell, and its histamine is sequestered in the cell's characteristic granules (Mota *et al.*, 1954; Hagen, Barnett and Lee, 1959; Lagunoff *et al.*, 1964). Whether or not the mast cell's mode of storing and releasing histamine is common to all cells with this capability is not known: the granules of the basophil leucocyte resemble in some ways those of the mast cell (Ackerman, 1963) and are in all probability the sites of localization of the histamine; the distribution of histamine within rabbit platelets is not known, but there are also granules present in these cell fragments (Schulz and Hiepler, 1959; Vasalli, Simon and Rouiller, 1964) which might serve as histamine binding sites; other cellular loci of histamine storage, in stomach mucosa, for instance, have not been definitely identified (Feldberg and Harris, 1953).

In order to avoid as far as possible poorly grounded conjecture, I will concentrate on the mast cell as a histamine binding cell and present some results of an analysis of the structure of its granules. This approach does not imply any assumption of general applicability of the suggested structure for histamine binding sites. There is sufficient diversity in fine structural morphology even among mast cell granules of various mammalian species to warrant extreme caution in generalizing even from rat mast cell granules to the homologous structures in human mast cells.

MATERIAL AND METHODS

All rats used in the study were Sprague–Dawley males weighing between 250 and 350 g. Standard sodium heparin prepared from pig intestine was

* Supported in part by a grant-in-aid from the United States Public Health Service (HE–03174).

generously provided by Dr. F. McIntyre of Abbott Laboratories and had an anticoagulant activity of 173 units/mg. Alpha-chymotrypsin was a crystalline preparation purchased from Worthington Biochemical Corporation, Freehold, N.J. Lysozyme was a crystalline preparation purchased from Armour Laboratories, Chicago, Ill. Histamine dihydrochloride was purchased from California Biochemical Corporation.

Crude mast cell granules were isolated from the complete cell population of rat peritoneal* fluids as previously described (Lagunoff *et al.*, 1964). For clean granule preparations, mast cells were separated from the other cell types by the method of Benditt and Arase or that of Glick, Bonting and Den Boer (1956). Completeness of sonic disruption of the cells was monitored by light microscopy. If intact cells or nuclei were present, the suspensions were resonicated. Nitrogen was determined by the method of Stegemann and Loeschke (1962). Digestion was performed in standard Kimble 5 ml ampoules heated in a distilled aluminium block mounted on a hot plate. N-sulfated hexosamine was determined by the method of Lagunoff and Warren (1962) and uronic acid by the carbazole method of Dische (1955). Chymase activity was measured as previously described (Lagunoff and Benditt, 1963). Histamine was determined by a modification of the method of Kremzner and Wilson (1961). All values are given as free base. The standard procedure of Lowry *et al.* (1951) was used to estimate protein but without precipitation by TCA.

RESULTS

1. *Solubility of Mast Cell Granules*

(*a*) Clean mast cell granules were collected from 30 male rats 250–350 g and washed several times in distilled water. The granule pellet was suspended in 0.05 M KCl, 0.01 M phosphate buffer pH 6.8 and aliquots distributed in equal volumes of buffered KCl to give final concentrations between 0.05 M and 2.0 M. The suspensions were mixed thoroughly and the apparent absorbence resulting from light scattering was measured at 450 mμ. The suspensions were then centrifuged and the Lowry modification of the Folin–Ciocalteau procedure performed on samples of clear supernatant and precipitates dissolved in 0.2 N NaOH, 2 per cent $NaCO_3$. Without prior TCA precipitation this reaction measures total content of aromatic amino acids and cysteine. Supernates were also assayed for histamine (Fig. 1). It is evident that although almost all the histamine is extracted from the granules at the lowest salt concentration tested (0.05 M KCl in 0.01 M phosphate buffer pH 6.8) the granule does not begin to dissolve until the salt

* In some experiments the rats were decapitated and pleural washings collected and pooled with the peritoneal washings.

concentration is increased to 0.5 M* and dissolution is not complete until about 2.0 M concentration.

In two experiments histamine, heparin and chymase were assayed in serial extracts of clean mast cell granules. Little heparin or chymase was extracted at salt concentrations below 1.0 M (Fig. 2).

Solubilization of heparin and chymase at salt concentrations which dissolve the granule suggests that the granule matrix is composed of a complex

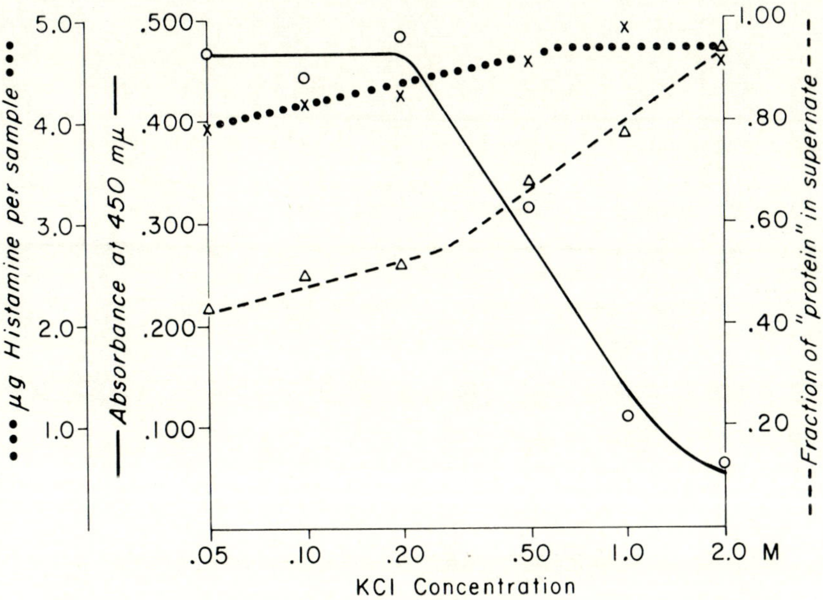

FIG. 1. Effect of salt on isolated mast cell granules. Apparent absorbence resulting from light scattering was employed as a measure of solubility of the granules. "Protein" value may include a non-protein contribution of undetermined magnitude.

between heparin and protein(s). In order to analyze the nature of the complex, crude granule preparations were washed with 0.2 M KCl to remove readily soluble protein and the insoluble granule material was dissolved in 1.0 M KCl or KBr and chromatographed on Sephadex G–100 (Granath and Flodin, 1961). When elution was carried out with 0.2 M NaCl, the chymase and heparin were eluted in a common fraction (Thursh, 1963). If however 1.0 M KCl or KBr was used for elution, heparin was separable from the chymase (Fig. 3). In the short columns used, heparin appeared in the void volume and chymase was retained.

* The decrease in apparent absorbence at 0.5 M KCl may represent swelling rather than dissolution.

Mast cell heparin unlike purified pig heparin consistently appeared in the void eluate from Sephadex G-100. When mast cell granule heparin was purified by phenol extraction and ethanol precipitation and then chromatogrammed on Sephadex G-100, this heparin also appeared in the void eluate. Mast cell granule heparin was also purified by dissolving in 1.0 M KCl, precipitating with cetylpyridinium chloride (CPC) and redissolving in

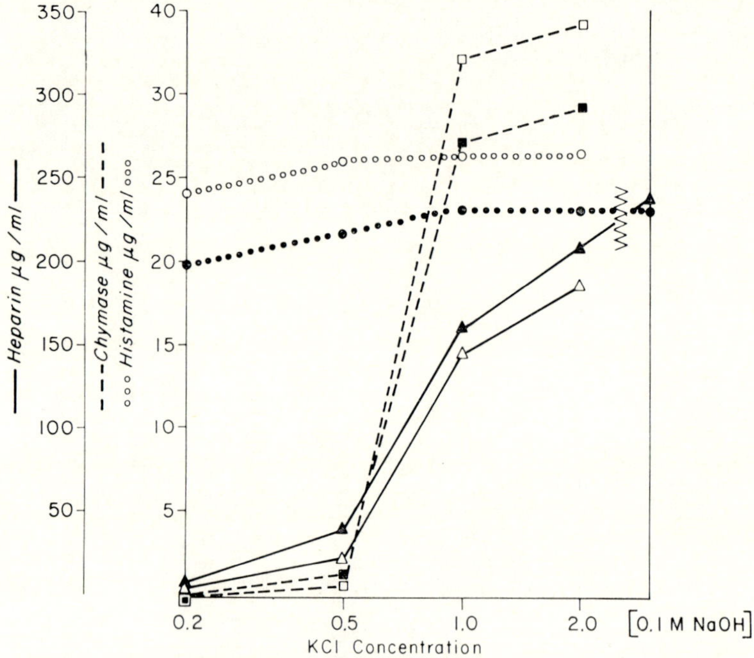

FIG. 2. Salt extraction of mast cell granule heparin, histamine and chymase. Two experiments were performed in which a single pellet of clean granules was serially extracted with increasing KCl concentration and the supernatant fluids after centrifugation assayed for histamine, heparin and chymase. The filled-in symbols represent one experiment, the empty symbols the other. In one experiment, the small residue left after 2.0 M KCl extraction was dissolved in 0.1 M NaOH and assayed for chymase and heparin.

2.0 M KCl; the CPC was then largely removed with KCNS (Korn, 1959). Mast cell heparin prepared in this manner was also largely excluded from Sephadex G-100 (Fig. 4).

Since preparations of heparin from pig mucosa did not behave anomalously on Sephadex G-100, the behavior of the mast cell granule heparin indicates a considerably larger molecular weight for rat heparin isolated without proteolytic digestion than for other heparin. The sedimentation velocity in a sucrose gradient (Martin and Ames, 1961) was approximately twice that of pig heparin. An S_{20w} of 2.3 for the latter was determined in the analytical

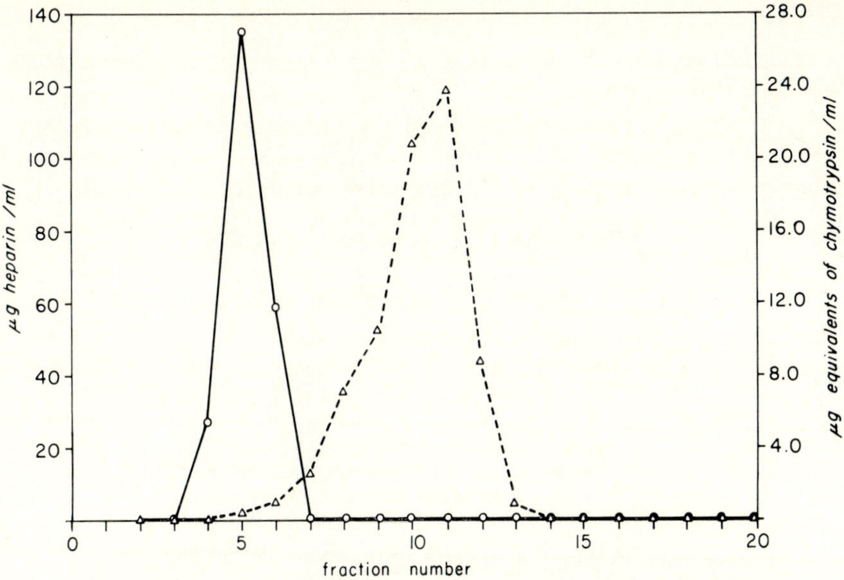

FIG. 3. Sephadex chromatography of a crude granule preparation. A sample of 1.0 M KBr extract of crude granule was applied to a 22.0 cm column of Sephadex G-100 with a total bed volume of 38.7 ml and a void volume of 10 ml. Volumes in fractions 1–5 were irregularly distributed but totalled 10 ml. Fractions thereafter were 2.5 ml. each. Elution was with 1.0 M KBr. Heparin was determined by N-sulfated hexosamine assay. Chymase activity against ATEE was measured in a pH-stat.

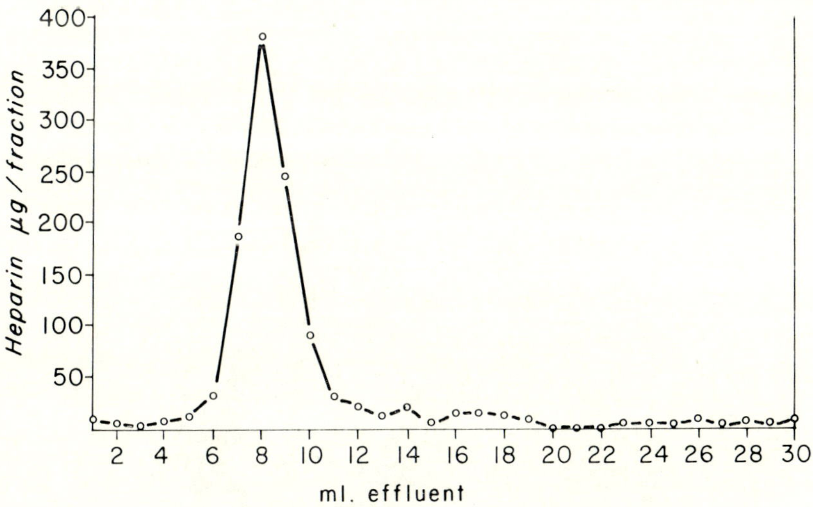

FIG. 4. Sephadex G-100 chromatography of a CPC purified preparation of granule heparin. Heparin was applied in 1.0 M KCl and eluted with 1.0 M KCl, 0.01 M phosphate buffer pH 6.8. The column was 21.5 cm long with a total calculated bed volume of 24.4 ml. and a void volume of 6 ml. One ml. fractions were collected. Residual CPC appeared in fraction 26.

ultracentrifuge at 1 per cent concentration in 1.0 M KCl* (cf. Barlow, Sanderson and McNeill, 1961).

Comparison of nitrogen and sulfated hexosamine determinations for mast cell heparin prepared by CPC precipitation followed by Sephadex chromatography indicated an excess of nitrogen in two preparations (Table 1).

TABLE 1. ANALYSES OF HEPARIN PREPARATIONS

Preparation	*Uronic acid*†	*N*†	S_{20W}
pig heparin	1.5	1.3	2.3
rat heparin* G129	—	1.9	—
G133	1.5	2.5	4.6‡

* Both samples of rat heparin were prepared from 1.0 M KCl extracts of crude granule preparations. They were purified by precipitation with CPC from 1.0 M KCl and chromatographed on Sephadex G-100 eluting with 1.0 M KCl buffered to pH 6.8 with 0.01 M phosphate.

† Expressed as molar ratios referable to N-sulfated hexosamine.

‡ Estimated from a comparison with pig heparin by sedimentation in a sucrose density gradient.

In order to determine the histamine binding capacity of mast cell granules, they were washed free of intrinsic histamine in 0.05 M KCl and then washed several times in distilled water. Aliquots of granules were then suspended in concentrations of histamine between 3.75 and 30 μg/ml in 0.001 M KCl, 0.0001 M phosphate buffer pH 6.5, and incubated 30 min at 25°C. The granules were collected by centrifugation and the histamine concentration of the supernatant measured. The precipitated granules were suspended in an aliquot of supernatant and the total histamine content determined and the value corrected for histamine in the supernatant phase (Table 2). The data

TABLE 2. HISTAMINE BINDING BY ISOLATED RAT MAST CELL GRANULES

Initial concentration of histamine μg/ml	*Final concentration of histamine μg/ml*	*μg histamine bound per μg heparin**
0	0	
2.5	0.66	0.027
5	1.36	0.050
10	3.57	0.082
20	12.7	0.099

* Each aliquot of granules contained 11.7 μg heparin as determined by N-sulfated hexosamine assay.

* Ultracentrifugation performed by Dr. E. Smuckler.

were plotted according to a Langmuir's adsorption equation (cf. Mathews, 1960) $1/b = K/c + C$, b is amount bound, K is the dissociation constant of the complex, c is the concentration in the soluble phase and C is the binding capacity; c was determined to be 0.127 μg histamine per μg heparin or 0.72 mole of histamine per mole of heparin disaccharide.

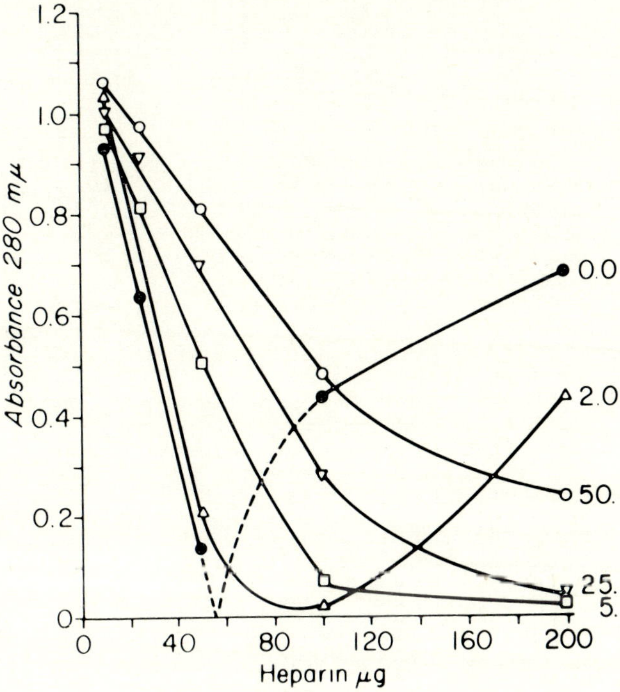

FIG. 5. Effect of $CaCl_2$ concentration on the formation of heparin–lysozyme complexes. The heparin was added to 0.5 mg of lysozyme in a total volume of 1 ml. at varying concentrations of $CaCl_2$; 0.0 = no added $CaCl_2$, 2·0 = 0.2 mM, 5 = 5 mM, 25 = 25 mM and 50 = 50 mM. All the solutions were 0.5 mM with respect to KCl and 0.05 mM with respect to phosphate buffer. The final pH of the supernates varied between 5.5 and 5.8. Absorbence at 280 mμ was measured after centrifugation at 15,000 × g in a Sorvall SS-1. The calculated charge equivalence was at 54 μg of heparin.

In a series of experiments chymotrypsin and lysozyme were tested for the formation of precipitates with heparin. Both proteins did precipitate with heparin under appropriate conditions. Chymotrypsin in 0.001 M KCl and concentrations of $CaCl_2$ from 0.0002 M to 0.05 M, however, did not precipitate quantitatively with amounts of heparin from about $\frac{1}{2}$ to 10 times the amount calculated for ionic equivalence. Lysozyme could be precipitated quantitatively in the presence of $CaCl_2$ with amounts of heparin which depended on the concentration of salt (Fig. 5). When precipitation was carried

TABLE 3. ANALYSIS OF LYSOZYME–HEPARIN PRECIPITATES

Heparin added μg	*Lysozyme** μg	*Heparin†* μg	*Heparin: Lysozyme‡*
0	0	0	—
25	154	22.5	1.36
50	284	43.5	1.43
75	410	64.4	1.46
100	485	86.0	1.65
200	487	147.0	2.82

To 1 mg/ml of lysozyme in 5 mM histamine, 0.001 M KCl, 0.0001 M phosphate buffer pH 6.8 was added the indicated amount of heparin; the tubes were allowed to stand for 1 hr at 25°C, and the precipitates collected at $15{,}000 \times g$ in a Sorvall SS-1. The supernatant was poured off and the tubes drained and the walls swabbed dry.

* Lysozyme was determined from the decrease in absorbence at 280 mμ in the supernate.

† Heparin was determined on the precipitate after dissolving it in 1.0 M KCl.

‡ Heparin: lysozyme ratios were calculated using a value of 181 as the equivalent weight of sodium heparin (634/3.5) and 1670 as the equivalent weight of lysozyme (Franklin and Schubert, 1960).

out in the presence of 5 mM histamine at pH 5.8, there was also a shift in the precipitation characteristics of the system almost identical with that produced by the same concentration of Ca^{++}. The composition of such precipitates at varying concentrations of heparin are given in Table 3. In a preliminary experiment the stability of precipitates prepared in the presence of 5 mM

TABLE 4. HISTAMINE BINDING BY HEPARIN–LYSOZYME–$CaCl_2$ COMPLEX

Initial conc. histamine $\mu g/ml$	*Final conc. histamine* $\mu g/ml$	μg *histamine bound per* μg *heparin*
0	0	—
2.5	1.10	0.020
5	3.3	0.027
10	6.5	0.043
25	13.8	0.065

Heparin–lysozyme–$CaCl_2$ complex was formed in presence of 4.6 fold excess of heparin in 5 mM $CaCl_2$. The complex was washed with water, suspended in 0.0001 M phosphate buffer pH 6.5 and incubated 30 min at 25°C. The complex was collected by centrifugation and the concentrations of histamine determined in the supernatant and an aliquot of supernatant in which the complex was resuspended. Each aliquot of complex contained 11.2 μg heparin. Total volume was 150 μl.

$CaCl_2$ at varying degrees of heparin excess were not significantly different, all dissolving in 0.2 M but not in 0.05 M KCl. A batch of lysozyme–heparin complex was prepared in 4.6 fold heparin excess in 5 mM $CaCl_2$ and washed in distilled water. The composition of the complex was determined and histamine binding measured (Table 4). The binding curve in this case did not fit a Langmuir adsorption isotherm.

DISCUSSION

Identification of the components of the mast cell granule is obviously of critical importance for an analysis of granule structure. Unfortunately it is only possible at present to account for about 70 per cent of granule dry weight (Lagunoff *et al.*, 1964). The data for heparin are the most satisfactory and indicate that this acidic polysaccharide makes up about 30 per cent of the dry granule weight. The most probable estimate for protein is also about 30 per cent. If it is assumed that all mast cell histamine is confined to the granules, then this amine would constitute about 3 per cent of the granule. Since TCA precipitable nitrogen and histamine account for only 60 per cent of the total granule nitrogen, there is an obvious gap in knowledge of the nitrogen-containing substances of the granule. Both taurine and cysteic acid have been found in neoplastic mouse mast cells (Green and Day, 1963) and may be present in rat granules. Low values for total phosphorus and chloroform: methanol soluble phosphorus rule out any significant amounts of nucleic acid, ATP or phospholipid (Lagunoff *et al.*, 1964), but no estimates of phosphorus-free lipid are available.

Although the analysis of granule constituents remains to be completed, it is possible to begin to investigate the way in which the known components are held together. One useful but not foolproof means of understanding what holds a structure together lies in discovering how it can be taken apart. Histamine can largely be removed from the granule merely by repeated washing in sucrose, and 0.05 M KCl will remove 90 per cent of the granule histamine. A large part of the histamine can be replaced by incubating the granules in a histamine-containing aqueous solution. When the concentration of salt reaches about 1.0 M the granule itself dissolves.

Manipulations which are known to release substances from membrane-bound structures such as lysozymes (deDuve, 1959) have little or no effect on mast cell granules; the non-ionic detergent Triton X-100, repeated freezing and thawing, and extremes of tonicity do not break up the granule matrix (Lagunoff *et al.*, 1964), and in electron micrographs there is no evidence of a membrane associated with isolated granules.

The several pieces of evidence are consistent with a hypothesis that the granule is a heparin–protein matrix to which histamine and perhaps other nitrogenous bases are attached electrostatically. Two general possibilities exist for the character of the heparin–protein complex. It might be analogous to the protein–polysaccharide complexes isolated from cartilage in which the protein is covalently bound to the chondroitin sulfate (Muir, 1958; Gregory, Laurent and Rodén, 1964) or the mast cell granule might represent an ionic complex of oppositely charged polyelectrolytes.

The evidence presented in this paper indicates that only a small portion of the granule protein is likely to be bound covalently to heparin since granule heparin isolated by mild procedures can be freed of all but about 2 per cent of non-hexosamine nitrogen. A small amount of protein covalently bound to heparin may be responsible for the unexpectedly large molecular size of the granule heparin. Unfortunately the availability of only small amounts of the macroheparin has interfered with an adequate study of this material. Lindahl and Rodén (1964) have recently provided evidence that at least in some species heparin is linked to a serine residue by the same glycosidic bond, a xylosyl-*O*-serine, that has been implicated in the chondroitin sulfate–protein complex of cartilage (Gregory, Laurent and Rodén, 1964).

When mast cell granules are dissolved in concentrated salt solution, heparin can largely be separated from both granule protein and non-granule contaminating protein by CPC precipitation at 1.0 M KCl followed by Sephadex G-100 chromatography. This behavior is compatible with the granule being a polyelectrolyte complex. The character of such polyelectrolyte complexes has been extensively studied by Bungenberg de Jong (1949) (see also Katchalsky, 1964).

Electrostatically determined heparin–protein complexes have been studied by Fischer (1935) and Jacques (1943) and more recently Godal (1960). In addition, a host of enzymes have been examined for inhibition by heparin (Engelberg, 1963), and Schubert and Franklin (1960) have provided an interesting investigation of a number of the properties of complexes formed between lysozyme and chondroitin sulfate, that closely resemble lysozyme–heparin complexes.

Chymotrypsin apparently cannot be quantitatively removed from solution at any concentration of heparin, whereas lysozyme can be completely precipitated by heparin at concentrations calculated to give net charge equivalence in very low salt concentrations. Although sufficiently low concentrations of non-valent cation do not dissolve the complex formed at equivalence, divalent cation including histamine tend to dissolve the complex so that the equivalence complexes would not seem a reasonable model for the mast cell granule. At very low concentrations of salt, an excess of heparin will tend to dissolve the complex with lysozyme and at high salt concentrations the complex is also soluble but at intermediate salt concentrations quantitative precipitation of protein occurs in the presence of an excess of heparin. Analysis of the flocculent precipitate formed under these conditions indicates that there is an excess of heparin in the complex as demonstrated by Schubert and Franklin (1960) for their system, and such complexes are capable of binding histamine.

Histamine binding by heparin has been studied by a number of workers (Parrot and Laborde, 1951; Parrot, Nicot and Laborde, 1960; Werle and Amann, 1956; Sanyal and West, 1959; Kobayashi, 1962; Phillips, 1965). In aqueous solutions these complexes are soluble and require the addition of ethanol to yield precipitates. It is interesting that the binding capacity of heparin as measured by equilibrium dialysis of 1 mole of histamine per mole of disaccharide heparin is closely approached by the binding capacity of the granules.

Little is known of the shape of heparin. Velluz (1959) has suggested the possibility of a helical structure and a recent investigation (Stone, 1964) of the interaction of metachromogenic dyes and heparin indicates a reasonable likelihood of helical structure under some conditions. Whether or not the helical structure occurs only in the presence of the dye counter-ion is not clear. One particularly interesting finding was that histamine unlike inorganic divalent cations may bind in a manner that does not disrupt the helix. In equilibrium dialysis experiments (Phillips, 1965) no evidence was found for any greater affinity of heparin for histamine than for other divalent cations, while trivalent inorganic cations were bound much more avidly than histamine.

One of the striking differences in mast cell granules from one species to another is their solubility. Granules of the rat and human are quite insoluble in water while those of the rabbit, guinea-pig and a number of fishes are notoriously soluble (Michel, 1938). It follows from the proposed model for the granule that differences in the protein or the polysaccharide portion of the complex which affect their interaction could easily give rise to differences in solubility. Further study of the influence on the formation of complexes of the configuration and charge of the protein as well as the polysaccharide should provide firmer ground for speculation. The task of isolating intact such highly soluble granules as occur in fish mast cells or guinea-pig circulating basophils will in all likelihood require a quite different approach from that utilized for isolating rat mast cell granules. The character of solubility may permit simple aqueous extraction of these polysaccharides and characterization of their interaction with proteins.

The sensitivity of granule-bound histamine to displacement by inorganic cations raises the important problem of how the histamine is retained by the granule intracellularly since the concentration of free intracellular potassium is probably not much lower than 0.2 M, a concentration which will release more than 90 per cent of granule histamine from isolated granules. One clue can be derived from electron micrographs of mast cells in which granules are frequently seen to be associated with a membrane (Fig. 6). This membrane may well function to protect the granule from intolerable ion concentrations. If this is the case, then release of granule histamine could result from either

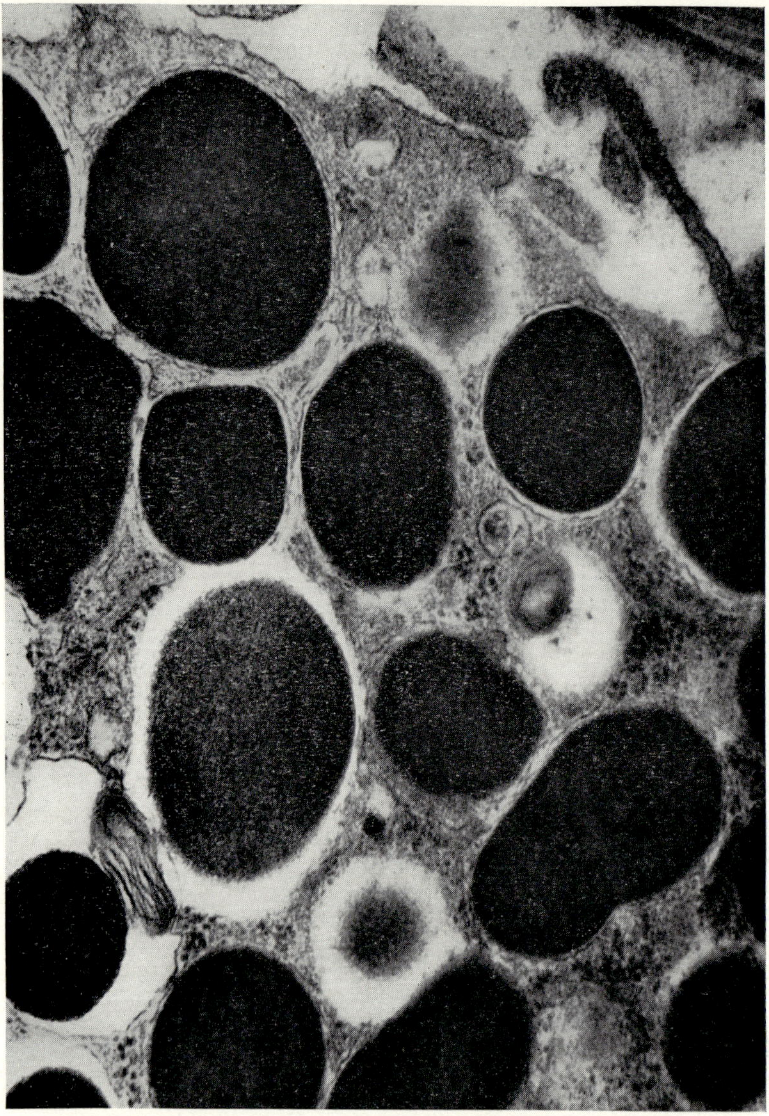

FIG. 6. Electron micrograph of a portion of a mast cell in the tongue of a rat. Most of the granules are seen to be associated with a dense peripheral membrane that in some instances appears to encompass the granule. These membranes are evident even in the absence of any halo formation about the granule. The tissue was fixed in cacodylate buffered 4 per cent glutaraldehyde, post-fixed in 2 per cent osmium tetroxide in collidine buffer and embedded in Epon 812. The electron micrograph was kindly provided by Dr. J. Combs and his co-workers (1965). × 68,000.

extrusion of the granule from its protective sac into the extracellular ionic milieu or from a breakdown of membrane function that allowed intracellular cations to reach the granule. Another possibility is that the membrane acts to maintain a high local concentration of free histamine.

There are numerous electron microscopic studies of mast cells (see Smith, 1963 for references prior to 1962, Bloom, 1963; Fernando and Movat, 1963). Although some granules in human mast cells (Steckenius, 1956) exhibit distinct internal structural complexity, the granules of rat peritoneal cells do not. Depending in part on techniques of preparation and in part on undefined conditions, the granules of the rat cells appear uniformly dense or show varying degrees of reticulation. These reticulated granules may correspond to the granules which light microscopists have described, particularly in the stimulated cell, as swollen. Low concentrations of salt which remove most of the histamine did not affect the appearance of isolated granules (Lagunoff *et al.*, 1964), but a more careful study of the effect of increased concentrations of salt on the structure of the granule as seen with the electron microscope is necessary.

Two sites of histamine binding, the mast cell and the basophil leucocyte, contain a highly acidic mucopolysaccharide well identified in the case of the former (Schiller and Dorfman, 1959; Bloom and Ringertz, 1960) and presumptively for the latter (Behrens and Taubert, 1952; Amann and Martin, 1961). Rabbit platelets are unique in their high content of histamine (Humphrey and Jacques, 1954), but platelets of many species seem capable of storing and releasing 5-hydroxytryptophan. Platelets of the rat have been demonstrated to contain a sulfated mucopolysaccharide (Odell and Anderson, 1957) that is a potential binding site for serotonin. The localization of the polysaccharide is not established so that it remains an untested hypothesis that the polysaccharide is present in the characteristic granules of the platelet. Recently a procedure has been developed for the isolation of platelet granules (Marcus and Zucker-Franklin, 1964), and these granules in the rabbit do apparently contain at least a portion of the serotonin (Wurzel and Marcus, 1965). Marcus has evidence that some elements present in the granule fraction from human platelets are lysosomal. There is a precedent for the presence of acidic mucopolysaccharide in lysosomes (Fedorko and Morse, 1965).

SUMMARY

The solubility properties of isolated rat mast cell granules have been investigated and it has been shown that 90 per cent or more histamine is extracted in 0.05 M KCl, but the granule matrix does not dissolve until the concentration is increased to 1.0 M KCl and the heparin–protein complex is dissolved. Precipitation of heparin with CPC followed by Sephadex chromatography yields heparin with 2 per cent or less excess nitrogen under conditions which are unlikely to break glycosidic or ester bonds. The heparin is, by sedimentation velocity in sucrose gradient and chromatography on Sephadex, considerably larger than heparin prepared from cow and pig by the conventional procedures, which involve proteolytic enzymes and/or alkaline conditions.

Precipitates of heparin and protein that form readily *in vitro* have been proposed as models for mast cell granule structure. These precipitates can bind histamine.

The almost complete release of histamine from granules *in vitro* by concentrations of salt that are well within the range expected intracellularly calls for a mechanism that isolates the granule from the ionic milieu of the cell sap. It is proposed that the perigranular membrane serves this function.

ACKNOWLEDGMENTS

Excellent technical assistance in carrying out these experiments was provided by Miss M. Arase and Miss C. Ryavec. I am also grateful to Dr. E. P. Benditt for counsel and encouragement. Dr. D. Thursh and Dr. M. T. Phillips have graciously allowed me to quote from unpublished work. Dr. J. Combs and his co-workers have generously provided the electron micrograph.

REFERENCES

ACKERMAN, C. A. (1963) Cytochemical properties of the blood basophilic granulocyte, *Ann. N.Y. Acad. Sci.* **103**, 376–93.

AMANN, R. and H. MARTIN (1961) Blutmastzellen und Heparin, *Acta Haemat.* (Basel) **25**, 209.

BARLOW, G. H. (1964) Macromolecular properties and biological action of heparin. III. Paper electrophoresis studies of histamine binding, *Biochim. Biophys. Acta* (Amst.) **83**, 122.

BARLOW, G. H., N. D. SANDERSON and P. D. MCNEILL (1961) Macromolecular properties and biological activity of heparin. *Arch. Biochem.* **94**, 518–25.

BEHRENS, M. and M. TAUBERT (1952) Der Nachweiss von Heparin in den Basophilen Leukocytes, *Klin. Wschr.* **30**, 76–8.

BLOOM, G. (1963) Electron microscopy of neoplastic mast cells: A study of the mouse mastocytoma mast cell, *Ann. N.Y. Acad. Sci.* **103**, 53–83.

BLOOM, G. and N. R. RINGERTZ (1960) Acid mucopolysaccharides of peritoneal mast cells of the rat and mouse, *Arkiv Kemi* **16**, 51–6.

BUNGENBERG DE JONG, H. G. (1949) Complex colloid systems, *Colloid Science* (Ed. H. R. Krugt), Amsterdam, Elsevier Publishing Co., Inc., pp. 335–432.

COMBS, J., L. CALDWELL and D. BOCKUS (1965) Effects of various fixatives on the electron microscopic appearance of rat mast cells. Unpublished.

DE DUVE, C. (1959) Lysosomes, a new group of cytoplasmic particles, *Subcellular Particles* (Ed. T. Hayashi), N.Y., The Ronald Press Co., pp. 128–58.

DISCHE, Z. (1955) New color reactions for the determination of sugars in polysaccharides, *Meth. Biochem. Anal.* **2**, 313–58.

ENGELBERG, H. (1963) *Heparin; Metabolism, Physiology and Clinical Application*, Springfield, C. C. Thomas, p. 44.

FEDORKO, M. E. and S. I. MORSE (1965) Isolation, characterization and distribution of acid mucopolysaccharides in rabbit leucocytes, *J. Exp. Med.* **121**, 39–48.

FELDBERG, W. and G. W. HARRIS (1953) Distribution of histamine in the mucosa of the gastro-intestinal tract of the dog, *J. Physiol.* (Lond.) **120**, 352–64.

FERNANDO, N. V. P. and H. Z. MOVAT (1963) The fine structure of connective tissue. III. The mast cell, *Exp. Molec. Path.* **2**, 450–63.

FISCHER, A. (1935) Die Bindung von Heparin an Eiweiss, *Biochem. Z.* **278**, 134–60.

GLICK, D., S. L. BONTING and D. DEN BOER (1956) Histochemistry XLIV. Use of density gradient for isolation of mast cells, *Proc. Soc. exp. Biol.* (N.Y.) **92**, 357–9.

GODAL, H. C. (1960) Precipitation of human fibrinogen with heparin, *Scand. J. Clin. Lab. Invest.* **12**, 56–65.

GRANATH, K. A. and P. FLODIN (1961) Fractionation of dextran by the gel filtration method, *Makromol. Chem.* **48**, 160–71.

GREEN, J. P. and M. DAY (1963) Biosynthetic pathways in mastocytoma mast cells in culture and *in vivo*, *Ann. N.Y. Acad. Sci.* **103**, 334–50.

GREGORY, J. D., T. C. LAURENT and L. RODÉN (1964) Enzymatic degradation of chondromucoprotein, *J. Biol. Chem.* **239**, 3312–20.

HAGEN, P., R. J. BARNETT and F.-L. LEE (1959) Biochemical and electron microscopic study of particles isolated from mastocytoma cells, *J. Pharmacol.* **126**, 91–108.

HUMPHREY, J. H. and R. JACQUES (1954) The histamine and 5-hydroxytryptamine content of platelets and leucocytes in various species, *J. Physiol.* (Lond.) **124**, 305–10.

JACQUES, L. B. (1943) The reaction of heparin with proteins and complex bases, *Biochem. J.* **37**, 189–95.

KAHLSON, G. (1960) A place for histamine in normal physiology, *Lancet* **I**, 67–71.

KATCHALSKY, A. (1964) Polyelectrolytes and their biological interactions. Connective Tissue: Intracellular Macromolecules. Proceedings of a symposium sponsored by the New York Heart Association. Boston, Little, Brown and Co., pp. 9–41.

KOBAYASHI, Y. (1962) Histamine binding by heparin, *Arch. Biochem.* **96**, 20–7.

KORN, E. D. (1959) The synthesis of heparin by slices of mouse mast cell tumor, *J. Biol. Chem.* **234**, 1321–4.

KREMZNER, L. T. and I. B. WILSON (1961) A procedure for the determination of histamine, *Biochim. Biophys. Acta* (Amst.) **50**, 364–7.

LAGUNOFF, D. and E. P. BENDITT (1963) Proteolytic enzymes of mast cells, *Ann. N.Y. Acad. Sci.* **103**, 179–97.

LAGUNOFF, D., M. T. PHILLIPS, O. A. ISERI and E. P. BENDITT (1964) Isolation and preliminary characterization of rat mast cell granules, *Lab. Invest.* **13**, 1331–44.

LAGUNOFF, D. and G. WARREN (1962) Determination of 2-deoxy-2-sulfoaminohexose content of mucopolysaccharides, *Arch. Biochem.* **99**, 396–400.

LINDAHL, U. and L. RODÉN (1964) The linkage of heparin to protein, *Biochem. Biophys. Res Commun.* **17**, 254–9.

LOWRY, O. H., N. J. ROSEBROUGH, A. L. FARR and R. J. RANDALL (1951) Protein measurement with the Folin phenol reagent, *J. Biol. Chem.* **193**, 265–75.

MARCUS, A. J. and D. ZUCKER-FRANKLIN (1964) Enzyme and coagulation activity of subcellular platelet fractions, *J. Clin. Invest.* **43**, 1241–2.

MARTIN, R. G. and B. N. AMES (1961) A method for determining the sedimentation behaviour of enzymes: Applications to protein mixtures, *J. Biol. Chem.* **236**, 1372–9.

MATHEWS, M. B. (1960) Trivalent cation binding of acid mucopolysaccharides, *Biochim. Biophys. Acta* (Amst.) **37**, 288–95.

MICHEL, N. (1938) The Mast Cells. *Downey's Handbook of Hematology*, Vol. 1, pp. 232–372. Reprinted *Ann. N.Y. Acad. Sci.* **103**, Appendix p. 491.

MOTA, I., W. T. BERALDO, A. G. FERRI and L. C. U. JUNQUEIRA (1954) Intracellular distribution of histamine, *Nature* (Lond.) **174**, 698.

MUIR, H. (1958) The nature of the link between protein and carbohydrate of a chondroitin sulfate complex from hyaline cartilage, *Biochem. J.* **69**, 198–204.

ODELL, T. T. and B. ANDERSON (1957) Isolation of a sulfated mucopolysaccharide from blood platelets of rats, *Proc. Soc. Exp. Biol.* (N.Y.) **94**, 151–6.

PARROT, J. L. and C. LABORDE (1951) Captation de l'histamine par l'héparine, *Comp. Rend. Soc. Biol.* **145**, 1047–51.

PARROT, J.-L., G. NICOT and C. LABORDE (1960) Captation de l'histamine par l'héparine. Influence du *p*H, *Comp. Rend. Soc. Biol.* **154**, 1426–8.

PHILLIPS, M. T. (1965) Histamine binding by heparin. Equilibrium dialysis. Unpublished.

SANYAL, R. K. and G. B. WEST (1959) The histamine–heparin complex, *J. Pharm. Pharmacol.* **11**, 548–52.

SCHAYER, R. W. (1963) Histidine decarboxylase in mast cells, *Ann. N.Y. Acad. Sci.* **103**, 164–78.

SCHILLER, S. and A. DORFMAN (1959) The isolation of heparin from mast cells of the normal rat, *Biochim. Biophys. Acta* (Amst.) **31**, 278–80.

SCHUBERT, M. and E. C. FRANKLIN (1960) Interaction in solution of lysozyme with chondroitin sulfate and its parent protein polysaccharide, *J. Amer. Chem. Soc.* **83**, 2920–5.

SCHULZ, H. and E. HIEPLER (1959) Über die Lokalisierung von gerinnungsphysiologischen Aktivitäten in submikroskopischen Strukturen der Thrombocyten, *Klin. Wschr.* **37**, 273–85.

SMITH, D. (1963) The tissue mast cell, *Int. Rev. Cytol.* **14**, 327–86.

STEGEMANN, H. and V. LOESCHKE (1962) Mikrobestimmung von Stickstoff als Indophenolblau nach Chloramin-T-oxydation, *Hoppe-Seyler's Zeitsch. Physiol. Chem.* **329**, 241–8.

STOECKENIUS, W. (1956) Zur Feinstruktur der Granuls menschlicher Gewebsmastzellen, *Exp. Cell Res.* **11**, 656–8.

STONE, A. L. (1964) Anomalous rotatory dispersion of metachromatic mucopolysaccharide-dye complexes. I. Heparin, *Biopolymers* **2**, 315–25.

TELFORD, J. M. and G. B. WEST (1961) The formation of histamine in the rat, *J. Pharm. Pharmacol.* **13**, 75–82.

THURSH, D. (1963) Sephadex chromatography of extracts of rat mast cell granules. Unpublished.

VASALLI, P., G. SIMON and C. ROUILLER (1964) Ultrastructural study of platelet changes initiated *in vitro* by thrombin, *J. Ultrastruct. Res.* **11**, 374–87.

VELLUZ, L. (1959) Recherches sur l'héparine les héparides antilipémiques, *Bull. Soc. Chim. Biol.* (Paris) **41**, 1–6.

WERLE, E. and R. AMANN (1956) Zur Physiologie der Mastzellen als träger des Heparins und Histamins, *Klin. Wschr.* **34**, 624–30.

WINQVIST, G. (1963) Electron microscopy of the basophil granulocyte, *Ann. N.Y. Acad. Sci.* **103**, 352–75.

WURZEL, M. and A. J. MARCUS (1965) Personal communication from A. J. Marcus.

DISCUSSION

CHAKRAVARTY: You mentioned that the mast cell granules could be depleted of histamine by potassium chloride solution and again reloaded with histamine by adding histamine to the medium. What is the mechanism of the uptake? Is it diffusion or active uptake or some form of chemical binding?

LAGUNOFF: The uptake and binding of histamine by mast cell granule heparin appears to be determined simply by ionic forces.

SOME ULTRASTRUCTURAL FEATURES OF RAT PERITONEAL MAST CELLS TREATED WITH COMPOUND 48/80

GUNNAR D. BLOOM and ÖSTEN HAEGERMARK*
Institute for Cell Research, and Dept. of Pharmacology,
Karolinska Institutet, Stockholm, Sweden.

IN connection with Dr. Lagunoff's presentation on mast granule structure we would like to give a brief survey of some recent work on mast cell and mast granule structure after treatment of the cells with the histamine liberator compound 48/80. Peritoneal fluid was collected from a number of male Sprague–Dawley rats in a buffered salt solution to which had been added glucose, human serum albumin and heparin. The cellular suspension was centrifuged and the sediments were resuspended in the buffered salt solution to which compound 48/80 had been added in three concentrations, 0·1, 0·25 and 0·50 μg/ml. As a control, cells were suspended in the buffered salt solution alone. After incubation at 37°C for 10 min the suspensions were centrifuged again. While supernatants and sediments were subjected to histamine assay, some sediments were fixed and processed for electron microscopy. The results of the histamine assay are shown in Fig. 1.

The electron microscopical examination reveals the following features: In the controls numerous mast cells are observed as completely intact elements and the mast granules appear as well-delimited, homogeneous, electron-dense structures (Figs. 2 and 6). At higher magnification a substructure of finely granular or reticular nature is faintly seen. The granules are surrounded by a distinct membrane, about 100 Å thick. The membrane may sometimes adhere so closely to the granule that it is indistinguishable from the granular matrix. More often, however, the membrane is separated from the periphery of the granule by an extremely narrow light space. Occasionally in these normal mast cells, granules are observed which differ from the majority. In such granules the general electron density is markedly diminished and the granular matrix clearly shows the substructure only faintly revealed in some of the normal dense granules. Furthermore, in such granules the space between periphery of the granule matrix and the surrounding perigranular membrane is widened. We have chosen to refer to such granules as "altered" mast granules.

* Supported in part by a grant from the United States Public Health Service (AM04063–05) and Konung Gustaf V: s 80-årsfond. A full report of this investigation is appearing in Exptl. Cell Research (1965).

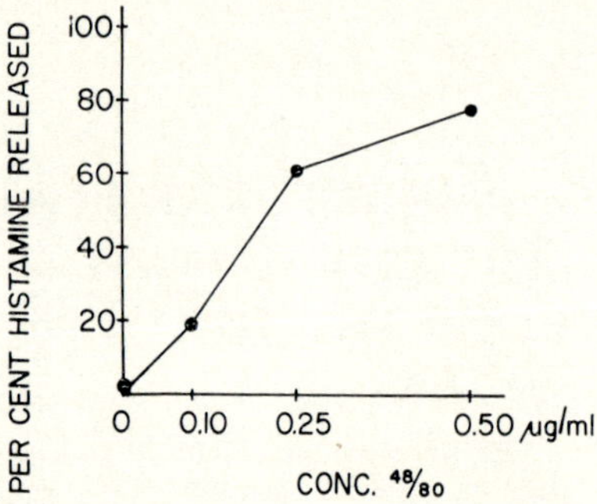

FIG. 1. Release of histamine from rat peritoneal mast cells induced by various concentrations of compound 48/80.

In specimens treated with the lowest dose of compound 48/80 (0·1 μg/ml) the majority of the mast cells are perfectly intact. However, a characteristic feature is that the number of "altered" granules is markedly increased. Furthermore, the perigranular space separating the granular matrix from its surrounding membrane appears to be widened as compared to the controls.

Incubation with 0·25 μg/ml of compound 48/80 leads to still more marked changes. The number of "altered" intracellular granules is significantly increased compared to both controls and those specimens treated with the lowest dose of compound 48/80. Furthermore, in this sample mast, granule release from the periphery of the mast cells is of frequent occurrence and all stages of this process may be observed. Larger vacuoles are frequently formed intracellularly in which several "altered" granules may lie, a phenomenon indicating rupture and disintegration of intervening perigranular

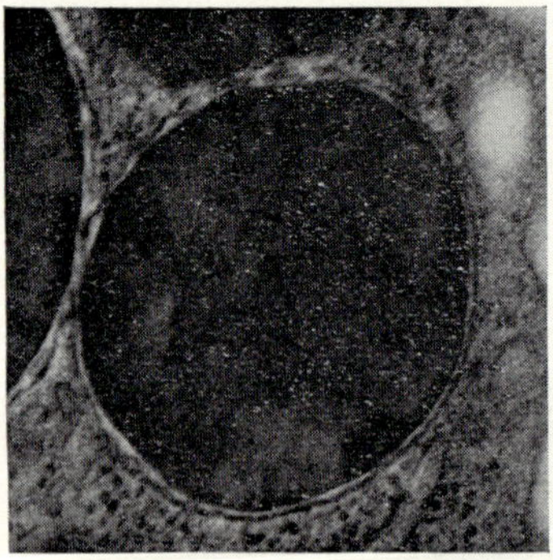

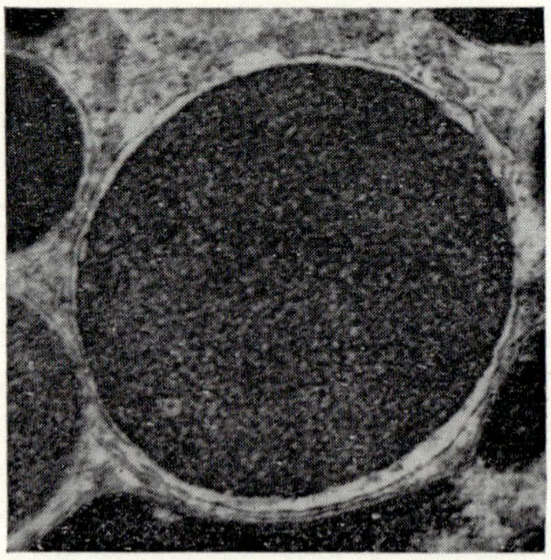

FIG. 2 (left). Normal mast granule. Note the dense, homogeneous appearance and the presence of a distinct perigranular membrane, here separated from the matrix by a light, narrow space. Electron micrograph. ×62,000.

FIG. 3 (right). "Altered" mast granule from rat peritoneal mast cell treated with 0·1 μg/ml. of 48/80. The granule is swollen, decreased in electron density and shows a peripheral halo between matrix and perigranular membrane. A fine substructure has appeared in the granule matrix. Electron micrograph. ×35,000.

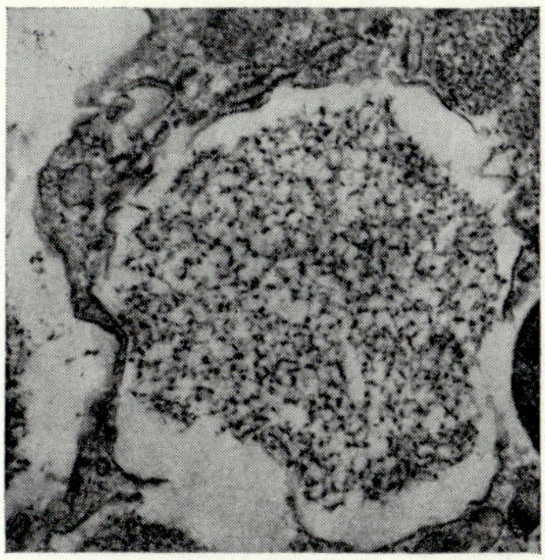

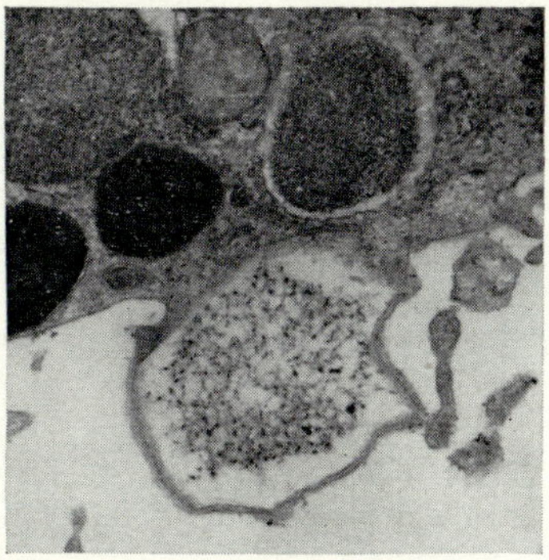

FIG. 4 (left). Grossly "altered" mast granule from specimens treated with 0·25 μg/ml. of 48/80. A wide space separates the reticulate matrix from surround-membrane. ×33,000.

FIG. 5 (right). Electron micrograph showing a protruding granule-containing vacuole prior to release of its contents. Note the double-walled structure of the vacuole and the reticulated appearance of the "altered" mast granule. Specimen treated with 0·25 μg/ml. of 48/80. ×21,400.

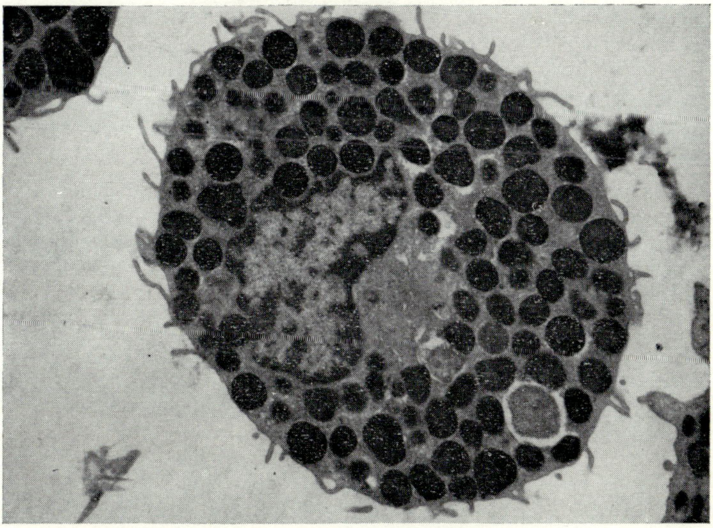

FIG. 6. Electron micrograph of a normal rat peritoneal mast cell showing the classical features of microvilli at the periphery of the cell and a content of numerous, well defined, homogeneous and electron dense mast granules. A Golgi apparatus and a centriole is also present in this section. To the lower left a typical "altered" mast granule is observed. Note its decreased electron density and the surrounding halo separating granule matrix from the perigranular membrane. ×8000.

membranes. Extracellular granules or granular aggregates are also frequently observed in these specimens. It is of interest to note that all granules observed extracellularly appear to be of the "altered" type. (Figs. 4, 5 and 7.)

Lastly, in the specimens treated with the highest dose of compound 48/80 (0·5 μg/ml) cellular changes are pronounced. Numerous mast cells are totally disrupted and appear as large conglomerates of granules in different stages

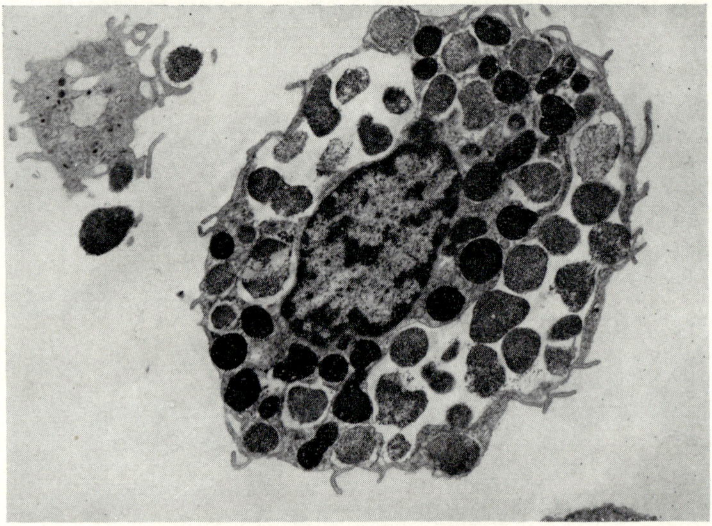

FIG. 7. Electron micrograph of rat mast cell treated with 0.25 μg/ml. of 48/80. The number of normal granules is strongly reduced while "altered" granules are very common. Large vacuolar formations containing numerous "altered" granules lacking surrounding membranes are seen as are freely lying, released mast granules. The latter are of "altered" type. $\times 8300$.

of alteration. However, seemingly intact cells are encountered. They contain numerous "altered" granules and show marked intracellular changes which appear to be of a pathologic nature.

It follows that there is a correlation between concentration of histamine liberator (compound 48/80) added to a suspension of rat peritoneal mast cells, and structural changes in the cells, especially the mast granules. There is also a correlation between amounts of histamine released and concentration of the liberator. Many of the numerous theories on the mechanism of histamine release from mast cells involve the interaction of the releasing agent or factor with cellular components far too minute to be resolved with the light microscope. Electron microscopy offers the possibility of visualizing cellular events on a much finer scale and may therefore be a fruitful approach to the study of the problem of amine release from the mast cell.

GENERAL DISCUSSION

ADAMS-RAY: With the before-mentioned fluorescence method Hillarp, Dahlström, Fuxe and I found *inter alia* in the skin of the hamster ear, cells with moderate and weak fluorescence, which was markedly reduced after reserpine.

After the administration of L-dopa these cells develop an intense and brilliant green fluorescence. If reserpine was given before L-dopa no or only slight increase of the fluorescence occurred. The intense fluorescence was also reduced or practically abolished by reserpine.

The decarboxylase inhibitor N.S.D. 1015 given 1 hr before L-dopa prevented the increase in fluorescence.

These cells were found to be mast cells. In the ear 80 per cent of the mast cells were of this type, but in the pouch none.

There is no or very little uptake of 5-HTP.

Dahlström, Fuxe, Häggendahl and I then found that in both normal and sympathectomized hamster after L-dopa administration the hamster ear tissue shows a unique capacity to form and store large amounts of DA for more than 24 hr, with maximum level after 2 hr. Heart and brain tissue on the other hand showed a peak of DA already after 30 min, after which the DA contents rapidly decreased and had after 24 hr reached normal levels. No significant changes were observed in the NA contents. Since only very few or practically no mast cells of this type are found in the brain and the heart the observed difference is in all probability due to the presence of abundant numbers of such mast cells in the skin of the ear and suggests that the primary catecholamine stored in the normal hamster ear is DA. The results also indicate that no or very low beta-hydroxylase activity is present in these mast cells, since no increase in the NA contents was found.

To conclude: in human granulation tissue we have found cells taking up dopa. We now are looking into the nature of these cells.

NORBERG: The histochemical fluorescence method of Falck and Hillarp has allowed the detection in various mammalian autonomic ganglia of synaptic structures formed by typical adrenergic nerve terminals. We have found basket-like structures of this type in sympathetic ganglia such as the inferior mesenteric and celiac ganglia of cat and rabbit and the superior cervical ganglion of the rabbit. The latter finding is of special interest in view of the physiological studies on this ganglion indicating an adrenergic mechanism influencing ganglionic transmission. Also certain parasympathetic ganglia, such as the ciliary ganglion of the cat and the intramural ganglia of the cat bladder and the intestine of various species contain such synaptic structures. Our denervation experiments in the cat have shown that the adrenergic synapses in the inferior mesenteric ganglion and the intramural ganglia of the bladder in all probability have their cell bodies in these same ganglia. The adrenergic terminals in the intramural ganglia of the colon have their cell bodies in the inferior mesenteric ganglion. The terminals in the small intestine in all probability have their origins in the celiac ganglia. It is reasonable to assume that the adrenergic synapses in the intestinal ganglia mediate the sympathetic inhibition of intestinal motility.

DAHLSTRÖM: I would like to say a few words about the effect of axotomy in monoamine neurons and the use of axone interruption.

A. *Above* the lesion there occurs, in both peripheral and central monoamine axons a large accumulation of fluorescent material. This accumulation is in all probability due to a damming up (by the obstacle of axotomy) of the specific so-called amine storage granules which are manufactured in the cell body and transported proximi-distally down the axon to the terminals, in accordance with the view of Weiss.

B. Also, below the site of axotomy there occurs during the first hours a small accumulation of amines just adjacent to the lesion. This accumulation disappears between 24–48 hr, and seems to be due to a retrograde axoplasmic flow occurring at least after axonal trauma.

C. Certain cell body reactions have been observed especially in the cell bodies in the brain stem. After 1–4 days can be seen a markedly increased content of fluorescent CA, the cells have a "swollen" appearance, and the nucleus is displaced towards the periphery.

As the normal concentration of amines in the axons is in most cases too low to be detected with the fluorescence method the use of axotomy is an important tool for the demonstration of CA fibres in peripheral adrenergic nerves and for mapping out amine neuron systems in the CNS. Due to the highly increased amine content after axotomy the

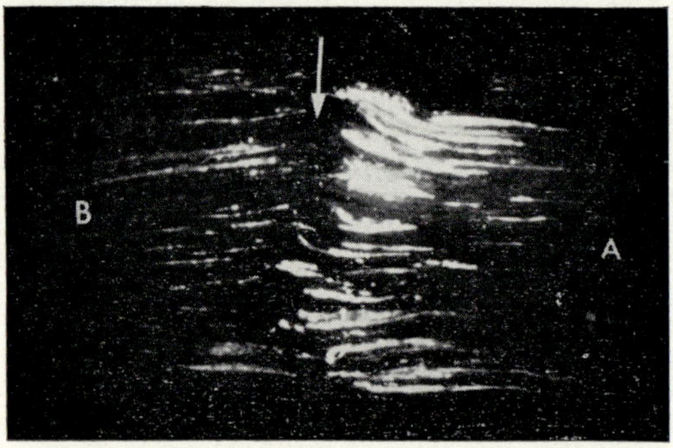

FIG. 1. Sciatic nerve of rat, 18 hr after compression (axotomy). Above (A) the lesion is seen large accumulations of strongly green-fluorescent material (accumulated NA, probably within storage granules) within distorted axons. Also, below (B) the compression is seen a small accumulation of weak to medium intensity. Site of lesion indicated by the arrow. ×80.

axons can be traced for considerable length, but seldom all the way to the cell body. But as the reactions here are very typical it is possible to state the origins of the transected axons.

Besides studying the organization of the peripheral adrenergic system (as done by Drs. Hamberger and Norberg) axotomy has been used to map out several amine systems in the CNS. Thus, by transections of the spinal cord we have been able to describe the large bulbospinal CA and 5-HT neuron systems. In close co-operation with Prof. Carlsson's group and Dr. Larsson in Gothenburg, combining biochemical and experimental approaches with Hillarp's fluorescence method, large nigro-neostrital dopamine systems have been detected. Also, several ascending neuron systems, arising in the mesencephalon, have been mapped

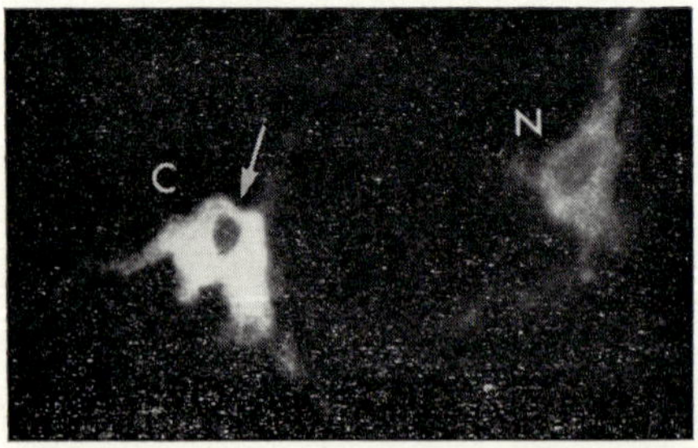

FIG. 2. From the medulla oblongata of a rat with a 24-hr-old transection of the spinal cord. Normal nerve cell body (N) and a nerve cell body (C), the axon of which is transected. This cell has a stronger fluorescence intensity than the normal cell, a "swollen" appearance and a displaced nucleus (→). ×310.

out. These systems give terminals to, for example, the limbic system, the preoptic area, the hypothalamus and to parts of the neocortex.

Axotomy is also used in order to give information of the syntheses, transport and turnover of the amine storage granules. Together with Dr. Häggendahl in Gothenburg we have started a series of experiments, combining biochemistry and axotomy with histochemistry in this purpose. Studies on the proximal accumulation in the sciatic nerve of rat for example show that the transport is very fast. 1 cm of a normal nerve contains about 1 ng of NA. 12 hr after axotomy the content has increased to about 8–10 ng/cm above lesion.

Thus, the use of axotomy is an important pool not only for mapping out amine systems but also for studies on the syntheses, transport and turnover of the "amine storage granules".

I would like to make some comments relating to Dr. Falck's lecture. He showed some pictures in which could be seen irregularities in the axons of neurons very close to the cell body. These irregularities were referred to by Dr. Falck as "varicosities". However, from the definition of a true varicosity under normal conditions there must be included the presence of *very* high amounts of the respective amine and thus a very strong fluorescence intensity should be seen using the present method, and, furthermore, the parts between of much lower intensity. The enlargements, often observed in both central and peripheral amine axons, exhibiting a weak to medium fluorescence intensity, might perhaps rather represent signs of some kind of peristaltic activity involved in the transport mechanism. This suggestion was made by Ochs (*Science* 1963) as an interpretation of the beaded appearance observed in myelinated peripheral axons after a certain treatment. Also Weiss (*Regional Neurochemistry* 1961) has stated that some sort of *active* mechanism must be involved in the proximo-distal transport of the axoplasm as judged from the appearance of the distorted axons above an applicated cuff. When a normal axon then shows about the same sort of "deformation" but to a *low* degree, these bulges might as well be signs of this active mechanism.

In our department we do not regard these enlargements as true "varicosities".

BIOSYNTHESIS AND RELEASE OF CATECHOLAMINES*

SIDNEY UDENFRIEND

Laboratory of Clinical Biochemistry, National Heart Institute,
National Institutes of Health, Bethesda, Maryland, U.S.A.

SYMPATHETIC storage particles are under active study by many groups and raise the question as to their significance. Do they exist merely to take up or store catecholamines or is storage just one of their functions? I would like to throw out the suggestion, that the physiologically significant particle of the sympathetic nerve, is in effect a synthesizing particle that holds the noradrenaline which it has formed until it is activated by some mechanism to release the transmitter.

Let us examine some of the evidence for such a hypothesis and consider its consequences. Two of the enzymes involved in noradrenaline synthesis, tyrosine hydroxylase (adrenal and brain) (Nagatsu, Levitt and Udenfriend, 1964) and dopamine-β-oxidase (adrenal) (Levin, Levenberg and Kaufman, 1960) have been found to be associated with particles which sediment between 15,000 and 20,000 $\times g$. Potter and Axelrod (1963) have presented suggestive evidence that dopamine-β-oxidase of heart is associated with the noradrenaline storage particles. We have similar evidence for dopamine-β-oxidase in brain. Only the decarboxylase appeared to be nonparticulate. However, when we examined the dopa decarboxylating activity of brain (dog, rat and guinea-pig), we found that on careful homogenization as much as 60 per cent of the activity could be found in synaptosomes (Levitt, Barchas and Udenfriend, to be published). De Robertis' findings (Rodriguez de Lores Aenaiz and De Robertis, 1964) with 5-hydroxytryptophan decarboxylation by brain synaptosomes are in agreement with these observations. It is our feeling that as suitable methods are developed, all three enzymes will be found in a particle (or vesicle) which may be schematically presented as in Fig. 1. The narrowest channel, representing tyrosine hydroxylase, is presumably the rate-limiting step (Levitt, Spector, Sjoerdsma and Udenfriend, 1965) and normally sets the pace for biosynthesis of the neurotransmitter. Such a vesicle which synthesizes the transmitter, releases it in response to nervous activity, and then recharges itself with transmitter, is not a new concept. Eccles (1965) has recently described it in a simplified

* The work reported in this paper represents studies carried out in collaboration with Dr. Sidney Spector and Dr. Albert Sjoerdsma.

form. However, the concept of a synthesizing vesicle raises questions which are not apparent from considering its storage properties alone. Is the rate of synthesis regulated? If so, how and at which step? Is there perhaps an end-product inhibition mechanism whereby excessive amounts of noradrenaline or dopamine inhibit further synthesis of the transmitter? Can one demonstrate allosteric sites which can respond to any of the components

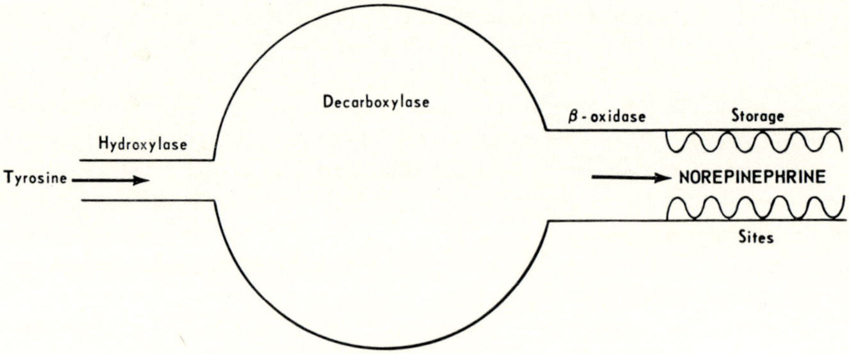

FIG. 1. Hypothetical norepinephrine synthesizing vesicles.

of a nerve impulse, i.e. ions, acetylcholine, noradrenaline, ATP, etc., by altering the rate of synthesis? Such a synthesizing particle also suggests that the three enzymes act in concert, that is, tyrosine enters and is rapidly converted to noradrenaline without the appearance of free intermediates.

The catalysts which may be included in such a particle are:

(1) tyrosine permease
(2) tyrosine hydroxylase
(3) dopa decarboxylase
(4) dopamine-β-oxidase
(5) noradrenaline binding protein

Although uptake of tyrosine into brain slices is a catalyzed process (Guroff, King and Udenfriend, 1961) it is not yet certain whether entry of tyrosine into the synthetic particle of sympathetic nerves is also catalyzed. The only evidence for this is rather indirect. In studies with the isolated heart, it was found that on perfusion with tyrosine-C^{14}, the isolated noradrenaline had a much higher specific activity than the tyrosine in the heart muscle (Spector, Sjoerdsma, Zaltman-Nirenberg, Levitt, and Udenfriend, 1963). One explanation is that tyrosine uptake into the sympathetic nervous system of the heart is more rapid than into heart muscle.

The properties of the three synthetic enzymes are now well established and have recently been summarized (Udenfriend, 1965). It should be pointed out, that although the observed properties of tyrosine hydroxylase make it the rate-limiting step with respect to noradrenaline synthesis, it is conceivable that uptake of tyrosine into the particle may be the actual rate-limiting step. If this is not true under normal conditions, it may be true when large amounts of other aromatic or long-chain amino acids occur as in phenylketonuria or branched-chain amino aciduria (Udenfriend, 1963).

This bring us to another very important point about the synthesizing particles. None of the catalysts is specific. As shown in Table 1, each of the

TABLE 1. SOME OF THE SUBSTRATES UTILIZED BY CATALYSTS OF THE SYMPATHETIC NERVOUS SYSTEM

*Tyrosine permease**	*Tyrosine hydroxylase*	*Decarboxylase*	*Dopamine-β-oxidase*	*Noradrenaline storage*
Tyrosine	Tyrosine	Dopa	Dopamine	Noradrenaline
Phenylalanine	Phenylalanine	Tyrosine	Tyramine	Norsynephrine
Leucine		Phenylalanine	Phenethylamine	Phenethanolamine
Isoleucine				
Valine				
Methionine				
5-HTP	α-CH_3-Tyrosine	α-CH_3-Dopa	α-CH_3-Dopamine	α-CH_3-Noradrenaline
		α-CH_3-*m*-Tyrosine	α-CH_3-*m*-Tyramine	Metaraminol
		Tryptophan	Paredrine	*p*-OH-Norephedrine
		5-HTP	Amphetamine	
		m-Tyrosine	*m*-Tyramine	*m*-Octopamine
		Histidine		

* Tyrosine permease data are based on data obtained with brain slices. 5-HTP, 5-hydroxytryptophan.

five catalytic systems has been shown to act on many synthetic and naturally occurring compounds, in addition to its normal substrate. This non-specificity has great significance. To the pharmacologist, it means that he can introduce a precursor of a sympathomimetic amine and have it converted to the active agent directly within the sympathetic machinery. The overall pharmacologic effects are quite different from those obtained with the amine itself. Such is the case with α-methyl-dopa and α-methyl-noradrenaline and with α-methyl-meta-tyrosine and metaraminol. With so many conversions possible within the sympathetic machinery, it becomes important to consider which is the active agent when administering a given drug. When tyramine is administered is norsynephrine responsible for some or all of the pharmacologic effects? Does dopamine owe any of its actions to newly synthesized noradrenaline? What about paredrine, *p*-hydroxy-norephedrine and related sympathomimetic agents?

Some interesting considerations concerning the significance of the catecholamine stores come from studies with inhibitors of tyrosine hydroxylase. One of these inhibitors, α-methyl-tyrosine, produces marked inhibition of endogenous noradrenaline synthesis, which starts with tyrosine. Noradrenaline synthesis is normal starting from dopa and the storage of catecholamines is unaffected. When α-methyl-tyrosine is administered to animals, catecholamine levels fall most rapidly in those tissues which synthesize them most rapidly. Brain noradrenaline and dopamine which turn over most rapidly are lowered first and to the greatest extent. Consistent with these observations the most prominent effects in animals are on the central nervous system. Although most tissue stores can be totally depleted with high doses of the inhibitor, marked effects are seen with lower doses when much of the peripheral noradrenaline is still present. The effects seen with the lowest doses are abolition of the shock-avoidance response and other disturbances in behavior (Stone-Merck, Sharp and Dohme). At higher doses animals become sedated and many of the recognized manifestations of lowered sympathetic activity appear (Spector, Sjoerdsma and Udenfriend, 1965). During the period of tyrosine hydroxylase inhibition animals are unable to survive stressful situations of various types. When rats are given α-methyl-tyrosine (150 to 200 mg per kg, i.p.) their body temperature falls rapidly, and if placed in a cold room they uniformly die within 6 to 8 hr. At death much of the noradrenaline is still present in peripheral organs and the amines in the adrenal medulla are only moderately lowered. The most marked depletion is seen in brain noradrenaline and dopamine. The levels in cold-stressed, untreated rats are not changed much. Those with drug alone show intermediate levels. The drug treated cold-stressed rats show normal vasoconstriction and apparently have sufficient peripheral noradrenaline to mobilize glycogen and fatty acids. All this would suggest that the inability to withstand stress may be centrally mediated.

It should be pointed out that α-methyl tyrosine does not lower catecholamine levels any more than do α-methyl-meta-tyrosine, metaraminol or reserpine. However, the latter compounds do not make rats anywhere near as susceptible to stress conditions. Furthermore, administration of dopa restores catecholamine levels and when given early enough protects the animals from α-methyl tyrosine effects. It appears, therefore, that at least in some tissues a rapid and continued synthesis of noradrenaline is required. In such tissues, at least, re-utilization of catecholamines by uptake mechanisms cannot be as important as *de novo* synthesis. The most fascinating conclusion suggested by the experiments with tyrosine hydroxylase inhibitors is that stimulation of the sympathetic nervous system, as in stress, increases the rate of catecholamine synthesis. If this is so, then one would expect the mechanism to involve induction or activation of the rate-limiting enzyme, tyrosine hydroxylase. There already exists other evidence which indicates that catecholamine synthesis can be increased by sympathetic stimulation. Studies on factors influencing tyrosine hydroxylase are certainly warranted.

SUMMARY

All three catalysts for the conversion of the dietary precursor, tyrosine to noradrenaline have now been characterized. Tyrosine hydroxylase appears to be the rate-limiting step in this overall conversion. Tyrosine hydroxylase and dopamine-β-oxidase are associated with dense particles. The dopa decarboxylating enzyme has also been shown to be associated with particulate material suggesting that all three enzymes are normally organized into a synthesizing organelle which may be the same as the noradrenaline storage vesicle. Such an organized system for synthesis and storage of noradrenaline may explain many observations concerning the sympathetic nervous system. It would also appear that a continued synthesis of noradrenaline is required for normal sympathetic function, irrespective of the concentration of noradrenaline in the tissue stores.

REFERENCES

ECCLES, J. (1965) The synapse, *Scientific American* **212**, 56–66.

GUROFF, G., W. KING and S. UDENFRIEND (1961) The uptake of tyrosine by rat brain *in vitro*, *J. Biol. Chem.* **236**, 1773–7.

LEVIN, E. Y., B. LEVENBERG and S. KAUFMAN (1960) The enzymatic conversion of dopamine to norepinephrine, *J. Biol. Chem.* **235**, 2080–6.

LEVITT, M., J. BARCHAS and S. UDENFRIEND (1965) Particle bound dopa decarboxylating activity (to be published).

LEVITT, M., S. SPECTOR, A. SJOERDSMA and S. UDENFRIEND (1965) Elucidation of the rate-limiting step in norepinephrine biosynthesis in the perfused guinea-pig heart, *J. Pharmacol.* (in press).

NAGATSU, T., M. LEVITT and S. UDENFRIEND (1964) Tyrosine hydroxylase: The initial step in norepinephrine biosynthesis, *J. Biol. Chem.* **239**, 2910–17.

POTTER, L. T. and J. AXELROD (1963) Properties of norepinephrine storage particles of the rat heart, *J. Pharmacol.* **142**, 299–305.

RODRIGUES DE LORES AENAIZ, G. and E. DE ROBERTIS (1964) 5-Hydroxytryptophan decarboxylase activity in nerve endings of the rat brain, *J. Neurochem.* **11**, 213–18.

SPECTOR, S., A. SJOERDSMA and S. UDENFRIEND (1965) Blockade of endogenous norepinephrine synthesis by α-methyl-tyrosine, an inhibitor of tyrosine hydroxylase, *J. Pharmacol.* **147**, 86–95.

SPECTOR, S., A. SJOERDSMA, P. ZALTZMAN-NIRENBERG, M. LEVITT and S. UDENFRIEND (1963) Norepinephrine synthesis from tyrosine-C^{14} in isolated perfused guinea-pig heart, *Science* **139**, 1299–1301.

UDENFRIEND, S. (1963) Factors in amino acid metabolism which can influence the central nervous system, *Amer. J. Clin. Nutr.* **12**, 287–90.

UDENFRIEND, S. (1965) Biosynthesis of the sympathetic neurotransmitter, norepinephrine, *Harvey Lect.* (in press).

DISCUSSION

KOPIN: Bhagat, B., E. K. Gordon, and I. J. Kopin (*J. Pharm. Exp. Ther.*, 1965, in press) showed increased rate of norepinephrine turnover which occurred during escape from tyramine tachyphylaxis in the cat. The "escape" could be prevented by inhibition of dopamine-β-oxidase with disulfiram, with prevention of the increased turnover seen during escape. These findings indicate that decrease in norepinephrine stores by tyramine administration is associated with increased rates of norepinephrine synthesis, α-CH_3-tyrosine would probably be a better drug to use to illustrate this.

PLETSCHER: (1) Did you compare the action of tetrahydropteridine with that of biopterin in activating tyrosine hydroxylase?

(2) Can you exclude that α-methyl-*p*-tyrosine causes a release of norepinephrine?

UDENFRIEND: (1) Dr. Seymour Kaufman has shown that dihydrobiopterin can be used as a cofactor for tyrosine hydroxylase.

(2) We have shown that α-methyl-tyrosine does not release norepinephrine.

EULER: How long does it take for the stores in different organs to become depleted after α-methyl-tyrosine?

UDENFRIEND: The rates of depletion parallel rates of turnover in the different tissues.

BRODIE: Since NA can itself block tyrosine oxidase, can one not think that normally the amine might act as a brake on its own synthesis. Accordingly, as the amine is released, the brake would be removed and synthesis would increase. Can one use this argument instead of postulating enzyme induction or activation?

UDENFRIEND: I have considered end-product inhibition as a control mechanism. Release of catecholamine could therefore stimulate synthesis.

UPTAKE AND STORAGE OF CATECHOLAMINES*

NORMAN KIRSHNER,† CAROLINE HOLLOWAY, W. JAMES SMITH‡
and ANNETTE G. KIRSHNER

Department of Biochemistry, Duke University Medical Center,
Durham, North Carolina

NORADRENALINE is released at the sympathetic nerve endings and adrenaline and noradrenaline are released from the adrenal gland following stimulation of the appropriate nerves. Adrenaline and noradrenaline can also be released from the intact or isolated gland by perfusion with acetylcholine, and noradrenaline can be partially released from sympathetic nerves by various sympathomimetic amines. Apart from this we know almost nothing of the steps between stimulation of the nerve and release of catecholamines.

Intimately associated with the release of catecholamines is the manner in which they are stored. The literature on the storage of catecholamines is too vast to be reviewed in detail here. However, biochemical (Blaschko and Welch, 1953; Hillarp, Lagerstedt and Nilson, 1953; Euler and Hillarp, 1956; Chrusciel, 1960; Potter and Axelrod, 1963) and electron microscopic studies (Lever, 1955; Sjöstrand and Wetzstein, 1956; de Robertis and Vaz Ferreira, 1957) have shown that the catecholamines are stored in granules and that the bulk of the catecholamines in various tissues can be recovered in a particulate form following homogenization in isotonic media. The studies of Hillarp, Högberg and Nilson (1955), Hillarp (1958a, 1958b), and Hillarp and Thieme (1959) on the granules of the adrenal medulla and of Schümann (1958), and Euler, Stjärne, and Lishajko (1963) on granules isolated from sympathetic nerves have indicated an important role for ATP and possibly the intragranular protein in the storage of adrenaline and noradrenaline.

More recently it has been found that the catecholamine-containing granules isolated from the adrenal medulla (Kirshner, 1962, 1962a; Carlsson, Hillarp and Waldeck, 1962, 1963) and from splenic nerves (Euler and Lishajko, 1963; Stjärne, 1964) can take up adrenaline and noradrenaline from the external medium by a mechanism which is temperature sensitive and stimulated by Mg^{++} plus ATP. The uptake is inhibited by EDTA, by reserpine and other drugs, and by a number of reagents which react with sulfhydryl

* This work was supported by USPHS grant AM–05427.
† Career Development Award K3–GM–15–184.
‡ Postdoctoral Fellow American Heart Association.

groups. The granules of the splenic nerve are similar in this respect to the adrenal granules but there are also some differences (Stjärne, 1964); splenic nerve granules are more stable to hypotonic media, there are quantitative differences in response to drugs, their catecholamine content turns over more rapidly and it is less tightly linked to ATP than that of adrenal granules. These differences may be important but do not necessarily indicate a basic difference in the uptake and storage mechanism.

In sharp contrast to the granules of the adrenal medulla and splenic nerve, the uptake of adrenaline or noradrenaline by the catecholamine-containing particulate fractions of cat heart (Potter and Axelrod, 1963) and cat brain (Mirkin and Gillis, 1963) is only moderately temperature sensitive, is not stimulated by Mg^{++} plus ATP and is not inhibited by low concentrations ofrese rpine.

To obtain further information on the storage and uptake of catecholamines, studies of the permeability of the adrenal granule membrane, of the ATPase activity of the granules, and of the granule soluble protein were undertaken.

Permeability of the Granule Membrane

Studies of the uptake of adrenaline and noradrenaline by adrenal granules indicated that a transport mechanism across the granule membrane might be involved. Specifically, the temperature dependence of the uptake, the stimulation by Mg^{++} and ATP, the kinetics of uptake, and the inhibition by reserpine and sulfhydryl group reagents indicated that an enzymatic, energy-dependent process was involved. Previously, Carlsson and Hillarp (1958), and Hillarp (1959) had reported that the granule membrane was permeable to catecholamines. However, in their studies, granules were incubated for much longer periods of time and with much higher concentrations of catecholamines than those used in measuring uptake. It was necessary, therefore, to determine whether catecchlamines at low external concentration could penetrate the granule membrane at sufficient rates to rule out membrane transport as the limiting factor. Since ATP is important for maximal uptake of catecholamines, the permeability of the granule membrane to this compound as well as to EDTA and urea was also measured.

The permeability of the granules to the above compounds was determined by measuring the distribution of isotopes between the intragranular water and incubation medium. From the concentrations of isotope in the two fluids, the intragranular space was calculated, using the ratio

$$\frac{\text{total cpm in intragranular water}}{\text{cpm/ml incubation medium}} = \text{volume of intragranular space}$$

The space is here defined as the volume necessary to contain the amount of intragranular isotope at concentrations equal to that in the incubation medium.

Duplicate samples of granules were incubated at 0° in 5 ml of 0.25 M sucrose containing 0.05 M tris pH 7.4, 0.1 μc of adrenaline-C^{14} (specific activity 1 μc/μmole), tritiated water (1×10^6 cpm), and 3 per cent dextran (average mol. wt. 200,000). At the end of the incubation the granules were centrifuged. The supernatant was assayed for catecholamines (Euler and Hamberg, 1949), tritium, carbon-C^{14}, and dextran (Seifter, Dayton, Novic and Muntwyler, 1950). The packed granules were resuspended in 5 ml of ice-cold isotonic sucrose and gently shaken in ice water for 1 hr. The granules were again sedimented and the supernatant assayed as above. One of the duplicates was lysed by suspension in distilled water, centrifuged, and the supernatant assayed for catecholamines, tritium and carbon-C^{14}. The second sample was used for wet and dry weight measurements. Similar procedures were used for the studies with ATP-C^{14}, urea-C^{14} and EDTA-C^{14}. Tritium and carbon-C^{14} were simultaneously determined using the Packard Tri-Carb liquid scintillation spectrometer. Paper chromatography of the C^{14}-compounds obtained from the various fractions showed them to be the same as the starting materials.

The total water in the pellets measured either by dry weight or T_2O were in excellent agreement and represented 75 ± 3 per cent of the wet weight. The extragranular water in all of the experiments averaged 16 ± 4 per cent, and the intragranular water 60 ± 2 per cent of the pellet weight. The per cent of extragranular water was about the same as previously reported by Carlsson and Hillarp (1958) but the intragranular water was about 10 per cent lower than that found by Hillarp (1959).

The data in Table 1 were obtained from three different granule preparations.

TABLE 1. PERMEABILITY OF CHROMAFFIN GRANULES TO ADRENALINE

Number of determinations	*Total pellet spaces*		*Extragranular pellet water*	*Intragranular space*			*Bound space*		
	T_2O	Adr.		T_2O	Adr.	Adr./T_2O	Adr.	T_2O	Adr./T_2O
5	0.233	0.182	0.059	0.174	0.123	0.70	0.160	0.042	4
3	0.244	0.205	0.048	0.196	0.157	0.80	0.186	0.007	27
8	0.199	0.163	0.051	0.148	0.112	0.75	0.083	0.005	16

The spaces were determined as described in the text. Granules were maintained in contact with the isotopes for 30 min at 0°C.

In these preparations the adrenaline space was equal to about 75 per cent of the intragranular water space. This by itself would indicate that the entire granule is not freely permeable to adrenaline. However, the data in Table 1 also show that amounts of adrenaline, about equal to the adrenaline space, were also bound. It was observed that if the pellet obtained after the first centrifugation was allowed to warm to room temperature before adding sucrose, only small amounts of adrenaline-C^{14} were diffusible, and the space thus measured was only 15 to 20 per cent of the T_2O space. Hence, one must conclude that the intragranular adrenaline and water spaces are equal.

The intragranular space was independent of external adrenaline concentration within the 20-fold range measured (25 to 460 μg/ml). The shortest incubation period was 5 min. At that time equilibration was complete and no changes were observed upon further incubation up to 30 min.

The data in Table 2 were obtained from parallel experiments using ethyl-

TABLE 2. RELATIVE GRANULE SPACE

Experiment	*Adrenaline*	*ATP*	*Urea*	*EDTA*
1	0.71	1.01	1.02	0.21
2	0.80	1.00	0.99	0.25
3	0.75	0.72	1.00	0.00

The spaces were determined as described in the text. The figures are relative to the T_2O space assigned a value of 1.0. Granules were maintained in contact with the isotopes for 30 min at 0°C.

enediaminetetra-acetate-C^{14} (EDTA), ATP-C^{14} and urea-C^{14}. It was expected that the granules would be permeable to urea-C^{14} and the good agreement between the T_2O space and the urea space affirms the reliability of the method.

From these data we have concluded that the granule membrane is freely permeable to adrenaline, ATP and urea but impermeable to EDTA. The latter is consistent with previous observations (Kirshner, 1962a) that EDTA did not inhibit dopamine-β-hydroxylase in intact granules but did inhibit the enzyme in lysed granules. These data, and the further observation that 10^{-5} M reserpine had no effect on the adrenaline space, indicate that transport across the granule membrane is not involved in the Mg^{++}-ATP-stimulated uptake of catecholamines. Stjärne (1964) has found splenic nerve granules to be permeable to noradrenaline and has expressed the possibility that the granules may contain internal structures, and that the catecholamines may be retained within intragranular sub-vesicles. Thus the possibility of an internal membrane transport mechanism cannot yet be excluded. Because the granules stain so intensely it has not yet been possible to detect any internal structure by electron microscopy.

ATPase Activity of Granule Membranes

Hillarp (1958, 1958b) has identified an ATPase activity as well as an adenylate kinase activity associated with the adrenal granules and proposed a possible functional role for these enzymes in the release of catecholamines.

However, at that time, no relationship between the storage or release of catecholamines and the enzymatic activities was demonstrated. Since then, the observations that the uptake of catecholamines by granules was inhibited by pre-incubation with N-ethylmaleimide (NEM), that this inhibition could be largely blocked if the pre-incubation was carried out in the presence of both Mg^{++} and ATP, and that NEM inhibited the ATPase activity of the granule preparations (Kirshner, 1965) prompted further investigation of the relationship of the ATPase activity to the uptake process.

It has been previously shown (Kirshner, 1962a) that EDTA prevented the uptake of catecholamines. Table 3 shows that EDTA also inhibited the

TABLE 3. THE EFFECT OF EDTA ON ATPASE ACTIVITY AND ADRENALINE-7-C^{14} UPTAKE

Additions	*Uptake* cpm Adrenaline-7-C^{14}	*Uptake* % of Mg^{++} value	*ATPase activity* μmoles Pi released	*ATPase activity* % of Mg^{++} value
None	4100	31	0.2	21
Mg^{++}	13,300	100	0.92	100
Ca^{++}	4040	30	0.67	72
$Mg^{++}+Ca^{++}$	12,500	94	0.84	91
EDTA	1637	12	0	0
EDTA$+Mg^{++}$	15,110	114	0.53	58
EDTA$+Mg^{++}+Ca^{++}$	11,880	90	0.78	85
EDTA$+Ca^{++}$	3380	25	0.50	54

Each reaction vessel contained: washed granules, 30 mg; Tris buffer pH 7.4, 50 μmoles; ATP, 10 μmoles; adrenaline-7-C^{14}, 0.08 μmoles, 75,000 c.p.m. Where indicated: EDTA, 5 μmoles; $MgCl_2$, 10 μmoles; and $CaCl_2$, 10 μmoles. The reaction mixture was made up to a final volume of 2 ml with 0.3 M sucrose. Incubated 15 min at 30°C.

ATPase activity, and that both ATPase activity and catecholamine uptake could be reactivated by addition of sufficient Mg^{++}. The addition of Ca^{++} to the medium reactivated the ATPase activity but not catecholamine uptake.

In attempts to demonstrate a further correlation between catecholamine uptake and Mg^{++}-stimulated ATPase activity the effects of increasing concentrations of Mg^{++} on both of these activities in the presence and absence of NEM were studied with granules that had been washed with 10^{-4} EDTA (Table 4). In the absence of NEM, the ATPase activity and the uptake increased with increasing concentration of magnesium to a maximum at 4.5 μmoles of Mg^{++}. The plot of adrenaline-7-C^{14} uptake versus μmoles of phosphate released was linear throughout this range. In the presence of NEM, the uptake of catecholamines was completely inhibited at all concentrations of added Mg^{++}. The ATPase activity was completely inhibited at the low Mg^{++} concentrations and 40 per cent inhibited at concentrations of Mg^{++} which, in the absence of NEM, gave maximal catecholamine uptake and

TABLE 4. THE EFFECTS OF Mg^{++}, Ca^{++}, AND NEM ON UPTAKE OF ADRENALINE-7-C^{14} AND ATPASE ACTIVITY

Additions μmoles	*ATPase activity μmoles Pi released* −NEM	+NEM	*Adrenaline-7-C^{14} uptake cpm* −NEM	+NEM
Mg^{++} None	0	0	775	600
0.25	0.20	0	2700	475
0.50	0.27	0	2900	425
1.5	0.45	0	3500	425
3.0	0.52	0.16	4300	450
4.5	0.62	0.37	4900	325
10.0	0.65	0.65	4500	335
Ca^{++} 0.5	0.34	0.34	830	870
1.5	0.51	0.57	1200	670
3.0	0.63	0.66	1000	680
4.5	0.63	0.75	1000	680
10.0	0.69	0.76	1000	600

Each reaction vessel contained: EDTA-treated granules, 30 mg; tris buffer pH 7.4, 50 μmoles; ATP, 10 μmoles; adrenaline-7-C^{14}, 0.08 μmoles, 75,000 cpm. The reaction mixture was made up to a final volume of 2 ml with 0.3 M sucrose. NEM, where indicated, was added to a final concentration of 1.5×10^{-4} M and all vessels were pre-incubated for 10 min at 30° in the absence of ATP and adrenaline-7-C^{14}. The addition of ATP and adrenaline-7-C^{14} initiated the reaction. Incubated 15 min at 30°C.

ATPase activity. In this preparation, ATPase activity was fully restored by the addition of 10 μmoles of Mg^{++}, however, in other preparations, full ATPase activity could not be restored with even larger amounts of Mg^{++}.

Increasing concentrations of Ca^{++} increased the ATPase activity but had an insignificant effect on adrenaline uptake. NEM had no effect on the Ca^{++}-stimulated ATPase activity.

The granules used in these studies contained varying amounts of mitochondria and it was not possible to distinguish the mitochondrial and granule ATPase activities. Hillarp (1958b) using a discontinuous sucrose gradient was able to separate almost completely, using succinic dehydrogenase activity as a criterion, the adrenal granules from mitochondria, and demonstrated an ATPase activity associated with the granule fraction which was not due to mitochondrial contamination. However, Fortier, Leduc and D'Iorio (1959) using similar methods, did not obtain clear separation of granules from succinic dehydrogenase activity, and reported that the ATPase activity in the granule fraction was probably due to mitochondrial contamination. Because of the possible relevance of ATPase activity to granule function, the localization of the ATPase activity in the granules was reinvestigated.

Granules were prepared as described by Hillarp (1958b). Three ml of the washed granule fraction (Gw), containing about 500 mg of granules in 0.25 M sucrose, was layered over 30 ml of a continuous linear sucrose gradient varying from 1.0 M to 2.25 M. The tubes were centrifuged at 25,000 rpm in a Spinco SW-25 swinging bucket rotor for 4 to 6 hr. Fractions of 100 drops each were collected by puncturing the bottom of the tubes. Identical fractions from the three centrifuge tubes were pooled, and diluted to 40 ml with distilled

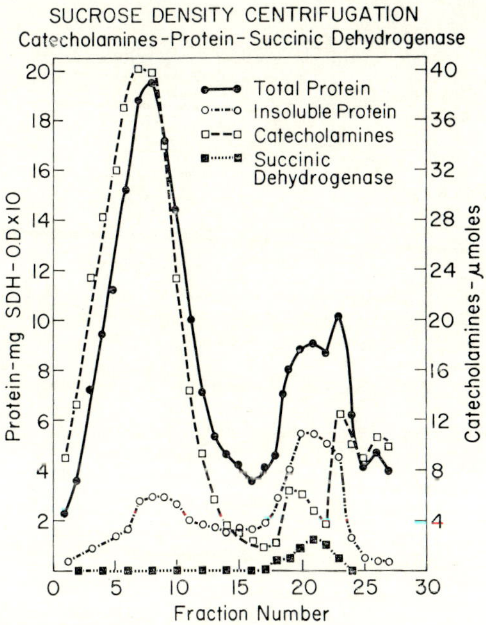

FIG. 1. Distribution of catecholamines, total and insoluble protein and succinic dehydrogenase activities. The preparation of granules, centrifugation procedure and assays are described in the text.

water to lyse the granules. Aliquots were removed for catecholamine (Euler and Hamberg, 1949) and total protein (Moore and Stein, 1954) analysis and the remainder centrifuged at $40{,}000 \times g$ for 30 min. The clear supernatants were discarded. The residues were suspended in 5 ml of distilled water using a glass–teflon homogenizer and aliquots of the suspension were used for measuring insoluble proteins (Moore and Stein, 1954), succinic dehydrogenase (Kamin, Gibbs and Merritt, 1957) and ATPase activities (Hillarp, 1958a).

Figures 1, 2 and 3 graphically present the data obtained. Fractions 1 to 16 contained no measurable succinic dehydrogenase activity and clearly indicated separation of mitochondria from the catecholamine-containing granules. In

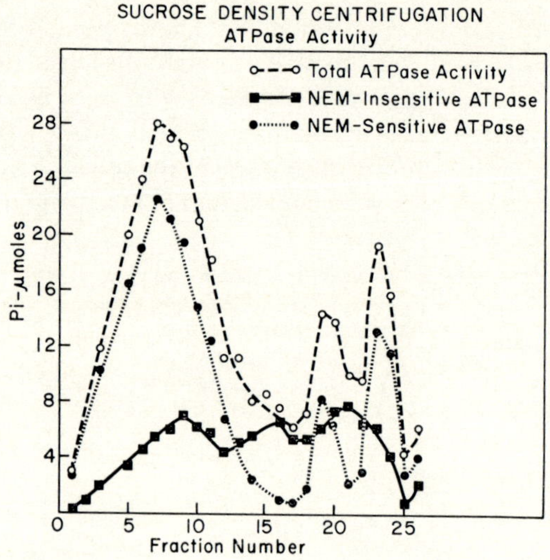

FIG. 2. Distribution of Mg^{++}-ATPase activity in the presence and absence of N-ethylmaleimide (NEM). The difference between these two activities was calculated and plotted as the NEM-sensitive ATPase activity.

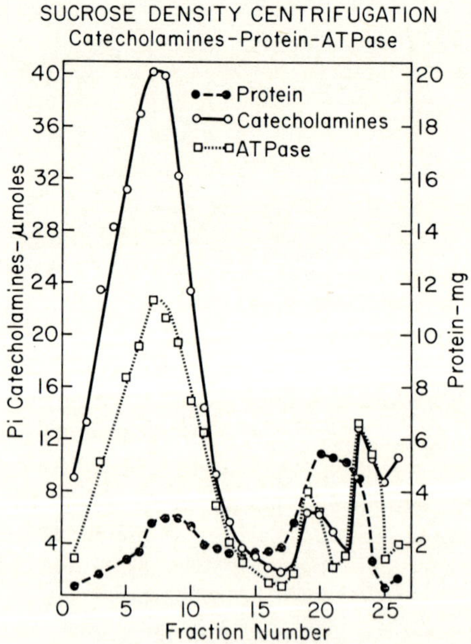

FIG. 3. Distribution of catecholamines, insoluble protein and NEM-sensitive Mg^{++}-ATPase activity.

fractions 1 to 12 the insoluble protein was about 16 per cent of the total protein, which is in good agreement with a previous report by Hillarp (1959) on the protein composition of the granules. In fractions 18 to 25 the insoluble protein was close to 50 per cent of the total protein.

The distribution of ATPase activity, especially the NEM-sensitive activity, correlated very well with the distribution of catecholamines. The peaks of ATPase activity and catecholamines in fraction 19 probably were due to trapped or damaged granules, or possibly, less dense, newly-formed granules; the peaks in fraction 23 probably were due to badly damaged or lysed granules. In fractions 1 to 12, NEM inhibited the ATPase activity an average of 79 per cent; in fractions 19 and 23 the inhibition was 57 and 60 per cent respectively. Although the distribution of the Ca^{++}-activated ATPase was similar to that of the Mg^{++}-activated ATPase, the former was only about one fourth as active as the latter. The CA^{++}-ATPase in these preparations was inhibited to the same extent as the Mg^{++}-ATPase, whereas in intact granules the CA^{++}-ATPase was not inhibited by NEM.

These studies confirm the observation of Hillarp (1958a) that adrenal granules contain an ATPase activity and provide a method for clearly separating the granules from mitochondria. The correlation between the inhibition of catecholamine uptake and ATPase activity by NEM warrants further investigation.

Soluble Granule Protein

A possible role in the storage of catecholamines for the soluble protein (Hillarp, 1958a) obtained upon water lysis of the granules, led us to investigate the properties of this protein.

Purification of the soluble protein was easily accomplished. Washed granules were suspended in distilled water and centrifuged at $20,000 \times g$ for 20 min. The supernatant was then lyophilized, redissolved in a small volume of water and centrifuged for 1 hr at $100,000 \times g$. The clear solution was desalted by passage through Sephadex G-25. The protein was then chromatographed on DEAE-cellulose at pH 6.8, using a 0.0 to 0.7 N sodium chloride gradient in 0.01 M phosphate. The protein peak was lyophilized, dialysed and passed through Sephadex G-200 in 0.01 M phosphate pH 6.8 (Fig. 4). During the early part of this work, chromatography on hydroxylapatite was also used but only small amounts of impurities were removed and it had no effect on the elution from Sephadex G-200. About 50 to 60 per cent of the total soluble protein obtained from the granules was recovered in the two peaks from the Sephadex columns. Figure 5 shows the sedimentation pattern in 0.05 M phosphate buffer, pH 6.8, of the protein eluted from the DEAE-cellulose column and Fig. 6 shows the two principal peaks obtained from the Sephadex G-200 column (larger peak on lower trace). The conditions were the same except that the slower moving fraction was present in

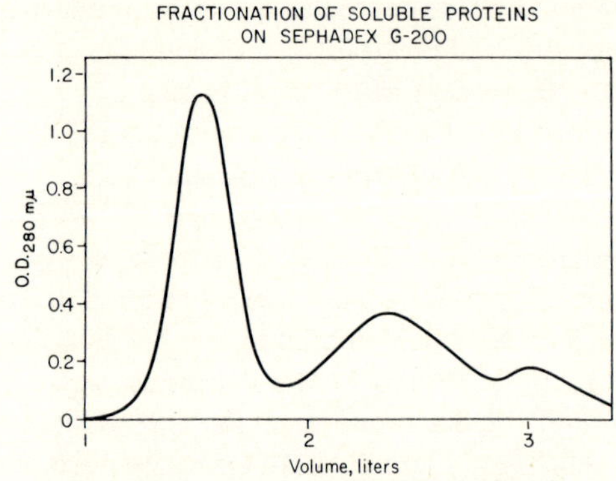

FIG. 4. The Sephadex column was equilibrated and eluted with 0.4 N NaCl in 0.01 M phosphate buffer, pH 6.8.

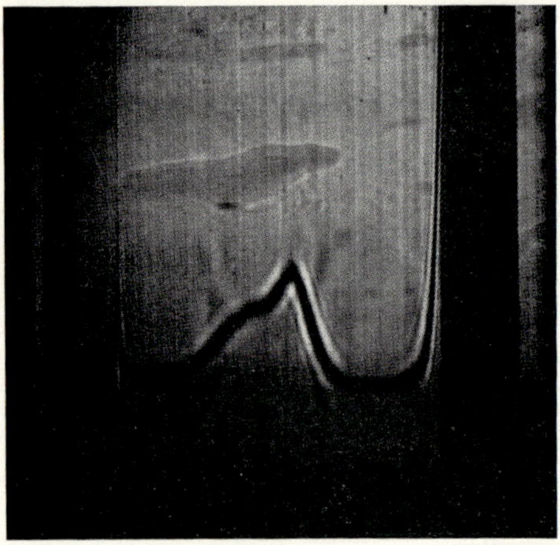

FIG. 5. The protein was centrifuged in a solution of 0.05 M phosphate, pH 6.8, at 25°C. The protein concentration was 5 mg/ml.

lower concentrations. The $S_{25,w}$ values of these fractions were approximately 1.4 and 2.8.

Further studies of these fractions showed that the sedimentation coefficients exhibited a marked and anomolous protein concentration dependence, and that the schlieren diagram exhibited a hypersharpening of the leading edge, which suggested an aggregating system of the type described by Gilbert (1959). It was found that in solutions of higher sodium chloride concentrations a more aggregated form could be stabilized. Addition of increasing

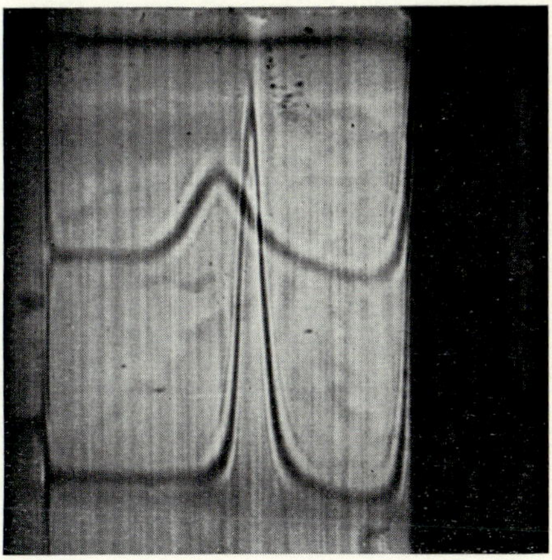

FIG. 6. The protein fractions were centrifuged in 0.05 M phosphate, pH 6.8 at 25°C. Upper curve: the second peak eluted from the Sephadex G-200 column. Protein concentration was 3 mg/ml. Lower curve: the first peak eluted from the Sephadex column. Protein concentration was 7 mg/ml.

amounts of sodium chloride to the 1.4 S fraction increased the $S_{25,w}$ to a maximum of 2.7 S (Fig. 7). Dialysis of the 2.7 S fraction overnight against distilled water changed its $S_{25,w}$ to approximately 1.6 S. The sedimentation velocity of the protein measured in 6 M guanidine containing 0.1 M mercaptoethanol gave an $S_{25,w}$ value of 1.7. Molecular weight determinations in high (0.5 M) and low (0.1 M) sodium chloride at a protein concentration of 0.2 mg per ml. gave weights of about 39,000 and 21,000 for the associated and dissociated systems. When the protein obtained from the DEAE-

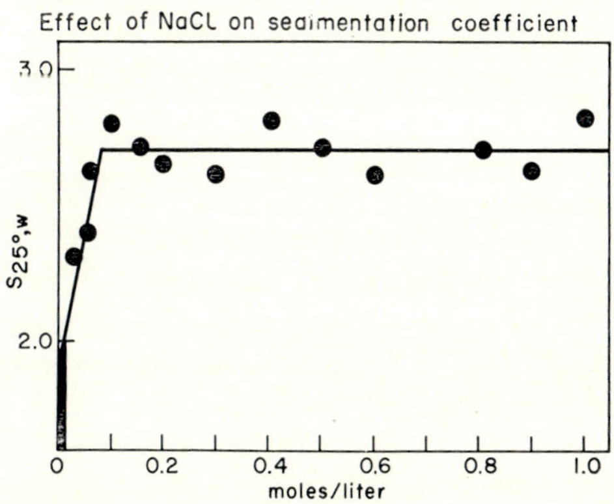

FIG. 7. The sedimentation coefficients were measured at a protein concentration of 5 mg/ml.

cellulose column was chromatographed on Sephadex G-200 in 0.01 M phosphate buffer pH 6.8 in 0.4 M sodium chloride only one fraction was obtained with an $S_{25,w}$ of 2.7.

The sedimentation behavior of the protein was much more sensitive to divalent cations, Mg^{++} and Ca^{++}, than it was to Na^{+}. In 0.001 M Mg^{++} or Ca^{++} the $S_{25,w}$ of the protein was 1.6 S; in 0.01 M Mg^{++} or Ca^{++} it rose sharply to 3.1 S. From the concentrations of Ca^{++} and Mg^{++} reported in the adrenal granule (Borowitz and Weiner, 1964) it would appear that the protein is present in an aggregated state. Blaschko and Helle (1963) in similar studies reported sedimentation coefficients of 1.4 and 1.8, depending upon the purification procedure, and that the sedimentation coefficients were raised to 2.4 upon addition of Mg^{++} plus ATP plus adrenaline. We believe the Mg^{++} by itself was sufficient to produce the changes.

TABLE 5. AMINO ACID COMPOSITION OF 2.7 S PROTEIN

Amino acid	*Residues per 40,000 mol.wt.*
Glutamic	82
Proline	32
Lysine	32
Alanine	30
Serine	29
Aspartic	29
Glycine	25
Leucine	22
Arginine	17
Valine	14
Threonine	10
Histidine	6
Methionine	6
Isoleucine	5
Phenylalanine	5
Tyrosine	3

The amino acid analysis was done as described by Spackman, Moore and Stein (1958). The figures for tryptophan and cysteine are not included.

Electrophoresis of the protein on cellulose acetate strips at pH 5.0 and on starch gel at pH 7.4 indicated a single species. Electrophoresis on polyacrylamide gels at pH 9.5 gave a major band with some trailing. Because of the low salt content of the gel the trailing may have been due to dissociation.

Chemical studies of the protein support the physical data which suggests that the 2.7 S protein is an aggregate of two peptides of slightly different molecular weights. The amino acid composition (Spackman, Stein and Moore,

1958) of the 2.7 S protein, calculated on the basis of a molecular weight of 40,000 is given in Table 5. These values are similar to those found by Weiner (1965). The values for cysteine and tryptophan are not included. Fingerprints of tryptic digests of both the aggregated and dissociated proteins gave identical patterns. About 40 distinct ninhydrin-staining spots were found. Since there are a total of 49 lysine and arginine residues one would expect about that number of peptides from a single polypeptide chain or from two different polypeptide chains. If the aggregate consisted of two identical chains one would expect about 25 ninhydrin-staining spots.

Preliminary binding studies by equilibrium dialysis indicated that the protein could bind between 2 and 3 moles of adrenaline per mole of protein and that this binding was readily reversible by dialysis and was not affected by Mg^{++} and ATP. This amount of binding, however, would account for only 1 or 2 per cent of the total catecholamines held within adrenal granules.

SUMMARY AND CONCLUSIONS

Studies of the permeability of adrenal granules to adrenaline and ATP indicate that the granule membrane is freely permeable to these compounds at 0°. Thus activation of a membrane transport mechanism for the Mg^{++}-ATP-stimulated uptake of catecholamines is unlikely. However, the possibility of intragranular structures must be considered and transport across membranes within the granule, if such exist, may be important in the storage and release of catecholamines.

Centrifugation of adrenal granules through a linear, continuous sucrose gradient has established that an ATPase activity is associated with the granules and that this activity is not due to mitochondrial contamination. The correlation between the inhibition of catecholamine uptake and of ATPase activity by N-ethylmaleimide indicates a possible relationship which warrants further investigation.

The mechanisms by which catecholamines are taken up, bound and released by granules of the adrenal medulla and other tissues remain obscure. The uptake in adrenal and splenic nerve granules is temperature sensitive, stimulated by Mg^{++} plus ATP, and inhibited by reserpine and other drugs and by reagents which react with sulfhydryl groups, indicating that an enzymatic reaction is involved in this process. The observations (Hillarp, 1958a) that catecholamines and adenine nucleotides are found in granules in molar ratios such that the total cationic charges of the catecholamines are balanced by the anionic charges of the adenine nucleotides has led to the hypothesis that these compounds react with the intragranular protein to form a non-diffusible complex which acts as a storage mechanism for the catecholamines. This postulate (Hillarp, 1958a) is not inconsistent with the requirement of an enzymatic reaction for the formation of the complex.

However, since lysis of the granule immediately liberates free catecholamines, and since all attempts to demonstrate a stable complex between catecholamines, ATP, Mg^{++} and intragranular protein have been unsuccessful, the nature of this complex is not clear and its existence has not been established.

On the other hand, it is equally feasible that the ATP present in the granules serves the purpose of supplying energy for the release of catecholamines. Hillarp (1958) has suggested how such a mechanism, which utilizes the ATPase activity of the granules, may act, but again, there is no evidence for such a mechanism other than the presence of ATP and ATPase activity.

It is hoped that further studies of the intragranular soluble protein and of the ATPase activity may lead to a better understanding of the nature of the uptake, storage and release of catecholamines.

REFERENCES

BLASCHKO, H. and K. B. HELLE (1963) *J. Physiol.* (Lond.) **169**, 120 P.

BLASCHKO, H. and A. D. WELCH (1953) *Arch Exp. Path. Pharmak.* **219**, 27.

BOROWITZ, J. L. and N. WEINER (1964) *Fed. Proc.* **23**, 350.

CARLSSON, A. and N.-Å. HILLARP (1958) *Acta Physiol. Scand.* **44**, 163.

CARLSSON, A., N.-Å. HILLARP and B. WALDECK (1962) *Med. Exp.* (Basel) **6**, 47.

CARLSSON, A., N.-Å. HILLARP and B. WALDECK (1963) *Acta Physiol. Scand.* **59**, Suppl. 215.

CHRUSCIEL, T. (1960) In: *Ciba Foundation Symposium on Adrenergic Mechanisms* (Edited by J. R. Vane, G. E. W. Wolstenholme and C. M. O'Connor. London, J. A. A. Churchill, Ltd.

EULER, U. S. VON and U. HAMBERG (1949) *Science* **110**, 561.

EULER, U. S. VON and N.-Å. HILLARP (1956) *Nature* (Lond.) **177**, 44.

EULER, U. S. VON and F. LISHAJKO (1963) *Acta Physiol. Scand.* **59**, 454.

EULER, U. S. VON, L. STJÄRNE and F. LISHAJKO (1963) *Acta Physiol. Scand.* **59**, 495.

FORTIER, A., J. LEDUC and A. D'IORIO (1959) *Rev. Can. Biol.* **18**, 110.

GILBERT, A. (1959) *Proc. Royal Soc.* (Lond.) Series A **250**, 377.

HILLARP, N.-Å. (1958) *Acta Physiol. Scand.* **42**, 144.

HILLARP, N.-Å. (1958a) *Acta Physiol. Scand.* **42**, 321.

HILLARP, N.-Å. (1958b) *Acta Physiol. Scand.* **43**, 82.

HILLARP, N.-Å. (1959) *Acta Physiol. Scand.* **47**, 271.

HILLARP, N.-Å., B. HÖGBERG and B. NILSON (1955) *Nature* (Lond.) **176**, 1032.

HILLARP, N.-Å., S. LAGERSTEDT and B. NILSON (1953) *Acta Physiol. Scand.* **29**, 251.

HILLARP, N.-Å. and G. THIEME (1959) *Acta Physiol. Scand.* **45**, 328.

KAMIN, H., R. H. GIBBS and D. A. MERRITT (1957) *Fed. Proc.* **16**, 202.

KIRSHNER, N. (1962) *Science* **135**, 107.

KIRSHNER, N. (1962a) *J. Biol. Chem.* **237**, 2311.

KIRSHNER, N. (1965) *2nd International Pharmacology Congress.* Pergamon Press.

LEVER, J. D. (1955) *Endocrinology*, **57**, 621.

MIRKIN, B. L. and C. N. GILLIS (1963) *Biochem. Pharmacol.* **12**, 1173.

MOORE, S. and W. H. STEIN (1954) *J. Biol. Chem.* **211**, 907.

POTTER, L. T. and J. AXELROD (1963) *J Pharmacol.* **142**, 291.

POTTER, L. T. and J AXELROD (1963) *J. Pharmacol.* **142**, 299.

ROBERTIS, E. DE and A. VAZ FERREIRA (1957) *Exp. Cell Res.* **12**, 568.

SCHÜMANN, H. J. (1958) *Arch Exp. Path. Pharmak* **233**, 296.

SEIFTER, S., S. DAYTON, B. NOVIC and E. MUNTWYLER (1950) *Arch. Biochem.* **25**, 191.

SJÖSTRAND, F. S. and R. WETZSTEIN (1956) *Experientia* (Basel) **12**, 196.

SPACKMAN, D. H., W. H. STEIN and S. MOORE (1958) *Anal. Chem.* **30**, 1190.
STJÄRNE, L. (1964) *Acta Physiol. Scand.* **62**, Suppl. 228.
WEINER, N. (1965) Personal communication.

SYNTHESIS, UPTAKE AND BINDING OF HISTAMINE AND 5-HYDROXYTRYPTAMINE IN MAST CELLS

JACK PETER GREEN

Department of Pharmacology, Yale University School of Medicine, New Haven, Conn.

As information about mast cells increases, it is becoming increasingly difficult to define them except in simplistic anatomic terms. They are large cells found in connective tissue, and their most striking property is the presence of basophilic, metachromatic granules (see Riley, 1959; Michel, 1963; Smith, 1963; Bloom, 1963; Spicer, 1963). However inexplicit this description, it is perhaps still too exclusive. The size and shape of mast cells vary with their location and differ among species (Michel, 1963; Spicer, 1963), the uterus of the mouse, for example, having small mast cells with relatively few granules and relatively little basophilia (Spicer, 1963). Differences among species are manifest biochemically, as exemplified by the esterases which in mast cells of the mouse and rat are homospecific with chymotrypsin whereas in man and dog the esterases resemble trypsin (see Lagunoff and Benditt, 1963; Hopsu and Glenner, 1963; Katayama and Ende, 1965).

What is more relevant to our interests is that there are differences among species in the amines found in their mast cells. Those of the rat and mouse contain both histamine and 5-hydroxytryptamine; those of other species examined, including guinea-pig, hamster, cow, and man contain histamine but lack 5-hydroxytryptamine (see Riley, 1959; West, 1959; Benditt, Holcenberg, and Lagunoff, 1963). Recently, mast cells in some regions of the cow and sheep have been shown to contain dopamine (see Phillips, Burch, and Hibbs, 1960; Coupland, 1963; Falck, Nystedt, Rosengren and Stenflo, 1964; Adams-Ray, Dahlström, Fuxe, and Hillarp, 1964).

SYNTHESIS OF AMINES

Virtually all the histamine present in mast cells is formed by the cells (Schayer, 1956). Histidine is taken up by an active process (Day and Green, 1962a) and decarboxylated by a pyridoxal-dependent enzyme which is found in the supernatant material after centrifugation of the cells. This decarboxylase shows high specificity for histidine, differing from the non-specific aromatic L-amino acid decarboxylase in its pH-optimum, its resistance to activation by benzene, and its failure to be inhibited by α-methyldihydroxyphenylalanine (Rothschild and Schayer, 1959; Weissbach, Lovenberg, and Udenfriend, 1961; Day and Green, 1962a). Inhibitors of histidine decarboxylase did not lower the levels of histamine in rat peritoneal mast cells (Levine, Sato, and Sjoerdsma, 1965), probably because of the slow turnover of the amine (Schayer, 1952).

5-Hydroxytryptamine in neoplastic mast cells of the mouse is formed from tryptophan which the cells take up by an active process, hydroxylate to form 5-hydroxytryptophan, which they then decarboxylate (Schindler, 1958; Day and Green, 1962a; Levine, Lovenberg, and Sjoerdsma, 1964). Presumably normal mast cells do the same, but only the decarboxylation has been shown (see Benditt, Holcenberg, and Lagunoff, 1963). The decarboxylase is the non-specific aromatic L-amino acid decarboxylase and is pyridoxal dependent. The absence of this enzyme in human mast cells accounts for the absence of 5-hydroxytryptamine in these cells (Demis, 1963).

In the formation of 5-hydroxytryptamine from the tryptophan by neoplastic mast cells of the mouse, negligible amounts of 5-hydroxytryptophan accumulate, the 5-hydroxytryptophan being decarboxylated as rapidly as it is formed (Schindler, 1958; Day and Green, 1962a, c); it can be concluded that under usual circumstances the uptake and hydroxylation of tryptophan is the rate-limiting step in the synthesis of 5-hydroxytryptamine.

However, in the formation of 5-hydroxytryptamine from administered 5-hydroxytryptophan, surprisingly large amounts of 5-hydroxytryptophan accumulate in these cells (Day and Green, 1962c). This persistence of *exogenously* formed 5-hydroxytryptophan in the cell may rest in part on the fact that the cells can take up 5-hydroxytryptophan more rapidly than they can decarboxylate it; but this explanation is not sufficient, since the exogenous 5-hydroxytryptophan persists for several days whereas the endogenous 5-hydroxytryptophan, i.e. that produced in the cell, is so rapidly decarboxylated that it can rarely be demonstrated. More likely the difference in rate of turnover of endogenous and exogenous 5-hydroxytryptophan is a reflection of different pools for the endogenous and exogenous 5-hydroxytryptophan, a compartmentation that has parallels in other cells (Furano and Green, 1963). It is possible that the decarboxylase is intimately associated with the hydroxylase so that the endogenously formed 5-hydroxytryptophan is more readily decarboxylated than that which is taken up by the cell (Day and Green, 1962a, c).

It should be emphasized that though mast cells and many other cells can take up 5-hydroxytryptophan, there is no evidence that this amino acid is found in blood except under pathological conditions.

The absence of dopamine in neoplastic (P-815) mast cells of the mouse was attributed to the absence of an enzyme to hydroxylate tyrosine (Day and

Green, 1962a). Recently hydroxylation of phenylalanine was demonstrated in the P-815 tumor (Levine, Lovenberg, and Sjoerdsma, 1964) in contrast with earlier findings (Day and Green, 1962a). The discrepant findings may be attributable to variants of the tumor. Or they may be yet another manifestation of the variation in metabolic behavior exhibited by these cells, as shown by the wide temporal fluctuations in their levels of histamine and 5-hydroxytryptamine, and in their changing capacities to take up and decarboxylate the precursor amino acids (Day and Green, 1962a).

The dopamine-containing mast cells of ungulates have not been studied except for the demonstration of the amine. Since dopamine is normally not found in significant concentrations in blood, the mast cells must synthesize the amine. It may be predicted that these mast cells can hydroxylate tyrosine, and then decarboxylate the product, 3,4-dihydroxyphenylalanine.

UPTAKE OF AMINES

Histamine and 5-hydroxytryptamine, both *in vivo* and *in vitro*, are taken up by mast cells of the peritoneal fluid of the rat (Furano and Green, 1964a) and by neoplastic mast cells of the mouse, the latter cells derived from both the Dunn–Potter and the Furth mastocytomas (Day and Green, 1962b; Green, 1963b; Furano and Green, 1964b). An indication of the capacity to concentrate amines is the distribution ratios of 5-hydroxytryptamine and histamine between the mast cells of the rat and the medium: at the end of 1 hr, the ratio of 5-hydroxytryptamine was 78, higher than any cell or tissue except the platelet; the corresponding value for histamine was 9.2, higher than any cell or tissue except the neoplastic mast cell (Furano and Green, 1964a). Despite this capacity, it is likely that under normal conditions, the levels of histamine and 5-hydroxytryptamine in mast cells are due mostly to synthesis rather than to uptake of amines, for under physiological conditions one does not attain blood levels of amines high enough to alter significantly the steady-state levels of amines in mast cells. But when high blood levels of amines are attained, as after the injection of histamine (Furano and Green, 1964a; Snyder, Axelrod and Bauer, 1964; Waton, 1964), it is reasonable to expect that a significant portion of the amine in the mast cell is attributable to uptake. Under these conditions, mast cells could serve as one of the sites of storage of unchanged amines. A more interesting implication of the capacity of mast cells to concentrate histamine is its possible physiological function in regulating the levels of free amines in tissues. When, for example, histamine synthesis is increased in tissues (Schayer, 1963), the nearly ubiquitously placed mast cells may take up some of the histamine and thereby reduce the amount of free histamine to which blood vessels are exposed.

The uptake of histamine by neoplastic mast cells is by the process of diffusion whereas 5-hydroxytryptamine is taken up by an active process as well as diffusion (Day and Stockbridge, 1964). In general, drugs that strongly inhibited the uptake of one amine had little effect on the uptake of the other. Uptake of histamine was decreased by norepinephrine, 5-hydroxytryptamine, and tryptamine; only tryptamine inhibited uptake of 5-hydroxytryptamine. Histamine uptake was sensitive to phenoxybenzamine and guanethidine whereas 5-hydroxytryptamine uptake was sensitive to cocaine, chlorpromazine, and imipramine. Surprisingly, substances that antagonize 5-hydroxytryptamine on smooth muscle (e.g. LSD-25) more powerfully inhibited uptake of histamine by these cells than uptake of 5-hydroxytryptamine. Analogously, antihistamines more effectively inhibited uptake of 5-hydroxytryptamine than of histamine (Day and Stockbridge, 1964).

With neoplastic mast cells, it was shown that the histamine and 5-hydroxytryptamine taken up by the cells have access to the granules containing the endogenous amines; but if the granules contain abundant quantities of amines, the additional amines are found in the clear cytoplasm (Furano and Green, 1964b). The compartmentation of the amines is reflected in the different rates of turnover of endogenous and exogenous amines (Day and Green, 1962c); and the rates of turnover of the amines in the cell depend on their intracellular localization which, in turn, depends on the amount of amine in the cell (Green and Furano, 1962; Furano and Green, 1964b). The best explanation to account for these observations on neoplastic mast cells is that most of the amines are held in the granules, but the granules contain a finite number of sites with which to bind the amines. When these sites are saturated, additional amine cannot enter the granular pool and hence remains in the clear cytoplasm. The amine in the clear cytoplasm turns over more rapidly than that in the granules. Comparable experiments have not been carried out on normal mast cells. These contain more granules than the neoplastic mast cells and have relatively little clear cytoplasm (see Bloom, 1963; Winqvist, 1963); it may prove difficult to saturate the granules with amines or to place significant amounts of amines in the sparse cell sap.

ELIMINATION OF AMINES

Unsuccessful attempts have been made to find catabolic products of histamine or of 5-hydroxytryptamine in peritoneal mast cells of the rat (Furano and Green, 1964a) or in neoplastic mast cells derived from both the Dunn–Potter and Furth tumors (Day and Green, 1962a, b; Furano and Green, 1964b). A variety of procedures failed to reveal either monoamine oxidase or diamine oxidase in these neoplastic cells (Day and Green, 1962a; Furano and Green, 1964b; cf. Hagen, 1961). Unchanged histamine and 5-hydroxytryptamine were also recovered in the medium of neoplastic cells growing in culture (Day and Green, 1962a, c). These amines are therefore eliminated by the cell, as is a small portion of the heparin (Green, 1963a).

BINDING OF AMINES

Most of the histamine and 5-hydroxytryptamine in the mast cell is found in dense, membrane-bound granules which can be isolated by differential and density gradient centrifugation (see Hagen, 1961; Green, 1962; Uvnäs, 1964; Lagunoff, 1965). In the peritoneal mast cells of the rat the histamine content varies from 10 to 40 μg per million cells; the corresponding figure for 5-hydroxytryptamine is 1 μg (Moran, Uvnäs, and Westerholm, 1962; Uvnäs, 1964; Furano and Green, 1964a). If it is assumed that each cell contains 500 granules (Uvnäs, 1964), that the average granule is 0.5 μ in diameter (Smith, 1963; Bloom, 1963) and is spherical, and that all the cellular histamine is confined to the granule, the concentration of histamine in the granule could be as high as 10 M; assuming that the diameter of each granule is 1.0 μ and that the concentration of histamine is 13 μg per million cells, Uvnäs (1964) arrived at an estimate of 0.6 M. Even if the lower estimate more nearly approaches the true value, and even if only 80 per cent of the histamine is granular-bound, it is clear that not all the histamine in the granule can exist in a freely diffusible form. The histamine and 5-hydroxytryptamine are assumed to be held in the granules by bonds of low energy since only mild conditions are necessary to cause the release of the amines (see Hagen, 1961; Green, 1962; Uvnäs, 1964). The presence of heparin (and perhaps other sulfomucopolysaccharides) in the granules suggested that the amines may be bound to heparin. Evidence, mostly circumstantial, supports this view (see Hagen, 1961; Green, 1962; Lagunoff, 1965; Uvnäs, 1965), but it should be noted that the granules may contain other acidic substances with the capacity to bind amines. Cerebroside sulfate, which forms complexes with amines (Green, Robinson, and Day, 1961) is found in neoplastic mast cells of the mouse (Green and Robinson, 1960), as are taurine, cysteic acid, and, of course, adenosine triphosphate (Green, Day, and Robinson, 1962), which also form complexes with amines (Green and Day, 1963). It is not known whether these or other substances that can bind amines are present in granules. In fact studies on granules have focused almost entirely on their content of a few enzymes, amines, and heparin.

Heparin: Properties that are Pertinent to Binding of Amines

Heparin consists of alternating units of hexosamine and uronic acid containing N-sulfate and O-sulfate groups (see Danishefsky and Steiner, 1965). There is a question whether heparin should be regarded as a single, chemically defined molecule or as a group of related molecules, analogous to the chondroitin sulfates (for references, see Green, 1962; Green and Day, 1963) and the keratan sulfates (Wortman, 1964). Supporting the latter view is the heterogeneity of preparations of heparin observed on ultracentrifugation (Barlow, Coen, and Mogen, 1964); the presence, in some preparations of heparin, of iduronic as well as glucuronic acid (Cifonelli and Dorfman, 1962) and acetylglucosamine (Yosizawa, 1964); and the reported differences among species in the properties of heparin (Bell and Jaques, 1956; Green, Roberts, and Day, 1963; Green and Day, 1963; Yosizawa, 1964; Lagunoff, 1965; cf., Barlow, Coen, and Mogen, 1964). Not all the differences can be attributed to differences in type and amount of contaminating impurities—which, perhaps, are themselves reflective of underlying differences in the heparins (Green, Day, and Roberts, 1963; Green and Day, 1963)—for the differences in anticoagulant activity are often not accompanied by differences in elemental analyses (Bell and Jaques, 1956; Green, Roberts, and Day, 1963; Barlow, Coen, and Mogen, 1964). Subtle differences in structure among heparins become all the more possible with the demonstration that heparin may have a helical structure (Stone, 1964).

There is no question that within the cell, heparin is covalently linked to protein. This fact is clear from the difficulty with which the bulk of the protein is removed from heparin, especially heparin from mouse and rat mast cells (Green, Roberts, and Day, 1963; Lagunoff, 1965) and by the presence, in highly purified commercial samples of heparin, of histidine and other amino acids (Green, Day, and Roberts, 1961), including serine (Lindahl and Rodén, 1964).

If the amines are bound to heparin, they are bound to the heparin–protein complex, which must have properties different from, or in addition to, those of heparin. Further, the heparin is held in the granule not in solution but as a gel, a state that would endow heparin (and heparin–protein) with properties different from those found in solution. In any attempt to simulate the binding of amines in mast cells, it would be appropriate to use the proteinsulfomucopolysaccharide, but studies have been carried out only with purified heparin (see Kobayashi, 1962; Green, Roberts and Day, 1963; Barlow, Coen, and Mogen, 1964). Although necessarily deficient, these studies may nevertheless yield some insight as to the mechanism of binding of amines by the mast cell and perhaps by other amine-rich cells containing sulfomucopolysaccharides (see Green, 1962; Robinson and Green, 1962). It is therefore fitting to review those properties of heparin that are important in the binding of amines (for references, see Green, 1962; Green, Roberts, and Day, 1963; Green and Day, 1963).

Of the characteristics that influence the capacity of a polyanion like heparin to bind cationic substances, the one most emphasized is the average electrostatic potential or, effectively, the number of dissociable acidic groups. In heparin these are the sulfate and carboxylate groups, both of which are dissociated at physiological pH. Since the amines exist as cations at this pH (see Albert, 1952), salt-like linkages between the positively charged amines and the polyanion have been inferred. It follows that in general the greater the number of free and dissociated amonic groups in the heparin molecule, the greater the amount of amines it can bind. The content of anionic groups, particularly sulfate, and their coulombic (i.e. salt-like) interactions with amines have been emphasized in studies of isolated granules (Uvnäs, 1965; Lagunoff, 1965) and of purified solutions of heparin (see Kobayashi, 1962; Green, Roberts, and Day, 1963; Barlow, Coen, and Mogen, 1964).

The acidic properties of heparin suggest an analogy to synthetic cation-exchange resins. And the fact that inorganic cations destroy the heparin–histamine complex (see Kobayashi, 1962) and release histamine from isolated granules (Uvnäs, 1965; Lagunoff, 1965) gives strength to the analogy. But this does not imply that only salt-like interactions are important in binding: evidence, recently reviewed (Green, 1962), has shown that in the binding of organic ions by synthetic ion-exchange resins, other factors such as hydrogen bonding, solubility, sieving, and Donnan equilibria must be considered. Similarly, studies on the binding by heparin of histamine, 5-hydroxytryptamine (Green, Roberts, and Day, 1963; Green and Day, 1963) and antibiotics (Gubernieva and Silaev, 1964) suggest that other forces, in addition to coulombic, are involved. These observations parallel those with other polyanions such as nucleic acids, which also form complexes with basic substances by interactions not attributable solely to coulombic forces (e.g. Moskowitz, 1963; Johns and Butler, 1964).

Although the ionization of the acidic groups in heparin is an important factor in binding, this fact does not imply that only salt-linkages are formed. With ionization, polyanions show an enhanced sieving effect (Laurent and Persson, 1964), a property of sulfomucopolysaccharides and protein-sulfomucopolysaccharides, which is manifest by the exclusion of protein and even small molecules from a portion of the (granular) water (see Gerber and Schubert, 1964; Laurent, 1964). The fact that inorganic salts destroy the heparin–histamine complex and release histamine from isolated granules may rest in part on disruption of the helical structure which heparin appears to have (Stone, 1964) rather than solely on disruption of an electrostatic bond between the anionic groups of heparin and histamine. Further, an increase in the ionization of heparin, while increasing the chances for coulombic interactions between heparin and amines, also tends to rupture hydrogen bonds (see Jacobson, 1963), which probably stabilize the helical structure. A rupture of intramolecular hydrogen bonds in heparin could make these bonds available for binding amines.

There is much opportunity for hydrogen bonding in heparin through the alcoholic hydroxyl, amide, and acidic groups. The importance of the alcoholic hydroxyl group has been emphasized in studies of the binding of cationic substances by glycopolymers (see Green, 1962, Puchtler, Sweat and Levine, 1962). Hydrogen bonds have been implicated in the binding by heparin of histamine, 5-hydroxytryptamine (Green, Roberts and Day, 1963; Green and Day, 1963) and basic antibiotics (Gubernieva and Silaev, 1964).

Besides being a possible linkage between the amines and heparin, hydrogen bonds have effects in the heparin molecule that can indirectly influence the binding of amines. Hydrogen bonding increases the ionization of acidic groups (Laskowski and Scheraga, 1954), which is a determinant of the extent of coulombic interactions between heparin and the amines. And hydrogen bonding influences the shape of heparin, which is a factor in its capacity to bind amines.

If heparin is regarded as a helix (Stone, 1964), it is expected that hydrogen bonds help to preserve the helix. It is likely that the helical structure would expose more binding groups than a random coil (see Jacobson, 1963). Also, a helical structure would limit the possible kinds of linkages and enhance the chances for specificity; it has been inferred that histamine can bind to heparin without disordering its helical configuration (Stone, 1964). If heparin is regarded as a non-helical polymeric chain, it would be expected that adjacent hydroxyl groups in a heparin molecule would tend to cause contraction and coiling of the molecule; this effect is opposed by the repulsion among the dissociated anionic groups, which causes stretching of the molecule: coiling reduces the accessibility of sites for binding cationic substances. It also reduces the sieving effect (Laurent and Persson, 1964) with a resultant increase in the amount of granular water, which could diminish the tendency for formation of intramolecular and intermolecular hydrogen bonds.

Other structural features of heparin can influence its capacity to bind cationic substances. In general, the more branched the molecule, the more functional groups are available for binding; but branching also reduces the sieving action (Laurent and Persson, 1964), which would thereby increase the granular water available to the amines and thus diminish binding. An increase in molecular weight increases the capacity to bind cationic substances; it also increases the sieving action (Laurent and Persson, 1964); it is of special interest that heparin isolated from the peritoneal mast cells of the rat has a greater molecular weight than ungulate heparin (Lagunoff, 1965). Finally, the fine structure of heparin is an important consideration in its capacity to bind. The type of uronic acid influences this capacity. The orientation of the sulfate groups is an important feature: sulfate groups in the axial position have a greater capacity to bind cations than do those in the equatorial position. The relative positions of the anionic groups may also be important, for if these sites are close enough, the bound amines may interact. Finally, consideration must be given to other substances that may be present in the granule and which can influence the capacity of heparin to bind amines. Among these are metals, some of which with acidic polysaccharides form complexes (Schweiger, 1964) which in turn can bind the amines (see below).

Properties of Histamine, 5-hydroxytryptamine and Dopamine that are Pertinent to Binding

A consideration of the mode of binding of amines requires a description of some of their properties. Only those properties will be considered that are relevant to the formation of reversible linkages of low energy to the granular material, presumably, although not necessarily, heparin.

The ethylamino side-chain, common to all three amines, suggests that all three can form a complex with heparin by ion–ion interactions between the polyanionic heparin and the cationic amino group. The pKa values of the primary amino groups of histamine (Leberman and Rabin, 1959) and 5-hydroxytryptamine (Rapport, Green and Page, 1948) are 9.8; dopamine has a pKa of 8.93 (Tuckerman, Mayer and Nachod, 1959). At physiological pH, these amino groups exist almost wholly as cations, the extent of ionization being, respectively, over 99 and 96 per cent. At physiological pH, almost all the carboxyl and all the sulfate groups (Albert, 1952) of heparin are ionized. It is not surprising then that the complex between heparin and an amine is attributed to the formation of a salt (see Jaques and Küttner, 1961; Kobayashi, 1962; Green, 1962; Green, Roberts and Day, 1963; Barlow, Coen and Mogen, 1964). As noted above, there are other interpretations. The aliphatic amino group can function as a donor or less likely, an acceptor in the formation of hydrogen bonds with heparin. Or the amino group may be linked to heparin through a metal (see below).

Even if salt-formation is a major type of interaction between the amines and heparin, there is no basis for assuming it is the only one. There is even less justification for disregarding the rest and major part of each amine molecule. All three contain functional groups besides a primary amino group, and all possess a conjugated system of electrons. In these regards there are differences among the molecules, and though it is likely that all three amines may share a similar site or sites or linkage to, say, heparin, other and additional sites may preferentially bind one of the amines, a situation analogous to that revealed in studies on the binding of dinitrophenol and dinitrotoluene to albumin (Carsten and Eisen, 1953). If the primary amino group is a major site of ignding to heparin, other sites may contribute to and stabilize the complex (Green, Roberts and Day, 1963). Some indication of these sites and the nature of the forces may be revealed by an examination of the structures of the three amines.

An immediate problem is to know which of the tautomeric forms of histamine to select, I or II.

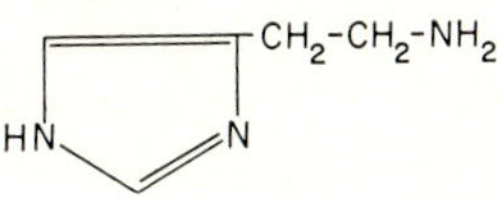

I

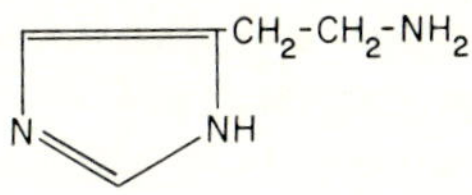

II

There is no significant difference in their resonance energies; the tautomers have eluded isolation; and attempts to determine which tautomer is dominant have been unsuccessful. A recent effort with nuclear magnetic spectroscopy yielded inconclusive results because the protons of the side-chain couple with the ring proton (Staab and Mannschreck, 1963; Staab, personal communication, 1964). Metabolism of histamine occurs mainly by methylation of the ring nitrogen farther away from the side-chain (Schayer, 1956), but this fact does not permit the conclusion that the dominant form of histamine contains the secondary ("pyrrole") nitrogen in this position. First, the site of methylation may rest solely on the specificity of the enzyme, which methylates only I, leaving II to be either catabolized by another route (Schayer, 1956) or excreted in unchanged form (Green, Fram and Kase, 1964). Second, if before methylation, the tertiary ("pyridine") nitrogen were to be protonated, the ring nitrogens would become indistinguishable (Pullman and Pullman, 1963), and any inference about the original position of the double-bond would not be possible.

Other studies, however, suggest that I exists, although they do not indicate its relative proportion. Imidazoles form complexes with metals and by different techniques, including nuclear magnetic spectroscopy, it has been shown that the chelate formed between histamine and metals involves the tertiary nitrogen and the primary amino group of the side-chain (Li, Doody and White, 1957; Leberman and Rabin, 1959; Martin and Edsall, 1960; Milnar and Pratt, 1962). The ring formed by the metal and these two groups in I has six atoms, whereas the ring in II would have seven; the latter structure is not likely from an examination of molecular models. It will be assumed for this discussion that histamine exists as I.

The work noted above and other observations (see Green, 1962; Andrews and Romary, 1964) demonstrating that histamine complexes with metals, including zinc, is pertinent to the possible mode of binding of histamine to metals. Zinc is present in mast cells. Together, heparin and zinc bind four times as much histamine as does heparin alone; the association constant of the ternary complex is more than twice as high as that obtained in the absence of zinc (Kerp, 1963). Zinc may attach to histamine through the tertiary nitrogen alone or through both the tertiary nitrogen and the primary amine (see Li, Doody and White, 1957). The complex may be completed by linkage of zinc to one or two molecules of heparin through the hydroxyl, carboxyl and/or sulfate groups, as may be deduced from studies of metal-binding by another acidic polysaccharide (Schweiger, 1964). Other combinations are obviously no less probable. 5-hydroxytryptamine and dopamine also form complexes with metals through which they may be linked to heparin (see Green, 1962).

One way of searching for possible sites of interaction in a molecule is to apply the techniques of quantum mechanics. The relative capacity of compounds to donate and accept electrons is often correlated with the energies of the highest filled and lowest empty molecular orbitals (see Mulliken and Person, 1962; Pullman and Pullman, 1962). On this basis, it may be concluded from the values shown in Table 1, obtained by the Hückel approximation,

TABLE 1. THE ENERGIES OF THE HIGHEST FILLED (H.F.M.O.) AND LOWEST EMPTY (L.E.M.O.) MOLECULAR ORBITALS OF 5-HYDROXYTRYPTAMINE, HISTAMINE, AND DOPAMINE

Substance	*H.F.M.O.*	*L.E.M.O.*
5-Hydroxytryptamine	0.46	−0.87
Histamine	0.61	−1.12
Dopamine	0·63	−1.05

Calculations were carried out by the Hückel approximation (see Pullman and Pullman, 1963) with a methyl group substituted for the ethylamino groups. Values are in β-units.

that of the three amines, 5-hydroxytryptamine is both the best donor and acceptor of electrons. Experimentally, 5-hydroxytryptamine and other indoles have been shown to form molecular complexes, some sort of which are of the charge–transfer type, with a variety of electron acceptors (Szent-Györgyi, 1961; Pullman and Pullman, 1963). The energies of the molecular orbitals of histamine and dopamine indicate that they are poor donors and acceptors. But the energy of the highest filled orbital does not always reflect the experimentally determined donor capacities, as shown by indoles, which are better donors than one would predict from the calculations (see Szent-Györgyi, 1961; Karreman, 1962; Pullman and Pullman, 1963). Catechols form complexes with flavins, one force in such complexing being charge–transfer (Fleischman and Tollin, 1965); and imidazoles and phenolic compounds have been shown to interact with nicotinamide when linked to nicotinamide by an ethylene bridge (Shifrin, 1964). The probability of the formation of donor–acceptor complexes is in part determined by the electron-accepting capacity of the acceptor (see Foster and Hanson, 1964) and the milieu. The capacity of a flavin to complex with election donors like indoles was influenced by the solvent, solvents capable of hydrogen-bonding increasing the interactions (Harbury, LaNone, Loach and Amick, 1959). Similarly, it was suggested that acridine shows some properties of a strong electron-acceptor in the presence of acetic acid but not of mineral acids (Nash and Allison, 1962).

Charge–transfer complexes have not been demonstrated between the amines and heparin. Studies of the interaction of basic dyes with heparin suggest that after the dyes are bound to heparin, the dye molecules stack, held together by forces between their π-electron systems (Bradley and Wolf, 1959), the propinquity of the anionic groups of heparin being one determinant of the extent of stacking (Stone, 1964). Such interactions are less probable with the amines. Although the possibility should be considered that π-electrons of the amines form hydrogen bonds with the electron-accepting groups in heparin—a type of linkage suggested between dyes and cellulose (Yoshida,

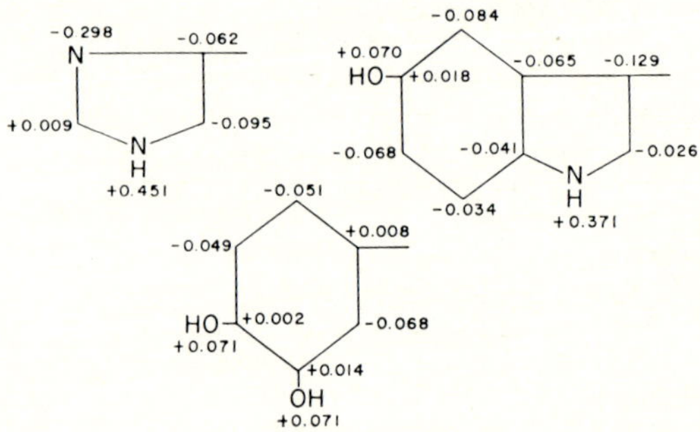

FIG. 1. The net charges of histamine, 5-hydroxytryptamine, and dopamine.

Osawa and Oda, 1964)—it is likely that the π-electrons make relatively little contribution. Even with strong electron acceptors, compounds like pyridine and ethylamine form complexes that are almost certainly attributable to the n-electrons (i.e. the "lone-pair") of the nitrogen (see Mulliken and Person, 1962; Hassel and Rømming, 1962). These complexes are more localized

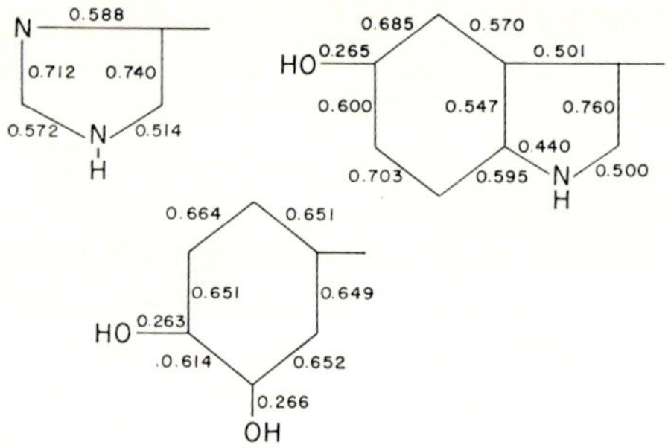

FIG. 2. The bond orders of histamine, 5-hydroxytryptamine, and dopamine.

and stronger than π-complexes (Mulliken and Person, 1962). It is the n-electrons on the tertiary and primary nitrogens of histamine that form bonds with the metals, as noted above. In histamine, 5-hydroxytryptamine and dopamine, there are not only n-electrons but other electron-rich atoms in the rings that can form complexes with electron-accepting groups in heparin. There are also atoms in the ring that could react with electron-donating atoms in heparin. An indication of these may be obtained by calculating the net charges by means of the Hückel approximation (Pullman and

Pullman, 1963). These charges are shown in Fig. 1. Figure 2 shows the estimates of bond-orders, from which may be approximated the bond lengths (see Pullman and Pullman, 1963).

Three items in histamine and 5-hydroxytryptamine are directly comparable: the carbon bearing the side-chain, the —NH group and between these, the —CH group. The lengths of the two bonds as estimated from the bond orders, are not significantly different in the two amines.

The carbon to which the side-chain is attached is negative in both molecules and relatively more so in 5-hydroxytryptamine. Although this carbon bears an ethylamine substituent, it is not necessarily inactive: small electrophilic species, such as a proton, can attack this position in skatole (Hinman and Shull, 1961). Szent-Györgyi (1961) has suggested that the complex formed between indoles and electron-acceptors may occur through a local interaction between this carbon and a specific point of the acceptor. The adjoining carbon in both molecules is negatively charged, histamine showing a relatively greater excess of electrons. In indoles this is the site of attack of large electrophilic species if the 3-position is substituted (Hinman and Shull, 1961), as it is in 5-hydroxytryptamine. Among the positively charged species that could react here is the nitrogen of the side-chain, which is protonated at physiological pH, and as shown with molecular models, free to abut against this —CH group. The —NH group, especially that in histamine, can readily react with electron donating groups. Formation of hydrogen bonds between this group and, say, the oxygen atoms in heparin would be expected.

Unlike the other two amines, histamine has a tertiary nitrogen carrying a strong negative charge. This atom would attract positive groups such as the charged amino group of the side-chain to form an intermolecular complex. And it can serve as an excellent proton acceptor for hydrogen bonding to heparin. The hydrogen-bonding tendencies of the ring-nitrogens in histamine is evident in the formation of oligomers of imidazole in non-polar solvents or in the crystalline state (Anderson, Duncan and Rossotti, 1961; Grimson, 1963; Brown and Aftergut, 1963). Even in aqueous solutions containing 0.1 M phosphate, histamine forms hydrogen-bonds, through its tertiary nitrogen, with atropine (Oroszlan and Maengwyn-Davis, 1962). Studies of the heparin of murine mast cells and its capacity to react with histamine and 5-hydroxytryptamine suggested hydrogen-bond formation between the amines and heparin (Green, Roberts and Day, 1963; Green and Day, 1963). 5-Hydroxytryptamine has no tertiary nitrogen, but it has a phenolic hydroxyl group which can form hydrogen bonds.

There are also two groups on the ring of dopamine to form hydrogen bonds. An appraisal of the probability of both forming hydrogen bonds with heparin is complicated by the likelihood of intramolecular hydrogen bonding between these phenolic hydroxyl groups (Tuckerman, Mayer and Nachod, 1959). From the calculation of bond orders, it may be estimated that the 4-hydroxyl group in dopamine is 5.5 Å from the carbon bearing the side-chain; in 5-hydroxytryptamine, the distance between the hydroxyl group and the analogous carbon is 5.6 Å. The 3-hydroxyl group however is farther away from the side-chain than is the —NH group in histamine and 5-hydroxytryptamine (4.1 compared with 2.8 Å). This difference may be lessened significantly by the free mobility of the side-chain $—CH_2NH_2$, which may permit conformation of the molecule to heparin. (It is interesting that the phenolic hydroxyl groups contribute little to the interaction between dopamine and ATP (Weiner and Jardetzky, 1964).)

Thus, all three amines have the same side-chain with a primary amino-group capable of forming a salt-linkage or a hydrogen bond with heparin. All have two other groups that can form hydrogen bonds. Histamine and 5-hydroxytryptamine have —NH groups similarly placed relative to the side-chain. Histamine has a strong electron-donating nitrogen capable of hydrogen bonding. Although 5-hydroxytryptamine has no such atom, it has a demonstrated capacity to function as a π-electron donor. In addition, 5-hydroxytryptamine has a phenolic hydroxyl group capable of hydrogen bonding. This is analogous to the 4-hydroxyl group of dopamine. The 3-hydroxyl group of dopamine may also form a hydrogen bond with heparin.

Environmental Factors influencing the Binding of Amines to Heparin

As the interaction between the amines and heparin takes place in the granule, an understanding of the nature of the binding requires information about the chemistry of the granule. This does not exist. What follows is a brief mention of some of the factors in the granule that are pertinent to the binding of amines to heparin; some of these considerations are also relevant to the binding of amines to substances other than heparin.

If the amines are primarily bound to heparin, other organic substances in the granule could influence the interaction. Among these are other acidic substances that form complexes with amines (see Green, 1962). An indication of how tenaciously some organic substances can hold cationic substances emerged from the observation that gangliosides (Robinson, Carlini and Green, 1963) and cerebroside sulfate (Green, Robinson and Day, 1961) even after extensive purification by solvent extraction and ion-exchange chromatography contained, respectively, norepinephrine and acetylcholine; the cerebroside sulfate appeared to be pure on elemental analysis. Although gangliosides are not in (neoplastic) mast cells, cerebroside sulfate is (Green and Robinson, 1960), and there is no reason to assume that there are not other anionic substances capable of competing with heparin. Conversely, cationic substances, such as basic amino acids, can compete for binding sites on heparin, the interaction of which with basic substances is well known (see Higginbotham, 1959; Green, 1962; Gubernieva and Silaev, 1964). If electron donor–acceptor interactions of any kind are involved in the binding of amines to heparin, the presence of other acceptors and donors would influence the probability of this interaction. Another effect of organic substances would be to alter the shape of heparin, which, as noted above, is an important consideration in its capacity to interact with cationic substances. Substances that interact with the alcoholic hydroxyl groups or the anionic groups would alter the shape. It has been shown for example that the presence of even trace amounts of basic organic substances in a polyanion can alter its shape (Scruggs and Ross, 1964).

Inorganic cations can also alter the shape of polyanions, and they can compete with cationic substances like histamine for binding to heparin. By decreasing the repulsion of the ionized groups, these cations can cause coiling of heparin. Salts have other effects such as causing shrinkage of gels (Katchalsky, 1964), interfering with the formation of coacervates, and disordering a helical configuration (Jacobson, 1963) including that of heparin (Stone, 1964). By forming a cloud of positive charges around the polyanion, inorganic cations can interfere with the accessibility of the anionic sites to the amines. Inorganic cations have been shown to release histamine from isolated granules (Uvnäs, 1965; Lagunoff, 1965) and to compete for the binding sites on heparin (see Kobayashi, 1962). Among the relatively specific effects of cations (see Katchalsky, 1964), special attention must be given the formation of complexes between metals and acidic polysaccharides, between metals and amines, and among zinc, heparin, and histamine; these have been mentioned above. It is perhaps pertinent that granules have an affinity for heavy metals (Selye, Gabbiani, and Serafimov, 1964; Gabbiani and Nielsen, 1964), and that a deficiency of magnesium causes a disruption of mast cells (Bois, 1963). One fact that should be noted is that when amines form complexes with cations, the molecular effects are not confined to the atoms linked to the metal: when for example histamine is bound to divalent cations, the secondary nitrogen, which is free, becomes more acidic (Leberman and Rabin, 1959).

Not only are the relative amounts and nature of the inorganic and organic substances important, but so are their concentrations in the granular water. With increasing concentration, the distance between molecules decreases and the frequencies of intermolecular collisions increase. This effect can enhance the probability of heparin-amine interactions, but it also would enhance the reactions of heparin and amines with other substances; these include inorganic cations which are drawn into the sphere of the polyanion and cause it to curl. The concentration of heparin also determines the probability of heparin–heparin interactions: the relative extent of inter- and intramolecular associations affects the shape of heparin. Shape is also affected by ionization of the acidic groups of heparin which is favored in dilute solution. The amount of granular water would determine the importance of the sieve action of heparin. Closely related to the water-content of the granules is the

CARSTEN, M. E. and H. N. EISEN (1953) The interaction of dinitrobenzene derivatives with bovine serum albumin, *J. Amer. Chem. Soc.* **75**, 4451–6.

CIFONELLI, J. A. and A. DORFMAN (1962) The uronic acid of heparin, *Biochem. Biophys. Res. Commun.* **7**, 41–5.

COUPLAND, R. E. (1963) Mast cells and chromaffin cells, *Ann. N.Y. Acad. Sci.* **103**, 139–50.

DANISHEFSKY, I. and H. STEINER (1965) Investigations on the chemistry of heparin. V. Disaccharides obtained after partial hydrolysis, *Biochim. Biophys. Acta* (Amst.) **101**, 37–45.

DAY, M. and J. P. GREEN (1962a) The uptake of amino acids and the synthesis of amines by neoplastic mast cells in culture, *J. Physiol.* (Lond.) **164**, 210–26.

DAY, M. and J. P. GREEN (1962b) The uptake of biogenic amines by neoplastic mast cells in culture, *J. Physiol.* (Lond.) **164**, 227–37.

DAY, M. and J. P. GREEN (1962c) Differences in turnover of endogenous and exogenous 5-hydroxytryptophan, 5-hydroxytryptamine, and histamine in neoplastic mast cells in culture, *Biochem. Pharmacol.* **11**, 1043–8.

DAY, M. and A. STOCKBRIDGE (1964) The effects of drugs on the uptake of amines by mast cells, *Brit. J. Pharmacol.* **23**, 405–19.

DEMIS, J. (1963) Mastocytosis syndrome: clinical and biological studies, *Ann. Int. Med.* **59**, 194–206.

FALCK, B., T. NYSTEDT, E. ROSENGREN and J. STENFLO (1964) Dopamine and mast cells in ruminants, *Acta Pharmacol.* (*Kbh.*) **21**, 51–8.

FLEISCHMAN, D. E. and G. TOLLIN (1965) Molecular complexes of flavins and phenols. III. Factors influencing spectra and stability, *Proc. Nat. Acad. Sci.* (Wash.) **53**, 38–46.

FOSTER, R. and P. HANSON (1964) Interaction of electron acceptors with bases. Part 12 Complexes of indoles with electron acceptor molecules, *Trans. Far. Soc.* **60**, 2189–95.

FURANO, A. V. and J. P. GREEN (1963) Differences in the disposition of endogenous and exogenous substances by cells, *Nature* (Lond.) **199**, 380–1.

FURANO, A. V. and J. P. GREEN (1964a) The uptake of biogenic amines by mast cells of the rat, *J. Physiol.* (Lond.) **170**, 263–71.

FURANO, A. V. and J. P. GREEN (1964b) The compartmentation and elimination of (^{14}C) histamine by neoplastic mast cells in culture, *Biochim. Biophys. Acta* (Amst.) **86**, 596–603.

GABBIANI, G. and K. NIELSEN (1964) Anaphylactoid inflammation and mastopexis, *Arch. Path.* **78**, 626–32.

GERBER, B. R. and M. SCHUBERT (1964) The exclusion of large solutes by cartilage protein-polysaccharide, *Biopolymers* **2**, 259–73.

GREEN, J. P. (1962) Binding of some biogenic amines in tissues, *Advances in Pharmacol.* **1**, 349–422.

GREEN, J. P. (1963a) Elimination and catabolism of ^{35}S-heparin by neoplastic mast cells in culture, *Biochem. Pharmacol.* **12**, 588–9.

GREEN, J. P. (1963b) Discussion, *Ann. N.Y. Acad. Sci.* **103**, 479–80.

GREEN, J. P. and M. DAY (1963) Biosynthetic pathways in mastocytoma cells in culture and *in vivo*, *Ann. N.Y. Acad. Sci.* **103**, 334–50.

GREEN, J. P., M. DAY and M. ROBERTS (1961) On the smooth-muscle stimulating activity of preparations of heparin, *J. Pharmacol.* **132**, 58–64.

GREEN, J. P., M. DAY and J. D. ROBINSON (1962) Some acidic substances in neoplastic mast cells and in the pineal body, *Biochem. Pharmacol.* **11**, 957–60.

GREEN, J. P., D. H. FRAM and N. KASE (1964) Methyl-histamine and histamine in the urine of women during the elaboration of oestrogen, *Nature* (Lond.) **204**, 1165–8.

GREEN, J. P. and A. V. FURANO (1962) Two pools for amines in neoplastic mast cells, *Biochem. Pharmacol.* **11**, 1049–53.

GREEN, J. P., M. ROBERTS and M. DAY (1963) The sulfomucopolysaccharides from a mast cell tumor of the mouse, *Biochem. Pharmacol.* **12**, 609–20.

GREEN, J. P. and J. D. ROBINSON (1960) Cerebroside sulfate (sulfatide A) in some organs of the rat and in a mast cell tumor, *J. Biol. Chem.* **235**, 1621–4.

GREEN, J. P., J. D. ROBINSON and M. DAY (1961) Interaction between cerebroside sulfate and amines, *J. Pharmacol.* **131**, 12–17.

GRIMISON, A. (1963) The deuterium isotope effect in the hydrogen bonding of imidazole in naphthalene solutions, *J. Phys. Chem.* **67**, 962–4.

GUBERNIEVA, L. M. and A. B. SILAEV (1964) The mechanism of interaction of heparin with antibiotics, *Biokhimiya* **29**, 831–6; *Chem. Abstracts* **62**, 1889 (1965).

HAGEN, P. (1961) Biosynthesis and storage of histamine, heparin, and 5-hydroxytryptamine in the mast cell, *Canad. J. Biochem. Pharmacol.* **39**, 639–46.

HARBURY, H. A., K. F. LA NONE, P. A. LOACH and R. M. AMICK (1959) Molecular interactions of isoalloxazine derivatives, II, *Proc. Nat. Acad. Sci.* **45**, 1708–17.

HASSEL, O. and C. RØMMING (1962) The structural evidence for weak charge transfer bonds in solids containing chemically saturated molecules, *Quart. Rev.* **16**, 1–18.

HIGGINBOTHAM, R. D. (1959) On the participation of micellophagosis in the resistance of tissue to injurious agents, *Int. Arch. Allergy* **15**, 195–222.

HINMAN, R. L. and E. R. SHULL (1961) The structure of diskatole, *J. Org. Chem.* **26**, 2339–45.

HOPSU, V. K. and G. G. GLENNER (1963) Observations on histochemical esterase and amidase activities with similarities to trypsin, *J. Histochem. Cytochem.* **11**, 520–8.

JACOBSON, A. L. (1963) Effect of a helix-coil transformation on magnesium binding to α-poly-L-glutaminic acid, *Biopolymers* **1**, 269–76.

JAQUES, R. and K. KÜTTNER (1961) Selektive Farbstoffverdrängung an biogenen Polyelektrolyten als Modell für die Liberation biogener Amine, *Helv. Physiol. Pharmacol. Acta* **19**, 335–43.

JOHNS, E. W. and J. A. V. BUTLER (1964) Specificity of the interactions between histones and deoxyribonucleic acid, *Nature* (Lond.) **204**, 853–5.

KARREMAN, G. (1962) Electronic aspects of quantum biology, *Ann. N.Y. Acad. Sci.* **96**, 1029–55.

KATAYAMA, Y. and E. ENDE (1965) Esterase studies on dog mast-cell tumours, *Nature* (Lond.) **205**, 190–1.

KATCHALSKY, A. (1964) Polyelectrolytes and their biological interactions, *Biophys. J.* **4**, 9–41.

KERP, L. (1963) Bedeutung von Zink für die Histaminspeicherung in Mastzellen, *Int. Arch. Allergy* **22**, 112–23.

KOBAYASHI, Y. (1962) Histamine binding by heparin, *Arch. Biochem. Biophys.* **96**, 20–7.

LAGUNOFF, D. and E. P. BENDITT (1963) Proteolytic enzymes of mast cells, *Ann. N.Y. Acad. Sci.* **103**, 185–98.

LAGUNOFF, D. (1965) The structure of mast cell granules. This Symposium, p. 79.

LASKOWSKI, M. and H. A. SCHERAGA (1954) Thermodynamic considerations of protein reactions. I. Modified reactivity of polar groups, *J. Amer. Chem. Soc.* **76**, 6305–18.

LAURENT, T. C. and H. PERSSON (1964) The interaction between polysaccharides and other macromolecules. VIII. The effect of various polymers on the sedimentation rates of serum albumin and α-crystallin, *Biochim. Biophys. Acta* (Amst.) **83**, 141–7.

LAURENT, T. C. (1964) Solubility of sodium urate in the presence of chondroitin-4-sulphate, *Nature* (Lond.) **202**, 1334.

LEBERMAN, R. and B. R. RABIN (1959) Metal complexes of histidine, *Trans. Far. Soc.* **55**, 1660–70.

LEVINE, R. J., W. LOVENBERG and A. SJOERDSMA (1964) Hydroxylation of tryptophan and phenylalanine in neoplastic mast cells of the mouse, *Biochem. Pharmacol.* **13**, 1283–90.

LEVINE, R. J., T. L. SATO and A. SJOERDSMA (1965) Inhibition of histamine synthesis in the rat by α-hydrazino analog of histidine and 4-bromo-3-hydroxybenzyloxyamine, *Biochem. Pharmacol.* **14**, 139–49.

LI, N. C., E. DOODY and J. N. WHITE (1957) Some metal complexes of glycine peptides, histidine, and related substances, *J. Amer. Chem. Soc.* **79**, 5859–63.

LINDAHL, U. and L. RODÉN (1964) The linkage of heparin to protein, *Biochem. Biophys. Res. Comm.* **17**, 254–9.

MARTIN, R. B. and J. T. EDSALL (1960) The association of divalent cations with acylated histidine derivatives, *J. Amer. Chem. Soc.* **82**, 1107–11.

MICHEL, N. A. (1963) The mast cells, *Ann. N.Y. Acad. Sci.* **103**, 491–628.

MILNER, R. S. and L. PRATT (1962) Proton resonance spectra of paramagnetic complexes in aqueous solution, *Disc. Faraday Soc.* **34**, 88–95.

MORAN, N. C., B. UVNÄS and B. WESTERHOLM (1962) Release of 5-hydroxytryptamine and histamine from rat mast cells, *Acta Physiol. Scand.* **56**, 26–41.

MOSKOWITZ, M. (1963) Differences in precipitability of nucleic acids with streptomycin and dihydrostreptomycin, *Nature* (Lond.) **200**, 335–7.

MULLIKEN, R. S. and W. B. PERSON (1962) Donor–acceptor complexes, *Ann. Rev. Phys. Chem.* **13**, 107–26.

NASH, T. and A. C. ALLISON (1962) Hydrogen bonding and charge transfer complexes, *Nature* (Lond.) **195**, 994.

OROSZLAN, S. I. and G. D. MAENGWYN-DAVIES (1962) Interaction of atropine with amino acids, *Biochim. Biophys. Acta* (Amst.) **63**, 117–27.

PHILLIPS, J. H., G. E. BURCH and R. G. HIBBS (1960) Significance of tissue chromaffin cells and mast cells in man, *Circulation Res.* **8**, 692–702.

PUCHTLER, H., F. SWEAT and M. LEVINE (1962) On the binding of congo red by amyloid, *J. Histochem. Cytochem.* **10**, 355–64.

PULLMAN, B. and A. PULLMAN (1962) *Quantum Biochemistry*. New York, Interscience.

RAPPORT, M. M., A. A. GREEN and I. H. Page (1948) Serum vasoconstrictor (serotonin). IV. Isolation and characterization, *J. Biol. Chem.* **176**, 1243–51.

RILEY, J. F. (1959) *The Mast Cells*. Edinburgh, Livingstone.

ROBINSON, J. D., E. A. CARLINI and J. P. GREEN (1963) The effects of ganglioside preparations on smooth muscle, *Biochem. Pharmacol.* **12**, 1219–23.

ROBINSON, J. D. and J. P. GREEN (1962) Sulfomucopolysaccharides in brain, *Yale J. Biol. Med.* **35**, 248–57.

ROTHSCHILD, A. M. and R. W. SCHAYER (1959) Characterization of histidine decarboxylase from rat peritoneal fluid mast cells, *Biochim. Biophys. Acta* (Amst.) **34**, 392–8.

SCHAYER, R. W. (1952) Biogenesis of histamine, *J. Biol. Chem.* **199**, 245–50.

SCHAYER, R. W. (1956) Formation and binding of histamine by free mast cells of rat peritoneal fluid, *Amer. J. Physiol.* **186**, 199–202.

SCHAYER, R. W. (1959) Catabolism of physiological quantities of histamine *in vivo*, *Physiol. Rev.* **39**, 116–26.

SCHAYER, R. W. (1963) Histidine decarboxylase in mast cells, *Ann. N.Y. Acad. Sci.* **103**, 164–78.

SCHINDLER, R. (1958) The conversion of C^{14}-labelled tryptophan to 5-hydroxytryptamine by neoplastic mast cells, *Biochem. Pharmacol.* **1**, 323–7.

SCHWEIGER, R. G. (1964) Complexing of Alginic acid with metal ions, *Kolloid-Zeitschrift. Zeitschrift für Polymere* **196**, 47–53.

SCRUGGS, R. L. and P. D. ROSS (1964) Viscosity study of DNA, *Biopolymers* **2**, 593–609.

SELYE, H., G. GABBIANI and N. SERAFIMOV (1964) Histochemical studies on the role of the mast cell in calcergy, *J. Histochem. Cytochem.* **12**, 563–9.

SHIFRIN, S. (1964) Noncovalent interactions between amino acid side chains and a coenzyme model, *Biochemistry* **3**, 829–33.

SMITH, D. E. (1963) Electron microscopy of normal mast cells under various experimental conditions, *Ann. N.Y. Acad. Sci.* **103**, 40–52.

SNYDER, S. H., J. AXELROD and H. BAUER (1964) The fate of C^{14}-histamine in animal tissues, *J. Pharmacol.* **144**, 373–9.

SPICER, S. S. (1963) Histochemical properties of mucopolysaccharide and basic protein in mast cells, *Ann. N.Y. Acad. Sci.* **103**, 322–33.

STAAB, H. A. and A. MANNSCHRECK (1963) Proton-resonance investigations of "4 (or 5)-substituted imidazoles", *Angewandte Chemie*, Int. Ed. **2**, 216–17.

STONE, A. L. (1964) Anomalous rotatory dispersion of metachromatic mucopolysaccharide-dye complexes. I. Heparin, *Biopolymers* **2**, 315–25.

SZENT-GYÖRGYI, A. (1961) The supra- and submolecular in biology, *J. Theoret. Biol.* **1**, 75–82.

TUCKERMAN, M. M., J. R. MAYER and F. C. NACHOD (1959) Anomalous pK_a values of some substituted phenylethylamines, *J. Amer. Chem. Soc.* **81**, 92–4.

UVNÄS, B. (1964) Release processes in mast cells and their activation by injury, *Ann. N.Y. Acad. Sci.* **116**, 880–90.

UVNÄS, B. and I. L. THON (1965) A physico-chemical model of histamine release from mast cells, this Symposium, p. 361.

WATON, N. G. (1964) Uptake and formation of histamine by cat tissues, *J. Physiol.* (Lond.) **172**, 475–81.

WEINER, N. and O. JARDETZKY (1964) A study of catecholamine nucleotide complexes by nuclear magnetic resonance spectroscopy, *Arch. Exp. Path. Pharmak.* **248**, 308–18.

WEISSBACH, H., W. LOVENBERG and S. UDENFRIEND (1961) Characteristics of mammalian histidine decarboxylating enzymes, *Biochim. Biophys. Acta* (Amst.) **50**, 177–9.

WEST, G. B. (1959) Tissue mast cells and tissue amines, *J. Pharm. Pharmacol.* **11**, 513–34.

WINQVIST, G. (1963) Discussion, *Ann. N.Y. Acad. Sci.* **103**, 461.

WORTMAN, B. (1964) Variability of keratan sulfate isolated from beef cornea, *Biochim. Biophys. Acta* (Amst.) **83**, 288–95.

YOSHIDA, Z., E. OSAWA and R. ODA (1964) Intermolecular hydrogen bond involving a π-base as the proton acceptor. I. Detection by the refractive index method, *J. Phys. Chem.* **68**, 2895–8.

YOSIZAWA, Z. (1964) A new type of heparin "ω-heparin" isolated from whale organs, *Biochem. and Biophys. Res. Commun.* **16**, 336–41.

DISCUSSION

UVNÄS: You suggested that histamine occurs in two pools in mastocytoma cells. Do you suggest that the one pool represents the free cell sap and histamine should occur 'free" in the cell?

GREEN: I think that all or almost all of the histamine in mast cells under ordinary conditions is associated with granules. The histamine is bound. But if the binding sites on the granules were saturated with amines, any additional amine that enters the cell would remain in the soluble portion of the cell.

RAŠKOVÁ: Could side effects of administrated heparin not be in connection with alterations of the formation of the protein–heparin complex?

GREEN: It is possible that the side effects that occasionally follow the injection of heparin are attributable to peptides that contaminate the heparin.

PLETSCHER: Is the uptake of 5-HT by mast cells *in vitro* dependent on the presence of glucose or other energetic substrates?

GREEN: The answer appears in the written version of the paper.

UDENFRIEND: The tryptophan hydroxylase of mouse mast cells has been solubilized and purified by Drs. Walter Lonenberg, Albert Sjoerdsma and Robert Levine (NIH).* The enzyme has a low K_m and requires a tetrahydropteridine cofactor.

HAEGERMARK: In this connection it might be of interest to mention that we (Cabut, Haegermark, Uvnäs) have been studying the uptake of radioactive histamine by rat peritoneal mast cells. The uptake is passive. The radioactive histamine is recovered in the granular fraction in the same proportion as the endogenous. It can also be released by compound 48/80 in the same proportions as the endogenous histamine.

GREEN: Uptake of histamine by neoplastic mast cells is also passive (Day, M. and A. Stockbridge, *Brit. J. Pharmacol.* **23**, 405–19, 1964). When these cells are low in endogenous amines, the radioactive histamine (exogenous) that is taken up is recovered in the granular fraction. But when the granules are rich in endogenous amines, the exogenous histamine is found in the soluble portion of the cell. We conclude from these experiments that the endogenous levels of amines determine the distribution of the additional amines taken up, for the number of free binding sites on the granules must be fixed and when these are saturated with amines, any additional amines can not be accommodated on (or in) the granules and therefore stay in the soluble cytoplasm.

If you saturate the granules in the normal mast cells by pre-incubating them with cold histamine and then incubate with radioactive histamine, the latter should be found mostly in the soluble cytoplasm, if our hypothesis is correct.

* *J. Biol. Chem.*, in press.

THE BINDING OF ACETYLCHOLINE BY BRAIN PARTICLES *IN VITRO**

V. P. Whittaker

Agricultural Research Council, Institute of Animal Physiology, Babraham, Cambridge, England

THE SIGNIFICANCE OF BOUND ACETYLCHOLINE IN BRAIN HOMOGENATES

Recent estimates made in this laboratory (Clementi and Whittaker, 1965) suggest that there are as many as 10^{11} cholinergic endings in the guinea-pig cortex accounting for a small, but nonetheless significant proportion by weight of the tissue. Comparable numbers occur in other mammalian brains. Here, then, is a rich source of cholinergic material with which to study the problems of the storage and release of acetylcholine. Although cholinergic endings are probably greatly outnumbered by non-cholinergic, the ready availability of the tissue in bulk and the ease with which it can be homogenized makes it in many ways more suitable for biochemical studies of the acetylcholine system than some other tissues with an exclusively cholinergic innervation.

The stability of the binding sites for acetylcholine in brain tissue is shown by the high proportion (70–75 per cent) of the transmitter which remains bound when the tissue is homogenized in isotonic sucrose. The bound acetylcholine is exclusively particulate in its localization; the significance of this became much clearer when E. G. Gray and I found (Whittaker, 1960; Gray and Whittaker, 1960, 1962) that a fraction rich in bound acetylcholine and other amines which had been separated from mitochondria and other subcellular particles by differential and density gradient centrifuging (Hebb and Whittaker, 1958; Whittaker, 1959) consisted almost entirely of pinched-off nerve endings. Evidently, then, bound acetylcholine was the fraction of the transmitter which was stored at cholinergic nerve endings; its survival was due to the survival of nerve endings as organized structures during homogenization and fractionation. Recently we have proposed the term *synaptosomes* for these pinched-off nerve endings (Whittaker, Michaelson and Kirkland, 1964) to emphasize their unique combination of synaptic origin and

* This work was supported, in part, by U.S.P.H.S. grant no. NB 03928 from the National Institute of Neurological Diseases and Blindness. Electron microscopic facilities were provided by the Wellcome Trust.

particulate character. The relationship between the nerve ending *in situ* and the synaptosome is illustrated in Fig. 1.

The origin of the 25–30 per cent of total brain acetylcholine which, in the presence of an anticholinesterase, is recovered in the high speed supernatant fraction consisting of cell sap diluted with sucrose (Whittaker, 1959) is not known. It may represent the proportion of cholinergic endings which do

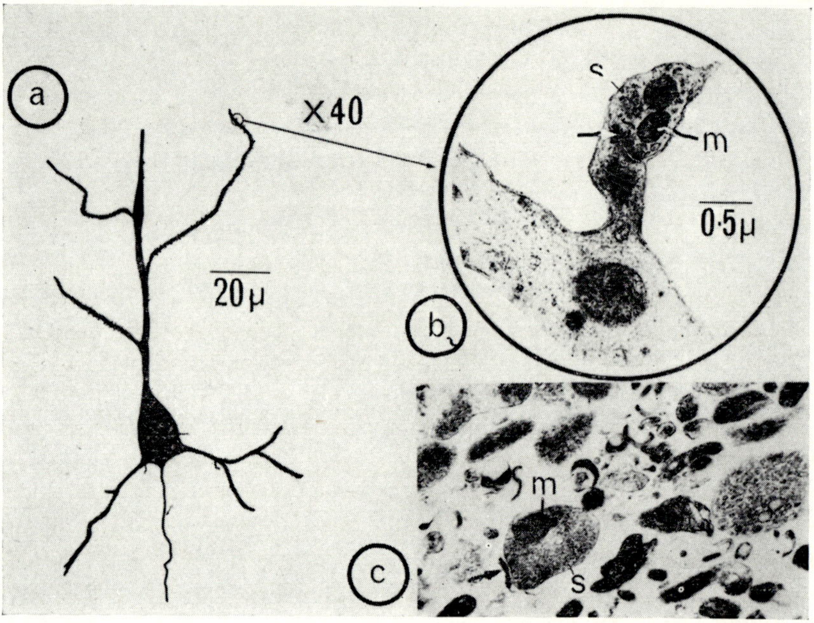

FIG. 1. The genesis of synaptosomes. (*a*) Drawing of a cortical neurone (silver preparation) showing dendritic spines which are the sites of axodendritic synapses. (*b*) Ringed area in (*a*) as seen in electron microscope showing details of spine and synaptic contact. (*c*) Electron micrograph of fraction of pinched-off nerve endings (synaptosomes) showing the same fine structure as the presynaptic nerve terminal in (*b*) including postsynaptic membrane (arrow), synaptic vesicles (s), mitochondria (m) and external membrane. (Electron micrographs by courtesy of Dr. E. G. Gray.)

not survive homogenization; however, it may equally well be derived from other portions of the cholinergic neurone, since there is much evidence for the presence of the enzyme which synthesizes acetylcholine, choline trans-acetylase, along the length of cholinergic neurones as well as at the nerve ending.

The synaptosomes are unstable structures; they readily autolyse on incubating at 37°C in 0·32 M sucrose (Whittaker, 1959) with the release of acetylcholine. Recent work in my laboratory (F. Clementi, unpublished) has shown, however, that if transferred to Krebs–Ringer solution and supplied with substrates, a proportion resists autolysis for up to 30 min at 37°C. Such preparations respire, and the contained mitochondria retain respiratory control (observations by K. Bicknell). Such preparations should provide an excellent *in vitro* system for the study of centrally-acting drugs. When non-radioactive (Whittaker, 1959) or radio-labelled (Burton, 1964) acetylcholine is added to synaptosomes a proportion is taken up, but there has so far been little study of the kinetics of such uptake.

Any treatment (exposure to acid pH, protein precipitants, organic solvents), which is disruptive to lipoprotein membranes, causes release of bound acetylcholine from the synaptosome fraction. However, several kinds of mildly disruptive treatment were found, such as freezing and thawing, and suspension in hypotonic media, which release only about 50 per cent of the bound acetylcholine (Whittaker, 1959). This led to the suggestion that there are two kinds of bound acetylcholine, a labile and a more stable fraction. It is now realised that the phenomenon of labile and stable forms of bound acetylcholine is more complex than was at first thought and the common end point (40–60 per cent) of the release is in part fortuitous. A better understanding of the release phenomena came from studies of the release of soluble cytoplasmic markers and of the morphological changes accompanying the release of acetylcholine (Johnson and Whittaker, 1962, 1963). A study of the morphological changes was made possible by the discovery (Horne and Whittaker, 1962) that negative staining could be used to study the morphology of intact and partially disrupted subcellular particles in the electron microscope. Conventional histological methods of positive staining and thin section are not well suited for this purpose because of the difficulty of preserving very fragile partially disrupted preparations from further disruption during positive staining and embedding.

Johnson and I found that in frozen and thawed preparations there were many intact synaptosomes: the ones which had been damaged appeared to be damaged from the inside. The soluble cytoplasmic marker lactate dehydrogenase was released *pari passu* with acetylcholine suggesting that the treatment had broken down a proportion of synaptosomes only, leaving about 50 per cent with their complement of lactate dehydrogenase and acetylcholine intact.

Synaptosomes suspended in hypotonic sucrose presented a very different picture: the outer membranes of almost all of them were disrupted, but the cytoplasmic organelles of the nerve endings, the synaptic vesicles and mitochondria, were intact. Lactate dehydrogenase had almost completely (80 per cent) leaked out reflecting the high proportion of disrupted synaptosomes, but about 50 per cent of the acetylcholine again remained bound. It seemed likely that in this case the more stable complement of bound acetylcholine represented intravesicular acetylcholine. To test this view further we set about isolating the synaptic vesicles from the soluble cytoplasm, broken external membranes and intra-terminal mitochondria.

This was achieved (Whittaker, Michaelson and Kirkland, 1964) by making use of a discontinuous density gradient consisting of sucrose layers varying in concentration from 0·4 to 1·2 M in 0·2 M steps. When the disrupted synaptosomes were centrifuged into such a gradient (Fig. 2), free synaptic vesicles (Fig. 3) were recovered as a hazy band in the 0·4 layer M (fraction D); external membranes (Fig. 4) were recovered in three particulate bands with

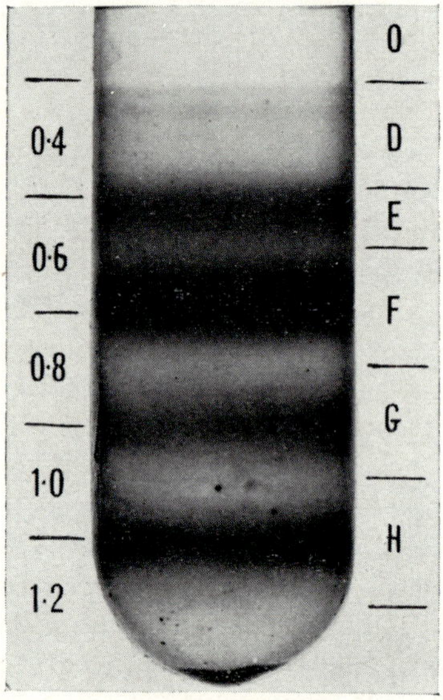

FIG. 2. Density gradient separation of component structures of synaptosomes disrupted by suspension in hypotonic media. *O*, soluble cytoplasm; *D*, synaptic vesicles; *E*, *F*, *G*, membrane fractions; *H*, incompletely disrupted synaptosomes; *I*, intra-terminal mitochondria. Figures at left are sucrose concentrations. Centrifuging was at 53,000 × *g* for 2 hr. (From Whittaker, Michaelson and Kirkland, 1963, 1964.)

densities equivalent to 0·4–0·6 M, 0·6–0·8 M and 0·8–1·0 M sucrose, intact and partially disrupted synaptosomes were collected between 1·0 and 1·2 M sucrose (fraction *H*) and intra-terminal mitochondria, as a pellet below 1·2 M sucrose (fraction *I*). A clear top layer (*O*) contained most of the soluble cytoplasmic constituents; control experiments with bovine serum albumin and other soluble substances showed that only about 20–25 per cent diffused out of this layer during centrifuging.

Figure 5 shows the distribution of the soluble cytoplasmic markers lactate dehydrogenase and potassium, the mitochondrial marker, succinate dehydrogenase, and the components of the acetylcholine system, cholinesterase, choline acetylase and acetylcholine itself, in the various fractions. The marker substances were distributed as expected from the morphological results: lactate dehydrogenase and potassium mainly in fraction *O*, succinate

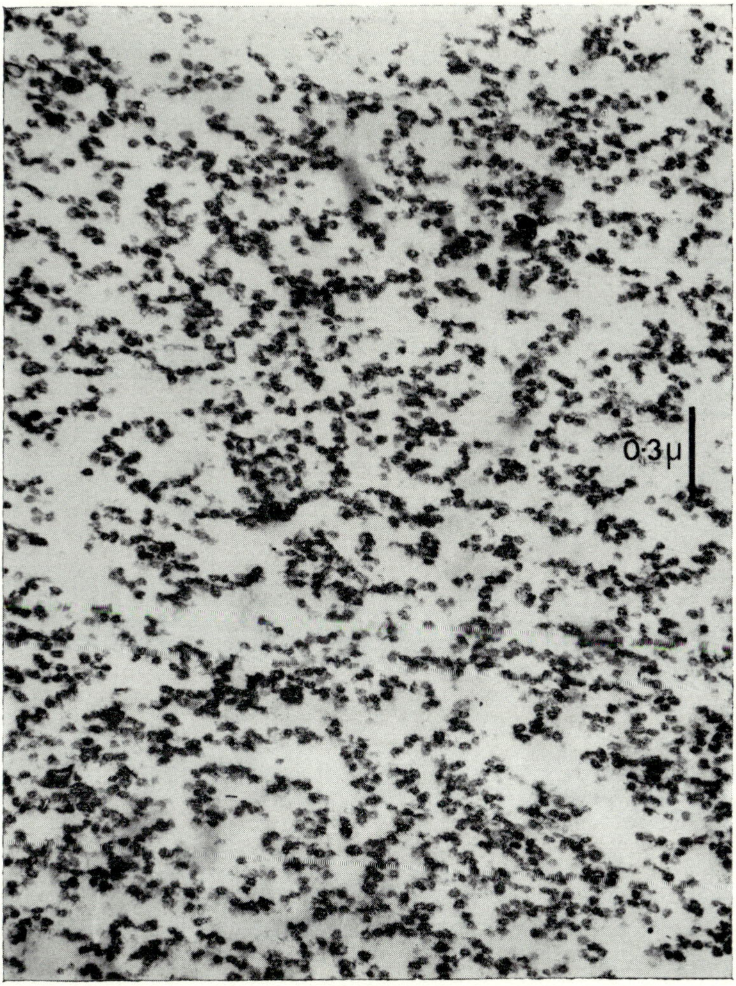

FIG. 3. Isolated synaptic vesicles of fraction *D* (electron micrograph by Dr. M. N. Sheridan). Permanganate staining. Note freedom from contamination by larger membrane fragments.

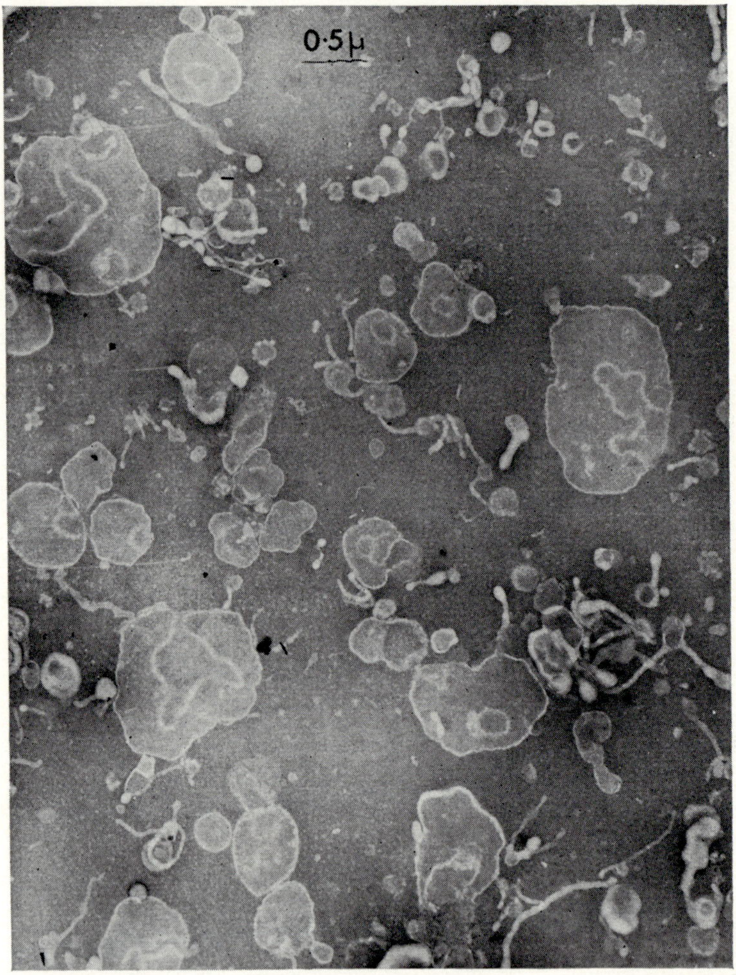

FIG. 4. External membranes of synaptosomes (fraction *G*). Negative staining with sodium phosphotungstate.

dehydrogenase mainly in fraction *I*. Cholinesterase was associated with external membrane fragments. Of chief interest was the distribution of choline acetylase and acetylcholine. The former behaved as a soluble

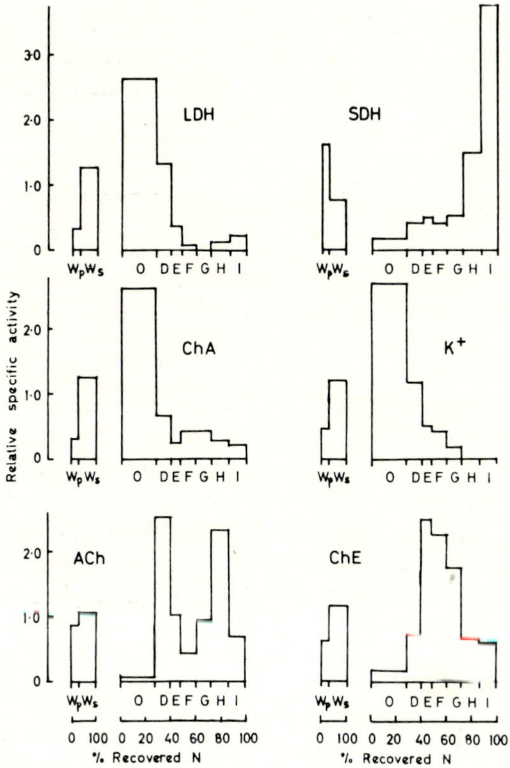

FIG. 5. Distribution of the cytoplasmic markers, lactate dehydrogenase (LDH) and potassium (K^+), the mitochondrial marker, succinate dehydrogenase (SDH) and the components of the acetylcholine system, cholinesterase (ChE), choline acetylase (ChA) and acetylcholine (ACh) in fractions of disrupted synaptosomes. For meaning of density gradient fractions see Fig. 2. The disrupted synaptosomes were centrifuged (left-hand block) to remove coarse particulate material (W_p) before submitting the supernatant (W_s) to density gradient separation. The ordinates represent the relative specific activity of the fractions, i.e. percentage of component recovered in fraction ÷ percentage of nitrogen recovered in fraction. (Graphs plotted from results of Whittaker, Michaelson and Kirkland, 1964.)

cytoplasmic component; the latter was *bimodally* distributed, with one peak in the isolated synaptic vesicle, and a second in another vesicle-rich fraction, that containing incompletely disrupted synaptosomes (fraction *H*). It thus seems clear that the synaptic vesicles are the ultimate storage sites of acetylcholine within the cholinergic ending, and that the acetylcholine is made not within the vesicle but in the cytoplasm. It is not clear whether the "labile" fraction of bound acetylcholine which disappeared during the initial hypotonic treatment represents a separate pool of acetylcholine in transit from the sites of synthesis in the cytoplasm to the storage sites in the vesicles or whether it is vesicular acetylcholine which leaked out during rupture. Experiments (Whittaker and Sheridan, 1965) in which eserine was added to the density gradient show that the vesicular acetylcholine is far from stable and leaks out during the density gradient separation.

THE ACETYLCHOLINE CONTENT OF ISOLATED SYNAPTIC VESICLES

It is well known that the spontaneous synaptic activity which can be recorded from neuromuscular junctions (Fatt and Katz, 1952, reviewed by del Castillo and Katz, 1956), spinal motoneurones (Katz and Miledi, 1963; Kuno, 1964), sympathetic ganglia (Blackman, Ginsborg and Ray, 1963) and the chick ciliary ganglion (Martin and Pilar, 1964) has been attributed to the quantized release of transmitter, of which the morphological counterpart could be the synaptic vesicle (del Castillo and Katz, 1956). Now that acetylcholine has definitely been shown to be associated with preparations of isolated synaptic vesicles of great purity, it would be of considerable interest to know if the number of acetylcholine molecules/vesicle as determined on a preparation of isolated vesicles agrees with physiological estimates of the number of acetylcholine molecules/quantum.

Unfortunately no accurate answer can be given to this question at the present time because different physiological estimates vary by more than a power of 10, and also because of uncertainties regarding the proportion of isolated vesicles which are derived from cholinergic endings. Nevertheless an estimate can be made of the average acetylcholine content of isolated vesicles (Sheridan and Whittaker, 1964; Whittaker and Sheridan, 1965) and this, when suitably corrected for the probable dilution of cholinergic vesicles by non-cholinergic, gives a figure within the range both of physiological estimates and theoretical estimates based on calculations of the number of molecules which can be inserted into a sphere of the same diameter as a synaptic vesicle.

The results are summarized in Table 1. In section I, estimates of the number of synaptic vesicles/g brain are obtained (A) from counts of synaptosomes and counts of the average number of vesicles contained within a synaptosome of average size (M. N. Sheridan and V. P. Whittaker, to be published), (B) from counts of isolated vesicles (Sheridan and Whittaker, 1964; Whittaker and Sheridan, 1965). Both counts made use of a polystyrene bead "tagging" technique. In section IIA, the average number of acetylcholine molecules/vesicle is calculated from the data in section I and the number of acetyl-

TABLE 1. ACETYLCHOLINE CONTENT OF CORTICAL SYNAPTIC VESICLES FROM THE GUINEA-PIG

I. Number of vesicles/g cortex	
A. From synaptosome counts	
a. Number of synaptosomes/g	4×10^{11}
b. Average number of vesicles/synaptosome	80
c. Number of vesicles/g ($=a\times b$)	$3{\cdot}2\times10^{13}$
B. From vesicle counts	
d. Number of vesicles/g in fraction *D*	$3{\cdot}8\times10^{12}$
e. Yield of vesicles in fraction (from acetylcholine content)	12%
f. Number of vesicles/g ($=100d/e$)	$3{\cdot}2\times10^{13}$
II. Number of acetylcholine molecules/vesicle	
A. Experimental	
g. Number of acetylcholine molecules/g in fraction *D* (by bioassay)	$1{\cdot}2\times10^{15}$
h. Average number/vesicle ($=g/d$)	306
i. Estimated proportion of cholinergic endings	15%
j. Corrected number/vesicle ($=100h/i$)	2050
B. Theoretical (for core of 310 Å)	
k. For isotonic (0·16 M) acetylcholine	1500
l. For solid acetylcholine	5×10^{4}
C. Physiological	
m. MacIntosh (1959)	900
n. Krnjević and Mitchell (1961)	1×10^{5}

choline molecules in the preparation as determined by bioassay. The figure (*h*) so obtained is revised upwards on the assumption that only a proportion of the vesicles are derived from cholinergic endings and so could contain acetylcholine. There is no means of knowing what this proportion is at present but it has been assumed to be the same as the proportion of cholinoceptive cells in the whole population. More recent results (Krnjević and Whittaker, 1965) have indicated a higher figure than 15 per cent, which would have the effect of reducing the corrected figure (*j*). However, the leakage of acetylcholine from the vesicles during preparation has almost certainly been underestimated.

In section IIB of the table, the number of molecules of acetylcholine which can be inserted into a core of 310 Å diameter (mean diameter of vesicle, 470 Å, less 2 × thickness of 80 Å thick wall) has been calculated on the basis of two different assumptions. On the likely assumption that the core is filled with isotonic acetylcholine, one obtains the figure (*k*) of 1500; on the rather unlikely assumption that the core is filled with solid acetylcholine the larger figure (*l*) of 5×10^4 is obtained. Figure *j* is satisfactorily close to the lower of these two theoretical estimates. It should perhaps be mentioned that there is no evidence that acetylcholine is in fact located in the core; it might be simply adsorbed on to the external surface. A synaptosome of average diameter could accommodate about $1{\cdot}4\times10^4$ molecules of acetylcholine as a surface monolayer.

The estimates (IIC) of the number of acetylcholine molecules/quantum from physiological data relate only to the neuromuscular junction and vary enormously: one of the lowest is MacIntosh's (1959) figure of 900, and the highest is Krnjević and Mitchell's (1961) figure of 10^5. However, Mitchell and Silver (1963) later thought that this figure was probably too high.

In spite of all the uncertainties in these calculations and comparisons, it seems fair to conclude that the observed acetylcholine content of isolated synaptic vesicles is consistent with the quantum hypothesis.

THE COMPOSITION OF THE VESICLE MEMBRANE

The lipid composition of a number of brain fractions has been studied by Eichberg, Whittaker and Dawson (1964). The results show that synaptic vesicles, synaptosomes and microsomes have a basically similar lipid composition, in contrast to myelin and mitochondria, each of which showed

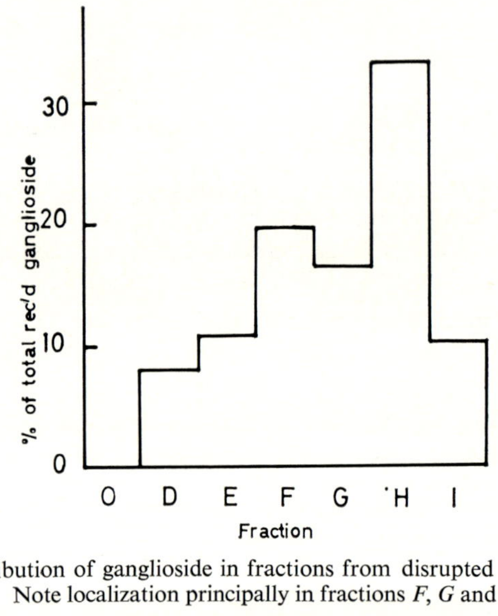

FIG. 6. Distribution of ganglioside in fractions from disrupted synaptosomes. Note localization principally in fractions *F*, *G* and *H*.

idiosyncrasies in lipid composition. However, differences in ganglioside and enzyme content have been found between the synaptic vesicles and external membranes which may have considerable functional significance.

The distribution of ganglioside in fractions of disrupted synaptosomes is shown in Fig. 6. It will be seen that the vesicle fraction accounts for a rather small proportion of the total ganglioside and that the richest fractions are those containing external membranes. The high ganglioside content of synaptic vesicles reported by Burton, Howard and Gibbons (1964) was probably due to the use of a preparation of synaptic vesicles heavily contaminated with external membranes and other structures (for discussion of their preparation see Whittaker, Michaelson and Kirkland, 1964).

The distribution of various adenosine triphosphatases (ATPase) in fractions of disrupted synaptosomes has been studied by Hosie (1964, 1965) working in my laboratory (Fig. 7). A DNP-stimulated ATPase was found to be localized in the mitochondrial (*I*) fraction. The external membrane fractions *E*–*G* and the synaptosome fraction *H* were rich in Na, K-stimulated ATPase,

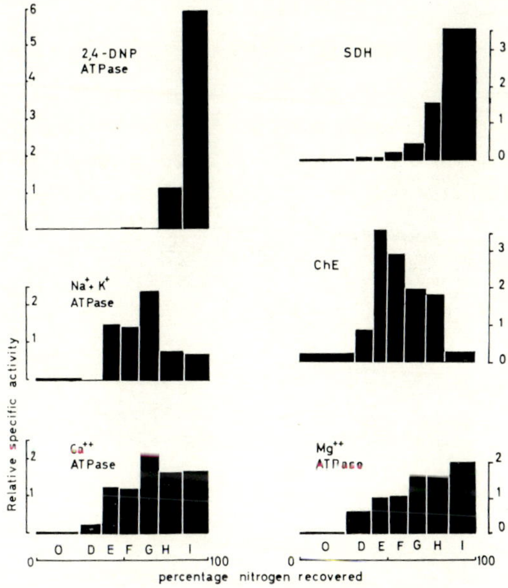

FIG. 7. Adenosine triphosphatase (ATPase) activity of fractions from disrupted synaptosomes. Note localization of 2:4-dinitrophenol (DNP)-stimulated enzyme (top left) in mitochondrial fraction like succinate dehydrogenase (SDH, top right), Na, K-stimulated ATPase (left, middle) in external membrane fractions with peak infraction *G* (cf. cholinesterase, middle right, which however, is more concentrated in the small membrane (microsomal) fragments in *E*) and Mg-activated ATPase which is the only significant type of ATPase activity seen in *D*.

maximum activity occurring in *G*. The synaptic vesicle fraction *D* contained a Mg-activated ATPase but was free from Na, K-stimulated ATPase. These results are in contrast to those of Campbell (1964) who reported Na, K-stimulated ATPase in a synaptic vesicle preparation prepared by sonic disintegration of synaptosomes suspended in hypotonic sucrose. However, in our hands this technique has led to synaptic vesicle preparations contaminated with fragments of external membrane.

The possibility was considered that the Mg-stimulated ATPase in the synaptic vesicle fraction was part of an ATP-requiring transport system for acetylcholine analogous to the well-known sodium transport system. However, no consistent stimulation of the ATPase could be obtained with added acetylcholine.

The ATPase results and the results summarized in Fig. 5 show that the synaptic vesicle membrane is neither derived from, nor significantly contaminated with membrane fragments derived from the external membrane or intraterminal mitochondria. This conclusion has certain consequences for theories of transmitter release.

MECHANISMS OF TRANSMITTER RELEASE

If one assumes, as now seems probable, that acetylcholine release involves the discharge of synaptic vesicles, there are several conceivable mechanisms, frequently discussed in recent years, by which this could be brought about. One possibility is that vesicles are simply extruded intact through the external

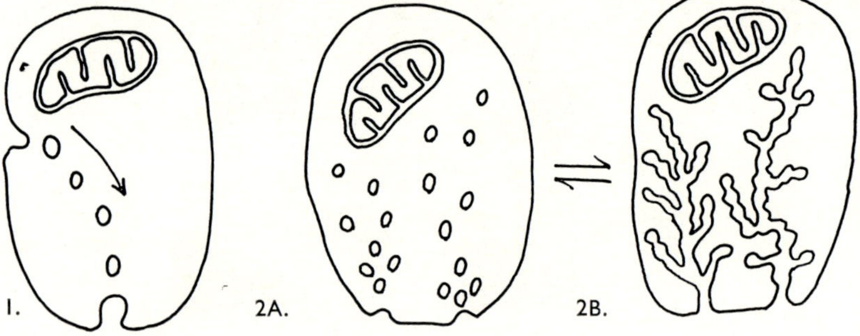

FIG. 8. Hypothesis of vesicle formation and transmitter release. 1. Vesicles are formed by withdrawal of external membrane material in regions distal to the area of synaptic contact, are charged with transmitter in the cytoplasm and fuse with the membrane in regions proximal to the region of synaptic contact. For reasons given in the text, the patches of membrane resulting from vesicle fusion must retain their identity. What maintains a unidirectional flow of vesicles? 2. Vesicles are present only in resting endings (A). On activation, they fuse to form tubules (B) which discharge transmitter accumulated during phase A. The vesicles readily reform from the tubules and the membrane of the tubule-vesicle system retains its identity. Possibly the vesicles are linked even in the resting phase by occluded lengths of tubule.

membrane and disintegrate in the synaptic cleft. Although there is evidence for this kind of release in the case of certain other storage granules, free, extracellular synaptic vesicles have not so far been convincingly demonstrated in tissue sections and the extracellular space round the ending (only about 200 Å wide) is hardly large enough to accommodate free vesicles (mean diameter 470 Å). Another possibility (shown diagrammatically in Fig. 8.1) is that synaptic vesicles fuse with the external membrane and discharge transmitter by a process of reversed micropinocytosis. There must, presumably, be some mechanism whereby material in the external membrane is withdrawn later to form fresh vesicles, otherwise the external membrane would extend indefinitely in area. In view of the differences in composition between the external membrane and the synaptic vesicles, the vesicular membrane, if it is incorporated into the external membrane, must retain its identity during this process of fusion and regeneration. It is rather difficult to see how this could be done.

Another possibility is that the vesicles represent merely the "resting" configuration of a two-phase system, the other active configuration of which is tubular in character (Fig. 8.2). The tubules may be envisaged as opening out through the external membrane and providing channels through which transmitter, taken up by vesicles in the resting phase, can rapidly diffuse out of the ending. When synaptic transmission ceased, the system would revert to its vesicle form. Although in contact with the external membrane at certain points, the vesicle–tubule membrane system would, in this scheme, retain its independent identity.

In examining large synaptosomes present in the nuclear fraction from the cerebellar cortex and probably originating from the large mossy-fibre endings of the granular layer, we noted the occurrence of one or more tubules as well as vesicles in the cytoplasm of most synaptosomes (Fig. 9a, arrow). In preparations which had been further fractionated to remove nuclei, myelin fragments and other contaminating structures (Israël and Whittaker,) 1965 synaptosomes were also seen (Fig. 9b) which had few or no synaptic vesicles but whose cytoplasm was filled with tubules. Slight warming (to 8–10°C) increased the number of such synaptosomes and also caused the appearance of appreciable numbers of disintegrating, badly autolysed synaptosomes. A continuous spectrum of forms left no doubt that structures of the type shown in Fig. 9b are synaptosomes. The tubules may in part be derived from mitochondria, but their presence in synaptosomes containing well preserved mitochondria but few or no vesicles strongly suggests that they are mainly derived from vesicles which have undergone intracytoplasmic fusion. The reaction may be an autolytic one, but the demonstration that intracytoplasmic fusion of vesicles can occur under these conditions raises the possibility, at least, that it may also occur under physiological conditions.

CONCLUDING REMARKS

The problem of bound acetylcholine in brain tissue has proved a complex one. An important step forward was the recognition that nerve endings are mechanically stronger than the structures which surround them; that on mincing or homogenizing brain tissue they are pinched off to form compact structures (synaptosomes) which behave like subcellular particles while

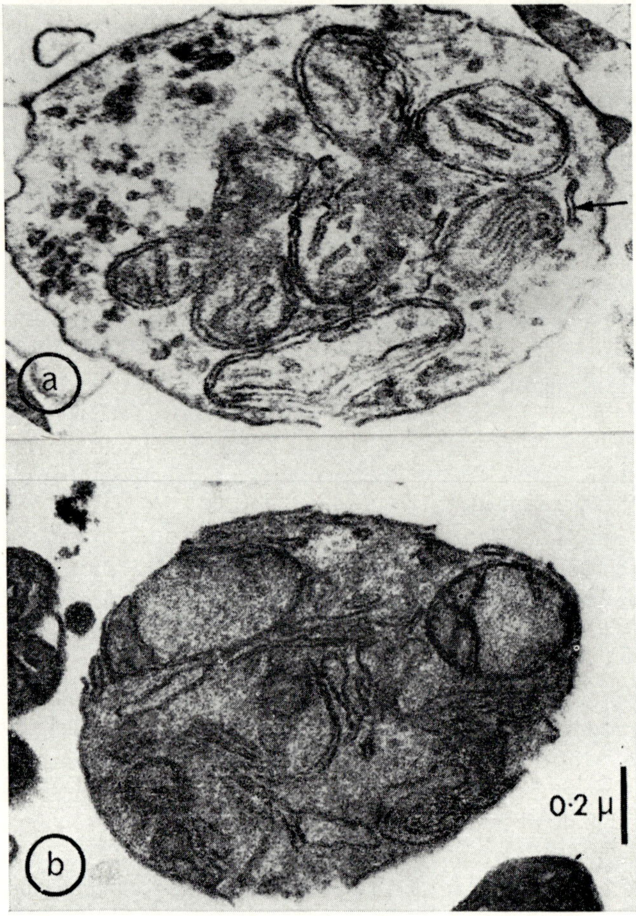

FIG. 9. (a) Normal large synaptosome, possibly derived from mossy fibre ending, isolated from cerebellar homogenate. Note mitochondria, vesicles and occasional tubules (arrow). (b) Second type of synaptosome, more abundant in preparations which have been slightly warmed, showing absence of vesicles and numerous tubules in cytoplasm. Permanganate fixation. (Electron micrographs by Dr. M. Israël.)

retaining the complex fine structure and transmitter content of the nerve ending *in situ*.

The second main advance has been the devising of a method which enables the component structures of the nerve ending to be isolated in highly purified form. This has made it possible to demonstrate that the synaptic vesicles are the ultimate storage sites of acetylcholine within the nerve ending and to determine the intraterminal compartmentation of functionally important components. Further work with the isolated organelles will be required before the detailed mechanisms of uptake, binding and release are fully understood.

SUMMARY

Subcellular fractionation of brain tissue suggests that there are three compartments containing acetylcholine in the cholinergic neurone: (*a*) that present in the cell body and axon, (*b*) that present in the nerve ending cytoplasm, (*c*) that present in the synaptic vesicles. A recently devised technique enables the component parts of isolated nerve endings, soluble cytoplasm, synaptic vesicles, external membranes and intraterminal mitochondria, to be separated thus permitting the compartmentation of transmitters and the enzymes metabolizing them to be determined. Evidence is presented indicating that acetylcholine generated in the neuronal cytoplasm is taken up by synaptic vesicles, thus enabling the local concentration of the transmitter to be raised at the nerve ending. "Tagging" techniques using polystyrene latex beads allow the average acetylcholine content of single nerve endings and vesicles to be determined. The membrane of the synaptic vesicles appears to be different from that of the external membrane. Under certain circumstances vesicular fusion may be observed to occur in isolated nerve endings, suggesting a possible release mechanism.

REFERENCES

BLACKMAN, J. C., B. L. GINSBORG and C. RAY (1963) On the quantal relase of the transmitter at a sympathetic synapse, *J. Physiol.* (Lond.) **167**, 402–15.

BURTON, R. M. (1964) Gangliosides and acetylcholine of the central nervous system—III. The binding of acetylcholine by subcellular particles of the brain, *Int. J. Neuropharmacol.* **3**, 13–22.

BURTON, R. M., R. E. HOWARD and J. M. GIBBONS (1964) Gangliosides and acetylcholine-containing synaptic vesicles of rat brain, Abstr. 6th Internat. Congr. Biochem. New York, 26 July–1 Aug., 1964. V–E–96.

CAMPBELL, W. J. (1964) The distribution of acetylcholine esterase in brain subcellular fractions and its relation to the "synaptic vesicle" fraction, Abstr. 6th internat. Congr. Biochem. New York, 26 July–1 Aug., 1964. V–E–97.

CLEMENTI, F. and V. P. WHITTAKER (1965) The yield of synaptosomes from the cerebral cortex of guinea pigs (to be submitted).

DEL CASTILLO, J. and B. KATZ (1956) Biophysical aspects of neuro-muscular transmission, *Progr. Biophys.* **6**, 121–70.

EICHBERG, J., V. P. WHITTAKER and R. M. C. DAWSON (1964) The distribution of lipids in subcellular particles of guinea-pig brain, *Biochem. J.* **92**, 91–100.

FATT, P. and B. KATZ (1952) Spontaneous subthreshold activity at motor nerve endings, *J. Physiol.* (Lond.) **117**, 109–28.

GRAY, E. G. and V. P. WHITTAKER (1960) The isolation of synaptic vesicles from the central nervous system, *J. Physiol.* (Lond.) **153**, 35–7*P*.

GRAY, E. G. and V. P. WHITTAKER (1962) The isolation of nerve endings from brain: an electron-microscopic study of cell fragments derived by homogenization and centrifugation, *J. Anat.* (Lond.) **96**, 79–88.

HEBB, C. O. and V. P. WHITTAKER (1958) Intracellular distribution of acetylcholine and choline acetylase, *J. Physiol.* (Lond.) **142**, 187–96.

HORNE, R. W. and V. P. WHITTAKER (1962) The use of the negative staining method for the electron-microscopic study of subcellular particles from animal tissues, *Z. Zellforsch.* **58**, 1–16.

HOSIE, R. J. A. (1964) The adenosine-triphosphatase activity of cerebral membranes of known morphological origin, *Biochem. J.* 13P.

HOSIE, R. J. A. (1965) The localization of adenosine triphosphatases in morphologically characterised subcellular fractions of guinea-pig brain, *Biochem. J.* **96**, 404–12.

ISRAËL, M. and V. P. WHITTAKER (1965) The isolation of mossy fibre endings from the granular layer of the cerebellar cortex, *Experientia* **21**, 325–6.

JOHNSON, M. K. and V. P. WHITTAKER (1962) Bound acetylcholine and entrapped cytoplasm in subcellular fractions of guinea-pig brain homogenates. Abstr. Int. Neurochem. Symp. on Enzyme Activity of the Central Nervous System, Göteborg. 17–21 June, 1962, pp. 60–1. Reprinted in *Acta Neurol. Scand.* **38**, Suppl. 1, 60.

JOHNSON, M. K. and V. P. WHITTAKER (1963) Lactate dehydrogenase as a cytoplasmic marker in brain, *Biochem J.* **88**, 404–9.

KATZ, B. and R. MILEDI (1963) A study of spontaneous miniature potentials in spinal neurones, *J. Physiol.* (Lond.) **168**, 389–422.

KRNJEVIĆ, K. and J. F. MITCHELL (1961) The release of acetylcholine in the isolatedr at diaphragm, *J. Physiol.* (Lond.) **155**, 246–62.

KRNJEVIĆ, K. and V. P. WHITTAKER (1965) Excitation and depression of cortical neurones by brain fractions released from micropipettes, *J. Physiol.* (Lond.) **197**, 288–322.

KUNO, M. (1964) Quantal components of excitatory synaptic potentials in spinal motoneurones, *J. Physiol.* (Lond.) **175**, 88–99.

MACINTOSH, F. C. (1959) Formation, storage and release of acetylcholine at nerve endings, *Canad. J. Biochem.* **37**, 343–56.

MARTIN, A. R. and G. PILAR (1964) Quantal components of the synaptic potential in the ciliary ganglion of the chick, *J. Physiol.* (Lond.) **175**, 1–16.

MITCHELL, J. F. and A. SILVER (1963) The spontaneous release of acetylcholine from the denervated hemidiaphragm of the rat, *J. Physiol.* (Lond.) **165**, 117–29.

SHERIDAN, M. and V. P. WHITTAKER (1964) Isolated synaptic vesicles: morphology and acetylcholine content, *J. Physiol.* (Lond.) **175**, 25–6*P*.

WHITTAKER, V. P. (1959) The isolation and characterization of acetylcholine-containing particles from brain, *Biochem. J.* **72**, 694–706.

WHITTAKER, V. P. (1960) The binding of neurohormones by subcellular particles of brain tissue. Proc. 4th Int. Neurochem. Symp. Varenna, 12–17 June, 1960. In *Regional Neurochemistry*: the Regional Chemistry, Physiology and Pharmacology of the Nervous System. Edited by S. Kety and J. Elkes. Oxford, Pergamon Press, pp. 259–63.

WHITTAKER, V. P., I. A. MICHAELSON and R. J. A. KIRKLAND (1963) The separation of synaptic vesicles from disrupted nerve ending particles, *Biochem. Pharmacol.* **12**, 300–2.

WHITTAKER, V. P., I. A. MICHAELSON and R. J. A. KIRKLAND (1964) The separation of synaptic vesicles from nerve ending particles ("synaptosomes"), *Biochem. J.* **90**, 293–303.

WHITTAKER, V. P. and N. M. SHERIDAN (1965) The morphology and acetylcholine content of isolated cerebral cortical synaptic vesicles, *J. Neurochem.* **12**, 363–72.

DISCUSSION

ROBERTSON: Have you performed an experiment in which membrane-limited vesicles derived from smooth endoplasmic reticulum in cells containing no acetylcholine are mixed with acetylcholine? If so does the acetylcholine bind to the membranes?

WHITTAKER: Yes, we have performed this kind of experiment with liver homogenates and mitochondrial fractions and with artificial lipid mixtures. In no case were we able to "synthesize" vesicles, nor did we get any evidence of acetylcholine binding. It would not, however, surprise me if one were to get evidence of binding to microsomes if sufficiently sensitive tests (e.g. with radiolabelled acetylcholine) were applied. Ganglioside and chondroitin sulphate are examples of negatively charged macromolecules which are known to bind amines non-specifically. It would then be a question of determining by affinity measurements and the use of inhibitors, whether the binding resembled that in the vesicle fraction. I suspect if it existed it would turn out to be non-specific.

UVNÄS: How can you be so sure that the acetylcholine-bearing structures are vesicles and that they have a membrane?

WHITTAKER: This follows (a) from the observed association between acetylcholine content and synaptic vesicles, (b) the morphological appearance of the synaptic vesicles in thin section and in negative staining after phosphate treatment.

ROBERTSON: You stated that at one stage of preparation the synaptosomes are ruptured by hypotonic shock. Why are not the vesicles also ruptured at this stage with release of acetylcholine if it is free within the vesicle lumen? Would you agree that your findings suggest that the acetylcholine is not free but bound to the vesicle membrane?

WHITTAKER: The vesicles are not entirely impervious to hypotonic conditions. If the vesicle preparation, after isolation, is further diluted, spun down and resuspended in water, acetylcholine is lost. However, the vesicles seem to be less sensitive to hypotonic conditions than the synaptosomes. I would agree that some sort of matrix may be present and that the acetylcholine may be bound to it.

KALOW: Dr. Whittaker: On inspection of electron-microscopic pictures of synapses, the accumulation of mitochondria is most striking. Would you care to comment whether this accumulation is the same in cholinergic and adrenergic endings, and whether it is possible to estimate what part of the metabolic activity of the brain has to be ascribed to that in synaptic regions?

WHITTAKER: There is some evidence from density gradient centrifugation that synaptosomes can be separated into "light" and "heavy" fractions, the latter having a larger proportion of their cytoplasm occupied by mitochondria. As far as I am aware, however, it has not been possible to demonstrate that these two fractions contain different transmitters. The synthesis and storage of acetylcholine and, no doubt, other transmitters requires metabolic energy, and I presume that the local concentration of mitochondria reflects this intense local metabolism.

METABOLISM OF MONOAMINES BY BLOOD PLATELETS AND RELATION TO 5-HYDROXYTRYPTAMINE LIBERATION

A. Pletscher, G. Bartholini and M. Da Prada
Medical Research Department,
F. Hoffmann-La Roche & Co. Ltd., Basle

Isolated blood platelets, which contain the biological equipment for the transport, storage, and metabolism of 5-hydroxytryptamine (5-HT) (membranes, granules, mitochondria) represent appropriate models for studying the various aspects of monoamine metabolism. In the present work, the metabolic transformation of 5-HT and dopamine as well as the relations between liberation and metabolism of 5-HT have been investigated.

I. METABOLISM OF MONOAMINES IN PHYSIOLOGICAL MEDIA

Isolated blood platelets of rabbits incubated in plasma or synthetic media like Tyrode solution* lose only small amounts of their 5-HT (1–2 per cent within 4–6 hr). In the incubation fluid, at best traces of 5-HT, but two major 5-HT metabolites are found. They show R_f values of 0·18 and 0·74 and exhibit the typical activation and fluorescence spectra of 5-hydroxyindoles (Fig. 1). The compound with the lower R_f value has the characteristics of 5-hydroxy-3-indoleacetic acid in paper and thin-layer chromatography. The derivative with the high R_f value behaves like 5-hydroxytryptophol (5-hydroxyindole-3-ethanol) or N-acetyl-5-HT which show almost identical properties in paper chromatography with various solvent systems. Using thin-layer chromatography (Silicagel, ethylacetate), however, the compound with the high R_f value is characterized as 5-hydroxytryptophol. Isolated human platelets to which exogenous 5-HT is added also form 5-hydroxytryptophol and 5-hydroxyindoleacetic acid (Bartholini and Pletscher, 1964; Bartholini *et al.*, 1964).

The *relation between 5-hydroxytryptophol and 5-hydroxyindoleacetic acid* formed from 5-HT depends on the amount of erythrocytes present in the incubation medium. Platelets containing low amounts of erythrocytes produce

* NaCl	7·60 g/l.	(0·130 M)	$NaHCO_3$	2·10 g/l.	(0·003 M)
KCl	0·42 g/l.	(0·006 M)	Glucose	2·00 g/l.	(0·011 M)
Versene	0·80 g/l.	(0·002 M)	Saccharose	4·50 g/l.	(0·010 M)
$NaH_2PO_4 . 2H_2O$	0·14 g/l.	(0·001 M)			

more 5-hydroxytryptophol than 5-hydroxyindoleacetic acid. The latter is, however, always present, even if the platelets have been repeatedly washed and recentrifuged. If high amounts of isolated erythrocytes (e.g. $\frac{1}{10}$ of the original content in blood) are added to the platelet suspension, the principal

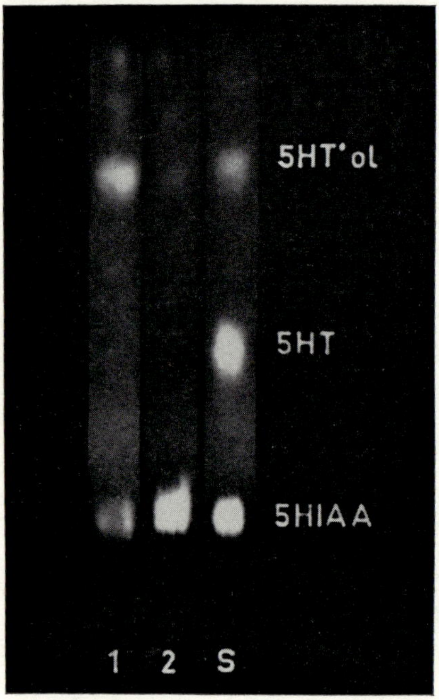

FIG. 1. Paper chromatography of the indolyl compounds formed during spontaneous release of 5-hydroxytryptamine from isolated rabbit platelets in presence and absence of erythrocytes. Incubation during 6 hr at 37°C in modified Tyrode buffer, pH 7·5. After centrifugation of the platelets, the buffer was extracted at pH 1 with ethyl ether. Schleicher and Schuell No. 2043; solvent system: N-propanol–ammonia 1 N (5:1). The fluorescence of the spots was developed by acidifying the paper with 4 N HCl and subsequent exposure to ultraviolet light (254 mμ). 1, thrombocytes; 2, thrombocytes+erythrocytes (5–10 per cent of the original content in the blood); S, standards: 5-HT'ol, 5-hydroxytryptophol (3 γ); 5-HT, 5-hydroxytryptamine (3 γ); 5HIAA, 5-hydroxy-indoleacetic acid (3 γ).

metabolite is 5-hydroxyindoleacetic acid and at best little formation of 5-hydroxytryptophol occurs (Fig. 1) (Bartholini and Pletscher, 1964). Pure erythrocytes do not transform added 5-HT into 5-hydroxyindoleacetic acid or 5-hydroxytryptophol, nor do they metabolize 5-hydroxytryptophol to 5-hydroxyindoleacetic acid. Furthermore, reserpinized platelets cause no major reduction of added 5-hydroxyindoleacetic acid to 5-hydroxytryptophol.

These findings indicate that isolated platelets transform released 5-HT into 5-hydroxyindoleacetaldehyde by oxidative deamination. In the absence of major amounts of erythrocytes, the aldehyde is reduced to 5-hydroxytryptophol, whereas in the presence of erythrocytes oxidation of the 5-hydroxyindoleacetaldehyde to the corresponding acid takes place. Therefore platelets, in contrast to erythrocytes, seem to contain monoamine oxidase (MAO) and a relatively active alcohol dehydrogenase, whereas aldehyde dehydrogenase is present in both platelets and erythrocytes (Table 1). Experiments with

TABLE 1. METABOLISM OF PLATELET 5-HT

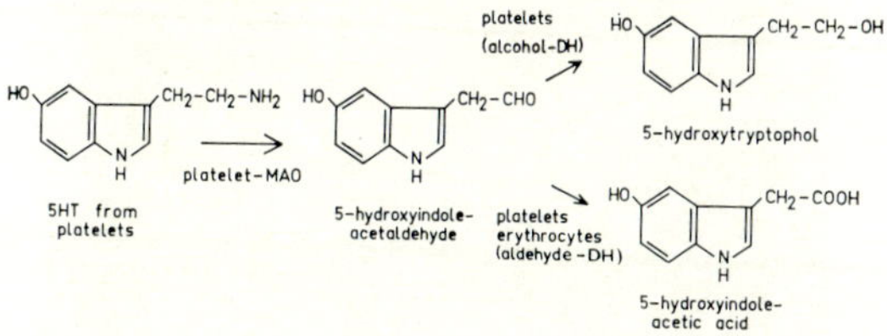

MAO inhibitors and indole-3-acetaldehyde, which is more stable than 5-hydroxyindoleacetaldehyde, confirm this hypothesis. Thus, the MAO inhibitors iproniazid and pargylin in appropriate doses completely inhibit the formation of 5-hydroxytryptophol and 5-hydroxyindoleacetic acid from released 5-HT. Furthermore, pure platelets metabolize indoleacetaldehyde* mainly to tryptophol (indole-3-ethanol)* and to a smaller extent to indoleacetic acid,* whereas with pure erythrocytes predominant formation of indoleacetic acid occurs (Fig. 2) (Bartholini *et al.*, 1964).

It may therefore be concluded that the 5-HT released from platelets under physiological conditions (e.g. in plasma or Tyrode solution) is metabolized by MAO and then preferentially transformed into 5-hydroxyindoleacetic acid or 5-hydroxytryptophol, depending on the presence or absence of erythrocytes.

Isolated platelets metabolize dopamine in a mode similar to 5-HT, i.e. the amine is transformed to 3,4-dihydroxyphenylethanol and 3,4-dihydroxyphenylacetic acid (Fig. 3).

II. EFFECT OF MONOAMINE LIBERATING AGENTS

The metabolism of platelet 5-HT has been studied in conditions under which the liberation of the amine is enhanced, i.e. by using (1) reserpine as well as a benzoquinolizine derivative, (2) thrombine, and (3) various sympathomimetic amines.

* Synthesized by Dr. H. Bruderer, F. Hoffmann-La Roche & Co. Ltd., Basle.

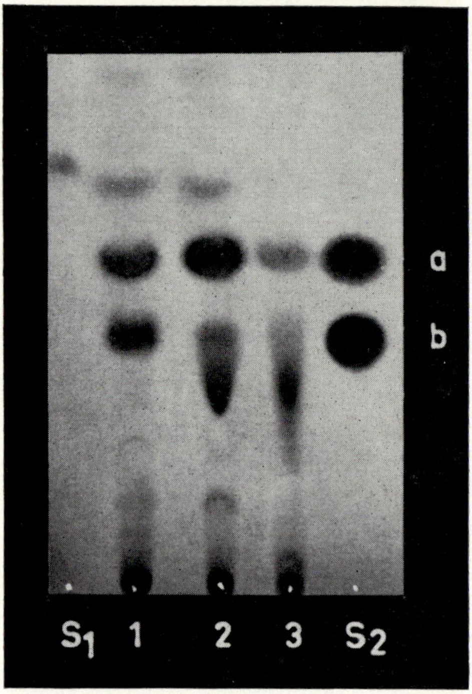

FIG. 2. Metabolism of indoleacetaldehyde (20 γ/ml.) by isolated thrombocytes and erythrocytes incubated in Tyrode solution at 37·5°C for 2 hr. Extraction of the buffer with ethyl ether at pH 1. Silicagel, solvent system: chloroform, acetic acid (95:5 v/v). S_1, indoleacetaldehyde; S_{2a}, indoleacetic acid; S_{2b}, indole-ethanol; 1, thrombocytes; 2, thrombocytes + erythrocytes; 3, erythrocytes.

1. *Reserpine-like Drugs*

The metabolism of the 5HT liberated by reserpine as well as by benzoquinolizine derivatives such as Ro 4-1284* (Pletscher *et al.*, 1962), is qualitatively similar to that of the spontaneously released amine. Two 5-HT metabolites, i.e. 5-hydroxytryptophol and 5-hydroxyindoleacetic acid, are present in the incubation medium (Fig. 4). Besides, considerable amounts of 5-HT appear indicating that only part of the released 5-HT is metabolized (Fig. 5). This metabolism seems to take place during or immediately after the liberation of 5-HT, since 5-hydroxytryptophol is formed at the very beginning of the incubation and since its relative amount (as compared to the total of released 5-HT) does not change with the time of incubation (Table 2) (Da Prada *et al.*, 1965).

* 2-hydroxy-2-ethyl-3-isobutyl-9,10-dimethoxy-1,2,3,4,6,7-hexahydro-11bH-benzo-[a]quinolizine.

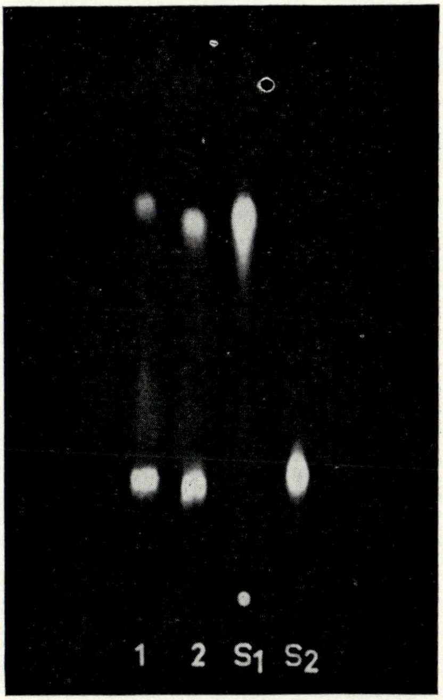

FIG. 3. Paper chromatography of the catechol metabolites formed by isolated rabbit platelets, incubated with 50 γ/cc dopamine in modified Tyrode buffer at 37·5°C for 2 hr. After centrifugation of the platelets, the medium was extracted at pH 3·5 with ethyl acetate. Paper and solvent system see Fig. 1. The fluorescence of the spots was developed by potassium ferricyanide 0·1 per cent in a 10 per cent aqueous solution of ethylene diamine (254 mμ). 1, thrombocytes, 16 hr after pretreatment of the rabbits with 5 mg/kg reserpine i.p. (no endogenous 5-HT); 2, thrombocytes after pretreatment of the rabbits with 150 mg/kg Disulfiram i.p. 16 hr before sacrifice; S_1, standard: 3,4-dihydroxyphenylethanol (3 γ); S_2, standard: 3,4-dihydroxyphenylacetic acid (3 γ).

The results show that due to the relatively slow 5HT release after reserpine (Carlsson *et al.*, 1957; Paasonen and Pletscher, 1959; Bartholini *et al.*, 1965) at least some of the liberated 5-HT gets into contact with MAO. It remains

TABLE 2. FORMATION OF 5-HYDROXYTRYPTOPHOL BY ISOLATED RABBIT PLATELETS INCUBATED IN TYRODE SOLUTION WITH 10 γ/ML RESERPINE OR 1 NIH UNIT/ML THROMBIN

Drugs	$\frac{1}{2}$ *hr*	2 *hr*
Reserpine	19·9 ± 4·8	21·7 ± 3·2
Thrombin	0·8 ± 0·0	12·1 ± 2·5

The figures are means ±S.E. of 4–7 experiments and indicate the 5-hydroxytryptophol formed in per cent of the released 5-hydroxytryptamine.

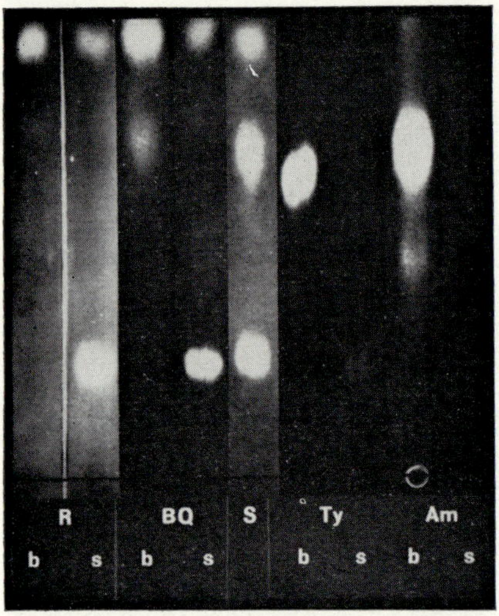

FIG. 4. Paper chromatography of 5-hydroxyindole derivative contained in Tyrode medium 2 hr after incubation of isolated rabbit platelets with various drugs. S, standard; above 5-hydroxytryptophol, middle 5-hydroxytryptamine, below 5-hydroxyindoleacetic acid; R, 2·5 γ/ml. reserpine; BQ, 50 γ/ml. benzoquinolizine derivative Ro 4-1284; Ty, 250 γ/ml. tyramine HCl; Am, 500 γ/ml. DL-amphetamine HCl; b, basic extract (pH 10 with ethyl acetate); s, acid extract (pH 1 with ethyl ether) (Pletscher and Bartholini, 1964).

to be elucidated whether platelet MAO is not sufficiently active for completely metabolizing the amine or whether part of the amine escapes the enzyme.

2. *Thrombin*

The 5-HT liberated by thrombine shows a different metabolic pattern from the amine released by reserpine. Thrombin causes an immediate marked liberation of 5-HT which is maximal after a few minutes of incubation of the platelets (Da Prada *et al.*, 1965). Initially (within 15–30 min), only negligible amounts of 5-HT metabolites are detected in the medium. After further incubation (1–2 hr), some 5-hydroxytryptophol and 5-hydroxyindoleacetic acid appear, however. Therefore, the relative amount of 5-hydroxytryptophol (as compared to the total of liberated 5-HT) increases with time (Fig. 5, Table 2). This indicates that the platelet MAO is still active, but that the 5-HT has probably escaped the MAO during its rapid liberation.

3. *Sympathomimetic Amines*

Sympathomimetic amines, such as amphetamine, para-chloro-N-methyl-amphetamine (Ro 4-6861) and tyramine, cause a marked liberation of platelet 5-HT which is rather more rapid than the reserpine-induced release of the

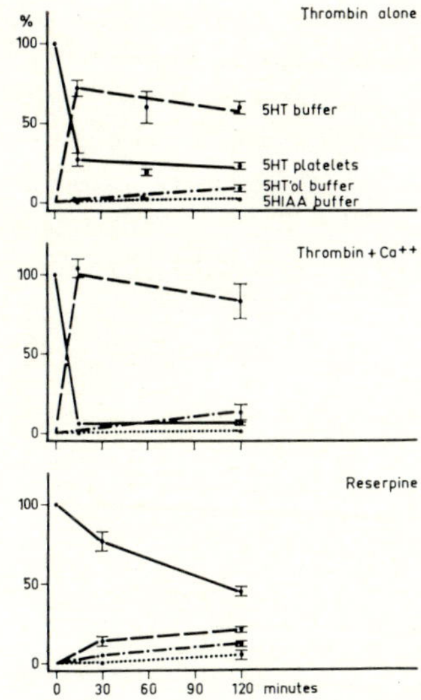

FIG. 5. Effect of thrombin (1 NIH unit/ml.),thrombin + Ca^{++}(0·3 g/l. = 0·0027 M), and reserpine (5 γ/cm^3) on the 5-hydroxytryptamine metabolism of isolated rabbit platelets in Tyrode buffer. Ordinate: 5-hydroxytryptamine (5-HT), 5-hydroxytryptophol (5-HT'ol) and 5-hydroxyindoleacetic acid (5-HIAA) in per cent of the 5-HT of the platelets before incubation. Abscissa: Incubation time. Each point represents an average of 4 experiments ± S.E. (Da Prada *et al.*, 1965).

amine. No 5-hydroxyindole metabolites appear, however, in the incubation medium even after 2 hr (Fig. 4) (Bartholini and Pletscher, 1964; Pletscher and Bartholini, 1964). The speed of 5-HT liberation alone does thus probably not explain the lack of 5-HT metabolism, because unlike with thrombin the liberated 5-HT is not metabolized on prolonged contact with the platelets. The following possibilities may be considered:

(*a*) Sympathomimetic amines displace the 5-HT from a different compartment than reserpine and Ro 4-1284. The 5 HT sensible to the sympathomimetic amines might, for instance, be located remote from metabolizing enzymes, whereas the reserpine-sensible pool possibly lies close to the MAO containing mitochondria. A similar theory has also been discussed for norepinephrine in sympathetic nerves and other tissues (Kopin, 1964). This pool hypothesis does, however, not seem to be very probable for the platelets, because almost all the 5-HT (90–95 per cent) can be liberated by sympathomimetic amines, such as Ro 4-6861, without 5-HT metabolites being formed.

(*b*) Sympathomimetic amines, in addition to their amine-liberating effect and contrary to reserpine and Ro 4-1284, interfere with the metabolism of liberated 5-HT by competitive (amphetamine, Ro 4-6861) or substrate (tyramine) inhibition. This hypothesis is supported by experiments with platelets from reserpinized animals containing virtually no endogenous 5-HT. Such platelets, even if they are agglutinated by thrombin and homogenized thereafter, still metabolize added 5-HT to 5-hydroxytryptophol and 5-hydroxyindoleacetic acid indicating that some MAO has remained active. The above mentioned sympathomimetic amines, in contrast to reserpine and Ro 4-1284, cause complete or marked inhibition of the formation of 5-HT metabolites which can hardly be explained by the pool theory, but rather by inhibition of MAO (Table 3).

TABLE 3. FORMATION OF 5-HYDROXYTRYPTOPHOL FROM EXOGENOUS 5-HYDROXYTRYPTAMINE BY ISOLATED PLATELETS OF RABBITS PRETREATED WITH 5 MG/KG RESERPINE I.P. 16 HR PRIOR TO THE EXPERIMENT

Drugs	*5-Hydroxytryptophol*
Controls	100±8
Tyramine	23±1
DL-Amphetamine	11±2
Ro 4–6861	3±2
Reserpine	117±8
Ro 4–1284	101±10

Incubation of the platelets in Tyrode solution for 2 hr at 37°C with 20 γ/ml exogenous 5-hydroxytryptamine. Addition of the drugs at the beginning of the incubation. Concentration of drugs: reserpine 10 γ/ml; Ro 4–1284 50 γ/ml; others equimolar to 100 γ/ml DL-amphetamine. The figures represent means ±S.E. of 3 experiments. They indicate the 5-hydroxytryptophol found in the incubation medium after 2 hr in per cent of the 5-hydroxytryptophol formed by platelets without supplementation of drugs. The sympathomimetic amines in the concentrations used do not inhibit the penetration of the exogenous 5-HT into platelets.

These findings indicate that the sympathomimetic amines, in addition to their monoamine liberating effect, also interfere with MAO whereby the metabolism of the liberated 5-HT is inhibited.

III. CONCLUDING REMARKS

The metabolism of platelet 5-HT depends on various factors, such as speed of 5-HT liberation, activity of MAO, relative activity of aldehyde and alcohol dehydrogenase. The findings have possibly some bearing on the *in vivo* metabolism of monoamines. Increased formation of aromatic alcohols due to absence or low activity of aldehyde dehydrogenase might, for instance, also take place in the intact organism. Thus, in arterial thrombosis, formation of platelet clots and liberation of 5-HT probably occurs in the absence of major amounts of erythrocytes. Thereby, 5-hydroxytryptophol might be formed and excreted. Furthermore, animals treated with Disulfiram eliminate considerable amounts of 5-hydroxytryptophol or 3,4-dihydroxyphenylethanol, especially after administration of 5-HT and 5-hydroxytryptophan or 3,4-dihydroxyphenylalanine (dopa) and dopamine respectively (Fig. 6). Patients

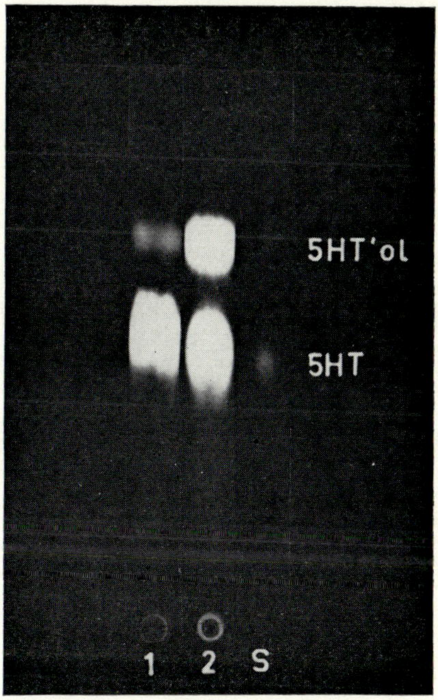

FIG. 6. Urinary elimination of 5-hydroxytryptophol after treatment of rats with Disulfiram + 5-hydroxytryptophan. 250 mg/kg Disulfiram were administered twice i.p. with an interval of 4 hr. After the second dose of Disulfiram, 100 mg/kg 5-hydroxytryptophan were injected i.p. Collection of urine during the following 12 hr. Basic extracts with butanol at pH 10. Paper, solvent system, and development of fluorescence see Fig. 1. 1, urine non-incubated with glusulase; 2, urine incubated with glusulase for 60 min; S, standards, 2 γ each.

treated with Disulfiram have also been found to excrete some 5-hydroxytryptophol in the urine.* It remains to be elucidated whether accumulation of endogenous aromatic alcohols in tissues (e.g. in brain) is of physiological importance. In preliminary experiments, exogenous 5-hydroxytryptophol and 3,4-dihydroxyphenylethanol, in contrast to the corresponding amines,

* Experiments performed together with Dr. R. Tissot, Psychiatric University Clinic Bel-Air, Geneva.

had at best little pharmacological effects on the cardiovascular, the autonomic and the central nervous system of animals.*

The above results might also explain some differences in the metabolic fate of monoamines liberated by reserpine and by sympathomimetic amines *in vivo*. Thus, in the brain of rats reserpine decreases the 5-HT and concomitantly increases the 5-hydroxyindoleacetic acid indicating that at least part of the released 5-HT is metabolized by MAO (Ashcroft and Sharman, 1962; Pletscher *et al.*, 1964a; Roos and Werdinius, 1962). In contrast, Ro 4-6861 causes a drop of both 5-HT and 5-hydroxyindoleacetic acid (Pletscher *et al.*, 1963; Pletscher *et al.*, 1964b). Based on the findings with platelets, this result may be due to a double effect of Ro 4-6861, i.e. monoamine liberation and MAO inhibition. The postulate of two 5-HT pools in order to explain the differences of reserpine and Ro 4-6861 is thus not imperative.

SUMMARY

The spontaneously released 5-hydroxytryptamine (5-HT) of isolated blood platelets is mainly converted to 5-hydroxyindole-3-ethanol (5-hydroxytryptophol) and to a minor extent to 5-hydroxyindole-3-acetic acid. The presence of erythrocytes enhances the formation of 5-hydroxyindole-3-acetic acid and diminishes that of 5-hydroxytryptophol. Isolated platelets also convert dopamine to the corresponding alcohol and acid. Isolated erythrocytes do not transform 5-HT into 5-hydroxytryptophol or 5-hydroxyindoleacetic acid. Indoleacetaldehyde is converted to indoleacetic acid by platelets as well as by erythrocytes and to indole-ethanol by platelets. These experiments indicate that platelets contain monoamine oxidase, an aldehyde oxidase and an alcohol dehydrogenase, whereas in erythrocytes only an aldehyde dehydrogenase seems to be present.

5-HT released from platelets by reserpine and the benzoquinolizine derivative Ro 4-1284 respectively is partly transformed into 5-hydroxytryptophol and 5-hydroxyindoleacetic acid. The amine seems to be metabolized during its relatively slow release. In contrast, platelets incubated with thrombin metabolize the 5-HT subsequently to its immediate liberation. The 5-HT liberated by sympathomimetic amines is not metabolized probably owing to competitive inhibition of the oxidative deamination of 5-HT.

REFERENCES

ASHCROFT, G. W. and D. F. SHARMAN (1962) Drug induced changes in the concentration of 5-OR indolyl compounds in cerebrospinal fluid and caudate nucleus, *Brit. J. Pharmacol.* **19**, 153–60.

BARTHOLINI, G. and A. PLETSCHER (1964) Two types of 5-hydroxytryptamine release from isolated blood platelets, *Experientia* (Basel) **20**, 376–8.

* Experiments carried out by Dr. H. Besendorf, Pharmacology Department, F. Hoffmann-La Roche & Co. Ltd., Basle.

BARTHOLINI, G., A. PLETSCHER and H. BRUDERER (1964) Formation of 5-hydroxytryptophol from endogenous 5-hydroxytryptamine by isolated blood platelets, *Nature* (Lond.) **203**, 1281–3.

BARTHOLINI, G., M. DA PRADA and A. PLETSCHER (1965) Inhibition of 5-hydroxytryptamine liberation from blood platelets by reserpine, *Nature* (Lond.) **205**, 400.

CARLSSON, A., P. A. SHORE and B. B. BRODIE (1957) Release of serotonin from blood platelets by reserpine *in vitro*, *J. Pharmacol.* **120**, 334–9.

KOPIN, I. J. (1964) Storage and metabolism of catecholamines: the role of monoamine oxidase, *Pharmacol. Rev.* **16**, 179–91.

PAASONEN, M. K. and A. PLETSCHER (1959) Increase of free 5-hydroxytryptamine in blood plasma by reserpine and a benzoquinolizine derivative, *Experientia* (Basel) **15**, 477.

PLETSCHER, A. and G. BARTHOLINI (1964) Pharmakologische Beeinflussung der 5-Hydroxytryptamin-Freisetzung aus Blutplättchen, *Helv. Physiol. Pharmacol. Acta* **22**, C84–C86.

PLETSCHER, A., G. BARTHOLINI, H. BRUDERER, W. P. BURKARD and K. F. GEY (1964b) Chlorinated aralkylamines affecting the cerebral metabolism of 5-hydroxytryptamine, *J. Pharmacol.* **145**, 344–50

PLETSCHER, A., A. BROSSI and K. F. GEY (1962) Benzoquinolizine derivatives: A new class of monoamine decreasing drugs with psychotropic action, *Int. Rev. Neurobiol.* **4**, 275–306.

PLETSCHER, A., W. P. BURKARD, H. BRUDERER and K. F. GEY (1963) Decrease of cerebral 5-hydroxytryptamine and 5-hydroxyindoleacetic acid by an aralkylamine, *Life Sci.* **2**, 828–33.

PLETSCHER, A., W. P. BURKARD and K. F. GEY (1964a) Effect of monoamine releasers and decarboxylase inhibitors on endogenous 5-hydroxyindole derivatives in brain, *Biochem. Pharmacol.* **13**, 385–90.

PRADA, M. DA, G. BARTHOLINI and A. PLETSCHER (1965) Formation of 5-hydroxytryptophol by blood platelets after thrombin and reserpine, *Experientia* (Basel) **20**, 498.

ROOS, B.-E. and B. WERDINIUS (1962) Effect of reserpine on the level of 5-hydroxyindoleacetic acid in brain, *Life Sci.* **1**, 105–7.

DISCUSSION

COSTA: 5-HT perfused in the isolated heart of rats given H^3-NE 18 hr previously, releases H^3-NE. Most of the radioisotope in the perfusate has basic properties, only a small fraction of the increased radioactivity in the perfusate has acidic properties. With high doeses of 5-HTP given to rats and rabbits one can observe a 50 per cent depletion of endogenous brain NE. These findings suggest caution in interpreting the pharmacological effects elicited by high doses of this monoamine precursor.

PLETSCHER: The first observation agrees with our experiments regarding 5-HT liberation from isolated platelets by sympathomimetic amines. Concerning the second remark, I also agree.

UVNÄS: (1) How is histamine and 5-HT stored in the thrombocytes? In vesicles? In granules? Or do they swim free within the platelet?

(2) Do the platelet granules contain heparin?

PLETSCHER: (1) As far as I know the 5-HT has been shown to be localized in a particulate fraction. Whether these are vesicles or granules remains to be shown.

(2) I do not know.

DISAPPEARANCE OF MONOAMINES AND OF ENZYMES RESPONSIBLE FOR THEIR FORMATION AFTER DENERVATION

NILS-ERIK ANDÉN

Department of Pharmacology, Göteborg, Sweden

BESIDES the histochemical technique the biochemically detectable disappearance of the monoamines noradrenaline (NA), dopamine (DA) and 5-hydroxytryptamine (5-HT) after surgical lesions has been of great value in localizing the amines at a cellular level and in mapping out the monoamine neurons in the central nervous system. It was first observed by Dr. Rosengren that both NA and 5-HT of the spinal cord had disappeared almost completely caudal to a transection ten days after the operation (Magnusson and Rosengren, 1963; Carlsson, Magnusson and Rosengren, 1963). Later on the time course of this NA and 5-HT disappearance in rabbits was investigated (Andén, Häggendal, Magnusson and Rosengren, 1964). There were no evident changes caudal to the transection during the first two days. Thereafter the two amines disappeared rather slowly and were not detectable in significant amounts seven days or more after the operation. In the part of the spinal cord cranial to the lesion taken as a whole there were no changes observed at any time. The loss of NA seemed to begin somewhat earlier and to be somewhat faster than that of 5-HT. A comparison between the NA disappearance from the spinal cord and from the iris after axotomy of the respective NA neurons was also performed. The iris was depleted between 16 and 24 hr after excision of the superior cervical ganglion, i.e. much earlier and faster than the spinal cord. It seems that the central nerve terminals are degenerating at a slower rate than the peripheral ones after axon damage. The disappearance of NA and 5-HT after the mentioned opercalons has been found to follow the same time course also by the histochemiufl atiorescence method (Carlsson, Falck, Fuxe and Hillarp, 1964; Malmfors and Sachs, 1965).

We have also observed by a combined histochemical and biochemical work that monoamines disappear from the rat brain after small and well localized lesions. After electrolytic lesions in the substantia nigra there was usually a pronounced reduction of the DA content in the corpus striatum, maximally by about 70 per cent. Lesions in the internal capsule produced an almost complete (more than 70 per cent) disappearance of the striatal DA. These effects are in all likelihood due to anterograde degeneration of the nigro-neostriatal DA neurons which have the cell bodies in the substantia nigra, the non-terminal axons in the internal capsule and the abundant axon terminals in the neostriatum (Andén, Carlsson, Dahlström, Fuxe, Hillarp and Larsson, 1964; Andén, Dahlström, Fuxe and Larsson, 1965a, and unpublished observations).

We have also placed lesions in the medial forebrain bundle (Andén, Dahlström, Fuxe and Larsson, 1965b). This structure has been found to contain many non-terminal monoamine fibres (Dahlström and Fuxe, 1964). Unilateral lesions of the medial forebrain bundle at the level of the anterior hypothalamus caused a reduction by about 50 per cent of the ipsilateral forebrain NA and 5-HT whereas after lesions at the junction between the mesencephalon and the diencephalon about 80 per cent of the two amines in the ipsilateral forebrain were lost. There were no significant changes on the side contralateral to the operation. These results and histochemical changes in the cell bodies and in the terminals as well (Andén, Dahlström, Fuxe and Larsson, 1965b) indicate that most if not all of the forebrain NA and 5-HT belongs to neurons whose cell bodies are localized in the lower brain stem and whose axons to a large extent ascend in the medial forebrain bundle.

We have also investigated the effect of lesions and subsequent degeneration of monoamine neurons on the activity of some enzymes participating in the monoamine metabolism. In this respect the L-DOPA decarboxylase has been most extensively studied (Andén, Magnusson and Rosengren, 1965). It was found that caudal to a spinal cord transection there was an almost complete disappearance of the decarboxylase activity both when determined *in vitro* and *in vivo*. Furthermore the enzyme activity was reduced at a similar rate as the amines. Also the iris was almost completely depleted of the decarboxylase activity after denervation, i.e. excision of the superior cervical ganglion. The disappearance rate is about the same as for NA. There is, thus, a faster loss of both NA and decarboxylase activity in the iris than in the spinal cord after denervation.

Some experiments have also been performed concerning the effect of sympathetic denervation on the activity of the enzyme DA-β-hydroxylase which normally converts DA to NA (Almgren, Andén and Waldeck, 1965). For technical reasons we have preferred to study the *in vivo* formation of octopamine from tyramine in rat salivary glands. There was practically no octopamine formation after an intravenous injection of ^{14}C-tyramine in a postganglionically sympathectomized gland whereas the formation was rather high in the innervated gland. This finding indicates that also the β-hydroxylase occurs in the NA neuron.

In conclusion: the disappearance of amines and of enzymes participating in their metabolism after denervation is a convenient biochemical method to determine the cellular localization of these compounds.

REFERENCES

Almgren, O., N.-E. Andén and B. Waldeck (1965) *Life Sci.* **4**, 121.
Andén, N.-E., A. Carlsson, A. Dahlström, K. Fuxe, N.-Å. Hillarp and K. Larsson (1964) *Life Sci.* **3**, 523.
Andén, N.-E., A. Dahlström, K. Fuxe and K. Larsson (1965a) *Am. J. Anat.* **116**, 329.
Andén, N.-E., A. Dahlström, K. Fuxe and K. Larsson (1965b) *Life Sci.* **4**, 1275.
Andén, N.-E., J. Häggendal, T. Magnusson and E. Rosengren (1964) *Acta Physiol. Scand.* **62**, 115.
Andén, N.-E., T. Magnusson and E. Rosengren (1965) *Acta Physiol. Scand.* **64**, 127.
Carlsson, A., B. Falck, K. Fuxe and N.-Å. Hillarp (1964) *Acta Physiol. Scand.* **60**, 112.
Carlsson, A., T. Magnusson and E. Rosengren (1963) *Experientia* (Basel) **19**, 359.
Dahlström, A. and K. Fuxe (1964) *Acta Physiol. Scand.* **62**, Suppl. 232.
Magnusson, T. and E. Rosengren (1963) *Experientia* (Basel) **19**, 229.
Malmfors, T. and C. Sachs (1965) *Acta Physiol. Scand.* **64**, 211.

SOME OBSERVATIONS ON THE DENERVATED NICTITATING MEMBRANE OF THE CAT DURING DEGENERATION OF THE ADRENERGIC INNERVATION

U. TRENDELENBURG, S. Z. LANGER and C. B. SMITH
Department of Pharmacology, Harvard Medical School, Boston, Massachusetts, U.S.A.

IMMEDIATELY after removal of the superior cervical ganglion (denervation) or section of the preganglionic fibers (decentralization) the nictitating membrane of the cat relaxes over the eye. Eighteen to 36 hr after denervation the nictitating membrane of the unanesthetized cat contracts for a period of 8 to 12 hr, while the decentralized nictitating membrane remains relaxed. This "degeneration contraction" appears to be due to the leakage of norepinephrine from degenerating nerve terminals (Trendelenburg and Langer, 1965).

The sensitivity to norepinephrine of the denervated nictitating membrane of spinal cats was determined at various stages of the "degeneration contraction". The earliest development of supersensitivity was found just before the "degeneration contraction" began; full supersensitivity was reached during relaxation from this contraction. At that time, the denervated nictitating membrane was 10 times more sensitive than the corresponding decentralized nictitating membrane, and for up to 14 days there was no further change in the difference in sensitivity to norepinephrine between the denervated and the decentralized nictitating membrane (Trendelenburg, 1963).

Full dose–response curves were obtained with intravenous injections of tyramine. The maximal response of the denervated nictitating membrane to tyramine remained normal during the beginning of the "degeneration contraction" and began to decline when the maximum of this contraction was reached. Immediately after the "degeneration contraction" had subsided the maximal response of the denervated nictitating membrane to tyramine was *ca.* 60 per cent of normal.

Fluorimetric determination of the norepinephrine content of the nictitating membrane showed that the norepinephrine stores declined during the "degeneration contraction". They were *ca.* 50 per cent of normal at the maximum of the contraction and *ca.* 10 per cent of normal immediately after the contraction.

The amount of H^3-norepinephrine was determined in the nictitating membrane of the spinal cat 60 min after the intravenous injection of 40 mc/kg of H^3-*dl*-norepinephrine (specific activity: 4.72 c/mM). A normal amount of H^3-norepinephrine was found until the "degeneration contraction" was subsiding. Immediately after this contraction only 30 per cent of the normal amount of H^3-norepinephrine was found.

These experiments show that the development of supersensitivity to norepinephrine coincides with the "degeneration contraction". Although this temporal relationship suggests that the development of supersensitivity to norepinephrine of the denervated nictitating membrane is causally related to the loss of some function of the degenerating nerve terminals, the results obtained with H^3-norepinephrine do not support the view that supersensitivity to norepinephrine is due solely to a breakdown of the mechanism responsible for the movement of norepinephrine into the nerve terminal.

REFERENCES

TRENDELENBURG, U. (1963) Time course of changes in sensitivity after denervation of the nictitating membrane of the cat, *J. Pharmacol.* **142**, 335–42.

TRENDELENBURG, U. and S. Z. LANGER (1965) The onset of denervation supersensitivity, *J. Physiol.* **178**, 54P.

GENERAL DISCUSSION

ERÄNKÖ: By electron microscopic examination of the superior cervical ganglion we failed to find many vesicles or dense-cored granules in the cell bodies, while there is much histochemically demonstrable noradrenaline. The same applies to adrenergic fibres of the iris, which show a formaldehyde-induced fluorescence throughout the fibre, while synaptic vesicles are only seen in the varicosities. Therefore, much of cytoplasmic or axonal noradrenaline does not appear to be bound into structures demonstrable by the present techniques of electron microscopy. This suggests that this NA is in a diffuse form in the axon.

MASUOKA: Normal rat brain stem has been subjected to differential centrifugation to provide evidence that the intensely fluorescent varicosities observed by the fluorescence method are correlated with the nerve ending particle or synaptosomes. Using the method of Gray and Whittaker, the P_2 subfractions A, B, C were isolated and examined by the fluorescence method. The B (synaptosome-rich) fraction was covered with many intensely fluorescent spots of the same size range as varicosities and the B and C fractions had relatively few. When brain stems from reserpine-pretreated rats were used, essentially all the fluorescent spots were absent.

COSTA: Experiments in progress in our laboratory (Dlabac, Costa and Brodie) using α-methyl-*p*-tyrosine, the inhibitor of NE biosynthesis (Spector *et al.*, 1965), have shown that rats given 200 mg/kg i.v. and then placed in a cold room several hours later, cannot maintain their body temperature. After 2 hr in the cold the body temperature declined and the rats did not shiver. At this time the animals were killed. Blood FFA, blood glucose, brain NE, brain dopamine and heart NE content were measured. The FFA and glucose levels were increased, the brain NE and dopamine content was depleted completely while heart NE content was decreased by 60 per cent.

These results suggest that brain NE or dopamine are essential to elicit shivering which in turn seems essential to utilize the energy substrates mobilized by the peripheral NE still present.

HANSON: In Gothenburg we have been using the specific tyrosine–hydroxylase inhibitor α-methyl-tyrosine (α-MT), mentioned by Dr. Udenfriend, in studying the correlation between biochemistry and behaviour measured as conditioned avoidance response (CAR). α-MT causes a selective depletion of brain catecholamines (CA), while leaving 5-hydroxytryptophane (5-HTP) essentially unaffected. The CAR experiments were performed on cats. Preliminary experiments had shown the effect on behaviour to have a slow onset. The full effect of α-MT was developed in about 15 hr as a rule.

The animal has been trained to a 100 per cent avoidance and has then been treated with 200 mg/kg α-MT i.p. and tested again 13 hr later. At that time we see a certain decrease of CAR, which is magnified at 14 hr. This mixture of CAR, escape and no response during one single session, which is the general course of the behavioural impairment after α-MT, was often seen and is interpreted as a "fatigue"-phenomenon. After treatment with L-DOPA 100 mg/kg i.p. there was a fast reversal of the α-MT effect on CAR. The DOPA-effect lasted in this case for about 8 hr. After that time the cat's performance successively decreased. The administration of L-DOPA to α-MT treated animals caused a rapid rise in dopamine and a slower and less pronounced but more sustained rise in noradrenaline.

Both the biochemical and behavioural effects of α-MT are slow in onset and may in fact coincide. However, most of the biochemical data given in the table are from animals not stressed by the testing procedure on CAR. Further investigations are needed to elucidate the possible role of stress in causing the biochemical and behavioural effects of α-MT. That such a connection exists is mentioned by Dr. Udenfriend, and the "fatigue"-phenomenon referred to earlier suggests it.

It would appear from the present data that the selective CA depletion in brain induced by α-MT leads to disruption of CAR and that restoration of CA levels by L-DOPA results in restoration of CAR. The data thus support the view that CA neurons in brain play a role in CAR.

SEDVALL: Together with Dr. Kernell, The Nobel Institute for Neurophysiology, Karolinska Institutet, Stockholm we have recently attempted to deplete adrenergic transmitter stores by long-term nerve stimulation. One of the lumbosacral sympathetic chains in cats was stimulated electrically and after different intervals the noradrenaline content in the gastrocnemius muscle of each hind limb was analysed. In non-stimulated control cats the noradrenaline content was roughly the same in the homologous muscles. Following the delivery of 50,000 impulses at a frequency of 5 impulses/sec there was a 50 per cent depletion in the muscles on the stimulated side—5 imp/sec is considered to be within the physiological range for the nerves. Stimulation with 12 imp/sec produced a 70 per cent depletion after 50,000 impulses. Similar results were obtained when the cervical sympathetic trunk in rats was stimulated with a frequency of 20/sec. Already after 10 min stimulation there was a 20 per cent depletion of the noradrenaline in the salivary gland, after 60 min stimulation the noradrenaline concentration reached a plateau level at 20–30 per cent of the normal content. Adrenergic function did not appear to be significantly impaired (pupillary response) by this depletion. Preliminary data indicate that after interruption of the stimulation the recovery curve for the noradrenaline content in the salivary gland has an initial rapid phase followed by a slow component which brings the content back to prestimulation levels within 5 hr. The results might suggest the occurrence of different mechanisms regulating the rate of net synthesis of the adrenergic transmitter in the absence of impulse activity.

MALMFORS: Exogenous NA is taken up by a mechanism in all probability located to the cell membrane over the whole neurone. The uptake is not affected by reserpine but is inhibited by cocaine and desmethylimipramine. In reserpinized animals the NA taken up is rapidly broken down by MAO intraneuronally, but after MAO inhibition it is possible to study uptake of exogenous NA in the adrenergic nerves of the dilator of the iris after injecting 10 μg/kg i.v. or 10^{-8} M in the anterior chamber of the eye. A further dose of reserpine or electric stimulation do not seem to affect the NA taken up in reserpinized animals which seems to be located extragranularly as judged, e.g., from the different intra neuronal distribution as compared with the distribution of the amine granules. Electric stimulation in an animal treated with the synthesis inhibitor H22/54 produces a remarkable decrease of the fluorescence intensity especially in the varicosities along the whole terminal.

WERDINIUS: Recent biochemical and histochemical investigations by the groups of Carlsson and Hillarp have shown that the 5-HT in the spinal cord is localized in descending pathways, the cell bodies of which are localized in the lower brain stem. After transection at the second thoracic segment the 5-HT disappears from the caudal part almost completely within a week.

We have now made a study on the time course of disappearance of 5-HIAA from the transected spinal cord. The lesions were performed in rabbits at the second to fourth thoracic segment. After various time intervals the contents of 5-HT and 5-HIAA were measured in the cranial and the caudal parts. The transection did not at any time alter the concentrations of the substances cranially to the lesion. The levels were for 5-HT about 0·33 μg/g, for 5-HIAA 0·20 μg/g.

Caudally to the site of transection, the concentrations of both substances were about 70–75 per cent of those cranially in the control cases. For the first two days after transection 5-HT was approximately unchanged, then it dropped substantially during the following days and had almost completely disappeared after 10 days.

5-HIAA, on the other hand, decreased with more than half the initial value in the first two days. Then there was a more gradual reduction, ending with almost complete depletion after 10 days. These results probably indicate that the metabolite level is a better indicator of the metabolic activity and turnover rate than the amine level after spinal cord transection.

RELEASE AND TURNOVER OF THE ADRENERGIC TRANSMITTER

A. G. H. BLAKELEY and G. L. BROWN
University Laboratory of Physiology, Oxford, England

WHEN the splenic nerves are excited at "physiological" frequencies, i.e. at 10/sec or less, very little of the liberated transmitter overflows into the blood stream. The amount escaping can be increased either by exciting the nerves at higher frequencies, i.e. up to 30/sec, or by giving a blocking agent like phenoxybenzamine (Brown and Gillespie, 1957; Brown, Davies and Ferry, 1961). We have recently made extensive use of the isolated spleen perfused with blood (Blakeley and Brown, 1963) and have been able, as a result, to continue giving trains of stimuli and collecting venous effluent for long periods, since the practical limit to the length of the experiment is the amount of blood available. The collection of blood during and just after stimulation involves a total loss to the system of any transmitter that overflows into this blood. When the transmitter loss is small, there is no decline in the overflow/stimulus, and at frequencies of stimulation of 10/sec the overflow/stimulus may even rise during successive trains of stimuli. When transmitter loss is high, there is an approximately exponential decline of overflow, the steepness of the decline being a function of the overflow/stimulus (Blakeley, Brown and Geffen, 1964). The response of the spleen, measured in terms of the area of pressor response follows closely the changes in overflow. It appears, therefore, that we were dealing with changes in transmitter liberation which are reflected in overflow. This suggested to us that the maintenance of transmitter liberation might depend upon the uptake and re-use by the nerves of the transmitter they liberate (Blakeley, Brown and Geffen, 1964).

Such an argument, however, takes no account of the possibility of synthesis of transmitter, and it appeared possible that an indication of the occurrence or otherwise of synthesis could be obtained by plotting the slopes of the decline of overflow against the overflow in the first sample. There proved, in the event, to be certain technical limitations to making a precise estimate of synthesis by this means, but Fig. 1 shows sufficiently clearly that synthesis is occurring. Assuming a linear relation between slope and initial overflow, an extrapolation of the line suggests that synthesis can maintain an overflow of transmitter equivalent to some 250 pg/stimulus; this is 68 per cent of the initial overflow at a frequency of stimulation of 30/sec in the

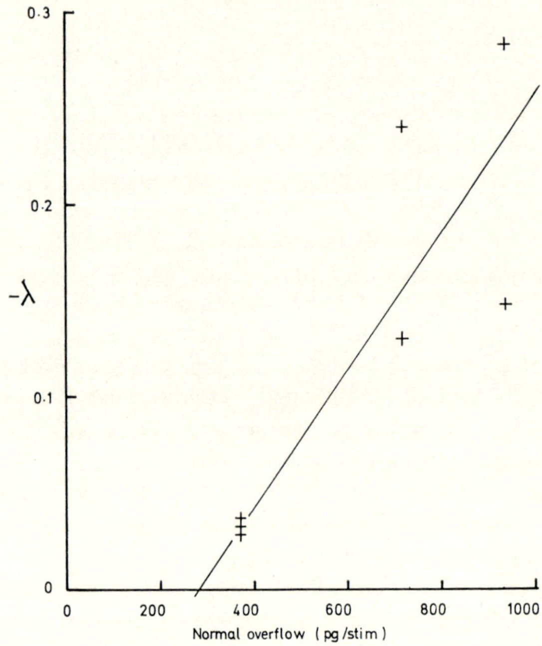

FIG. 1. Slope of decline of transmitter overflow in seven experiments, plotted against normal overflow of transmitter under the conditions of the experiments.

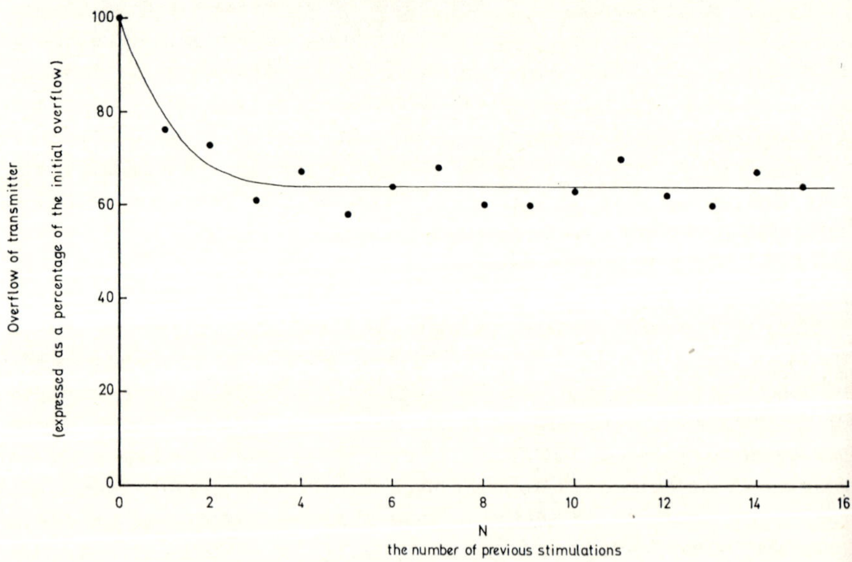

FIG. 2. Overflow of transmitter from blood-perfused spleen in response to successive trains of 200 stimuli to nerve at 30/sec.

perfused spleen. Under the conditions of these experiments, in which trains of 200 stimuli were given at 10 min intervals, this would mean a rate of synthesis of 0·3 μg/hr.

If synthesis occurs, continued stimulation and collection of the overflowing transmitter might be expected to produce a decline of overflow, followed eventually by a steady overflow at a rate corresponding to the level of synthesis. Recent experiments have shown that this occurs. In Fig. 2, stimulation at 30/sec with trains of 200 stimuli was continued for 2 hr 40 min, and after the initial decline, a steady level of overflow was maintained at 64 per cent of the initial overflow.

It appears therefore that, at physiological frequencies of excitation (less than 10/sec), the nerve endings of the spleen maintain their output of transmitter by uptake and re-use of what is liberated. If loss of transmitter is brought about by any means, transmitter release is maintained at a lower level by synthesis.

REFERENCES

BLAKELEY, A. G. H. and G. L. BROWN (1963) A method for perfusion of the isolated spleen, *J. Physiol.* (Lond.) **169**, 66–8P.

BLAKELEY, A. G. H., G. L. BROWN and L. B. GEFFEN (1964) Uptake and re-use by sympathetic nerves of the transmitter they liberate, *J. Physiol.* (Lond.) **173**, 22–3P.

BROWN, G. L., B. N. DAVIES and C. B. FERRY (1961) The effect of neuronal rest on the output of sympathetic transmitter from the spleen, *J. Physiol.* (Lond.) **159**, 365–80.

BROWN, G. L. and J. S. GILLESPIE (1957) The output of sympathetic transmitter from the spleen of the cat, *J. Physiol.* (Lond.) **138**, 81–102.

DISCUSSION

ROSELL: I would like to comment upon your finding that there is a greater overflow of the transmitter at 30/sec than at 10/sec. In experiments on the gracilis muscle of dogs Renkin and I showed that activation of sympathetic vasoconstrictor nerves reduced the transcapillary transport of Rb^{86} (Renkin and Rosell, *Acta Physiol. Scand.* **54**, 223, 1962). This effect may be due to a reduced number of open capillaries and as a consequence a reduced capillary surface area over which diffusion occurs. Accordingly, in the same preparation Axelrod, Kopin and I (*Am. J. Physiol.* **205**, 317–21, 1963) demonstrated that during sympathetic nerve stimulation the released transmitter substance is more or less trapped in the muscle depending on the degree of the vasoconstrictor response. Could your finding in part be due to a smaller vasoconstrictor response when you stimulate with a frequency of 30/sec during about 6 sec than with 10/sec during 20 sec.

BROWN: There is no doubt that the overflow of transmitter is greatly affected by bloodflow through the tissue, but the differences between overflows of 10/sec and 30/sec cannot be due entirely to this. In our experiments the perfusion was done with a constant output pump, so the rise in pressure and its duration are a good index of the degree of vasoconstriction. The response is, in fact, always *greater* at 30/sec than at 10/sec. It is interesting that as transmitter liberation declines, the duration of the response declines more than the peak pressure attained.

FOLKOW: In this connection I will briefly mention some observations that, on the whole, agree with Sir Lindor's findings and interpretation. While studying the physiological discharge range of the sympathetic nervous system some 15 years ago I was struck by the fact that upon interruption of stimulations at physiological frequencies (up to some 6–8 imp/sec) the blood vessels relaxed immediately, while after stimulations at higher, "supraphysiological" rates vascular relaxation was markedly retarded. The characteristics of these events strongly suggested that normally the transmitter is rapidly eliminated by some specific mechanism, which is not able to cope entirely with the excess release occurring at supraphysiological discharge rates, then leading to diffusion and a significant "overflow". This was subsequently supported by observations on the remote effects on denervated, supersensitive smooth muscles. However, the exact nature of the assumed specific elimination mechanism was unknown to us by then; an enzymatic breakdown appearing perhaps most likely in analogy with the situation for the cholinergic nerve fibres.

With the specific blocking agents now available, studies along these lines have again been taken up in collaboration with Dr. A. Carlsson and Dr. J. Häggendal. It was observed that the normally prompt vascular relaxation upon cessation of low-frequency sympathetic stimulation was markedly prolonged after administration of *desmethylimipramin* (DMI), 2–3 mg/kg, mimicking the situation when the vasoconstrictor fibres have been driven at definitely supraphysiological rates. No such effects were seen when COMT and MAO inhibitors were given, even in huge amounts. These findings suggest that normally the vasoconstrictor fibre transmitter is eliminated predominantly via re-uptake in the fibre endings and that this is, in fact, a necessary prerequisite in case a proper nervous vascular control is to be exercised with prompt onset and interruption of the effector adjustments.

Further, Häggendal's technique was used for measuring the extent of transmitter "overflow" to the blood stream when the muscle vasoconstrictor fibres were stimulated at physiological rates (usually 6/sec). No really significant overflow took place as long as the vasoconstrictor fibres in the normal way restricted the muscle blood supply. However, exactly the same "physiological" sympathetic stimulation produced a considerable overflow if simultaneously muscle work and exercise hyperaemia was induced. Presumably this overflow was a reflection of the by now greatly improved chances for the released transmitter to escape the otherwise evidently efficient re-uptake mechanism and to be eliminated partly via the enhanced capillary blood flow. In any case, these findings make it doubtful whether, on the whole, the content of noradrenaline and its breakdown products in blood and urine can be trusted as reliably reflecting the extent of sympathetic discharge, as evidently the relationship between transmitter re-uptake and transmitter overflow can be so markedly affected by the current level of regional blood perfusion. It also may explain why in exercise there is often a rapid rise in noradrenaline in the blood, while massive sympathetic discharge in other situations often hardly affects the blood concentration of noradrenaline.

CONTROL OF CATECHOLAMINE AND INDOLEAMINE METABOLISM BY SYMPATHETIC NERVES

JULIUS AXELROD

Laboratory of Clinical Science, National Institute of Mental Health,
Bethesda, Maryland, U.S.A.

THE concept that the sympathetic nerves secrete a physiologically active compound was first proposed by Elliott in 1905. On the basis of the sympathomimetic actions of adrenaline, he suggested that sympathetic nerves release a pressor substance which could then act on an effector organ. Many years later, Loewi (1921) provided experimental evidence that stimulation of sympathetic nerves results in the release of an adrenaline-like compound. The identity of this transmitter substance was established by Euler (1948) who provided conclusive evidence that it was noradrenaline. The development of specific and sensitive chemical methods for the measurement of catecholamines, the use of adrenergic drugs, the introduction of radioactive catecholamines of high specific activity and the application of biochemical techniques have added to our knowledge concerning the sympathetic nervous system. This report will describe some of our work on the role of the sympathetic nervous system in the uptake, storage, release and metabolism of catecholamines and indoleamines.

The uptake, storage, release and metabolism of catecholamines in sympathetic nerves. The possibility that the sympathetic nerves may be involved in the selective uptake and retention of catecholamines stemmed from tissue distribution studies using the radioactive amines (Axelrod, Weil-Malherbe and Tomchick, 1959; Whitby, Axelrod and Weil-Malherbe, 1961). After the administration of relatively small amounts of adrenaline or noradrenaline, the unchanged hormones were found to disappear from the whole mouse in a multiphasic fashion. In the first 5 min there was a rapid metabolism of about half the injected dose, mainly by O-methylation. The remaining 50 per cent disappeared slowly, indicating that it was taken up and stored in tissues in such a manner that it could not be destroyed by enzymes. An examination of the distribution of H^3-norepinephrine in different organs showed that it was selectively taken up by the heart, spleen and salivary gland organs that have a rich sympathetic innervation. Because the catecholamines were taken up and retained in a physiologically inactive form it was suggested that binding was an important mechanism for the inactivation of these hormones (Axelrod, 1960; Whitby, Axelrod and Weil-Malherbe, 1961). Following these findings there have been many other reports showing that certain peripheral tissues can accumulate and retain catecholamines (Strömblad and Nickerson, 1961; Harvey and Pennefather, 1962).

The uptake and retention of the administered catecholamines by tissues having an abundant adrenergic supply suggested that the sympathetic nerves might be the site of uptake. When sympathetically innervated structures are cut, the catecholamines are depleted (Euler and Purkhold, 1951). Presumably, this is due to the subsequent degeneration of the nerves. Denervation experiments provided an opportunity to examine the importance of sympathetic fibers in the uptake and storage of noradrenaline (Hertting, Axelrod, Kopin and Whitby, 1961). The superior cervical ganglia from cats were removed unilaterally and about 5 days were allowed to elapse to insure degeneration of the nerve fibers. H^3-noradrenaline and H^3-adrenaline were administered intravenously, the cats were killed 1 hr later and the H^3-catecholamine content was examined in the innervated and denervated salivary gland (Fig. 1). There was a profound decrease in the ability of the denervated salivary gland to take up and retain the circulating noradrenaline and adrenaline. These results indicated that the sympathetic nerves are the site of the uptake and retention of circulating catecholamines. The supersensitivity to catecholamines that results from denervation may be due, in part, to the destruction of an important inactivating mechanism for these hormones; binding by sympathetic nerves. In the denervated tissue the concentration of unbound physiologically active catecholamines could be expected to persist for longer periods of time to interact with the effector cells.

The ability to take up and store catecholamines by sympathetic nerves is not confined solely to noradrenaline and adrenaline. The injection of H^3-tyramine resulted in the accumulation of a radioactive product in the salivary glands which was identified as the β-hydroxylated metabolite octopamine (Fischer, Musacchio, Kopin and Axelrod, 1964). Sympathetic denervation of the salivary gland caused a striking reduction in the storage of octopamine (Fig. 1). Using a similar approach several other amines such as dopamine, α-methyl octopamine, were also found to be taken up by sympathetic nerves (Musacchio, Fischer and Kopin, 1965).

In a detailed study on the uptake of catecholamines, Iversen (1963) has demonstrated that this mechanism is saturable and is probably located in the axonal membrane in the terminal regions of the postganglionic sympathetic nerves. This uptake mechanism is selective for the physiologically active L-isomers of noradrenaline and adrenaline. Noradrenaline appears to be transported across the cell membrane more rapidly than adrenaline. Once the catecholamines enter the sympathetic nerves they undergo a further intracellular redistribution and enter storage sites in which they are retained in such a way as to prevent rapid leakage out of the axon or metabolism by degradative enzymes.

Studies on the storage of catecholamines in the sympathetic nerves have been also facilitated by the use of the electron microscope and the centrifuge. Several investigators have shown that noradrenaline is stored in particulate material after homogenizing and centrifuging the splenic nerves (Euler and Hillarp, 1956), heart (Wegmann and Kako, 1961), and brain (Weil-Malherbe and Bone, 1957). Work in our laboratory demonstrating that radioactive noradrenaline can selectively accumulate in the sympathetic nerves (Hertting, Axelrod, Kopin and Whitby, 1961) gave us an opportunity to examine the precise intracellular localization of noradrenaline by electron microscopic radioautography. One hour after the intravenous injection of H^3-noradrenaline the pineal gland, an organ rich in sympathetic nerves (Kappers, 1960), was examined by electron microscopy and radioautography (Wolfe, Potter,

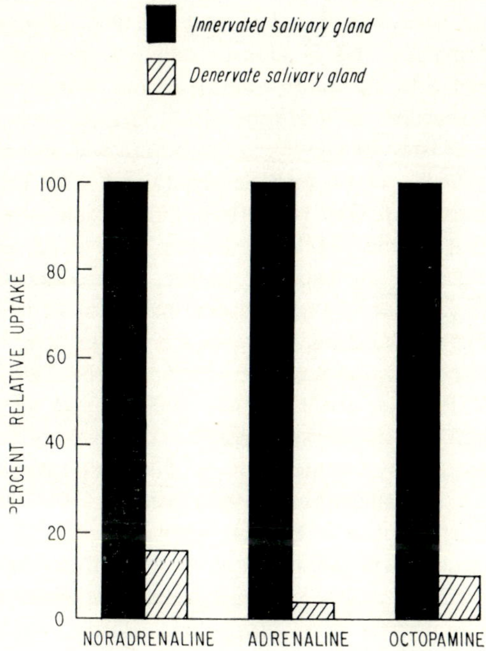

FIG. 1. Uptake of noradrenaline and other sympathomimetic amines in sympathetically innervated and denervated salivary glands. The right superior cervical ganglia were removed from cats or rats. After 5 to 50 days cats were given H^3-noradrenaline or H^3-adrenaline intravenously and 1 hr later the H^3-catecholamine content was assayed in the innervated and denervated salivary glands. H^3-tyramine was administered intravenously to unilaterally denervated rats and radioactivity measured in both salivary glands 0.5 hr later. All the radioactivity was found to be present in the salivary gland as octopamine.

Richardson and Axelrod, 1962). A marked localization of photographic grains was found only in the non-myelinated axons which contained dense core granules 40 to 50 mμ wide. These experiments enabled us to label the storage particles and study their subcellular distribution (Potter and Axelrod, 1963a, b). H^3-norepinephrine was administered intravenously to rats and one hour later the hearts were removed, homogenized and layered on a continuous sucrose gradient. The subcellular fractions were separated by a single centrifugation and assayed for radioactive and endogenous noradrenaline. Most of the endogenous and H^3-noradrenaline were found in the lightest particles associated with the microsomal fraction. After centrifugation in a modified sucrose gradient, almost pure noradrenaline storage particles were obtained which consisted of small vesicles 50 mμ wide (Snyder, Michaelson and Musacchio, 1964). These particles contained ATP in a similar molar ratio to noradrenaline as has been found in adrenal medullary granules (Hillarp, 1958; Potter and Axelrod, 1963b). The norepinephrine storage particles isolated from sympathetic nerves also contained the enzyme, dopamine β-oxidase, that converts dopamine to noradrenaline. H^3-noradrenaline was found to be taken up by the storage particles by a temperature dependent process (Potter and Axelrod, 1963b; Stjärne, 1964) and this uptake is enhanced by ATP (Euler and Lishajko, 1963). These storage particles are presumably encased by a membrane since they have negligible pressor activity unless they are lysed in hypotonic solution (Potter and Axelrod, 1963b).

Pharmacological (Trendelenburg, 1961; Stjärne, 1961) and biochemical studies (Axelrod, Hertting and Patrick, 1961; Potter and Axelrod, 1963c) indicate that the norepinephrine is stored in sympathetic nerves in a heterogeneous fashion. Part of the store is easily releasable with certain drugs such as tyramine and another is more firmly held (Potter, Axelrod and Kopin, 1962). Only a small fraction of the stored noradrenaline appeared to be required for the functioning of the heart (Crout, Muskus and Trendelenburg, 1962). It has not been established whether these pools of norepinephrine are stored in a separate compartment within the nerves or whether they correspond to storage granules situated near the surface of nerve fibers and thus readily released by nerve impulses or accessible to drugs such as tyramine (Potter and Axelrod, 1963a).

The capacity of the newborn rat heart to take up and retain H^3-noradrenaline shows dynamic changes in the developing animal. For several days after birth the ability of the newborn rat heart to take up and retain H^3-norepinephrine is about one-fifth that of the adult heart (Glowinski, Axelrod, Kopin and Wurtman, 1964). Between six and fifteen days of age there is a rapid increase in the uptake and binding of the catecholamine followed by a slower increase until adult levels are reached at about 5 weeks of age. These results suggest that the sympathetic innervation of the newborn heart has not reached adult levels or that the uptake mechanism in sympathetic nerves is not fully developed. The administration of an antiserum to the nerve growth factor to newborn rats results in the atrophy of sympathetic ganglia (Levi-Montalcini and Booker, 1960). The administration of the antiserum to the nerve growth factor presumably blocks the development of the sympathetic nerves in the newborn where they have not yet formed. Adult rats which had been treated with the nerve growth factor antiserum at birth were totally unable to take up circulating H^3-norepinephrine into the heart and certain other tissues (Iversen, Glowinski and Axelrod, unpublished observations).

The uptake of H^3-norepinephrine by the sympathetic nerves made it possible to introduce the neurohumor into the nerve. Because of the high specific activity of the H^3-norepinephrine it was possible to study the fate of this hormone when it is discharged from the nerve either spontaneously or by electrical stimulation of the nerve (Hertting and Axelrod, 1961). H^3-noradrenaline was administered to cats and several hours later the spleen which had taken up considerable quantities of radioactive neurohumor was perfused with blood from a donor cat. The splenic nerve was stimulated and H^3-noradrenaline and its metabolites were assayed in the venous outflow before and after stimulation. There was a marked increase in the concentration of H^3-noradrenaline in the efferent outflow after each stimulation (Fig. 2). There was also a small amount of H^3-noradrenaline released spontaneously before stimulation. After each stimulation there was an increase in the O-methylated metabolite, normetanephrine, in the venous outflow. There was a direct relationship between the amount of H^3-noradrenaline discharged and the increase in H^3-normetanephrine. The amount of radioactive deaminated metabolites appearing in the splenic outflow was small and was not elevated during stimulation. This experiment provided evidence that the neurohumor discharged from sympathetic nerves is inactivated by several processes: diffusion into the blood stream, O-methylation locally by catechol-O-methyl transferase, and binding by re-uptake into the sympathetic nerves. Rebinding appears to be the most important mechanism for the rapid inactivation of the locally released hormone.

The importance of re-uptake of noradrenaline by sympathetic nerves as a mechanism of inactivation has been obtained in a similar experiment to that described above (Rosell, Kopin and Axelrod, 1963). The dog skeletal muscle was perfused with blood containing H^3-noradrenaline at constant flow rate. The venous blood was then examined for noradrenaline and its metabolites under resting conditions and during stimulation of the sympathetic vasomotor nerves. Under resting conditions H^3-noradrenaline represented about 25 per cent of the radioactivity and normetanephrine, 3-methoxy-4-hydroxymandelic acid and 3-methoxy-4-hydroxyphenylglycol accounted for 25 per cent each. Less than 2 per cent of deaminated catechols were found. After stimulation of the sympathetic nerves, there was a delay in the outflow of H^3-noradrenaline and a fall in the amount of metabolites. Just before stimulation stopped, there was a sharp increase in the outflow of H^3-noradrenaline followed by a slow fall so that the pre-stimulatory levels of perfusion pressure and of H^3-noradrenaline were reached at about the same time. To determine the role of vasoconstriction that might be caused

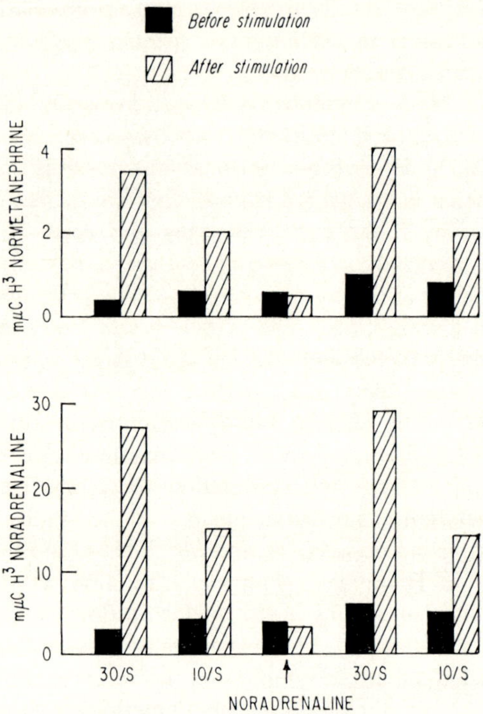

FIG. 2. Release and metabolism of H^3-noradrenaline after stimulation of the cat splenic nerve. Cats were given H^3-noradrenaline intravenously and the spleen was isolated and perfused with blood from a donor cat. Splenic nerves were stimulated at a frequency of 30 or 10 imp/sec. Blood was collected for 2 min before and 2 min during and after stimulation, and was assayed for H^3-noradrenaline and its metabolites. Non-radioactive noradrenaline was injected into the splenic artery.

by the discharged noradrenaline, the dogs were treated with phenoxybenzamine to block this constriction. After treatment with this drug, stimulation of the sympathetic nerves caused an immediate and greater rise in H^3-noradrenaline. Phenoxybenzamine might also act by inhibiting the uptake of the noradrenaline which is released (Hertting, Axelrod and Whitby, 1961). From these results it was concluded that noradrenaline released from sympathetic nerves causes vasoconstriction which then traps the hormone in the vascular bed. This permits the released noradrenaline to return to the sympathetic nerve endings in the same manner as exogenously administered noradrenaline. In similar experiments it was shown that sympathetic nerves can take up and discharge compounds such as adrenaline (Rosell, Axelrod and Kopin, 1964) and octopamine (Fischer, Horst and Kopin, 1965).

After the administration of H^3-noradrenaline, the catecholamine disappears from sympathetic nerves in the heart over many days (Axelrod, Hertting and Patrick, 1961), indicating a spontaneous release and metabolism of the stored neurohumor. When impulses to the sympathetic nerves were interrupted by decentralization or treating animals with ganglionic blocking agents (Hertting, Potter and Axelrod, 1962) the release was slowed in the denervated side (Table 1). The spontaneous release of noradrenaline appeared to be partially dependent on nerve impulses.

TABLE 1. EFFECT OF DECENTRALIZATION OF THE SPONTANEOUS RELEASE OF H^3-NOREPINEPHRINE IN TISSUE

Tissue	*Innervated*	*Decentralized*
Submaxillary	62	98
Parotid	25	32
Striated eye muscle	13	23
Retractor bulbae muscles	5.6	9.2
Iris	88	123

Cats were given H^3-noradrenaline intravenously. Thirty minutes later the superior cervical ganglia was decentralized on one side by cutting the preganglionic fibers. Twenty-four hours later the cats were killed and the innervated and decentralized tissues were analysed for H^3-noradrenaline. Results are expressed as $m\mu c$ H^3-noradrenaline/g tissue.

Noradrenaline in the sympathetic nerves can undergo different types of metabolic transformation depending on how it is discharged (Kopin and Gordon, 1962, 1963). During the initial rapid phase of noradrenaline release it is mainly metabolized by catechol-O-methyl transferase and during the slower phase of release it is metabolized mainly by monoamine oxidase. Further evidence for this type of disposition of noradrenaline in the sympathetic nerves has been obtained by the use of the drugs tyramine and reserpine. When tyramine is administered it causes a release of small amounts of noradrenaline and a marked rise in blood pressure, while reserpine produces a considerable depletion of tissue norepinephrine without producing a rise in blood pressure. The noradrenaline released by tyramine is mainly metabolized by O-methylation while the noradrenaline released by reserpine is metabolized mainly by deamination (Kopin and Gordon, 1962). Tyramine appears to produce its effect by rapidly discharging noradrenaline in a physiologically active form (Axelrod, Gordon, Hertting, Kopin and Potter, 1962). The released hormone is then metabolized by catechol-O-methyl transferase outside the nerve. Tyramine presumably releases noradrenaline from the storage vesicles close to the synaptic terminal or from the noradrenaline bound to cytoplasm components in the neurone. Reserpine appears to slowly release noradrenaline from the vesicles deep in the nerve so that the noradrenaline is deaminated by monoamine oxidase in the mitochondria which are present in the nerves. As a result the neurohumor leaves the sympathetic nerves as an inactive deaminated metabolite.

Evidence that monoamine oxidase is present in sympathetic nerves has been obtained by denervation experiments (Snyder, Fischer and Axelrod, 1965). Since a considerable portion of the total mass of the pineal gland is composed of sympathetic nerves (Kappers, 1960), a large ratio of intraneural monoamine oxidase to extraneural monoamine oxidase would be expected. Accordingly, the effect of sympathetic denervation on monoamine oxidase activity in the rat pineal gland was examined. After chronic interruption of the sympathetic nerves to the pineal there was a 50 per cent reduction in monoamine oxidase activity in the pineal gland (Table 2). Sympathetic

TABLE 2. MONOAMINE OXIDASE (MAO) ACTIVITY IN THE PINEAL AND SALIVARY GLAND AFTER SUPERIOR CERVICAL GANGLIONECTOMY

	Pineal MAO activity	*Salivary MAO activity*
Innervated	1.01 ± 0.06	4.67 ± 0.23
Denervated	0.53 ± 0.02*	3.32 ± 0.35†

MAO activity is expressed as mμmoles C^{14}-acetic acid formed from C^{14} tryptamine per mg tissue per hour ± standard error of the mean. Rats were subjected to unilateral superior cervical ganglionectomy and both salivary glands were assayed for monoamine oxidase one week later.

* $P < 0.001$. † $P < 0.01$.

denervation of the submaxillary gland resulted in a smaller but significant decrease in monamine oxidase activity in this organ.

The control of the metabolism, storage and release of indoleamines by sympathetic nerves in the pineal gland. For many years the pineal gland had been thought to be a vestige of the photoreceptor complex the "third eye" of cold-blooded animals. Investigators who observed precocious sexual development in young boys with pineal tumors had believed that the pineal was an endocrine gland. However, no convincing experimental evidence was provided to support such a conclusion. It was also thought that the pineal gland served no useful function in man since it becomes hardened by calcification at puberty.

In recent years the pineal gland has been the subject of considerable investigation by workers in many different disciplines. The main stimulation for this activity was the discovery by Lerner, Case and Heinzelman (1959), that the pineal contained a unique indole derivative, melatonin (5-methoxy-N-acetyltryptamine).

This compound, isolated from the bovine pineal gland, was shown to be the most potent skin-lightening agent in amphibians known. Melatonin did not have effects on mammalian pigmentation, where it was found, and it could not be detected in amphibians where it exerted its effects. Another important discovery about the pineal gland was made by Kappers (1960), who showed that the pineal gland of the adult rat was innervated by sympathetic nerves. The sympathetic nerves originated in the superior cervical ganglia and the nerve endings actually terminated on pineal cells (Wolfe, Potter, Richardson and Axelrod, 1962), instead of on blood vessels or smooth muscle cells. It was also found that the pineal cells of amphibians but not of mammals contained photoreceptive elements which were indistinguishable from those found in the cone cells of the retina (Kelly, 1962). It was also found that the pineal glands were rich in serotonin and histamine (Giarman and Day, 1958). Using a specific fluorescent microscopic method it was observed that the serotonin in the pineal glands was partly localized in the parenchymal cells and partly in the sympathetic nerves (Bertler, Falck and Owman, 1963). Denervation of the superior cervical ganglia produces more than a ninety per cent depletion of noradrenaline in organs innervated by nerves whose cell bodies arise in superior cervical ganglia but only a 50 per cent decline in the serotonin content of the pineal gland (Pellegrino de Iraldi, Zieher and de Robertis, 1963). Pineal cells of rat fetus were also shown to contain an unidentified secretory material (Owman, 1961).

Interest in our laboratory in the pineal gland came from the observation of Lerner, Case and Heinzelman, 1959, that melatonin contained an O-methyl group. We have been interested in enzymatic O-methylation reactions and had described an enzyme that can O-methylate catechols (Axelrod, 1957). The presence of an O-methyl group on melatonin suggested that it might arise from S-adenosylmethionine. An extract of bovine pineal gland was incubated with C^{14}-methyl S-adenosylmethionine and N-acetyl-serotonin (Axelrod and Weissbach, 1961). This resulted in the formation of a radioactive product that had the same R_f as authentic melatonin on paper chromatographs. The high concentration of serotonin in the pineal suggested that the amine might form N-acetylserotonin. When an extract of bovine pineal gland was incubated with serotonin together with acetyl co-enzyme A, N-acetylserotonin was formed (Weissbach, Redfield and Axelrod, 1960). These results showed that the pineal gland contained all the enzymes and substrates necessary for the formation of melatonin which proceeded in 2 steps as follows:

$$\text{serotonin} \xrightarrow{\text{acetyl Co. A}} \text{N-acetylserotonin} \xrightarrow{\text{S-adenosylmethionine}} \text{melatonin}$$

The distribution of melatonin-forming enzyme (hydroxyindole-O-methyl transferase) was examined in a number of species such as monkey, cow, goat, man, rat and hen (Axelrod and Weissbach, 1961; Axelrod, Wurtman and Winget, 1964). It was found only in the pineal gland. In the frog (*Rana pipiens*), however, it was found in all areas of the brain and in the eye (Axelrod and Quay, unpublished observations). Hydroxyindole-O-methyl transferase is relatively nonspecific since it can O-methylate many hydroxy-indoles including serotonin, bufotenine, 5-hydroxyindole acetic acid and 5,6-dihydroxyindole (Axelrod and Weissbach, 1961; Axelrod and Lerner, 1963). N-acetylserotonin, however, was the best substrate by far, being about 10 times more active than any other examined.

In contrast to amphibians the mammalian pineal gland does not respond to light, directly. However, many studies have shown that it is markedly influenced by environmental lighting. Rats exposed to continuous light show a decrease in pineal weight (Fiske, Bryant and Putnam, 1960), cyto-plasma RNA and nucleolar size (Roth, Wurtman and Altschule, 1962). When rats are exposed to continuous light the ovaries become enlarged and the incidence of estrus increases (Fiske, Bryant and Putnam, 1960); both effects are inhibited by the administration of pineal extracts (Jöchle, 1956; Wurt-man, Roth, Altschule and Wurtman, 1961). The active principle in the pineal gland that blocks the effects of light on the ovaries has been identified as melatonin (Wurtman, Axelrod and Chu, 1963). The effects of melatonin were specific since neither its immediate precursor N-acetylserotonin nor its metabolite 6-hydroxymelatonin had appreciable blocking action (Chu, Wurtman and Axelrod, 1964). However, 5-methoxytryptophol has been shown to decrease ovarian weight in the rat and to decrease the incidence of estrus (McIsaacs, Taborsky and Farrell, 1964).

The administration of H^3-melatonin intravenously resulted in the selective uptake of the compound in the ovaries (Wurtman, Axelrod and Potter, 1964). Melatonin is presumably secreted by the pineal gland since it has been identified in nervous tissues (Lerner, Case, Mori and Wright, 1959) which cannot synthesize it, and in the human urine (Lerner and Takahashi, unpublished observations). On the basis of these observations melatonin appears to satisfy the classical criteria for a hormone. It is produced by a specialized structure; only the pineal gland has the enzyme for its synthesis. It is released into the circulation, since it can be found in the urine. It has an effect on a distant target organ, the ovaries, and it can be concentrated by this organ from the circulation.

The mechanism whereby light alters the estrous cycle in the rat was explored next. One way in which light may have its effect on the gonads would be on the formation and release of melatonin, since this compound can inhibit the estrous cycle. It also seemed likely that the enzyme that synthesizes melatonin was rate limiting because of the large amounts of serotonin present and the low activity of hydroxyindole O-methyl transferase in the rat pineal gland. Rats were placed in constant dark and light environments for several days after which the pineal glands were examined for hydroxyindole O-methyl transferase activity (Wurtman, Axelrod and Phillips, 1963). Rats kept in a constant dark environment for 5 days showed a 3- to 10-fold increase in hydroxyindole O-methyl transferase activity compared to those kept in continuous light. In addition, the weight of the pineal gland was significantly heavier when rats were kept in the dark. Thus it appeared that light was increasing the incidence of estrus in the rats by reducing the synthesis of melatonin, a compound that inhibits the estrous cycle. Since the effects of illumination could be blocked by agents that interfere with protein synthesis such as puromycin (Axelrod, Wurtman and Snyder, 1965), it appeared that light was influencing the rate of formation of the enzyme protein itself.

There are several possible routes in which information about lighting can reach the pineal gland. Light can penetrate the skull and impinge upon the pineal. This is not an unlikely possibility since it has been shown that sufficient light can pass through the sheep skull to activate photoelectric cells implanted within the brain (Ganong, Shepherd, Wall and van Brunt, 1963). To examine this possibility, rats were blinded, placed in light or dark environments and after several days their pineal hydroxyindole O-methyl transferase activity was examined (Wurtman, Axelrod and Fischer, 1964). Removal of both eyes resulted in the complete loss of the ability of the pineal gland to respond to light and dark with the accompanying changes in weight and hydroxyindole O-methyl transferase activity (Table 3). This experiment demonstrated that the action of light on the rat pineal gland was mediated via retinal receptors. Light could act on the pineal gland through the hypothalamus and pituitary body or other endocrine organs. To determine whether pituitary or ovarian hormones are required for light to affect the hydroxyindole O-methyl transferase activity, adult males, immature females, adult females in which both ovaries had been removed and hypophysectomized adult females were placed in continuous light and darkness. In all treated animals, light produced an inhibitory effect on pineal weight and the melatonin synthesizing enzyme. Finally, information about lighting could be transmitted to the pineal gland via a neural route. Kappers (1960) had already shown that the pineal gland was innervated by sympathetic nerves. These nerves were interrupted by the removal of both superior cervical ganglia (Wurtman, Axelrod and Fischer, 1964), and the denervated rats were placed in continuous light and darkness. After sympathetic denervation,

TABLE 3. EFFECT OF LIGHT ON HYDROXYINDOLE O-METHYL TRANSFERASE ACTIVITY AND ITS CONTROL BY SYMPATHETIC NERVES

		Pineal wt. (*mg*)	*HIOMT activity*	*Per cent in estrus*
Control	Dark	1.29	14.0	51
	Light	0.97*	6.4*	95*
Blinded	Dark	1.27	9.9	35
	Light	1.20	11.5	44
Ganglionectomy	Dark	1.05	7.9	58
	Light	1.10	9.1	71

Rats were placed in constant light or darkness 1 day after the operation and killed 11 days later. Hydroxyindole-O-methyl transferase activity (HIOMT) is expressed as $\mu\mu$moles melatonin formed from N-acetyl serotonin per gland per hour.

* $P < 0.001$.

rats placed in constant darkness no longer responded with decreased pineal weight or hydroxyindole O-methyl transferase activity (Table 3). These experiments demonstrated that the transmission of information about lighting to the pineal gland required the presence of an intact sympathetic innervation. Thus, the sympathetic nervous system is directly concerned in the regulation of melatonin synthesis in the pineal gland. The actions of the sympathetic nervous system on pineal gland melatonin synthesis are cued by environmental lighting. Light could change the rate of release of a neurotransmitter from sympathetic nerve endings and this transmitter could then influence the activity of the melatonin synthesizing enzyme. It is well established that the sympathetic nerves release noradrenaline (Euler, 1956). Preliminary results indicated that the repeated administration of noradrenaline to rats prevented the decline in hydroxyindole O-methyl transferase activity when rats were exposed to light (Axelrod and Wurtman, unpublished observations).

If light influences the rat estrous cycle by affecting the synthesis and secretion of melatonin in the pineal, it would be expected that sympathetic denervation of the pineal should abolish the effects of light on the gonads. When the superior cervical ganglia were removed from rats, exposure to light no longer affected the estrous cycle in rats (Wurtman, Axelrod and Chu, 1964). This procedure also eliminated the light-induced increase in ovary and uterine weight. These results demonstrated that the sympathetic nervous system plays a role in the regulation of the estrous cycle in rats via the control of melatonin synthesis in the pineal gland.

Other biochemical processes in the pineal have been found to be influenced by environmental lighting and by the activity of the sympathetic nerves.

The mammalian pineal gland has a high concentration of serotonin (Quay and Halevy, 1962; Bertler, Falck and Owman, 1963), as well as 5-hydroxytryptophan decarboxylase activity (Snyder and Axelrod, 1964). Since serotonin serves as a precursor to melatonin (Axelrod and Weissbach, 1960) the influence of light and sympathetic denervation on the serotonin-forming enzyme 5-hydroxytryptophan decarboxylase was examined (Snyder, Axelrod, Fischer and Wurtman, 1964, 1965) (Fig. 2). Rat pineal glands were denervated by bilateral superior cervical ganglionectomy and then kept in diurnal lighting conditions for 7 days. Following ganglionectomy there was about a 2-fold increase in 5-hydroxytryptophan decarboxylase activity. There was no significant change in pineal weight. Other investigators have shown about a 50 per cent decline in serotonin in a denervated rat (Bertler, Falck and Owman, 1963). There was no change in 5-hydroxytryptophan decarboxylase in sympathetically denervated salivary gland or intestines. Rats were then kept in continuous light or dark environments and their pineals examined for 5-hydroxytryptophan decarboxylase activity. There was twice as much enzyme activity when rats were kept in light for 5 days as those kept in darkness (Fig. 3). Previous work had shown that the pathway by which light affects hydroxyindole O-methyl transferase activity in the pineal involves the eyes and the superior cervical ganglia. Consequently, the eyes or the superior cervical ganglia were removed from groups of rats which were then placed in constant light or darkness. Blinding or the removal of the superior cervical ganglia prevented the elevation in pineal 5-hydroxytryptophan activity of rats kept in constant light (Fig. 3). The effect of drugs that interfere with sympathetic nerve impulses on the responses of pineal 5-hydroxytryptophan decarboxylase to constant light was also examined. Rats placed in continuous light or dark environments were given bretylium or guanethidine. Bretylium prevented the rise in pineal 5-hydroxytryptophan decarboxylase in constant light but guanethidine had no effect. Both drugs, however, prevented the decrease in pineal weight normally induced by exposure to constant light. Quay and Halevy (1962) have shown that the serotonin content of the pineal glands of rats exposed to long periods of constant light is reduced by half. Denervation also reduces the serotonin content by about the same amount (Bertler, Falck and Owman, 1963; Pellegrino de Iraldi, Zieher and de Robertis, 1963). In rats exposed to continuous light or after denervation, there is a doubling of 5-hydroxytryptophan decarboxylase activity suggesting an inverse relationship between serotonin content and the activity of the enzyme that forms this amine.

The results described above showed that environmental lighting affects two enzyme systems in the pineal gland, hydroxyindole O-methyl transferase which is decreased by lighting and 5-hydroxytryptophan decarboxylase which is increased by lighting. Exposure to light also reduces pineal gland weight (Fiske, Bryant and Putnam, 1960). Blinding and superior cervical ganglionectomy have been shown to prevent the actions of light on the pineal gland (Wurtman, Axelrod and Fischer, 1964). Bretylium and guanethidine, drugs that block impulses in the sympathetic nerves, also block the effect of light on pineal weight. It is unlikely that changes in serotonin and melatonin content and the enzymes involved in their metabolism would influence pineal weight to a measurable degree. The observation that drugs which affect the

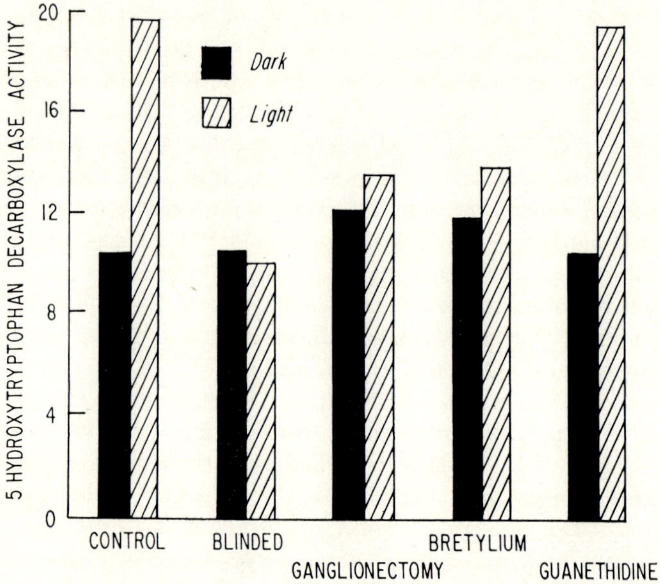

FIG. 3. Effect of ganglionectomy and drugs on the response of rat pineal 5-hydroxytryptophan decarboxylase activity to constant light and darkness. Bilaterally ganglionectomized and blinded rats were placed in constant light or darkness for 30 days. Some rats received bretylium (20 mg/kg) or guanethidine (20 mg/kg) for 30 days in the experimental lighting conditions. 5-hydroxytryptophan decarboxylase activity was expressed as mμg serotonin formed from 5-hydroxytryptophan per mg pineal per hour.

sympathetic nervous system also affect pineal weight suggests that the sympathetic innervation of the pineal gland may influence many other biochemical parameters in this organ.

The pineal has been shown to respond under unusual lighting conditions, continuous light or dark. In nature, animals rarely live under such environments. Consequently, the biochemical response of the pineal to circadian 24-hr light and dark cycles were examined. Rats were kept under controlled lighting conditions (lights on 7 a.m. to 7 p.m.) and the pineals were assayed for hydroxyindole O-methyl transferase activity at 6 a.m., noon, 6 p.m. and midnight (Axelrod, Wurtman and Snyder, 1965). By 5 hr after the onset of darkness (midnight), enzyme activity increased 2- to 3-fold (Table 4). In

TABLE 4. EFFECT OF GANGLIONECTOMY AND BLINDING ON THE DIURNAL CHANGES IN HYDROXYINDOLE-O-METHYL TRANSFERASE IN THE RAT PINEAL GLAND

	HIOMT activity	
	6 p.m.	*Midnight*
Treatment		
Control	12.3	20.5*
Ganglionectomy	12.6	11.7
Control	17.7	45.9*
Blinded	25.5	27.9

Groups of rats were blinded or had their superior cervical ganglia removed. One week later the rats were killed at 6 p.m. (lights off from 7 p.m. to 7 a.m.) and at midnight and the pineals assayed for hydroxyindole-O-methyl transferase (HIOMT) activity. Enzyme activity is expressed as $\mu\mu$moles C^{14}-melatonin formed after incubating N-acetylserotonin and C^{14}-methyl S-adenosylmethionine with rat pineal gland.

* $P < 0.001$.

addition, there was a significant increase in pineal weight during this period. All circadian rhythms studied thus far have been found to be endogenous. They can persist in the absence of environmental cues. To determine whether changes in pineal hydroxyindole O-methyl transferase were the consequence of an intrinsic rhythm or were regulated by changes in environmental lighting, rats were subjected to various lighting regimens. Animals kept in continuous darkness had considerable more enzyme activity than rats kept in light, but there was no change in the melatonin synthesis in either group between 6 p.m. and midnight. Another group of rats were deprived of one photoperiod of darkness by being placed in light instead of darkness from 7 p.m. to midnight. Controls consisted of a group of rats which were killed at 6 p.m. and another at midnight after the usual 5 hr of darkness. The control group showed the expected rise in hydroxyindole O-methyl transferase between 6 p.m. and midnight. When a single photoperiod of darkness (5 hr) was omitted the rise in hydroxyindole O-methyl transferase was abolished. Thus, the diurnal rhythm in pineal hydroxyindole O-methyl transferase appears to be exogenous and entirely dependent on environmental lighting changes. To see whether the information concerning diurnal light reaches the pineal gland via the eyes and the sympathetic nervous system the following experiments were done (Axelrod, Wurtman and Snyder, 1965): Both eyes were removed and the animals were kept in diurnal lighting conditions. After one week the hydroxyindole O-methyl transferase was examined at 6 p.m. and midnight. There was no rise in the enzyme activity at midnight as had been observed with control rats (Table 4). In other groups of rats the pineal was denervated by the removal of the superior cervical ganglia. This procedure also eliminated the rise in enzyme activity in darkness (Table 4).

Quay (1963) has shown the presence of a circadian rhythm in serotonin content in the rat pineal with a peak concentration at noon and a minimal concentration about 10 p.m. To examine the possible mechanisms controlling the circadian rhythm in serotonin content (Snyder, Zweig and Axelrod, 1965; Snyder, Zweig, Axelrod and Fischer, 1965), groups of rats were kept under diurnal lighting conditions and were killed at 1 p.m. (6 hr of light) and 11 p.m. (4 hr of darkness) and the pineal assayed for serotonin by an ultra micromethod (Snyder, Axelrod and Zweig, 1965). Serotonin levels were found to be 2 to 3 times higher in pineal glands of rats killed at 1 p.m. than in those of rats killed at 11 p.m., essentially confirming the results of Quay (1963) (Table 5). Groups of rats were then transferred to continuous dark or light environments. After one week the animals were killed at 1 p.m. and at 11 p.m. The circadian changes in pineal serotonin level were abolished in rats kept in constant light. Under these conditions the serotonin was elevated to the 1 p.m. values found in diurnal lighting. However, the circadian rhythm persisted unchanged in rats kept in continuous darkness or in blinded rats. These results indicated that the circadian serotonin rhythm in the rat pineal gland is endogenous but can be suppressed by continuous light exposure. To establish how rapidly continuous light abolishes the serotonin

TABLE 5. CONTROL OF THE CIRCADIAN RHYTHM IN PINEAL SEROTONIN LEVELS BY THE SYMPATHETIC NERVES

	Pineal serotonin mμg/mg	
	1 p.m.	*11 p.m.*
Diurnal light	72	34*
Continuous light	66	62
Continuous darkness	68	32*
Blinded (diurnal light)	64	23*
Ganglionectomy (diurnal light)	45	42
Decentralization (diurnal light)	42	44

Groups of rats were kept in continuous light or darkness or under diurnal lighting conditions. Ganglionectomized, decentralized or blinded rats were kept in diurnal light for 5 days after the operation. Rats were killed at 1 p.m. or 11 p.m. and examined for serotonin content.

* $P < 0.001$.

rhythm, one group of rats was placed in a room in which the lights remained on from 7 p.m. to 11 p.m. In contrast to rats kept in darkness during this period, there was no decrease in the serotonin level at 11 p.m. in the pineals of rats which were exposed to an additional photoperiod of illumination.

Previous work showed that the circadian rhythm of eosinophil counts requires the intact adrenal gland (Halberg, Visscher and Bittner, 1953). To examine the possible effects of endocrine organs on circadian rhythms in pineal serotonin, adrenalectomized, thyroidectomized, oophorectomized and hypophysectomized rats were kept in diurnal lighting conditions and one week later the pineals were examined for serotonin at 1 p.m. and 11 p.m. The removal of these glands had no influence on the fall of the serotonin level in the pineal at 11 p.m. The effect of sympathetic denervation on the circadian pineal rhythm was examined next. The circadian rhythm in pineal serotonin was completely abolished in rats whose superior cervical ganglia were removed (Table 5). The 1 p.m. values were lowered and the 11 p.m. values elevated.

There are several possible explanations for the suppressive action of ganglionectomy on the circadian serotonin rhythm. It may be intrinsic to the pineal gland but require intact sympathetic nerves for its expression. The controlling mechanism may be in the superior cervical ganglia. The control might be neural but localized elsewhere in the body and communicated to the pineal gland via preganglionic fibers. To examine this latter possibility the preganglionic nerves to the superior cervical ganglia were cut bilaterally in groups of rats that were kept in diurnal lighting for 6 days before assaying their pineals for serotonin. This treatment completely abolished the circadian serotonin rhythm. These results indicated that the serotonin rhythm is extrinsic to the pineal gland and information is communicated from the central nervous system to this organ via preganglionic fibers to the superior cervical ganglia. Light presumably acts as an external synchronizer. The pineal gland thus appears to contain at least two biological clocks: one (hydroxyindole O-methyl transferase) completely dependent upon environmental lighting and the other (serotonin) originating within the central nervous system but cued by changes in lighting.

SUMMARY

Circulating noradrenaline and adrenaline are selectively taken up and retained by sympathetic nerves in tissues. This uptake and binding serves as an important mechanism for the rapid inactivation of these hormones. The noradrenaline is stored in small vesicles within the nerves which can be separated by centrifugation. Biochemical and pharmacological studies show that there is more than one bound form of noradrenaline; an easily releasable store and a firmly bound store. When noradrenaline is discharged from the nerve it is inactivated by O-methylation, diffusion and re-uptake by sympathetic nerves. Evidence is presented showing the presence of monoamine oxidase within sympathetic nerves. This enzyme inactivates noradrenaline within the sympathetic nerves.

The sympathetic nervous system has been shown to be important in regulating the biochemical and physiological activity of the rat pineal gland. This gland contains an enzyme (hydroxyindole O-methyl transferase) that can synthesize melatonin, a compound that inhibits the estrous cycle. Light affects the estrous cycle by inhibiting the melatonin-synthesizing enzyme in the pineal. Information about lighting reaches the pineal via the superior cervical ganglia and sympathetic nerves. Light also regulates 5-hydroxytryptophan decarboxylase activity in the pineal by means of the sympathetic nerves. There are diurnal rhythms in the serotonin content and melatonin synthesis in the pineal. The former is endogenous and receives information from the central nervous system; the latter is exogenous and receives information from environmental lighting. Both rhythms are controlled by sympathetic nerves.

REFERENCES

AXELROD, J. (1957) O-methylation of epinephrine and other catechols *in vitro* and *in vivo*, *Science* **126**, 400–1.

AXELROD, J. (1960) The fate of adrenaline and noradrenaline. In: *Ciba Foundation Symposium on Adrenergic Mechanisms*, J. R. VANE, G. E. W. WOLSTENHOLME and M. O'CONNOR (eds.), London, England, J. & A. Churchill, Ltd., pp. 28–39.

AXELROD, J., E. GORDON, G. HERTTING, I. J. KOPIN and L. T. POTTER (1962) On the mechanism of tachyphylaxis to tyramine in the isolated rat heart, *Brit. J. Pharmacol.* **19**, 56–63.

AXELROD, J., G. HERTTING and R. W. PATRICK (1961) Inhibition of H^3-norepinephrine release by monoamine oxidase inhibitors, *J. Pharmacol.* **134**, 325–8.

AXELROD, J. and A. B. LERNER (1963) O-methylation in the conversion of tyrosine to melanin, *Biochim. Biophys. Acta* (Amst.) **71**, 650–5.

AXELROD, J., P. D. MACLEAN, R. W. ALBERS and H. WEISSBACH (1961) Regional distribution of methyl transferase enzymes in the nervous system and glandular tissues. In: *Regional Neurochemistry*, S. S. Kety and J. Elkes (Eds.). New York, N.Y., Pergamon Press, pp. 307–11.

AXELROD, J., H. WEIL-MALHERBE and R. TOMCHICK (1959) The physiological disposition of H^3-epinephrine and its metabolite metanephrine, *J. Pharmacol.* **127**, 251–6.

AXELROD, J. and H. WEISSBACH (1960) Enzymatic O-methylation of N-acetylserotonin to melatonin, *Science* **131**, 1312.

AXELROD, J. and H. WEISSBACH (1961) Purification and properties of hydroxyindole-O-methyl transferase, *J. Biol. Chem.* **236**, 211–13.

AXELROD, J., R. J. WURTMAN and S. H. SNYDER (1965) Control of hydroxyindole-O-methyl transferase activity in the rat pineal gland by environmental lighting, *J. Biol. Chem.* **240**, 949–54.

AXELROD, J., R. J. WURTMAN and C. M. WINGET (1964) Melatonin synthesis in the hen pineal gland and its control by light, *Nature* (Lond.) **201**, 1134.

BERTLER, A., B. FALCK and C. OWMAN (1963) Cellular localization of 5-hydroxytryptamine in the rat pineal, *Kgl. Fysiograf. Sällsk. Lund. Förh.* **33**, 13–16.

CHU, E. W., R. J. WURTMAN and J. AXELROD (1964) An inhibitory effect of melatonin on the estrous phase of the estrous cycle of the rodent, *Endocrinology* **75**, 238–42.

CROUT, J. R., A. J. MUSKUS and U. TRENDELENBURG (1962) Effect of tyramine on isolated guinea-pig atria in relationship to their noradrenaline stores, *Brit. J. Pharmacol.* **18**, 600–11.

ELLIOTT, T. R. (1905) The action of adrenaline, *J. Physiol.* (*Lond.*) **32**, 401-7.

EULER, U. S. VON (1948) Identification of the sympathomimetic ergone in adrenergic nerves of cattle with laevo noradrenaline, *Acta Physiol. Scand.* **16**, 63-74.

EULER, U. S. VON (1956). *Noradrenaline*, Charles C. Thomas, Springfield, Ill., pp. 133–54.

EULER, U. S. VON and N.-Å. HILLARP, (1956) Evidence for the presence of noradrenaline in submicroscopic structure of adrenergic axon, *Nature* (Lond.) **177**, 44.

EULER, U. S. VON and F. LISHAJKO (1963) Effect of adenine nucleotides on catecholamine release and uptake in isolated nerve storage granules, *Acta Physiol. Scand.* **59**, 454–61.

EULER, U. S. VON and A. PURKHOLD (1951) Effect of sympathetic denervation on the noradrenaline and adrenaline content of the spleen, *Acta Physiol. Scand.* **24**, 212–17.

FISCHER, J. E., D. W. HORST and I. J. KOPIN (1965) *β*-Hydroxylated sympathomimetic amines as false neurochemical transmitters, *Brit. J. Pharmacol.* **24**, 477–84.

FISCHER, J. E., J. MUSACCHIO, I. J. KOPIN and J. AXELROD (1964) Effects of denervation on the uptake and *β*-hydroxylation of tyramine in the rat salivary gland, *Life Sci.* **3**, 413–17.

FISKE, V. M., G. K. BRYANT and J. PUTNAM (1960) Effect of light on the weight of the pineal in the rat, *Endocrinology* **66**, 489–90.

GANONG, W. F., M. D. SHEHPERD, J. R. WALL, E. E. VAN BRUNT and M. T. CLEGG (1963) Penetration of light into the brain of mammals, *Endocrinology* **72**, 962–3.

GIARMAN, N. J. and M. DAY (1958) Presence of biogenic amines in the bovine pineal body *Biochem. Pharmacol.* **1**, 235.

GLOWINSKI, J., J. AXELROD, I. J. KOPIN and R. J. WURTMAN (1964) Physiological disposition of H^3-norepinephrine in the developing rat, *J. Pharmacol.* **146**, 48–53.

HALBERG, F., M. B. VISSCHER and J. J. BITTNER (1953) Eosinophil rhythm in mice, *Amer. J. Physiol.* **174**, 109–22.

HARVEY, J. G. and J. N. PENNEFATHER (1962) Effect of adrenaline infusion on the catecholamine content of rat and cat tissue, *Brit. J. Pharmacol.* **18**, 183–9.

HERTTING, G. and J. AXELROD (1961) Fate of tritiated noradrenaline at the sympathetic nerve endings, *Nature* (Lond.) **192**, 172–3.

HERTTING, G., J. AXELROD, I. J. KOPIN and L. G. WHITBY (1961) Lack of uptake of catecholamines after chronic denervation of sympathetic nerves, *Nature* (Lond.) **189**, 66.

HERTTING, G., J. AXELROD and L. G. WHITBY (1961) Effect of drugs on the uptake and metabolism of H^3-norepinephrine, *J. Pharmacol.* **134**, 146–53.

HERTTING, G., L. T. POTTER and J. AXELROD (1962) Effect of decentralization and ganglionic blocking agents on the spontaneous release of H^3-norepinephrine, *J. Pharmacol.* **136**, 289–92.

HILLARP, N.-Å. (1958) Adenosine phosphates and inorganic phosphate in the adrenaline- and noradrenaline-containing granules of the adrenal medulla, *Acta Physiol. Scand.* **42**, 321–32.

IVERSEN, L. L. (1963) The uptake of noradrenaline by the isolated perfused rat heart, *Brit. J. Pharmacol.* **4**, 523–37.

JÖCHLE, W. (1956) Ueber die Wirkung eines Epiphysenextraktes auf sexual Entwicklung und sexual Cyclus junger weiblicher Ratten unter normalen Holtungsbedingungen und Dauerbeleuchtung, *Endokrinologie* **33**, 287.

KAPPERS, J. (1960) The development, topographical relations and innervation of the epiphysis cerebri in the albino rat, *Z. Zellforsch* **52**, 163–215.

KELLY, D. E. (1962) Pineal organs: photoreception, secretion and development, *American Scientist* **50**, 597–625.

KOPIN, I. J. and E. K. GORDON (1962) Metabolism of norepinephrine-H^3 released by tyramine and reserpine, *J. Pharmacol.* **138**, 351–9.

KOPIN, I. J. and E. K. GORDON (1963) Metabolism of administered and drug released norepinephrine-H^3 in the rat, *J. Pharmacol.* **140**, 207–16.

LERNER, A. B., J. D. CASE and R. V. HEINZELMAN (1959) Structure of melatonin, *J. Amer. Chem. Soc.* **81**, 6084–5.

LERNER, A. B., J. D. CASE, W. MORI and M. R. WRIGHT (1959) Melatonin in peripheral nerve, *Nature* (Lond.) **183**, 1821.

LEVI-MONTALCINI, R. and B. BOOKER (1960) Destruction of the sympathetic ganglia in mammals by an antiserum to the nerve growth factor, *Proc. Nat. Acad. Sci.* (Wash.) **46**, 384–91.

LOEWI, O. (1921) Ueber humorale Uebertragbarkeit der Herznervenwirkung, *Arch. ges. Physiol.* **189**, 239–42.

MCISAACS, W. M., R. G. TABORSKY and G. FARRELL (1964) 5-Methoxy tryptophol: effect on estrus and ovary weight, *Science* **145**, 63–4.

MUSACCHIO, J., J. E. FISCHER and I. J. KOPIN (1965) Effect of chronic sympathetic denervation on the subcellular distribution of some sympathomimetic amines, *Biochem. Pharmacol.* **14**, 898–900.

OWMAN, C. (1961) Secretory activity of the fetal pineal gland of the rat, *Acta Morph. Neerl. Scand.* **3**, 367–94.

PELLEGRINO DE IRALDI, A., L. M. ZIEHER and E. DE ROBERTIS (1963) The 5-hydroxytryptamine content and synthesis in normal and denervated pineal gland, *Life Sci.* **9**, 691–6.

POTTER, L. T. and J. AXELROD (1963a) Subcellular localization of catecholamines in tissue of the rat, *J. Pharmacol.* **142**, 291–8.

POTTER, L. T. and J. AXELROD (1963b) Properties of norepinephrine storage particles of the rat, *J. Pharmacol.* **142**, 299–305.

POTTER, L. T. and J. AXELROD (1963c) Studies on the storage of norepinephrine and the effect of drugs, *J. Pharmacol.* **140**, 199–206.

POTTER, L T., J. AXELROD and I. J. KOPIN (1962) Differential binding and release of norepinephrine and tachyphylaxis, *Biochem. Pharmacol.* **11**, 254–6.

QUAY, W. B. (1963) Circadian rhythm in rat pineal serotonin and its modification by estrous cycle and photo period, *Gen. Comp. Endocr.* **3**, 473–9.

QUAY, W. B. and A. HALEVY (1962) Experimental modification of the rat pineal content of serotonin and related indole amines, *Physiol. Zool.* **35**, 1.

ROSELL, S., J. AXELROD and I. J. KOPIN (1964) Release of tritiated epinephrine following sympathetic nerve stimulation, *Nature* (Lond.) **201**, 301.

ROSELL, S., I. J. KOPIN and J. AXELROD (1963) Fate of H^3-noradrenaline in the skeletal muscle before and following sympathetic stimulation, *Amer. J. Physiol.* **205**, 317.

ROTH, W., R. J. WURTMAN and M. D. ALTSCHULE (1962) Morphological changes in the pineal parenchymal cells of rats exposed to continuous light or darkness, *Endocrinology* **71**, 888–92.

SNYDER, S. H. and J. AXELROD (1964) A sensitive assay for 5-hydroxytryptophan decarboxylase, *Biochem. Pharmacol.* **13**, 803–6.

SNYDER, S. H., J. AXELROD, J. E. FISCHER and R. J. WURTMAN (1964) Neural and photic regulation of 5-hydroxytryptophan decarboxylase in the rat pineal gland, *Nature* (Lond.) **203**, 981.

SNYDER, S. H., J. AXELROD, R. J. WURTMAN and J. E. FISCHER (1965) Control of 5-hydroxytryptophan decarboxylase activity in the rat pineal by sympathetic nerves, *J. Pharmacol.* **147**, 371–5.

SNYDER, S. H., J. AXELROD and M. ZWEIG (1965) A specific and senstitive assay for serotonin in tissues, *Biochem. Pharmacol.* **14**, 831–5.

SNYDER, S. H., J. FISCHER and J. AXELROD (1965) Evidence for the presence of monoamine oxidase in sympathetic nerves, *Biochem. Pharmacol.* **14**, 363–5.

SNYDER, S. H., I. A. MICHAELSON and J. MUSACCHIO (1964) Purification of norepinephrine storage granules from the heart, *Life Sci.* **3**, 965–70.

SNYDER, S. H., M. ZWEIG and J. AXELROD (1965) Control of the circadian rhythm in serotonin content of the ratipineal gland, *Life Sci.* **3**, 1175–9.

SNYDER, S. H., M. ZWEIG, J. AXELROD and J. E. FISCHER (1965) Control of the circadian rhythm in serotonin content of the rat pineal gland, *Proc. Nat. Acad. Sci.* (Wash.). **53**, 301–5.

STJÄRNE, L. (1961) Tyramine effects of catecholamine release from spleen and adrenal in the cat, *Acta Physiol. Scand.* **51**, 224–9.

STJÄRNE, L. (1964) Studies on catecholamine uptake, storage and release mechanisms, *Acta Physiol. Scand.* **62**, Suppl. 228.
STRÖMBLAD, B. C. R. and M. NICKERSON (1961) Accumulation of epinephrine and norepinephrine by some rat tissues, *J. Pharmacol.* **134**, 154–9.
TRENDELENBURG, U. (1961) Modifications of the effects of tyramine by various agents and procedures, *J. Pharmacol.* **134**, 8–11.
WEGMANN, A. and K. KAKO (1961) Particle-bound and free catecholamines in dog hearts and the uptake of injected norepinephrine, *Nature* (Lond.) **192**, 978.
WEIL-MALHERBE, H. and A. BONE (1957) Intracellular distribution of catecholamines in the brain, *Nature* (Lond.) **180**, 1050–1.
WEISSBACH, H., B. G. REDFIELD and J. AXELROD (1960) Biosynthesis of melatonin: enzymatic conversion of serotonin to N-acetylserotonin, *Biochim. Biophys. Acta* (Amst.) **43**, 352–3.
WHITBY, G. L., J. AXELROD and H. WEIL-MALHERBE (1961) The fate of H^3-norepinephrine in animals, *J. Pharmacol.* **132**, 193–201.
WOLFE, D. E., L. T. POTTER, K. C. RICHARDSON and J. AXELROD (1962) Localizing tritiated norepinephrine in sympathetic axons by electron microscopic autoradiography, *Science* **138**, 440–1.
WURTMAN, R. J., J. AXELROD and E. W. CHU (1963) Melatonin a pineal substance: effect on the rat ovary, *Science* **141**, 277–8.
WURTMAN, R. J., J. AXELROD, E. W. CHU and J. E. FISCHER (1964) Mediation of some effects of illumination on the rat estrous cycle by the sympathetic nerves, *Endocrinology* **75**, 266–72.
WURTMAN, R. J., J. AXELROD and J. E. FISCHER (1964) Melatonin synthesis in the pineal gland: effect of light mediated by the sympathetic nervous system, *Science* **143**, 1328–9.
WURTMAN, R. J., J. AXELROD and L. S. PHILLIPS (1963) Melatonin synthesis in the pineal gland: control by light, *Science* **142**, 1071–2.
WURTMAN, R. J., J. AXELROD and L. T. POTTER (1964) The uptake of H^3-melatonin in endocrine and nervous tissues and the effects of constant light exposure, *J. Pharmacol.* **143**, 314–18.
WURTMAN, R. J., W. ROTH, M. D. ALTSCHULE and J. J. WURTMAN (1961) Interaction of the pineal and exposure to continuous light on organ weights of female rats, *Acta Endocr.* (Kbh.) **36**, 617–24.

DISCUSSION

MEYERSON: I would like to ask Dr. Axelrod if the effects on estrus he reported in his elegant experiments are restricted to the peripheral vaginal response or if he also had studied the effects on estrus behaviour. We have found that the estrus behaviour (lordosis response) is inhibited by increased cerebral serotonin content (Meyerson, B. J., *Acta Physiol. Scand.* **63**, suppl. 241, 1964).

AXELROD: We have not examined the effects of melatonin on estrus behaviour, but we did find that it blocked the effects of light in increasing ovary weight and it also antagonized the action of the follicle stimulating hormone on increasing uterine weight.

GREEN: The findings on rats in estrus remind me that we have recently shown in women treated with follicle stimulating hormones that the ratio of methylhistamine to histamine in urine is correlated with the quantity of estrogens (Green, J. P., Fram, D. H. and Kase, N. *Nature*, 1964).

CATECHOLAMINES IN NERVE AND ORGAN GRANULES

U. S. von Euler

Department of Physiology, Karolinska Institutet, Stockholm

NORADRENALINE STORES IN THE ADRENERGIC NEURON

The demonstration of noradrenaline (NA) in characteristic amounts in adrenergic nerves and in organs supplied with such nerves was the first indication of special NA stores (cf. Euler, 1951). Since the stores disappeared after degeneration of the corresponding adrenergic nerves, it was concluded that the NA was intimately connected with nerve structures. The amount of NA in the stores therefore seemed to reflect the supply of adrenergic nerves to the organ, and consequently any organ supplied with adrenergic nerves, for instance vessels, should contain NA. In agreement with this concept no catecholamines were found in the placenta which is devoid of nerves. These results have since been amply confirmed, also with labelled amines distributed in the stores (Axelrod, Weil-Malherbe and Tomchick, 1959; Hertting, Axelrod, Kopin and Whitby, 1961).

So far no other natural stores of NA have been found than those which are due to the adrenergic nerves in the tissue or possibly to chromaffin cells. It was also recognized early that all parts of the neuron contained the transmitter, which led to the question of the distribution of NA within the axon. Quantitative studies on the NA content in sympathetic ganglia, splenic nerves and the spleen made it clear that the NA was not evenly distributed in the neuron. The finding that the nerve trunk contained several times as much NA as the ganglia and much less NA per unit weight of nerve than the organ itself, directly indicated that the transmitter must be accumulated in the peripheral parts of the neurons. The appearance of special knobs along the terminal portions, noted by Hillarp (1946), made it appear possible that these contained the transmitter in concentrated form. Garven and Gairns (1952) suggested "that the small beads on the course of the finest fibrils represent the actual release points of the humoral agents within the cytoplasmic continuum provided by cells other than the neurones". This view has been proved correct by more recent studies using fluorescence microscopy (Falck, 1962) and electron microscopy (Lever and Esterhuizen, 1961; Richardson, 1962) which clearly demonstrated not only the postulated accumulation of NA at the endings but also showed the presence of junctions between the peripheral nerve and the effector cell.

SPECIFIC STORAGE GRANULES

The specific localization of NA in the terminal portions of the adrenergic neuron suggested that the transmitter was bound to some special kind of structure (Euler, 1954). The occurrence of specific storage granules for catecholamines in the adrenal medullary cells, indicated by Cramer (1919) and demonstrated by Hillarp *et al.* (1953) and by Blaschko and Welch in the same year, made it appear possible that similar structures were present in the adrenergic neurons. This proved to be the case and a particle-bound fraction of NA could be prepared from adrenergic nerves as well as from organs like the spleen (Euler and Hillarp, 1956). In the rabbit heart the proportion of bound NA (high-speed sediment after homogenization) was about 50 per cent of the total amount less the coarse fraction. The proportion of bound NA was 60–70 per cent in homogenates of the vesicular gland of the bull, which is very rich in NA. The specific NA storage granules of the adrenergic nerves have been demonstrated in electron micrographs from suspensions after isolation from nerves in 1957 (cf. Euler, 1961; Euler and Swanbeck, 1964) and in adrenergic nerves (Lever and Esterhuizen, 1961; Richardson, 1962; Wolfe *et al.*, 1962; de Robertis, 1964). The finer structure of these granules, which have a diameter of 0·03–0·20 μ, is still incompletely known.

CAPACITY OF STORES

As mentioned above, normal organs contain amounts of NA which are approximately constant and characteristic for each organ. This observation suggests that during normal conditions the stores are filled to near capacity. Isolated nerve granules obtained either from the adrenergic nerve trunk or from the organ itself similarly take up only small amounts of exogenous NA after preparation from normal material. Above a certain maximal capacity level no more amines seem to be taken up. Assuming that some kind of equilibrium normally exists between the specific stores and the extragranular stores the total capacity for storing the transmitter in the organs would thus be limited under physiological conditions.

Notwithstanding this general rule, it is a common finding that the amounts of noradrenaline in an organ can be moderately increased by various means, probably indicating a certain normal deficit, presumably as a result of continuous activity and some lag in the refilling of the stores. Thus Ryd (1962) observed that after bretylium the stores increased by some 30–50 per cent in several organs in the guinea pig. It is conceivable that "neuronal rest" caused this increase, which is supported by the finding of Sedvall (1964) that denervation causes an initial rise in the NA content of skeletal muscle.

Some apparently controversial results have been obtained after injecting large amounts of catecholamines into animals. Large increases in the NA content of some organs have been noted after injections, but only for a brief time and after large doses. Wegmann and Kako (1961) showed that this increase did hardly affect the specific storage granules in the dog heart but was present in the soluble fraction. Similar results were obtained by Bhagat (1963). The nature of this temporary pool is still unknown and it may include several different components, some presumably of unspecific character. A rebinding of excess transmitter at specific binding sites should aid in terminating its action (Axelrod, 1964) and also be useful in transmitter economy (Stjärne, 1964).

Some attempts have been made to influence the amine content of animal organs by administering precursors. After insulin depletion of the suprarenals Van Arman (1951) observed a 30 per cent increase in the amine content after DOPA. Euler and Uddén (1951) found an increase in the total NA content of heart, spleen and liver in the cat after infusion of DOPA.

The granules isolated from bovine splenic nerves and from rabbit heart have essentially the same appearance (Euler and Lishajko, 1965a). Potter and Axelrod (1963), Schümann, Schnell and Philippu (1964) and Snyder, Michaelson and Musacchio (1964) have succeeded in obtaining very pure fractions of granules from organs.

SPECIFICITY OF THE TRANSMITTER

Differential estimation of catecholamines has shown that although noradrenaline is the predominant catecholamine in all organs, adrenaline (Euler, 1949) and dopamine (Goodall, 1951) are constantly found in small amounts. The dopamine amounts vary greatly in different organs and do not seem to be directly connected with the neurotransmitter system in mammals, except in the CNS. The existence of special dopamine cells (Falck, Hillarp and Torp, 1959) seems to offer an explanation for its occurrence in many organs for instance in the lungs of ruminants. While noradrenaline disappears to a large extent after denervation and degeneration of the adrenergic nerves, the adrenaline content seems to be little or not at all affected (Euler and Purkhold, 1951; Strömblad, 1960). It therefore appears that adrenaline is bound to some other structure than the nerve axons, and it has been suggested that it occurs in chromaffin cells in the tissues. Support for this concept has recently been obtained by Sjöstrand (1965).

PROPERTIES OF NA-STORING GRANULES

Most of the studies of the specific NA-storing granules have been carried out on material obtained from bovine splenic nerves. By squeezing the nerves between rollers (Euler, 1958) a press juice is obtained which contains a suspension of granules. After dilution with isotonic ice-cold sucrose or potassium phosphate larger particles may be removed by centrifugation at $9000 \times g$ for 10 min, leaving a supernatant which contains most of the NA-storing

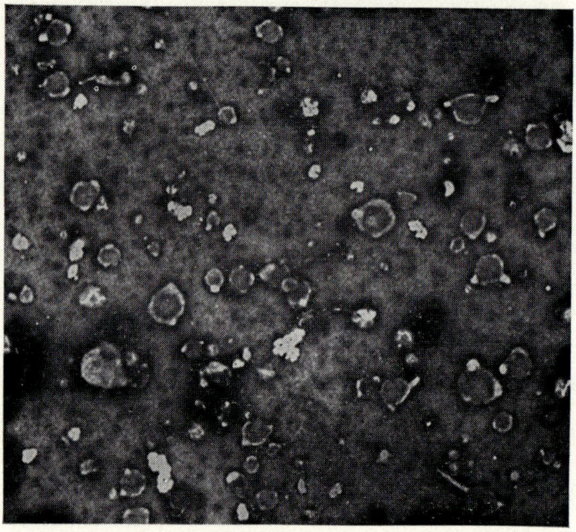

FIG. 1. Electron micrograph of isolated granules from bovine splenic nerves. Negative staining with phosphotungstic acid. ×36,000. (Euler and Swanbeck, unpubl. observations.) (Note small knobs on membrane.)

granules or vesicles. Centrifugation for 30 min at $50{,}000 \times g$ yields a semi-translucent small pellet which consists largely of particles with a diameter of 300–2000 Å (Fig. 1). These particles may be resuspended in sucrose or phosphate buffer with only a small loss of activity. The particles contain the NA in a biologically inactive form. Passage of the particles through an alumina column does not cause adsorption of NA, nor is the NA contained in the granules oxidized by exposure to MnO_2.

At low temperature the particles retain their NA for several hours with little loss. When incubated at 20°C in phosphate buffer the half-life is about 1 hour at pH 7·5. At 37°C the release rate is greatly increased, the half-life being some 10 minutes. In this respect as in several others (cf. Stjärne, 1964) the nerve granules differ from the adrenal medullary granules which have a half-life of more than 100 minutes at 37°C.

The nerve granules are sensitive to acids and rapidly lose their NA at pH 4 or lower. A number of drugs influence the release rate in either direction (Euler and Lishajko, 1965b), a few examples of which will be mentioned below.

In the presence of NA 10^{-4} M no net loss of NA occurs during incubation, which has been shown to depend on a simultaneous uptake, as demonstrated

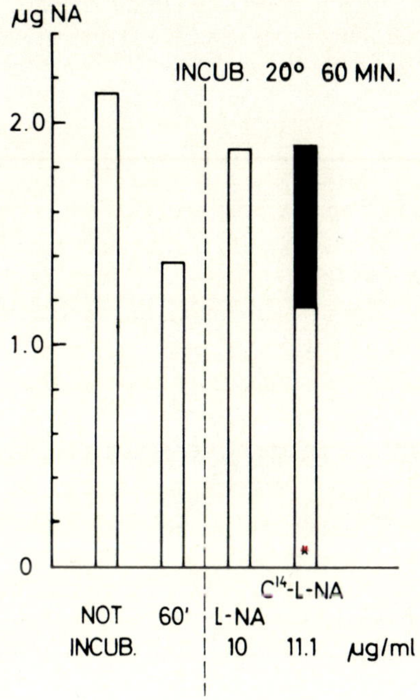

FIG. 2. Isolated splenic nerve granules incubated 60 min in isotonic potassium phosphate at pH 7·5. Noradrenaline (NA), μg per tube. Bars from left (1) not incubated. (2)–(4) incubated 60 min at 20°, (2) no addition, (3) and (4) with l-NA and C^{14}-l-NA. Black part of bar: radioactive NA (Euler and Stjärne, unpubl. observations).

with the aid of labelled amines (Euler, Stjärne and Lishajko, 1963) (Fig. 2).

The "protective" effect of ATP (and ADP) is likewise due to concomitant uptake of NA from the incubation medium into the granules, although this occurs at a NA concentration level of 2×10^{-6} and higher. AMP lacks this effect (Euler and Lishajko, 1963b).

When incubated with adrenaline (A) in concentrations of 10^{-4} M, NA is released at about the normal rate from granules while A is taken up to approximately equivalent amounts (Euler and Lishajko, 1963a).

UPTAKE IN PARTIALLY DEPLETED GRANULES

After partial depletion at 20° or 37°C, addition of NA or A to the incubation medium causes an uptake of these amines. This uptake is favoured by shifting the pH in the incubation medium to more alkaline reaction, and is greatly enhanced by addition of ATP and Mg both at low (10^{-5} M) and high (10^{-4} M) concentration of NA or A (Euler and Lishajko, 1963b), as previously shown for adrenal medullary granules by Kirshner (1962) and by Carlsson, Hillarp and Waldeck (1963).

STEREOSPECIFICITY OF UPTAKE AND RELEASE

The "protective" effect of NA on isolated nerve granules described above is markedly higher for the *l*-isomer of NA than for the *d*-isomer or the racemic compound, indicating that the uptake process is to some extent stereospecific (Euler and Lishajko, 1964). A comparison of the uptake of unlabelled isomers in combination with tracer amounts of the H^3-*dl*-NA has shown that the uptake of the labelled *d*-isomer is only about $\frac{1}{5}$ of that of the *l*-isomer (Stjärne and Euler, 1965).

In order to compare the release rate of the two isomers, isolated nerve granules were incubated with *l*- and *d*-A respectively. After uptake (exchange) of the isomers the granules were washed and the release rate of the two isomers determined over a period of 90 min. It then emerged that the release rate was identical for the two isomers, indicating that while the uptake process shows a well marked stereospecificity, the two isomers, once incorporated in the granules, leave them at the same temperature-dependent rate.

EFFECT OF SOME DRUGS ON THE NA RELEASE AND UPTAKE IN ISOLATED GRANULES

A large number of chemical substances and drugs have been tested with regard to their action on the release and uptake of amines in nerve granules. In some instances rather marked effects have been noticed both in inhibitory and enhancing direction.

The release rate is increased by a number of drugs which also cause various degrees of NA depletion *in vivo*, such as tyramine, α-methyl-*m*-tyramine, prenylamine (Segontin) and decaborane. These drugs either substitute the normal transmitter or cause some alteration in the ability of the granules to retain NA. Other drugs, such as reserpine, prenylamine in low concentrations, phenoxybenzamine and nethalide decrease the release rate (Euler and Lishajko, 1965b; Euler, Stjärne and Lishajko, 1964). Particularly striking is the inhibitory effect of reserpine on the spontaneous release rate which is obvious even after pretreatment of isolated granules with reserpine in 10^{-8} M concentration (see Euler and Lishajko, 1965b).

In most cases the mechanism for these actions is unknown, but the fact that both phenoxybenzamine, DCI and nethalide strongly inhibit the release and uptake rates suggests the possibility that granules are supplied with a structure bearing relationships to effector cell receptors. Some drugs which may cause a release *in vivo* do not seem to influence the NA release rate from granules, such as acetylcholine, nicotine, 5-HT, bretylium and guanethidine. It therefore seems that these compounds act on other nervous structures than the storage granules.

The NA depleting effect of *d*- and *l*-A on organs *in vivo* seems to be due to a competitive exchange reaction, since A is taken up in approximately equivalent amounts (Andén, 1964; Westfall, 1965). This kind of effect occurs with several amines or amine precursors and can be readily demonstrated in the isolated granule system (Euler and Lishajko, 1963a).

NA DEPLETION AND UPTAKE *IN VIVO*

As shown by Schöne and Lindner (1962) the drug Segontin causes a loss of NA from various organs after administration *in vivo*. This effect has been confirmed by Mackenna (1965) who also showed that injection of NA or A could refill the depleted stores to their normal content. The amines thus taken up are retained only for a brief time and then leave the stores. In homogenates of the heart of Segontin-treated rabbits the amine taken up was distributed in an approximately normal fashion, indicating that the uptake mechanism was not lost while the ability to retain the amines was reduced.

Recently another drug, the boron hydride $B_{10}H_{14}$ (decaborane) has been shown to deplete the NA stores in the rat brain (Merritt, Schultz and Wykes, 1964). In rabbits a dose of 4 mg/kg causes an 80–90 per cent depletion of the NA in the heart, spleen and kidneys and 60–70 per cent depletion in the brain (Euler and Lishajko, 1965c). The depletion proceeds gradually and is most marked between 24 and 48 hr after the injection of the drug. If NA or A are administered intravenously or intramuscularly in adequate doses in the depleted animal the amines are taken up by the organs (Fig. 3), and remain for several hours, whereafter they gradually leave the organs over a period of about 20 hr. The proportion of bound NA appears to be constant, irrespective of the total amount of NA in the organ (cf. Stjärne, 1964), suggesting a dynamic equilibrium between the granule-bound and the "free" pool. Although about 40 per cent of the adrenaline still remains in the adrenals after decaborane treatment no uptake of circulating adrenaline from the adrenals into the depleted organs is noted.

While the uptake of *l*-A is about 80 per cent of that of *l*-NA the uptake of the *d*-isomer of A is consistently less and amounts to about $\frac{1}{2}$–$\frac{1}{3}$ of that of

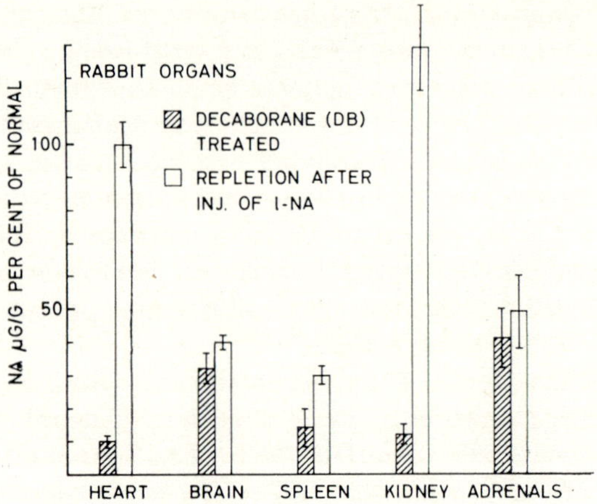

FIG. 3. NA content in rabbit organs in per cent of normal after depletion with decaborane 24 hr previously (hatched bars) and 1–2 hr after 0·3–0·4 mg/kg Na intramuscularly (empty bars) (Euler and Lishajko, 1965c).

the *l*-isomer (Euler and Lishajko, 1965d). A stereospecificity of the uptake process has been previously noted in non-depleted organs by Nash, Costa and Brodie (1963), Kopin and Bridgers (1963), Maickel, Beaven and Brodie (1963) and in the perfused heart by Iversen (1963).

GRANULAR STORAGE IN RELATION TO NERVE TRANSMISSION

While part of the NA occurring in the peripheral stores (nerve terminals) occurs in a granule-bound and relatively stable form, another part occurs less firmly bound. This portion probably corresponds to some extent to the part appearing as free transmitter after homogenization or other treatment of the nerve or tissue used for the separation of granules. The exact proportions between these two "pools" cannot at present be ascertained but even with rough mechanical disintegration of the tissue (Ultra-Turrax homogenizer) the proportion of firmly bound NA is as high as 70 per cent of the total amount in the vesicular gland of the bull after removal of the coarse, incompletely homogenized fraction.

It may be assumed that the first step in the transmitter release associated with nerve stimulation is a mobilization of the loosely bound NA, presumably in connection with the depolarization of the nerve membrane. Its exact localization is not known at present, but one might speculate that some membrane system closely related to the granules and axon membrane is involved (cf. Kopin, this Symposium). In this connection it might be mentioned that adrenergic nerve phospholipids—which occur in membranes—

have a high binding capacity for NA, which is readily reversible by even slight ionic concentration changes (Euler and Lishajko, to be published). A binding of this kind may be particularly suited for a temporary storage and rapid release.

The disequilibrium ensuing after the release induced by nerve activity may be postulated to enhance a release from the storage granules which has been shown to occur at decreased NA concentration in the medium. This, again, must be assumed to trigger the resynthesis process, since physiological nerve activity hardly affects the total amount present in the stores. How this autonomic regulation of the resynthesis rate is organized is not known but several possibilities present themselves, such as product inhibition, limited storage space and others. This general scheme is of course not incompatible with associated mechanisms such as rebinding of released transmitter and uptake of circulating amines, since granules and presumably other binding structures readily take up amines as far as their binding capacity permits. Whether or not the storage granule represents a complete system for biosynthesis (cf. Udenfriend, this Symposium) is still not definitely known, even though an array of enzymes in the granule as a synthesizing unit carrying out the various steps of synthesis in a kind of "moving belt" system is an attractive thought.

SUMMARY

Some of the present concepts of the storage of noradrenaline in adrenergic neurons and terminals in organs are briefly reviewed and may be summarized as follows.

1. Peripheral stores of noradrenaline are directly associated with adrenergic axons and disappear after degeneration of the nerves.

2. The transmitter occurs in the whole neuron but is concentrated in the terminal parts.

3. Part of the transmitter is bound to granules or vesicles, 0·03–0·2 μ in diameter, which can be isolated and studied *in vitro*.

4. The noradrenaline store, represented by granules, has a limited capacity which is normally utilized to a high extent.

5. A transient extragranular uptake in organs can be observed after administration of noradrenaline and other amines.

6. Partially depleted terminals as well as isolated granules can take up noradrenaline and also other amines on a competitive basis.

7. Various drugs enhance the spontaneous release of isolated granules (tyramine, prenylamine 3×10^{-4} M, decaborane) or inhibit the release (reserpine 10^{-7}–10^{-5} M, phenoxybenzamine, nethalide, prenylamine 10^{-5} M).

8. After depletion of the peripheral stores in the rabbit heart by decaborane, administration of noradrenaline causes a rapid uptake to almost normal values. The amines taken up leave the stores over a period of several hours.

9. The proportion of noradrenaline found in the granules before or after decaborane depletion is approximately constant, irrespective of the total amount in the rabbit heart.

10. The uptake of adrenaline in the decaborane-depleted rabbit heart and in isolated granules is partly stereospecific, in contrast to the release which occurs at equal rates for the *l*- and *d*-isomer both *in vivo* and *in vitro*.

11. The function of the storage granules in relation to transmitter release is briefly discussed.

ACKNOWLEDGMENTS

Most of the work presented in this review has been sponsored by the *Air Office of Scientific Research* under Grant AF EOAR 65-52 through the European Office of Aerospace Research (OAR) United States Air Force, by Grant NA 04432-02 from the United States Public Health Service and by the Swedish Medical Research Council, project nr 14X-97-01.

REFERENCES

ANDÉN, N.-E. (1964) Uptake and release of dextro- and laevo-adrenaline in noradrenergic stores, *Acta Pharmacol.* (Kbh.) **21**, 59–75.

AXELROD, J. (1964) The uptake and release of catecholamines and the effect of drugs In: *Progress in Brain Research* **8**, 81–9. Eds H. E. Himwich and W. A. Himwich. Elsevier Publ. Company, Amsterdam.

AXELROD, J., H. WEIL-MALHERBE and R. TOMCHICK (1959) The physiological disposition of H^3-epinephrine and its metabolite metanephrine, *J. Pharmacol. Exp. Ther.* **127**, 251–6.

BHAGAT, B. (1963) Effect of noradrenaline injection on the catecholamine content of the rat heart, *Arch. Int. Pharmacodyn.* **146**, 47–55.

BLASCHKO, H. and A. D. WELCH (1953) Localization of adrenaline in cytoplasmic particles of the bovine adrenal medulla, *Arch. Exp. Path. Pharmak.* **219**, 17–22.

CARLSSON, A., N.-Å. HILLARP and B. WALDECK (1963) Analysis of the Mg^{++}-ATP dependent storage mechanism in the amine granules of the adrenal medulla, *Acta Physiol. Scand.* **59**, Suppl. 215.

CRAMER, W. (1919) Observations on the functional activity of the suprarenal gland in health, and in disease, *Sci. Rept. Imp. Cancer Research Fund* **6**, 1–23.

EULER, U. S. VON (1949) The distribution of sympathin N and sympathin A in spleen and splenic nerves of cattle, *Acta Physiol. Scand.* **19**, 207–14.

EULER, U. S. VON (1951) The nature of adrenergic nerve mediators, *Pharmacol. Rev.* **3**, 247–77.

EULER, U. S. VON (1954) Adrenaline and noradrenaline: Distribution and action, *Pharmacol. Rev.* **6**, 15–22.

EULER, U. S. VON (1958) The presence of the adrenergic neurotransmitter in intra-axonal structures, *Acta Physiol. Scand.* **43**, 155–66,

EULER, U. S. VON (1961) Neurotransmission in the adrenergic nervous system, *The Harvey Lectures*, Series 55, 43–65.

EULER, U. S. VON and N.-Å. HILLARP (1956) Evidence for the presence of noradrenaline in submicroscopic structures of adrenergic axons, *Nature*, *Lond.* **177**, 44–5.

EULER, U. S. VON and F. LISHAJKO (1963a) Catecholamine release and uptake in isolated adrenergic nerve granules, *Acta Physiol. Scand.* **57**, 468–80.

EULER, U. S. VON and F. LISHAJKO (1963b) Effect of adenine nucleotides on catecholamine release and uptake in isolated adrenergic nerve granules, *Acta Physiol. Scand.* **59**, 454–61.

EULER, U. S. VON and F. LISHAJKO (1964) Uptake of L- and D-isomers of catecholamines in adrenergic nerve granules, *Acta Physiol. Scand.* **60**, 217–22.

EULER, U. S. VON and F. LISHAJKO (1965a) Free and bound noradrenaline in the rabbit heart, *Nature, Lond.* **205**, 179–80.

EULER, U. S. VON and F. LISHAJKO (1965b) Effect of drugs on the storage granules of adrenergic nerves. In: Pharmacology of cholinergic and adrenergic transmission, *Proc. 2nd Int. Pharmacol. Meeting*, Prague, 20–23 August 1963. Czechoslovak Medical Press, Praha, pp. 245–59.

EULER, U. S. VON and F. LISHAJKO (1965c) Catecholamine depletion and uptake in adrenergic nerve vesicles and in rabbit organs after decaborane, *Acta Physiol. Scand.* In press.

EULER, U. S. VON and F. LISHAJKO (1965d) Stereospecific catecholamine uptake in rabbit hearts depleted by decaborane, *Int. J. Neuropharmacol.* **4**, 273–80.

EULER, U. S. VON and A. PURKHOLD (1951) Effect of sympathetic denervation on the noradrenaline and adrenaline content of the spleen, kidney, and salivary glands in the sheep, *Acta Physiol. Scand.* **24**, 212–17.

EULER, U. S. VON, L. STJÄRNE and F. LISHAJKO (1963) Uptake of radioactively labelled *dl*-catecholamines in isolated adrenergic nerve granules with and without reserpine, *Life Sci.* **2**, 878–85.

EULER, U. S. VON, L. STJÄRNE and F. LISHAJKO (1964) Effects of reserpine, Segontin and phenoxybenzamine on the catecholamines and ATP of isolated nerve and adreno-medullary storage granules, *Life Sci.* **3**, 35–40.

EULER, U. S. VON and G. SWANBECK (1964) Some morphological features of catecholamine storing nerve vesicles, *Acta Physiol. Scand.* **62**, 487–8.

EULER, U. S. VON and P. UDDÉN (1951) Increase in noradrenaline content of organs after administration of DOPA in the cat, *Experientia* (Basel) **7**, 465–6.

FALCK, B. (1962) Observations on the possibilities of the cellular localization of mono-amines by a fluorescence method, *Acta Physiol. Scand.* **56**, Suppl. 197.

FALCK, B., N.-Å. HILLARP and A. TORP (1959) A new type of chromaffin cells, probably storing dopamine, *Nature, Lond.* **183**, 267–8.

GARVEN, H. S. D. and F. W. GAIRNS (1952) The silver diammine ion staining of peripheral nerve elements and the interpretation of the results: with a modification of the Biel-schowsky-Gros method for frozen sections, *Quart. J. Exp. Physiol.* **37**, 131–42.

GOODALL, MCC. (1951) Studies of adrenaline and noradrenaline in mammalian heart and suprarenals, *Acta Physiol. Scand.* **24**, Suppl. 85.

HERTTING, G., J. AXELROD, I. J. KOPIN and L. G. WHITBY (1961) Lack of uptake of catechol-amines after chronic denervation of sympathetic nerves, *Nature, Lond.* **189**, 66.

HILLARP, N.-Å. (1946) Structure of the synapse and the peripheral innervation apparatus of the autonomic nervous system, *Acta Anat.* (Basel) **2**, Suppl. 4.

HILLARP, N.-Å., S. LAGERSTEDT and B. NILSON (1953) The isolation of a granular fraction from the suprarenal medulla, containing the sympathomimetic catechol amines, *Acta Physiol. Scand.* **29**, 251–63.

IVERSEN, L. L. (1963) The uptake of noradrenaline by the isolated perfused rat heart, *Brit. J. Pharmacol.* **21**, 523–37.

KIRSHNER, N. (1962) Uptake of catecholamines by a particulate fraction of the adrenal medulla, *J. Biol. Chem.* **237**, 2311–17.

KOPIN, I. J. and W. BRIDGERS (1963) Differences in D- and L-norepinephrine-H^3, *Life Sci.* **2**, 356–62.

LEVER, J. D. and A. C. ESTERHUIZEN (1961) Fine structure of the arteriolar nerves in the guinea pig pancreas, *Nature, Lond.* **192**, 566–7.

MACKENNA, B. R. (1965) Uptake of catecholamines by the hearts of rabbits treated with Segontin, *Acta Physiol. Scand.* **63**, 413–22.

MAICKEL, R. P., M. A. BEAVEN and B. B. BRODIE (1963) Implications of uptake and storage of norepinephrine by sympathetic nerve endings, *Life Sci.* **2**, 953–8.

MERRITT, J. H., E. J. SCHULTZ and A. A. WYKES (1964) Effect of decaborane on the nor-epinephrine content of rat brain, *Biochem. Pharmacol.* **13**, 1364–6.

NASH, C. W., E. COSTA and B. B. BRODIE (1963) Stereospecificity in the release of H^3-NE from rat hearts by D- and L-isomers of NE, *Pharmacologist* **5**, 258.

POTTER, L. T. and J. AXELROD (1963) Properties of norepinephrine storage particles of the rat heart, *J. Pharmacol.* **142**, 299–305.

RICHARDSON, K. C. (1962) The fine structure of autonomic nerve endings in smooth muscle of the rat vas deferens, *J. Anat.* (Lond.) **96**, 427–42.

ROBERTIS, E. DE (1964) *Histophysiology of Synapses and Neurosecretion.* Pergamon Press, New York, N.Y.

RYD, G. (1962) Protective effect of bretylium on noradrenaline stores in organs, *Acta Physiol. Scand.* **56**, 90–3.

SCHÜMANN, H. J., K. SCHNELL and A. PHILIPPU (1964) Subcelluläre Verteilung von Noradrenalin und Adrenalin im Meerschweinchenherzen, *Arch. Exp. Path. Pharmak.* **249**, 251–66.

SCHÖNE, H.-H. and E. LINDNER (1962) Über die Wirkung von N-(3′-Phenyl-propyl-(2′))-1,1-diphenyl-propyl-(3)-amin auf den Katecholamin-Stoffwechsel, *Klin. Wschr.* **40**, 1196–1200.

SEDVALL, G. (1964) Noradrenaline storage in skeletal muscle, *Acta Physiol. Scand.* **60**, 39–50.

SJÖSTRAND, N. (1965) The adrenergic innervation of the vas deferens and the accessory male genital glands, *Acta Physiol. Scand.* **65**, Suppl. 257.

SNYDER, S. H., I. A. MICHAELSON and J. MUSACCHIO (1964) Purification of norepinephrine storage granules from rat heart, *Life Sci.* **3**, 965–70.

STJÄRNE, L. (1964) Studies of catecholamine uptake storage and release mechanisms, *Acta Physiol. Scand.* **62**, Suppl. 228.

STJÄRNE, L. and U. S. VON EULER (1965) Stereospecificity of amine uptake mechanism in nerve granules, *J. Pharmacol.*

STRÖMBLAD, B. C. R. (1960) Adrenaline–noradrenaline content of the submaxillary gland of the cat, *Experientia* (Basel) **16**, 417–18.

VAN ARMAN, C. G. (1951) Amino acids and amines as precursors of epinephrine, *Amer. J. Physiol.* **164**, 476–9.

WEGMANN, A. and H. KAKO (1961) Particle-bound and free catecholamines in dog hearts and the uptake of injected norepinephrine, *Nature, Lond.* **192**, 978.

WESTFALL, T. C. (1965) Uptake and exchange of catecholamines in rat tissues after administration of *d*- and *l*-adrenaline, *Act Physiol. Scand.* **63**, 336–42.

WOLFE, D. E., L. T. POTTER, K. C. RICHARDSON and J. AXELROD (1962) Localizing tritiated norepinephrine in sympathetic axons by electron microscopic autoradiography, *Science* **138**, 440–2.

DISCUSSION

BROWN: (1) Richardson described 3 types of vesicles as being characteristic of sympathetic endings, did you see this?

(2) We have some shaky evidence that after depletion of stores by nerve stimulation, there is a latency of some 30 min after an infusion of noradrenaline before the amount liberated by nerve stimulation is restored.

EULER: (1) We have not looked in detail for different types, except for the size.

(2) When a tissue has been depleted with decaborane, the uptake of noradrenaline is extremely fast. After intravenous injection of noradrenaline, the stores may be completely filled in a minute or two.

ERÄNKÖ: What was the concentration of NA in the incubation medium used to replenish the NA stores in the granules? Do you consider this uptake an active or passive process?

EULER: The concentration of NA used for replenishment is of the order of 10^{-4} M when no ATP is added. In the presence of ATP-Mg uptake occurs with much lower concentrations of NA. We consider the latter uptake as an active uptake process, which may not be the case with high concentrations.

UVNÄS: Can you influence the release of noradrenaline from your granules?

EULER: The release rate may be greatly influenced by a number of drugs, and chemical agents, both in enhancing and retarding direction. Some drugs like reserpine, Segontin, and phenoxybenzamine have very marked actions on the release rate.

STEREOSPECIFICITY OF AMINE UPTAKE MECHANISMS IN NERVE GRANULES

L. STJÄRNE, U. S. VON EULER and F. LISHAJKO
Department of Physiology, Karolinska Institutet, Stockholm

THE question of the stereospecificity of the amine uptake mechanisms of noradrenergic nerves has attracted much interest during recent years. One major reason for this is that the radioactively labelled NE which has been commercially available and widely used for studies of the amine uptake mechanisms as well as of other functions of this neuron, and of the metabolic fate of circulating and locally released NE, has consisted of a racemic mixture of the two stereoisomers.

In recent experiments with isolated granules from bovine splenic nerves we have made observations, which clearly show that these granules distinctly prefer the *l*-isomer. We have ventured to express this preference quantitatively as an "affinity ratio" *l*-NE/*d*-NE of about 6/1.

The technique used involved a two-step incubation of the isolated granules in isotonic potassium phosphate, pH 7·5, at 20°C. In the first, the incubation was carried out in the presence of 42–48 nanomoles (i.e. 7–8 μg) per ml of nonlabelled *l*-, *dl*- or *d*-NE, together with a tracer dose of tritiated *dl*-NE. After this the granule suspension was centrifuged and the granules were resuspended in NE-free potassium phosphate for a second incubation period of 120 min.

I should like to demonstrate these results. Figure 1 shows one typical experiment from each of the *l*-, *dl*- and *d*-NE series. The left ordinate shows the NE content of the granules, in per cent of the starting level. In all three experiments the NE content of the granules in the control tubes, which were incubated in the absence of added exogenous NE, during 60 min at 20°C fell by about 45 per cent. However, in the presence of 48 nanomoles/ml of added *l*-NE this fall was strongly counteracted, while the same amount of *dl*-NA was less efficient, and *d*-NE virtually inactive in this respect. Thus in agreement with previous observations at this laboratory *l*-NE was shown to have the best "protective effect" of the two stereoisomers. After reincubation of the granules in a NE-free medium the intragranular NE decreased at a rate largely identical with that of the controls from the first incubation.

The right ordinate shows the specific activity ratios, defined as the ratio or the specific activity of the NE in the granules in cpm/nanomol of NE, over the specific activity of the original suspension medium. The uptake of the tritiated *dl*-NA was quite low in the *l*-NE experiment, considerably higher in the *dl*- and excessively high, even reaching the value of unity, in the *d*-NE experiment.

This is further established on the next figure (Fig. 2). This shows the means and range of the specific activity ratios during the second incubation period. The uptake of the labelled *dl*-NE in the *d*-NE experiments was twice as high as in the *dl*- and three times as high as in the *l*-NE experiments. The main-

NE IN MEDIUM (nanomol/ml)

	a	b	a	b	a	b
l-NE	6	54	6	24	6	6
d-NE	0	0	0	24	0	48
l-NE+d-NE	6	54	6	48	6	54

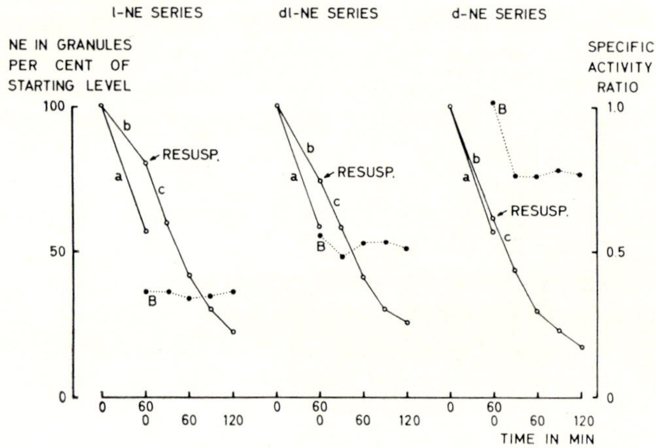

FIG. 1.

tenance, at least in the *l*- and *dl*-series, of the initial specific activity ratio throughout the second incubation period in a NE-free medium, in spite of a substantial fall in the total intragranular NE, shows that in these cases the exogenous NE taken up into the granules was subsequently released largely at the same rate, and therefore presumably in the same manner, as the endogenous NE.

Figure 3 shows the "protective efficiency", defined as the "protective effect" per nanomol of *l*-, *dl*- or *d*-NE. The "protective efficiency" of *l*-NE was about 5 times as high as that of *d*-NE.

One may thus conclude that the isolated nerve granules, under the experimental conditions used, distinctly discriminate between the two stereoisomers of NE.

The total amount of NE present in the granules at the end of an incubation period consists partly of endogenous NE molecules, which have remained

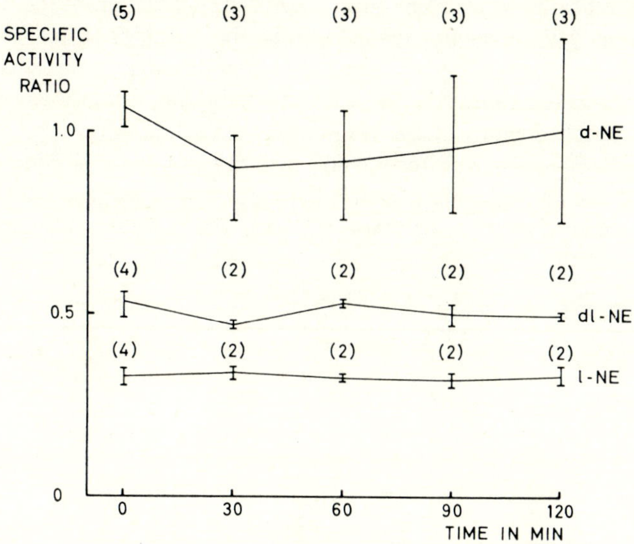
SPECIFIC ACTIVITY RATIO
(5)
(3)
(3)
(3)
(3)
(4)
(2)
(2)
(2)
(2)
(4)
(2)
(2)
(2)
(2)
1.0
0.5
0
d-NE
dl-NE
l-NE
0
30
60
90
120
TIME IN MIN

FIG. 2.

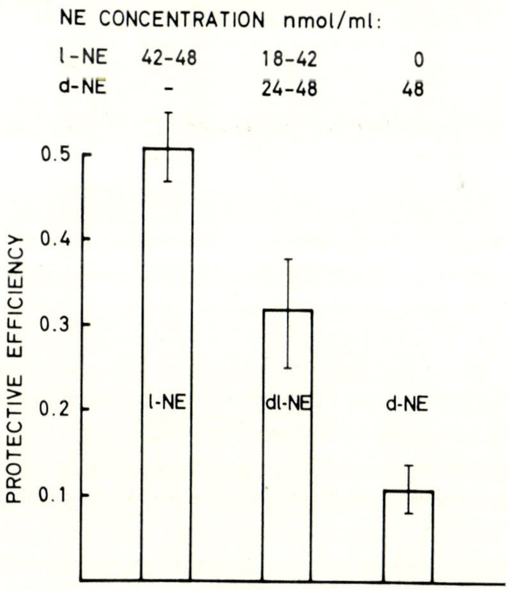
NE CONCENTRATION nmol/ml:
l-NE 42-48 18-42 0
d-NE - 24-48 48
0.5
0.4
0.3
0.2
0.1
PROTECTIVE EFFICIENCY
l-NE
dl-NE
d-NE

FIG. 3.

in the granules throughout the experiment, partly of endogenous NE molecules released into the medium and recaptured into the granules, and finally of exogenous NE molecules taken up into the granules from the medium, in exchange for some of the endogenous NE molecules lost. Calculation of the amount of exogenous NE taken up into the granules in the *dl*-NE experiments, by determination of the fraction of the total radioactive NE added to the medium which was recovered from the granules, showed that 32·5 per cent of the endogenous NE originally present in the granules remained there at the end of the first incubation period. On the basis of the assumption

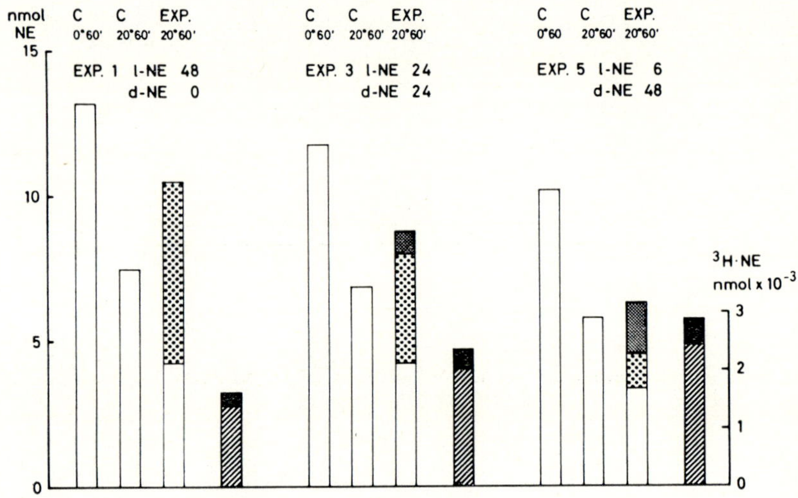

FIG. 4.

that the release of NE from the granules into the medium is spontaneous and independent of the presence or absence of NE in the medium, at least at the concentrations used in the present series, this figure was used to give a first approximation of remaining endogenous NE in the *l*- and *d*-NE experiments as well.

Under this premise it was possible to calculate the uptake of labelled and unlabelled *l*- and *d*-NE in the three types of experiment by comparing the specific activities of the NE in the medium with those of the NE taken up into the granules. It turned out that the "affinity ratio" of *l*-NE to *d*-NE was 5·9, i.e. the *l*-isomer was taken up and/or retained about 6 times as efficiently as the *d*-isomer.

Figure 4 shows the results from one experiment from each of the *l*-, *dl*- and *d*-NE series. The left ordinate shows the total NE content of the granules. The first column in each group shows the NE content in the controls kept at 0°C, the second column that of the controls incubated at 20°C in the absence of added exogenous NE but with about 6 nanomoles of endogenous *l*-NE present in the medium. The third column shows the total amount of NE fluorimetrically determined in the granules incubated at 20°C in the presence of the indicated amounts of exogenous *l*-, *dl*- or *d*-NE. The heavily dotted part of this column indicates exogenous *l*-NE, the finely dotted part exogenous *d*-NE. The fourth column in each group refers to the right ordinate which shows the tritiated *dl*-NE taken up into the granules in picomoles. The striped part of those columns indicates the tritiated *l*-NE, the filled part the tritiated *d*-NE.

The ratio of labelled *l*- to *d*-NE in all experiments, independently of the absolute uptake of tritiated NE, was about 6 to 1. In the *l*-NE series the exogenous NE in the granules consisted almost exclusively of the *l*-form, while *l*-NE accounted for about 85 per cent of the total exogenous NE in the *dl*-series and about 40 per cent in the *d*-NE series.

It may be significant that the figure for the "affinity ratio" *l*-NE/*d*-NE arrived at in our experiments with isolated nerve granules coincides with that which can be calculated for the isolated rat heart from the K_m values for *l*- and *d*-NE presented by Iversen. It does suggest that the preference in noradrenergic neurons for the *l*-isomer, reported by several authors, although contested by others, may be of this order of magnitude.

However, the exact implications for interpretation of data obtained with the use of racemic tracer require additional experiments, particularly concerning the retention of the *d*-NE taken up.

BIOCHEMICAL ASPECTS OF STORAGE AND RELEASE OF BIOGENIC AMINES FROM SYMPATHETIC NERVES

IRWIN J. KOPIN

Laboratory of Clinical Science
National Institute of Mental Health
U.S. Department of Health, Education and Welfare
Public Health Service
Bethesda, Md. U.S.A.

THE concept of chemical transmission of a stimulus between nerve cells and at neuroeffector junctions was first proposed nearly a century ago (Dubois-Reymond, 1875). The similarity of responses to sympathetic nerve stimulation and epinephrine administration led Elliott (1905) to suggest that this substance might be the chemical stimulant liberated at the sympathetic nerve ending. The release of a sympathomimetic substance was demonstrated by Loewi (1921) and the work of Cannon and his coworkers which is summarized in a classic monograph (Cannon and Rosenblueth, 1937). Although a number of investigators had suggested its chemical structure, the identity of the adrenergic transmitter was not established until 1946 (Euler, 1946). It is now generally accepted that norepinephrine is the major transmitter substance released at sympathetic nerve endings (Euler, 1956).

A neurochemical transmitter must be released rapidly in response to a nerve impulse, act on a receptor, and have its action rapidly terminated. Following release from the neuron, the transmitter must be replaced so rapidly that subsequent impulses will be effective. The microstructure of the nerve ending and receptor, the intracellular distribution of the transmitter and various enzymes responsible for its synthesis and metabolism, and the mechanism for transmitter storage, release, transport, and interaction with a receptor are all important aspects of neurochemical transmission. These have been studied by a variety of techniques.

CATECHOLAMINE METABOLISM

The study of tritiated catecholamines of high specific activity, administered in tracer quantities, was initiated by Axelrod and his coworkers. Initial studies of the metabolism of epinephrine-7-H^3 in man and animals showed that O-methylation is the predominant pathway for metabolic inactivation of administered catechols (Axelrod, 1959). A portion of the administered norepinephrine-H^3 is taken up into the tissues, retained in an inactive form, and slowly released (Whitby, Axelrod, and Weil-Malherbe, 1961; Iversen and Whitby, 1962).

There are a number of observations which indicate that norepinephrine-H^3 which is taken up into the tissues enters the sympathetic nerves and mixes with the endogenous catecholamine. The uptake of labeled catecholamine is markedly diminished after chronic sympathetic denervation (Hertting, Axelrod, Kopin, and Whitby, 1961). When depleted by reserpine, endogenous and labeled norepinephrine disappear at the same rate (Hertting and Hess, 1962). Labeled norepinephrine can be released from the perfused cat spleen (Hertting and Axelrod, 1961), skeletal muscle (Rosell, Kopin and Axelrod, 1963), and dog heart (Chidsey and Harrison, 1963) by sympathetic nerve stimulation. Autoradiography has provided direct evidence for the presence of norepinephrine-H^3 in sympathetic nerves (Wolfe, Potter, Richardson and Axelrod, 1962; Marks, Samorajski and Webster, 1962). Labeled norepinephrine taken up by the nerve endings may therefore be considered a valid tracer for the endogenous catecholamine.

The metabolic fate of intravenously administered norepinephrine appears to depend on the extent of tissue binding. Norepinephrine which is destroyed immediately following intravenous administration is metabolized predominantly by O-methylation, presumably in the liver and kidney. The labeled catecholamine bound in the nerve endings and slowly released is destroyed mainly by deamination (Kopin and Gordon, 1963).

Following administration of a monoamine oxidase (MAO) inhibitor, norepinephrine in brain increases, followed by the accumulation of normetanephrine. These observations suggested to Carlsson (1960) that MAO is close to the site of norepinephrine synthesis and that the rise in normetanephrine indicates overflow of free catecholamine, with consequent excitation.

Direct evidence for this hypothesis has been obtained *in vivo* and in the isolated perfused rat heart. Ten hours after intravenous administration of norepinephrine-H^3 to a rat, urinary excretion of radioactivity decreases in a single exponential phase. If a drug which depletes tissue of norepinephrine is administered, the rate of excretion of radioactivity increases. Examination of the metabolic products of norepinephrine in the urine reveals the metabolic fate of the catecholamine released (Kopin and Gordon, 1962, 1963). Reserpine, which depletes catecholamine stores without inducing much sympathomimetic response, causes a marked increase in excretion of deaminated catechols. Tyramine, other sympathomimetic agents, and ganglionic stimulants which result in increased sympathetic nerve activity release free norepinephrine into the circulation. Like administered catecholamines, these are mostly O-methylated. An estimate of the proportion of depleted norepinephrine which reaches the circulation can be made from a comparison of norepinephrine which is excreted unchanged after drug administration and after intravenous infusion of norepinephrine. After reserpine, less than 15 per cent of the catecholamine lost from the tissue reaches the circulation. When tyramine is administered, however, over 60 per cent of the norepinephrine is released into the circulation (Kopin and Gordon, 1962). These results have been confirmed by observations of the metabolites released during drug depletion of catecholamine stores from the isolated perfused rat heart (Nash, Costa, and Brodie, 1964).

When norepinephrine-H^3 is infused into an isolated rat heart, the amine is taken up and stored (Kopin, Hertting, and Gordon, 1962; Iversen, 1963). If the heart of a reserpinized animal is used, some uptake still occurs, but the amine is not retained and storage is markedly diminished. While there is no change in the proportion of norepinephrine which is O-methylated, there is a large increase in deaminated products. These results suggest that reserpine does not greatly interfere with entry of the catecholamine into the cells, but does interfere with storage. The amine taken up is not protected from oxidation by intraneuronal MAO (Kopin, Hertting, and Gordon, 1963).

Similar results have been obtained in the rat brain. Glowinski and Axelrod (1965) have shown that norepinephrine-H^3 introduced into the lateral ventricle of reserpinized rats is taken up and then is deaminated to a greater extent than normal. This uptake can be blocked by pretreatment with amphetamine; when this is done, O-methylated products increase. It appears therefore, that intraneuronal destruction by monoamine oxidase and extraneuronal metabolism by catechol-O-methyl transferase occurs in the brain, as suggested by Carlsson (1960), as well as in the peripheral adrenergic nervous system.

When tyramine is administered, norepinephrine is released, probably by displacement from its storage site. Subsequent doses of tyramine have a diminishing effect as tachyphylaxis results, and less norepinephrine is released despite the presence of large stores of the catecholamine (Potter, Axelrod, and Kopin, 1962; Axelrod, Gordon, Hertting, Kopin and Potter, 1962). It has been demonstrated recently (Pöch and Kopin, 1965) that the catecholamine released by tyramine is replaced by octopamine and that when tachyphylaxis has developed, octopamine rather than norepinephrine is selectively released by tyramine (Fig. 1). The presence of octopamine appears to protect norepinephrine stores from further depletion by tyramine.

SYNTHESIS OF FALSE NEUROCHEMICAL TRANSMITTERS

Octopamine formation from tyramine does not occur after chronic sympathetic denervation (Carlsson and Waldeck, 1964; Fischer, Musacchio, Kopin and Axelrod, 1964). When dopamine-β-oxidase is inhibited with disulfiram (Goldstein, Anagnoste, Lauber and McKeregan, 1964), octo-

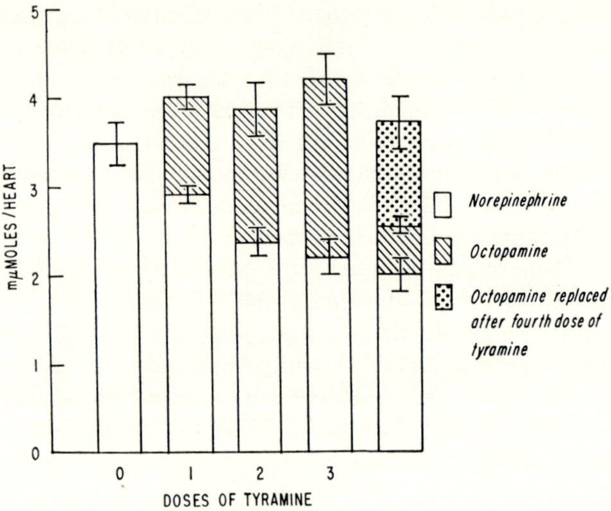

FIG. 1. Replacement of norepinephrine by octopamine during tachyphylaxis to tyramine. Tyramine (10 mg/kg) labeled with tritium (50 μc/μM) was administered intramuscularly to rats at intervals of 15 min for one, two, or three doses. One other group of rats received three doses of labeled tyramine and a fourth dose of unlabeled tyramine (Group 4A) while another group received three doses of unlabeled tyramine and a fourth dose of labeled tyramine (Group 4B). Fifteen minutes after the last dose of tyramine the rats were killed and the hearts examined for norepinephrine and octopamine-H^3. The octopamine-H^3 in the hearts of animals in Group 4A is "old" octopamine, while that in Group 4B is "new" octopamine.

pamine does not form and tyramine is rapidly destroyed by monoamine oxidase (Musacchio, Kopin and Snyder, 1964).

Replacement of a neurotransmitter at its storage site by a closely related molecule which can function as a relatively ineffective false transmitter has been considered for both cholinergic and adrenergic nerve endings. Burgen, Burke, and Desbarats-Schonbaum (1956) found that triethylcholine can be acetylated by choline acetylase, and Bowman and Rand (1961) suggested that the relatively inactive triethylacetylcholine might replace acetylcholine and be responsible for the myasthenia-like picture produced in animals treated with triethylcholine. When Carlsson and Lindqvist (1962) showed that α-methylnorepinephrine could form from α-methyldopa it was suggested that the α-methylcatecholamine might replace norepinephrine and be a false transmitter. Day and Rand (1963) restored the actions of tyramine in reserpinized animals by administering α-methyldopa. This restoration does not occur when dopamine-β-oxidase is inhibited by pretreatment with disulfiram (Musacchio, Bhagat and Kopin, 1965). Muscholl and Maitre (1963) demonstrated release of α-methyl-norepinephrine by nerve stimulation of the isolated perfused heart of a rabbit pretreated with α-methyldopa.

These findings and the formation of octopamine in sympathetic nerves indicated earlier prompted Fischer, Horst, and Kopin (1965) to examine a number of sympathomimetic amines with regard to their possible role as false neurochemical transmitters. Octopamine, but not tyramine, could be released by sympathetic nerve stimulation from the isolated perfused cat spleen (Fischer, Horst and Kopin, 1965) and heart (Fischer and Kopin, 1965). Similarly the β-hydroxylated derivatives of *m*-tyramine and α-methyltyramine, but not the amines themselves, could be released by nerve stimulation. Dopamine and α-methyldopamine, lacking the β-hydroxyl group, could also be released by sympathetic nerve stimulation (Musacchio, Fischer and Kopin, 1965). While this work was in progress, Crout and Shore (1964) reported that metaraminol could be released by sympathetic nerve stimulation. Thus, a number of phenylethylamine derivatives having two of the three hydroxyl groups of norepinephrine may act as false neurochemical transmitters (Table 1). The amines released by nerve stimulation are retained,

TABLE 1. AMINE RETENTION IN THE "MICROSOMAL" FRACTION OF TISSUE HOMOGENATES AND THEIR RELEASE BY SYMPATHETIC NERVE STIMULATION

	% *in particulate*	*Release by nerve stimulation*
Norepinephrine	48.1±1.1	+
α-CH_3 Norepinephrine	51.0±2.7	+
Dopamine	38.0±2.2	+
α-CH_3 Dopamine	34.0±3.5	+
m-Octopamine	30.0±3.4	+
α-CH_3 *m*-Octopamine	—	+
Octopamine	25.5±1.4	+
α-CH_3 Octopamine	27.0±3.1	+
Tyramine	0	—
α-Methyl-Tyramine	6	—
m-Tyramine	0	—

Tritiated amines were administered intravenously with or without disulfiram pretreatment. Hearts were removed, homogenized in cold isotonic sucrose and the microsomal fraction separated and assayed for the tritiated amine and its β-hydroxylated derivative (Musacchio, Kopin and Weise, 1965).

at least partially, in the particulate fractions. Epinephrine-H^3 has also been shown to be taken up by sympathetic nerve fibers in skeletal muscle and released by sympathetic nerve stimulation.

Replacement of norepinephrine by a false transmitter has been implicated in the diminished sympathetic responsiveness seen after treatment with α-methyldopa (Day and Rand, 1963), metaraminol (Crout and Shore, 1964), monoamine oxidase inhibitors and tachyphylaxis to sympathomimetic amines (Kopin, Fischer, Musacchio and Horst, 1964).

NOREPINEPHRINE STORAGE AND RELEASE

There are a number of similarities in synthesis, storage, and release mechanisms of cholinergic and adrenergic neurohumors. In both cases, the enzyme responsible for synthesis is located at the storage sites. Subcellular particles which contain both choline-acetylase and acetylcholine were described by Hebb and Whittaker (1958). Structures containing dopamine-β-oxidase and catecholamines have been isolated from the adrenal medulla (Kirshner, 1957) and the rat heart (Potter and Axelrod, 1963a). Perry (1953), and Birks and MacIntosh (1957) interpreted observations on release of acetylcholine from the isolated perfused superior cervical ganglion of the cat in terms of "available" and "bound" stores of acetylcholine. Trendelenburg (1961) first postulated such variably available stores of norepinephrine. The limited release of norepinephrine by tyramine (Potter, Axelrod and Kopin, 1962; Potter and Axelrod, 1963a), the multiphasic changes in specific activity of norepinephrine in the rat heart after administration of a tracer dose of the labeled catecholamine (Axelrod, Hertting and Patrick, 1961; Montanari, Costa, Beaven and Brodie, 1963; Kopin, Hertting and Gordon, 1962), and the difference in specific activity of tyramine-released and residual norepinephrine in the rat heart (Potter, Axelrod and Kopin, 1962) provide direct evidence for a division of catecholamine stores.

Recent studies of the miniature end-plate potentials in sympathetically innervated guinea pig vas deferens (Burnstock and Holman, 1962) suggest that, as in cholinergic systems (Fatt and Katz, 1952), neurochemical transmitter substance is released in quanta. Vesicular storage may represent an anatomical basis for quantal release. The simultaneous release of the contents of numerous vesicles at the synaptic membrane (Palay, 1958) could provide sufficient numbers of quanta to result in receptor activation.

Norepinephrine-containing subcellular particles have been obtained from the adrenal medulla (Hillarp, Lagerstedt and Nilson, 1953), adrenergic nerves (Euler and Hillarp, 1956), brain (Weil-Malherbe and Bone, 1959), and sympathetically innervated structures (Potter and Axelrod, 1963b).

The chemical composition of these structures, their enzymatic content, their catecholamine-binding properties *in vitro* and the effect of drugs and ions on these have been examined in a number of laboratories. These properties have been reviewed (Stjärne, 1964) and are the subject of other communications at this symposium. The particles contain dopamine-β-oxidase and are probably involved in the synthesis as well as storage of the transmitter. Because dopamine-β-oxidase is relatively nonspecific (Creveling, Daly, Witkop and Udenfriend, 1962), phenylethylamine derivatives other than dopamine can be β-hydroxylated and replace norepinephrine in the storage sites. The enzyme can synthesize false transmitters.

Musacchio, Kopin and Weise (1965) examined the subcellular distribution of some phenylethylamine derivatives and their β-hydroxylated derivatives formed *in vivo*. When *p*-tyramine, *m*-tyramine, or α-methyltyramine are administered, their β-hydroxylated derivatives are rapidly formed and are retained in the "microsomal" fraction, but not as well as norepinephrine (Table 1). When β-hydroxylation is prevented by treatment with disulfiram, the amines are almost wholly restricted to the soluble fraction. Dopamine and α-methyl dopamine are retained in the particulate fraction almost as efficiently as norepinephrine, though they lack a β-hydroxyl group (Musacchio and Kopin, unpublished observations). That different proportions of the amines are retained in the particulate fraction indicates that release into the soluble fraction is not merely a consequence of physical disturbance of subcellular structures by homogenization. Musacchio, Kopin and Weise (1965) related their subcellular distribution findings obtained *in vitro* to the release of amines by drugs *in vivo*. It was found that greater retention of an amine by the particulate fraction correlated with greater resistance to depletion by reserpine, tyramine, and guanethidine (Table 2). There appears to be a relation of structure to strength of binding.

TABLE 2. AMINE RETENTION IN THE MICROSOMAL FRACTION COMPARED WITH RETENTION AFTER RELEASE BY VARIOUS CATECHOLAMINE-DEPLETING AGENTS

	% *in particulate*	*Retention** in vivo *after administration of:*		
		reserpine	*guanethidine*	*tyramine*
Norepinephrine-H^3	48±1.1	12.4	64	50
Dopamine-H^3 (+Disulfiram)	38±2.2	2.9	14.5	13.4
m-Octopamine-H^3	30±3.4	1.9	3.1	8.7
Octopamine-H^3	25±1.4	0.86	1.5	4.1

* Per cent control.

There is considerable evidence that norepinephrine stores in the sympathetic nerves are divided into several functional compartments as well as storage compartments. The undiminished response to tyramine after a major portion of the norepinephrine stores are depleted by reserpine (Trendelenburg, 1961), the limited depletion of catecholamine stores by tyramine (Potter, Axelrod and Kopin, 1962), and the multiphasic release of labeled norepinephrine from the heart (Axelrod, Hertting and Patrick, 1961; Montanari, Costa, Beaven and Brodie, 1963) indicate at least two major pools of norepinephrine in tissues. Stjärne (1964) presented evidence for this division at the level of the storage granules.

When norepinephrine stores are depleted by reserpine, nerve stimulation fails to produce a sympathetic response but responses are restored by administration of norepinephrine (Burn and Rand, 1960), although only minute quantities of the catecholamine are retained in the reserpinized tissue (Crout, Muskus and Trendelenburg, 1962). These findings suggest a small pool of norepinephrine readily released by nerve stimulation which is resistant to the action of reserpine. Sedvall and Thorson (1963) showed that after acute sympathetic denervation of cat skeletal muscle, reserpinization did not prevent the transmission of a limited number of nerve impulses. Presumably this reserpine-resistant store is also preserved following ganglionic blockade or decentralization, both of which prevent depletion of norepinephrine by nerve impulses (Hertting, Potter and Axelrod, 1962).

INTERACTION OF BRETYLIUM WITH SYMPATHETIC NERVE ENDINGS

Of the drugs which affect release of catecholamines at the nerve ending, bretylium is of particular interest. This quarternary amine is taken up by sympathetic nervous tissue (Boura, Copp, Duncombe, Green and McCoubrey, 1960) and blocks the effects of sympathetic nerve stimulation (Boura and Green, 1959) without interfering with conduction of nerve impulses (Exley, 1960). The transient adrenergic affects of bretylium appear to be related to release of norepinephrine (Gillis and Nash, 1961; Gokhale, Gulati and Kelkar, 1963). Bretylium causes a transient release of norepinephrine-H^3 from the perfused cat spleen (Hertting, Axelrod and Patrick, R. W., 1962). The labeled catecholamine appears to be derived from the reserpine-resistant norepinephrine store which can be released by nerve stimulation (Fischer, Weise and Kopin, 1965). Subsequent release of norepinephrine-H^3 by nerve stimulation is blocked. When large doses of bretylium are administered over a prolonged period, norepinephrine stores can be depleted (Benmiloud and Euler, 1963). Bretylium can also release bound octopamine and α-methyloctopamine, the α-methyl derivative being more susceptible to the action of bretylium (Fig. 2). The mechanism of this releasing action is currently being investigated.

Burn and Rand (1959) suggested that the impulse passing down the post-ganglionic fiber releases acetylcholine, which in turn releases norepinephrine. Brandon and Rand (1961) demonstrated acetylcholine in the cat spleen and found that chronic post-ganglionic denervation caused its disappearance to the same extent as the disappearance of norepinephrine. They showed that a cholinergic-like substance was released into the venous effluent during stimulation of the splenic nerves of the spleen of a reserpinized cat and showed that hemicholinium produced failure of contraction of the spleen in response to nerve stimulation. This effect of hemicholinium was reversed by choline.

Acetylcholine was shown to contract the spleen and this effect was diminished by reserpine treatment and by adrenolytic drugs. All of these findings support the hypothesis presented by Burn and Rand (1959).

Ferry (1963) found that acetylcholine caused a centripetal discharge of the C fibers of the splenic nerve. These findings suggested that sympathetic postganglionic nerves had been excited somewhere near their endings. Since hexamethonium does not abolish splenic contraction when the splenic nerves are stimulated, Ferry (1963) concluded that acetylcholine stimulation of the

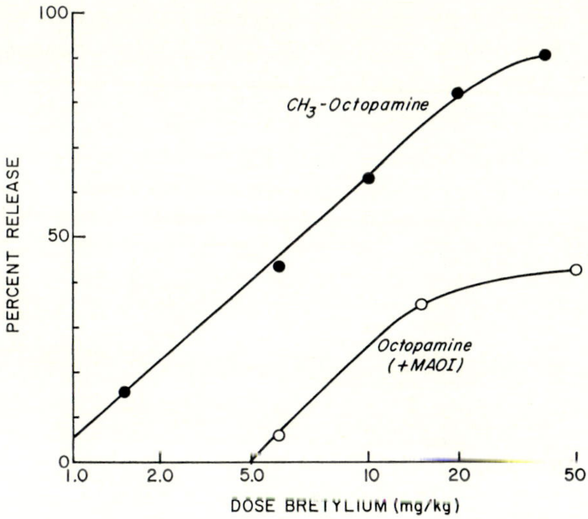

FIG. 2. Dose-response curves for release of octopamine-H^3 and α-methyl-octopamine-H^3 by bretylium. Rats received 20 μc tyramine-H^3 or α-methyl-tyramine-H^3 intravenously. Rats given tyramine were pretreated with pheniprazine (10 mg/kg) one hr earlier. Two hrs after administration, when the labeled tissue amines were mainly β-hydroxylated (Musacchio, Kopin and Weise, 1965), bretylium was administered subcutaneously. The per cent depletion of the β-hydroxylated amines in the heart 1 hr later is indicated.

postganglionic nerves caused norepinephrine release. Hexamethonium blocks the actions of both acetylcholine and bretylium, presumably by interfering with their entry into the nerve (Burn and Gibbons, 1964). In a bath containing bretylium, isolated strips cut from spleen failed to contract in the presence of acetylcholine (Brandon and Rand, 1961). Huković (1960) showed that the adrenergic action of acetylcholine on the rabbit ear was prevented by bretylium, but on washing out the drug this action was restored. Bretylium in high concentration may interfere with entry of acetylcholine into the nerve.

Fischer, Weise and Kopin (1965) and Hertting and Widhalm (1965) showed that acetylcholine releases norepinephrine-H^3 and some of the false transmitters from the isolated perfused cat spleen. Bretylium, in high concentrations, prevents the release of norepinephrine-H^3 by acetylcholine or by nerve stimulation. When the drug is washed out, however, release of norepinephrine-H^3 by acetylcholine is restored, but the effects of nerve stimulation are still blocked. The dissociation of the effects of nerve stimulation and acetylcholine after bretylium treatment suggest that postganglionic nerve stimulation by acetylcholine (Ferry, 1963) is not the mechanism of norepinephrine release.

One day after administration of bretylium-H^3 its release from the isolated perfused cat spleen could be demonstrated (Fischer, Weise and Kopin, 1965). There are a number of differences between bretylium and the other transmitters. Acetylcholine fails to release bretylium. The relationship of release of bretylium to frequency of nerve stimulation is different from the relationship observed for the other transmitters (Fischer, Weise and Kopin, 1965). Furthermore, bretylium does not appear to be bound in the norepinephrine storage granules (Costa, Chang and Brodie, 1964).

These observations suggest that bretylium cannot be considered a typical false adrenergic transmitter. Bretylium appears to interact with the sympathetic neuron at the level of the release mechanism triggered by the nerve impulse, but does not block the action of acetylcholine in releasing norepinephrine. Koelle (1962) has summarized the evidence for a possible action of acetylcholine in liberating another neurohumoral transmitter from the presynaptic site. If release of acetylcholine were essential to release of norepinephrine, as suggested by Burn and Rand (1959), and bretylium interfered with release of acetylcholine, then its adrenergic blocking action might be explained.

In addition to blocking adrenergic transmission, bretylium can also block neuromuscular conduction (Boura and Green, 1959). This blocking action does not appear to be the result of competition or depolarization (Dixit, Gulati and Gokhale, 1961). These observations are consistent with the view that bretylium may interfere with acetylcholine release. The release of bretylium-H^3 by splenic nerve stimulation suggests that bretylium may replace a substance usually released during sympathetic nerve stimulation. The possibility that bretylium replaces acetylcholine or a similar quarternary ammonium compound is currently being investigated.

After nerve degeneration consequent to chronic sympathetic denervation, small amounts of norepinephrine (Hertting, Axelrod, Kopin and Whitby, 1961) and other amines (Carlsson and Waldeck, 1964; Fischer, Musacchio, Kopin and Axelrod, 1964) can still be bound. This binding is not due to incomplete denervation since it is insensitive to reserpine (Fischer, Kopin and Axelrod, 1965). The significance of extraneuronal binding has not been established.

It had been previously suggested that there is a mechanism for active transport of catecholamines into the sympathetic neuron (Dengler, Spiegel and Titus, 1961; Furchgott, Kirpekar, Riker and Schwab, 1963; Iversen, 1963). Recently Iversen (1965) has described two distinct processes for norepinephrine uptake by the rat heart which are affected differently by sympathomimetic amines and other drugs (Burgen and Iversen, 1965). Re-uptake into the nerve appears to be a major means of terminating the action of the released transmitter (Hertting and Axelrod, 1961). The mechanism of active amine transport into the neurons, however, appears to be different from the mechanism of intracellular binding.

HYPOTHETICAL STRUCTURE OF THE SYMPATHETIC NERVE ENDING

While it is generally accepted that norepinephrine is stored in subcellular structures in the neighborhood of the sympathetic neuro-effector junction, it is also recognized that some of the catecholamine must be stored in a form which readily enters the soluble portion in tissue homogenates. Radioactive norepinephrine is taken up into sympathetic ganglia and stored, but there is no evidence of particulate binding of the catecholamine in these structures (Fischer and Snyder, 1965). Reserpine releases the labeled catecholamine indicating that granules or vesicles, as usually described, are not the only locus for norepinephrine binding, nor the only site at which reserpine acts.

Although reserpine in small doses releases norepinephrine *in vivo*, low concentrations strongly inhibit spontaneous release of catecholamines from isolated bovine splenic nerve granules (Euler and Lishajko, 1961) and inhibit exchange of granular norepinephrine with catecholamines in solution (Euler and Lishajko, 1963). Only in high concentrations does reserpine release norepinephrine from these structures.

A hypothetical functional model of the sympathetic neuron must correlate anatomical, biochemical, physiological, and pharmacological findings. This model must account for the presence of vesicles in the terminal axon, but their absence from the perikaryon. It should explain the quantal release of the transmitter. The multiphasic disappearance of labeled norepinephrine and its different metabolic fates must be explained. One day after administration of norepinephrine-H^3, reserpine will cause depletion, by deamination, of labeled and unlabeled norepinephrine at the same rate, while tyramine given at this time preferentially releases the unlabeled catecholamine into the circulation intact. The differential effect of tyramine on labeled and unlabeled norepinephrine does not appear to be related to the division of the stores into soluble and particulate fractions (Iversen and Whitby, 1963). These findings must be accounted for by the neuron model. The apparently paradoxical action of low concentrations of reserpine on isolated bovine splenic nerve particles must also be clarified.

A possible model of the sympathetic neuron would include a hypothetical pre-vesicular structure formed in the perikaryon and extending into the axon (Fig. 3). This structure, containing dopamine-β-oxidase, has qualitatively different binding sites separated by a membrane crossed by an active transport system. It is proposed that near the neuro-effector junctions this membrane forms vesicles which are later incorporated into a terminal cell membrane during its depolarization (Katz, 1962).

During homogenization the membrane might form some additional vesicles, the extent of vesicle formation varying with the homogenization

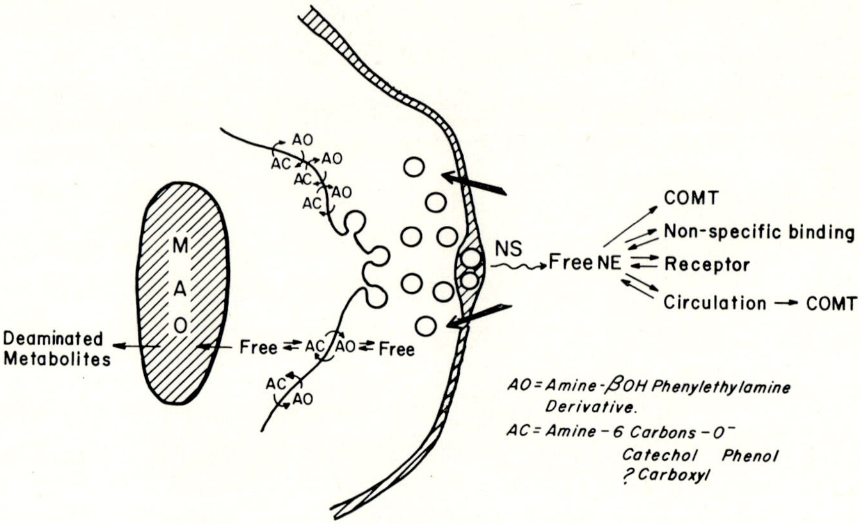

FIG. 3. Hypothetical model of the catecholamine storage sites in the sympathetic nerve. The membrane is shown as a smooth structure for sake of clarity, but would be highly convoluted in the cell. The details are discussed in the text.

technique, but would also liberate some bound norepinephrine into the soluble fraction. The binding sites on the outside of the vesicles and on the membrane surface protected from the mitochondria are postulated to require a phenylethylamine structure for entry, and a β-hydroxyl group for retention at the binding site. The side of the membrane exposed to the mitochondria corresponds to the inside of the vesicle. The binding sites here may require both a positively charged amine and an acidic group separated from the amine by about six carbon atoms. This site would firmly bind the strongly acidic catecholamines but only weakly bind the phenolic amines. The free amines in equilibrium with this side of the hypothetical membrane would be destroyed by mitochondrial MAO. The free amines on the other side of the membrane would be protected from monoamine oxidase and could escape from the cell, intact, if their concentration were elevated and the uptake mechanism blocked by a drug such as tyramine. Metabolism by catechol-O-methyltransferase predominates outside the neuron. This pathway also predominates in the interval immediately after exogenous norepinephrine-H^3 reaches the area of the nerve ending.

Administered norepinephrine-H^3 taken into the cell first labels the "outer" catecholamine binding sites. These have a rapid turnover rate and after the labeled molecules have disappeared from the "outer" side of the membrane and vesicles, the remaining labeled molecules on the "inside" are slowly metabolized by monoamine oxidase. The molecules bound on one side of the membrane are not all equally likely to enter a given empty binding site on the other side. While this model does account for two pools, they cannot be considered mathematical compartments in the usual sense. Kinetic analysis of this model yields a non-linear set of equations. The solution of these will yield a sum of exponentials, but this cannot be considered *evidence* for or against this model or the true open compartment system suggested by Montanari, Costa, Beaven and Brodie (1963).

When reserpine interferes with the binding mechanism on the "inside" of the membrane and vesicles, free norepinephrine is liberated. The molecules on the outside may be transferred to the empty binding site "inside" but are then released from the reserpine blocked sites or are released "outside" and diffuse "in" or outside the cell. Thus "inner" labeled molecules and "outer" unlabeled molecules on the membrane disappear at about the same rate. Amines in the vesicles already formed (their number may vary with the rate of nerve stimulation) before reserpine acts are liberated into the interior and are still available for release by nerve stimulation (an apparent reserpine-resistant store). Eccles (1964) estimates that the reserve supply of vesicles at the synaptic knob in the skeletal neuromuscular junction is sufficient for only 10,000 nerve impulses. This is remarkably similar to the size of the reserpine-resistant store of norepinephrine in cat skeletal muscle (Sedvall and Thorson, 1963). This scheme also explains the inhibition by reserpine of spontaneous release of catecholamines from isolated bovine splenic nerve vesicles.

Following administration of tyramine, the norepinephrine on the outer surface of the membrane and vesicles is replaced by octopamine. Because these binding sites are occupied, transfer of the norepinephrine molecules from the "inside" to the "outside" does not occur. It is the rapidly turning over store which is preferentially released from the neuron; its return is blocked by the phenolic amine and it reaches the circulation or is destroyed by catechol-O-methyltransferase. On subsequent homogenization the specific activity of the "soluble" and particulate bound norepinephrine are similar, but higher than before depletion of unlabeled catecholamine by tyramine.

While this model does appear to account for a number of biochemical and pharmacological observations which are not easily explained if all of the norepinephrine in the sympathetic nerve endings were stored in vesicles, it must be emphasized that there is as yet no anatomical evidence for the hypothetical membrane postulated to bind catecholamines or β-hydroxylated amines. The concept of intracellular membranes and their possible involvement in the "import, export and intracellular circulation" of various substances is not new (Palade, 1956). The formation of subcellular particles from this membrane *in vivo* or artifactually has also been discussed (e.g. Robertson, 1959; Rothschild, 1963). It would not be surprising, therefore, if such a membrane did exist.

SUMMARY

The metabolic fate of norepinephrine in sympathetic nerves varies with its site of destruction. When destroyed in the neuron spontaneously or during reserpine depletion, deamination predominates. When released in an active form outside the neuron as a consequence of nerve stimulation or displacement by a sympathomimetic amine, O-methylation is the major pathway for metabolic inactivation. A number of amines lacking only one of the three hydroxyl groups of norepinephrine are retained in the particulate fraction of tissue homogenates to varying degrees. Retention in the particulate fraction correlates with release by sympathetic nerve stimulation for these substances. Betylium releases octopamine and α-methyloctopamine. This drug blocks the release of norepinephrine-H^3 by sympathetic nerve stimulation, but not release by acetylcholine. Acetylcholine releases false neurochemical transmitters, but does not release bretylium-H^3. Bretylium-H^3, however, is released by nerve stimulation. These findings are considered in relation to the cholinergic-link hypothesis of Burn and Rand (1959). A model of the sympathetic nerve ending which includes a pre-vesicular membrane with distinct binding sites on either side, crossed by an active transport system, is discussed. This hypothetical model is useful in explaining several apparently paradoxical observations in literature, such as the different metabolic fates of norepinephrine, the differential effect of tyramine on labeled and unlabeled norepinephrine while reserpine depletes these at the same rate, and the strong inhibition of spontaneous release of norepinephrine from bovine splenic nerve granules by low concentrations of reserpine.

REFERENCES

Axelrod, J. (1959) Metabolism of epinephrine and other sympathomimetic amines, *Physiol. Rev.* **39**, 751–76.

Axelrod, J., E. Gordon, G. Hertting, I. J. Kopin and L. T. Potter (1962) On the mechanism of tachyphylaxis to tyramine in the isolated rat heart, *Brit. J. Pharmacol.* **19**, 56–63.

Axelrod, J., G. Hertting and R. W. Patrick (1961) Inhibition of H^3-norepinephrine release by monoamine oxidase inhibitors, *J. Pharmacol.* **134**, 325–8.

BENMILOUD, M. and U. S. VON EULER (1963) Effects of bretylium, reserpine, guanethidine and sympathetic denervation on the noradrenaline content of the rat submaxillary gland, *Acta Physiol. Scand.* **59**, 34–42.

BIRKS, R. and MACINTOSH (1957) Acetylcholine metabolism at nerve endings, *Brit. Med. Bull.* **13**, 157–61.

BOURA, A. L. A. and A. F. GREEN (1959) The actions of bretylium: adrenergic neurone blocking and other effects, *Brit. J. Pharmacol.* **14**, 536–48.

BOURA, A. L. A., F. F. COPP, W. G. DUNCOMBE, A. F. GREEN and A. MCCOUBREY (1960) The selective accumulation of bretylium in sympathetic ganglia and their post-ganglionic nerves, *Brit. J. Pharmacol.* **15**, 265–70.

BOWMAN, W. C. and M. J. RAND (1961) Actions of triethylcholine on neuromuscular transmission, *Brit. J. Pharmacol.* **17**, 176–95.

BRANDON, K. W. and M. J. RAND (1961) Acetylcholine and sympathetic innervation of the spleen, *J. Physiol., Lond.* **157**, 18–32.

BURGEN, A. S. V., G. BURKE and M. L. DESBARATS-SCHONBAUM (1956) Specificity of brain choline acetylase, *Brit. J. Pharmacol.* **11**, 308–12.

BURGEN, A. S. V. and L. L. IVERSEN (1965) Inhibition of norepinephrine uptake by sympathomimetic amines in the isolated rat heart, *Brit. J. Pharmacol.* **25**, 18–33.

BURN, J. H. and W. R. GIBBONS (1964) The sympathetic postganglionic fibre and the block by bretylium: the block prevented by hexamethonium and imitated by mecamylamine, *Brit. J. Pharmacol.* **22**, 549–57.

BURN, J. H. and M. J. RAND (1959) Sympathetic postganglionic mechanism, *Nature, Lond.* **184**, 163–5.

BURNSTOCK, G. and M. E. HOLMAN (1962) Spontaneous potentials at sympathetic nerve endings in smooth muscle, *J. Physiol., Lond.* **160**, 446–60.

CANNON, W. B. and A. ROSENBLUETH (1937) *Autonomic Neuro-effector Systems.* New York, Macmillan Co.

CARLSSON, A. (1960) Discussed in *CIBA Found. Symp. on Adrenal Mechanisms* (Edited by J. R. Vane, G. E. W. Wolstenholme and M. O'Connor). London, J. & A. Churchill, Ltd.

CARLSSON, A. and M. LINDQVIST (1962) *In vivo* decarboxylation of α-methyl metatyrosine, *Acta Physiol. Scand.* **54**, 37–94.

CARLSSON, A. and B. WALDECK (1964) β-hydroxylation of tyramine *in vivo*, *Acta Pharmacol.* (Kbh.) **20**, 371–4.

CHIDSEY, C. A. and D. C. HARRISON (1963) Studies on the distribution of exogenous norepinephrine in the sympathetic neurotransmitter store, *J. Pharmacol.* **140**, 217–23.

COSTA, E., C. CHANG and B. B. BRODIE (1964) Interaction of guanethidine with adrenergic neurons at a molecular level, *Pharmacologist* **6**, 174.

CREVELING, C. R., J. W. DALY, B. WITKOP and S. UDENFRIEND (1962) Substrates and inhibitors of dopamine-β-oxidase, *Biochim. Biophys. Acta* (Amst.) **64**, 125–34.

CROUT, J. R., A. J. MUSKUS and U. TRENDELENBURG (1962) Effect of tyramine in isolated guinea pig atria in relation to their noradrenaline stores, *Brit. J. Pharmacol.* **18**, 600–11.

CROUT, J. R. and P. A. SHORE (1964) Release of metaraminol (aramine) from the heart by sympathetic nerve stimulation, *Clin. Res.* **12**, 180.

DAY, M. D. and M. J. RAND (1963) A hypothesis for the mode of α-methyldopa in relieving hypertension, *J. Pharm. Pharmacol.* **15**, 221–4.

DENGLER, H. J., H. E. SPIEGEL and E. O. TITUS (1961) Uptake of tritium labeled norepinephrine in brain and other tissues of the cat *in vitro*, *Science* **133**, 1072–3.

DIXIT, B. N., O. D. GULATI and S. D. GOKHALE (1961) Action of bretylium and guanethidine at the neuromuscular junction, *Brit. J. Pharmacol.* **17**, 372–9.

DUBOIS-REYMOND, E. (1875) *Gesammelte Abhandlungen für allgemeine Muskel-und Nervenphysik*, Viet u. Comp., Leipzig.

ELLIOTT, T. R. (1905) The action of adrenalin, *J. Physiol., Lond.* **32**, 401–67.

ECCLES, J. C. (1964) *The Physiology of Synapses.* New York, Acad. Press, Inc., p. 92.

EULER, U. S. VON (1946) A specific sympathomimetic ergone in adrenergic nerve fibres (sympathin) and its relations to adrenaline and noradrenaline, *Acta Physiol. Scand.* **12**, 73–97.

EULER, U. S. VON (1956) *Noradrenaline*. Springfield, Ill., U.S.A., Charles C. Thomas.
EULER, U. S. VON and N.-Å. HILLARP (1956) Evidence for the presence of noradrenaline in subcellular microscopic structures of adrenergic axone, *Nature, Lond.* **177**, 44–5.
EULER, U. S. VON and F. LISHAJKO (1961) Effect of reserpine on the release of catecholamines from isolated nerve and chromaffin cell granules, *Acta Physiol. Scand.* **52**, 137–45
EULER, U. S. VON and F. LISHAJKO (1963) Uptake of radioactively labeled DL-catecholamines in isolated adrenergic nerve granules with and without reserpine, *Life Sci.* **11**, 878–85.
EXLEY, K. A. (1960) The persistence of adrenergic nerve conduction after TM 10 or bretylium in the cat, *Adrenergic Mechanisms* (Edited by J. R. Vane, G. E. W. Wolstenholme and M. O'Connor), pp. 158–61.
FATT, P. and B. KATZ (1952) Spontaneous subthreshold activity at motor nerve endings, *J. Physiol., Lond.* **117**, 109–28.
FERRY, C. B. (1963) The sympathomimetic effect of acetylcholine on the spleen of the cat, *J. Physiol., Lond.* **167**, 487–504.
FISCHER, J. E., W. D. HORST and I. J. KOPIN (1965) β-hydroxylated sympathomimetic amines as false neurotransmitters, *Brit. J. Pharmacol.* **24**, 477–84.
FISCHER, J. E. and I. J. KOPIN (1965) Release of false neurochemical transmitters from the heart. In preparation.
FISCHER, J. E., I. J. KOPIN and J. AXELROD (1965) Evidence for extraneuronal binding of norepinephrine, *J. Pharmacol.* **147**, 181–5.
FISCHER, J. E., J. M. MUSACCHIO, I. J. KOPIN and J. AXELROD (1964) Effects of denervation on the uptake and β-hydroxylation of tyramine in the rat salivary gland, *Life Sci.* **3**, 413–19.
FISCHER, J. E. and S. H. SNYDER (1965) Norepinephrine in sympathetic ganglia. In *J. Pharmacol.* In press.
FISCHER, J. E., V. WEISE and I. J. KOPIN (1965) Interactions of bretylium at sympathetic nerve endings and its release by stimulation. *J. Pharmacol.* In press.
FURCHGOTT, R. F., S. M. KIRPEKAR, M. RIKER and A. SCHWAB (1963) Actions and interactions of norepinephrine, tyramine, and cocaine on aortic strips of rabbit and left atria of guinea pig and cat, *J. Pharmacol.* **142**, 39–58.
GILLIS, C. N. and C. W. NASH (1961) The initial pressor actions of bretylium tosylate and guanethidine sulfate and their relation to relase of catecholamine, *J. Pharmacol.* **134**, 1–7.
GLOWINSKI, J. and J. AXELROD (1965) Drugs and H^3-norepinephrine uptake in brain. In *J. Pharmacol.* **149**, 43–9.
GOKHALE, S. D., O. D. GULATI and V. V. KELKAR (1963) Mechanisms of the initial adrenergic effects of bretylium and guanethidine, *Brit. J. Pharmacol.* **20**, 362–77.
GOLDSTEIN, M., B. ANAGNOSTE, E. LAUBER and M. R. MCKEREGAN (1964) *Life Sci.* **3**, 763–8.
HEBB, C. O. and V. P. WHITTAKER (1958) Intracellular distributions of acetylcholine and choline acetylase, *J. Physiol., Lond.* **142**, 187–96.
HERTTING, G. and J. AXELROD (1961) Fate of tritiated noradrenaline at the sympathetic nerve endings, *Nature, Lond.* **192**, 172–3.
HERTTING, G., J. AXELROD, I. J. KOPIN and L. G. WHITBY (1961) Lack of uptake of catecholamines after chronic denervation of sympathetic nerves, *Nature, Lond.* **189**, 66.
HERTTING, G., J. AXELROD and R. W. PATRICK (1962) Action of bretylium and guanethidine on the uptake and release of H^3-norepinephrine, *Brit. J. Pharmacol.* **18**, 161–6.
HERTTING, G. and S. M. HESS (1962) The site of binding of injected norepinephrine-H^3, *Experientia*, (Basel) **18**, 214–15.
HERTTING, G., L. T. POTTER and J. AXELROD (1962) Effect of decentralization and ganglionic blocking agents on the spontaneous release of H^3-norepinephrine, *J. Pharmacol.* **136**, 289–92.
HERTTING, G. and S. WIDHALM (1965) Über den Mechanismus der Noradrenalin-Freisetzung aus Sympathischen Nervenendigungen, *Arch. Exp. Path. Pharmak.* In press.
HUKOVIC, S. (1960) The action of sympathetic blocking agents on isolated and innervated atria and vessles, *Brit. J. Pharmacol.* **15**, 117–12.
IVERSEN, L. L. (1963) The uptake of noradrenaline by the isolated perfused rat heart, *Brit. J. Pharmacol.* **21**, 523–37.

IVERSEN, L. L. (1965) Uptake of catecholamines at high perfusion concentrations in the isolated rat heart—a novel catecholamine uptake process, *Brit. J. Pharmacol* **25**, 18–33.

IVERSEN, L. L. and L. G. WHITBY (1962) Retention of injected catecholamines by the mouse, *Brit. J. Pharmacol.* **19**, 355–64.

IVERSEN, L. L. and L. G. WHITBY (1963) The subcellular distribution of catecholamines in normal and tyramine-depleted mouse hearts, *Biochem. Pharmacol.* **12**, 582–4.

KATZ, B. (1962) The transmission of impulses from nerve to muscle and the subcellular unit of synaptic action, *Proc. Roy. Soc.* B. **155**, 455–79.

KIRSHNER, N. (1957) Pathway of noradrenaline formation from dopa, *J. Biol. Chem.* **226**, 821–5.

KOELLE, G. (1962) A new general concept of the neurohumoral functions of acetylcholine and anticholinesterase, *J. Pharm., Lond.* **14**, 65–95.

KOPIN, I. J., J. E. FISCHER, J. MUSACCHIO and W. D. HORST (1964) Evidence for a false neurochemical transmitter as a mechanism for the hypotensive effect of monoamine oxidase inhibitors, *Proc. Nat. Acad. Sci.*, (Wash.) **52**, 716–21.

KOPIN, I. J. and E. K. GORDON (1962) Metabolism of norepinephrine-H^3 released by tyramine and reserpine, *J. Pharmacol.* **138**, 351–9.

KOPIN, I. J. and E. K. GORDON (1963) Metabolism of administered and drug-released norepinephrine-7-H^3 in the rat, *J. Pharmacol.* **140**, 207–16.

KOPIN, I. J., G. HERTTING and E. K. GORDON (1962) Fate of norepinephrine-H^3 in the isolated perfused rat heart, *J. Pharmacol.* **138**, 34–40.

LOEWI, O. (1921) Über humorale Übertragbarkeit der Herznerven wirkung, *Pflügers Arch. ges. Physiol.* **189**, 239–42.

MARKS, B. H., T. SAMORAJSKI and E. J. WEBSTER (1962) Radioautographic localization of norepinephrine-H^3 in the tissues of mice, *J. Pharmacol.* **138**, 376–81.

MONTANARI, R., E. COSTA, A. BEAVEN and B. B. BRODIE (1963) Turnover rates of norepinephrine in hearts of intact mice, rats, and guinea pigs, using tritiated norepinephrine, *Life Sci.* **4**, 232–40.

MUSACCHIO, J. M., B. BHAGAT and I. J. KOPIN (1965) Effect of dopamine-β-oxidase inhibition on restoration of the actions of tyramine, *J. Pharmacol.* In press.

MUSACCHIO, J. M., J. E. FISCHER and I. J. KOPIN (1965) Release of dopamine and α-methyl dopamine by sympathetic nerve stimulation, *J. Pharmacol.* In press.

MUSACCHIO, J. M., I. J. KOPIN and S. H. SNYDER (1964) Effects of disulfiram on tissue norepinephrine content and subcellular distribution of dopamine, tyramine, and their β-hydroxylated metabolites, *Life Sci.* **3**, 769–75.

MUSACCHIO, J. M., I. J. KOPIN and V. K. WEISE (1965) Subcellular distribution of some sympathomimetic amines and their β-hydroxylated derivatives in the rat heart, *J. Pharmacol.* **148**, 22–8.

MUSACCHIO, J. M., V. K. WEISE and I. J. KOPIN (1965a) Some observations on the mechanisms of norepinephrine binding, *Nature, Lond.* **205**, 606–7.

MUSCHOLL, E. and L. MAITRE (1963) Release by sympathetic nerve stimulation of α-methyl noradrenaline stored in the heart after administration of α-methyl dopa, *Experientia* (Basel) **19**, 658.

NASH, C. W., E. COSTA and B. B. BRODIE (1964) Actions of reserpine, guanethidine and metaraminol on cardiac catecholamine stores, *Life Sci.* **3**, 441–9.

PALADE, G. E. (1956) The endoplasmic reticulum, *J. Biophys. Biochem. Cytol.* **2**, Suppl. 4, 85–98.

PALAY, S. L. (1958) The morphology of synapses in the central nervous system, *Exp. Cell Res. Suppl.* **5**, 275–93.

PERRY, W. L. M. (1953) Acetylcholine release in the cat's superior cervical ganglion, *J. Physiol., Lond.* **119**, 439–54.

PÖCH, G. and I. J. KOPIN (1965) Replacement of norepinephrine by octopamine during tachyphylaxis to tyramine, *Biochem. Pharmacol.* In press.

POTTER, L. T. and J. AXELROD (1963a) Studies on the storage of norepinephrine and the effect of drugs, *J. Pharmacol.* **140**, 199–206.

POTTER, L. T. and J. AXELROD (1963b) Subcellular localization of catecholamines in tissues of the rat, *J. Pharmacol.* **142**, 291–8.

POTTER, L. T. and J. AXELROD (1963c) Properties of norepinephrine storage particles of the rat heart, *J. Pharmacol.* **142**, 299–305.

POTTER, L. T., J. AXELROD and I. J. KOPIN (1962) Differential binding and release of norepinephrine and tachyphylaxis, *Biochem. Pharmacol.* **11**, 254–6.

ROBERTSON, J. D. (1959) The ultrastructure of cell membranes and their derivatives, *Biochem. Soc. Symposia* **16**, 3–43.

ROSELL, S., J. AXELROD and I. J. KOPIN (1964) Release of tritiated epinephrine following sympathetic nerve stimulation, *Nature, Lond.* **201**, 301.

ROSELL, S., I. J. KOPIN and J. AXELROD (1963) Fate of H^3-norepinephrine in skeletal muscle before and following sympathetic nerve stimulation, *Amer. J. Physiol.* **205**, 317–21.

ROTHSCHILD, J. (1963) Isolation of microsomal membrane, *Biochem. Soc. Symposia* **22**, 4–31.

SEDVALL, G. (1964) Short-term effects of reserpine on noradrenaline levels in skeletal muscle, *Acta Physiol. Scand.* **62**, 101–8.

SEDVALL, G. and J. THORSON (1963) The effect of nerve stimulation on the release of noradrenaline from a reserpine resistant transmitter pool in skeletal muscle, *Biochem. Pharmacol.* **12**, Suppl. 65–6.

STJÄRNE, L. (1964) Studies of catecholamine uptake, storage, and release mechanisms, *Acta Physiol. Scand.* **68**, Suppl. 28.

TRENDELENBURG, U. (1961) Modification of the effect of tyramine by various agents and procedures, *J. Pharmacol.* **134**, 8–18.

WEIL-MALHERBE, H. and A. D. BONE (1959) The effect of reserpine on the intracellular distribution of catecholamines in the brain stem of the rabbit, *J. Neurochem.* **4**, 251–363.

WHITBY, L. G., J. AXELROD and H. WEIL-MALHERBE (1961) Fate of norepinephrine-H^3 in animals, *J. Pharmacol.* **132**, 193–201.

WOLFE, D. E., L. T. POTTER, K. C. RICHARDSON and J. AXELROD (1962) Localizing tritiated norepinephrine in sympathetic axons by electron microscopic autoradiography, *Science* **138**, 440–2.

DISCUSSION

PLETSCHER: Do you have to assume two NE pools in order to explain the difference of reserpine and tyramine? Could the effect of tyramine not be due to competitive inhibition of the oxidative deamination of liberated NE?

KOPIN: If MAO was inhibited, a decrease in deaminated products would be expected. This does not occur *in vivo* or *in vitro*. Furthermore, when tyramine releases norepinephrine, there may be a rise in specific activity of residual norepinephrine. This could occur only if there were more than one pool.

BROWN: Acetylcholine has a non-specific effect in C-fibers, afferent or efferent. Ferry showed that liberation of noradrenaline by injected ACh was associated with centripetal nerve impulses. It would be interesting to find out whether the liberation of noradrenaline by ACh in the preparation treated with bretylium (when the nerves are ineffective) was associated with centripetal impulses.

KOPIN: I agree. Ferry is in the best position to answer this important question and I hope he is willing to do this.

RELEASE OF CATECHOLAMINES FROM THE HEART

E. Muscholl

Pharmakologisches Institut der Universität Mainz, Deutschland

It is well known that a variety of drugs is capable of releasing endogenous catecholamines from sympathetically innervated organs. Prototypes of depleting agents are reserpine, guanethidine, sympathomimetic amines and α-methylated amino acids such as α-methyldopa and α-methyl-*m*-tyrosine. The pertinent literature has been reviewed by many authors and under different aspects. Recent reviews are those by Green (1962), Shore (1962), Brodie and Beaven (1963), Fawaz (1963), Axelrod (1964), Kopin (1964) and Zaimis (1964).

This paper presents a summary of experimental work carried out in our laboratory during the last two years. The main purpose was to study the release of adrenergic transmitters by sympathetic stimulation from the perfused heart and the effects of drugs on the release. The quantity of transmitter released from the nerve endings cannot be estimated directly. The transmitter output of the heart is not identical with the amount of transmitter released since part of it is taken up into the amine store and part may be metabolized. Under our experimental conditions an estimate of the quantity taken up and/or metabolized can be obtained by infusing transmitter substance and measuring both disappearance from the perfusion fluid and net uptake into the heart. Therefore balance studies were carried out with noradrenaline and α-methyl noradrenaline, both of which can serve as adrenergic transmitters.

Since some of the experimental results have been published elsewhere details will be presented only for those observations not yet published.

METHODS

The isolated heart of the rabbit with the sympathetic nervous supply from the right stellate ganglion intact was set up according to Huković and Muscholl (1962). The ganglion was stimulated for 30 sec at supramaximal voltage with rectangular impulses of 3·5 msec duration at a frequency of 10 shocks/sec. Each stimulation period was followed by a resting period of 30 sec. This schedule was usually repeated for 6 min. The effects of drugs were studied by infusing them into the aortic cannula 20 min after the control stimulation 5 min before and during the second stimulation experiment.

Noradrenaline outputs of consecutive stimulation experiments were corrected for the experimental decay observed in the controls.

Removal of amines from the perfusion fluid by the isolated rabbit heart was studied using the method of Lindmar and Muscholl (1964).

For estimation of catecholamines rabbit hearts were homogenized in 25 ml of 5 per cent trichloroacetic acid (TCA) with an "Ultra-Turrax". Extraction was carried out for 20 min, the extract spun and the precipitate re-extracted with the same amount of TCA. The supernatants were combined and an aliquot of 20 ml was taken for adsorption of the catecholamines on alumina. The amines contained in TCA-extracts of plasma and adrenal glands or in heart perfusates were adsorbed on alumina (Euler and Orwén, 1955). If the amines were assayed fluorimetrically or biologically, the alumina was eluted with 5 ml 0·25 N HCl. If paper chromatography was used for separation of the amines, the alumina was eluted with two portions of 0·5 N H_2SO_4, adding up to 2·0 ml. The pH was brought to 4 with solid $NaHCO_3$, the precipitate removed by centrifugation and 6 vol. acetone added. After 2 hr at 4°C the precipitate was removed, 50 mg NaCl added to the acetone extract and the volume reduced to dryness *in vacuo*. The residue was leached with several portions of small amounts of an acetone–ethanol mixture (1 : 1) and the evaporation to a small volume repeated. After spinning, the extract was applied to acid-washed Whatman paper No. 1. Chromatograms were run by the descending technique at 26°C. The following solvents were used: n-butanol–N HCl (2 : 1) and *m*-cresol–glacial acetic acid–H_2O (10 : 0·4 : 9·6, vol./vol.) After 24 hr the papers were dried. *m*-cresol was removed by washing the papers in benzene. After spraying with 0·1 per cent potassium ferricyanide in 5 per cent aqueous ethylenediamine and drying at 50°C (Ellman, 1958) the fluorescence was examined. For the mean travelling distances of catechol amines see Table 1.

TABLE 1. SEPARATION OF CATECHOLAMINES. MEAN TRAVELLING DISTANCES OF AMINES FROM BASE-LINE IN MM. DESCENDING CHROMATOGRAMS RUN FOR 26 HR AT 26°C ON WHATMAN NO. 1 PAPER (COMPOSITION OF Bu–HCl AND *m*-CRESOL–AA–H_2O SEE METHODS)

	Bu–HCl	*m-Cresol-AA-H_2O*
Noradrenaline	90	38
Adrenaline	111	122
Dopamine	140	103
α-Methyl noradrenaline	144	72
Dihydroxy ephedrine	163	185
Isoprenaline	200	250
α-Methyl dopamine	215	163

Differential fluorimetric estimation of adrenaline and noradrenaline with the trihydroxy indole (THI) method and biological assay of noradrenaline, α-methyl noradrenaline or adrenaline on the pithed rat were described previously (Lindmar and Muscholl, 1964, 1965).

3,4-Dihydroxy ephedrine was determined fluorimetrically utilizing a strong fluorescence with the THI method of similar characteristics as adrenaline. Therefore, differential estimation of mixtures with noradrenaline could also be performed. Another extremely sensitive test was the depressor effect on the pithed rat. Most preparations responded to 1 ng. Recoveries of known amounts added to heart and adrenal extracts and to perfusates were about 80 per cent.

The following substances were used as hydrochlorides: *l*-noradrenaline, *l*-adrenaline, *l*-α-methyl noradrenaline (Corbasil, Farbwerke Hoechst, Frankfurt), *d,l*-dihydroxy ephedrine and *d,l*-dihydroxy pseudoephedrine (both C. H. Boehringer Sohn, Ingelheim a. Rhein). Doses and concentrations refer to the bases. Doses of *l*-cocaine, dichloroisoproterenol and 1-(3-methylphenoxy)-2-hydroxy-3-isopropylamino-propane, MHIP (Kö 592, C. H. Boehringer Sohn, Ingelheim) refer to the hydrochlorides.

RESULTS AND DISCUSSION

Release of noradrenaline. The effects of a number of drugs on the noradrenaline output of the heart have recently been summarized by Muscholl (1964). The following points need further comment.

Drugs which are known to inhibit uptake of exogenous noradrenaline (cocaine, tyramine, dichloroisoproterenol) increase noradrenaline output during sympathetic stimulation. Secondary effects can arise if impulse propagation is affected, e.g. by local anaesthetic agents. The inhibition of noradrenaline uptake by dichloroisoproterenol found in the rat *in vivo* (Muscholl, 1961) has been confirmed in this species by Axelrod, Hertting and Potter (1962) and Iversen (1965). A seemingly contradictory finding was reported by Williams and Mayer (1964). These authors injected a dose of 4 μg dichloroisoproterenol into perfused rat hearts and studied the uptake of tritiated noradrenaline rapidly administered in doses of 1 and 5 μg 8 and 30 min later. There was no inhibition of noradrenaline uptake. In the perfusion experiments by Lindmar and Muscholl (1964) on rabbit hearts and Iversen (1965) on rat hearts noradrenaline was infused continuously at concentrations of 10–40 ng/ml. Dichloroisoproterenol inhibited uptake if present during the noradrenaline infusion in concentrations ranging from 0·6–10 μg/ml. Obviously the noradrenaline concentration in the experiments by Williams and Mayer (1964) was too high for demonstrating an uptake inhibition.

The question arises whether the inhibition of noradrenaline uptake by dichloroisoproterenol is causally related to its property to block adrenergic β-receptors. Brown and Gillespie (1957) observed an increase in noradrenaline output from the spleen after blockade of the adrenergic receptors and sympathetic stimulation. They suggested that the transmitter outflow increased because noradrenaline was prevented from combining with the receptors. Alternative explanations of this phenomenon are as follows: (1) The blocking agents prevent the decrease of blood flow following sympathetic stimulation. Hence, re-uptake of transmitter (which in the spleen is dependent on the flow rate) is delayed and the transmitter output increases (Hertting and Schiefthaler, 1963). (2) Adrenergic blocking agents inhibit noradrenaline uptake into stores independently of their action on the receptors

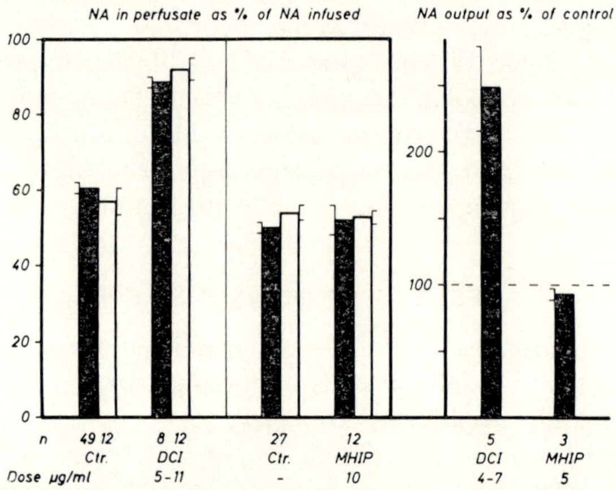

FIG. 1. Effect of dichloroisoproterenol (DCI) and methylphenoxy-2-hydroxy-3-isopropylaminopropane (MHIP) on the removal of noradrenaline (NA) from the perfusate of the isolated rabbit heart (left panel) and the NA output during sympathetic stimulation (right panel). Solid columns result of bioassay, open columns fluorimetric estimations. In infusion experiments all values corrected for individual recoveries. n, number of estimations. NA infused at 10–40 ng/ml.

(Farrant, Harvey and Pennefather, 1964; Thoenen, Hürlimann and Haefely, 1964). This idea is strongly supported by the lack of correlation between intensity of adrenergic blockade and increase of transmitter output (Thoenen, Hürlimann and Haefely, 1964). Apparently no adrenergic blocking drug which is completely devoid of the property to inhibit noradrenaline uptake has yet been investigated on its effect on transmitter release. Such a drug is 1-(3-methyl-phenoxy)-2-hydroxy-3-isopropylamino-propane (MHIP). It inhibits adrenergic β-receptors but does not exert sympathomimetic effects (Engelhardt, 1965). In contrast, dichloroisoproterenol (Moran and Perkins, 1958) has intrinsic sympathomimetic activity. Dr. Weber of our laboratory has studied the effect of MHIP on the removal of noradrenaline from the perfusate of the isolated rabbit heart. Figure 1 shows that the removal was not affected by this drug. The dose of 10 μg/ml is sufficient to block β-receptors completely. For the purpose of comparison the effects of dichloroisoproterenol are included in the graph. Both dichloroisoproterenol and MHIP in the doses indicated blocked the increase in heart rate and force of contraction usually seen during sympathetic stimulation. After dichloroisoproterenol the noradrenaline output rose to 250 per cent of the control output, whereas MHIP did not alter the noradrenaline output significantly.

These results suggest that blockade of adrenergic receptors is not causally related to inhibition of noradrenaline uptake or increase of transmitter output. Rather, the inhibition of amine uptake and increase of transmitter output caused by dichloroisoproterenol seem to be related to its sympathomimetic property. Various sympathomimetic amines have been found to inhibit noradrenaline uptake (Iversen, 1964).

Release of adrenaline. Significant concentrations of adrenaline are present in the hearts of several species (Hökfelt, 1951; Goodall, 1951; Muscholl, 1959). Synthesis of noradrenaline has been shown to occur in the heart itself (Spector, Sjoerdsma, Zaltzman-Nirenberg, Levitt and Udenfriend, 1963) and only a small proportion of the amine stored was found to be taken up from the circulation (Kopin and Gordon, 1963). However, heart adrenaline seems to be replenished by uptake from the circulation of amine released by the adrenal medulla (Strömblad, 1961). Sympathetic stimulation of the isolated rabbit heart released a small amount of adrenaline, 3·6 per cent of the noradrenaline, into the perfusate (Huković and Muscholl, 1962). Sympathetic stimulation releases also tritiated adrenaline from vasomotor nerves which previously was taken up from the blood stream (Rosell, Axelrod and Kopin, 1964). These are examples of liberation of an amine not synthesized at the site of release.

Release of α-methyl noradrenaline. Administration of α-methyldopa depletes heart and brain noradrenaline (Goldberg, Da Costa and Ozaki, 1960; Hess, Connamacher, Ozaki and Udenfriend, 1961; Porter, Totaro and Leiby, 1961). Carlsson and Lindqvist (1962) found that α-methyldopamine and α-methyl noradrenaline accumulated in the brain after injections of α-methyldopa. In the hearts of rats and guinea pigs the noradrenaline is partly replaced by α-methyl noradrenaline after administration of α-methyldopa (Maitre and Staehelin, 1963). This interesting finding was followed up in experiments by Muscholl and Maitre (1963). We found that α-methyl noradrenaline which is incorporated into the amine store after doses of α-methyldopa is released by sympathetic stimulation or by the nicotinic drug dimethyl-phenyl-piperazinium iodide (DMPP). This seems to be the first direct evidence that a drug metabolite can be liberated as a "false transmitter". Such a possibility has previously been discussed by Carlsson and Lindqvist (1962) and by Day and Rand (1964). Release by sympathetic stimulation of metaraminol incorporated into the amine stores of the heart by an infusion was recently demonstrated by Crout, Alpers, Tatum and Shore (1964).

In the rabbit four doses of 50 mg/kg *l*-α-methyldopa decreased the noradrenaline concentration in auricles, right and left ventricles by 38–60 per cent. An equimolar amount of α-methyl noradrenaline was detected in the tissues (Lindmar and Muscholl, 1965). It was also shown that reserpine and guanethidine in doses which lower but do not block completely the binding capacity of the amine stores cause a greater loss of noradrenaline than of α-methyl noradrenaline. Balance experiments carried out on isolated rabbit hearts indicate that monoamine oxidase activity is responsible for the preferential loss of noradrenaline.

α-Methyl noradrenaline is removed from the perfusion fluid of the isolated heart by an uptake process similar to the process which terminates the biological actions of noradrenaline (Lindmar and Muscholl, 1965). Dr. Weber of our laboratory demonstrated recently that cocaine inhibited the removal of α-methyl noradrenaline from the perfusion fluid of the rabbit heart. In a control series $54{\cdot}5 \pm 2{\cdot}8$ per cent of the α-methyl noradrenaline infused at a concentration of 10 ng/ml was recovered in 16 perfusates. In the presence of 10 μg/ml cocaine the percentage recovered in 16 perfusates rose to $85{\cdot}5 \pm 2{\cdot}7$. Cocaine strongly potentiated the ino- and chronotropic action of α-methyl noradrenaline. On the nictitating membrane of the cat cocaine was found to potentiate α-methyl noradrenaline in the same manner as noradrenaline (Fleckenstein and Bass, 1953). It appears, therefore, that cocaine causes supersensitivity to both noradrenaline and α-methyl noradrenaline by inhibiting their uptake into storage sites.

Release of α-methyl adrenaline. In the preceding paragraph evidence was quoted for *in vivo* formation of α-methyl noradrenaline in animals treated with α-methyldopa. Biosynthesis of α-methyl adrenaline has not been reported so far. It was thought that it may occur if α-methyl noradrenaline was retained in an organ capable of N-methylating noradrenaline.

For theoretical reasons it has been assumed that α-methyl noradrenaline formed biologically from L-α-methyldopa consists of one isomer with the configuration 1R : 2S (Van Rossum and Hurkmans, 1963; Lindmar and Muscholl, 1965). R and S are notations of absolute configuration according to the sequence rule by Cahn, Ingold and Prelog (1956). In his review article, Van Rossum (1963) has quoted the evidence from rotation dispersion techniques showing that levorotatory cobefrine and metaraminol have the 1R : 2S configuration. The same configuration is present in (−) ephedrine, whereas (+) pseudoephedrine is 1S : 2S. According to the Merck Index, *d,l*-cobefrine has the same (*erythro*) configuration as *d,l*-ephedrine. Thus it may be reasonable to assume that biosynthesis of α-methyl adrenaline, if it occurs, will lead to a substance with the 1R:2S configuration like (−) 3,4-dihydroxy ephedrine. The diastereomer, (*threo*) 3,4-dihydroxy pseudoephedrine, should be expected to differ in its biological actions from the actions of α-methyl adrenaline.

In order to find out whether α-methyl adrenaline is a further metabolite of α-methyl noradrenaline, experiments were carried out using the rabbit adrenal gland. In the adrenal medulla of this species the ratio adrenaline/nor-adrenaline is higher than in other laboratory animals (Euler, 1956). In rabbits a ratio of 50/1 was found after separation of the medullary amines by paper chromatography (Muscholl and Vogt, 1958). Moreover, the efficiency of the methylation process in the rabbit adrenal medulla was demonstrated by the observation that after a dose of reserpine the percentage of methylated amine was normal before the total amine content was restored. Therefore in this organ the rate of N-methylation seems to be comparatively high.

In a preliminary experiment it was checked whether α-methyl adrenaline was taken up and retained by the adrenal gland of a normal rabbit. An infusion of 120 μg dihydroxy ephedrine per kg body weight and min was administered intravenously for 25 min to a rabbit anaesthetized with pento-barbital sodium. The animal was bled 10 min after termination of the in-infusion. Extracts of heart, spleen, adrenals and plasma were prepared. The catechol amines were adsorbed on alumina, eluted, applied on paper and chromatographed with butanol–HCl. Concentrations of dihydroxy ephedrine exceeding those of endogenous noradrenaline were found in heart and spleen. The plasma contained a much smaller concentration of dihydroxy ephedrine than either heart or spleen. In the adrenal gland there was a normal concentration of adrenaline and noradrenaline. The ratio of the amines was 32/1 (Table 2). The eluate of the paper strip corresponding to

TABLE 2. CATECHOLAMINES IN ADRENAL GLANDS OF THE RABBIT

		μg/adrenals				
		A	N	A/N	aM-N	aM-A
Control		184	0	—	0	0
Reserpine + aM-DOPA						
1 mg/kg	2×50 mg/kg	12·1	4·6	2·6	0·450	0·050
2 mg/kg	3×50 mg/kg	10·8	3·4	3·2	0·230	0·025
2 mg/kg	4×50 mg/kg	45	7·0	6·5	0·300	0·025
2 mg/kg	5×50 mg/kg	15·6	1·9	8·2	0·250	0·010
2 mg/kg	6×50 mg/kg	99	12·5	7·9	1·310	0·160
Infusion of aM-A		107	3·3	32	0	0·250

A, adrenaline; N, noradrenaline. aM-N and aM-A are α-methylated amines.

the R_f-value of dihydroxy ephedrine was tested on the blood pressure of the pithed rat. Like dihydroxy ephedrine it caused a depressor effect. The total amount in both adrenal glands was 250 ng. Rechromatography of the eluate and biological assay revealed depressor activity only in the paper strip corresponding to the R_f-value of dihydroxy ephedrine. This experiment showed that dihydroxy ephedrine was taken up from the circulation by heart, spleen, and to a lesser degree, also by the adrenal glands and retained in these organs against a concentration gradient.

In a number of additional rabbits the adrenal glands were depleted of the catecholamines by a single intravenous dose of reserpine (1 or 2 mg/kg). Beginning on the day after the administration of reserpine intravenous doses of 50 mg/kg L-α-methyldopa were given daily. Two to six doses of α-methyldopa were administered and the rabbits killed 20 hr after the last injection. The catecholamines of the adrenal glands were purified as described under Methods. Adrenaline and noradrenaline were estimated fluorimetrically in a small aliquot of the alumina eluate. About $\frac{1}{8}$ of the final acetone-ethanol extract was applied on paper as a spot, the remainder was applied on a line. Both samples were chromatographed with butanol-HCl on the same paper together with reference substances. The part of the paper carrying the spot sample and the reference substances was sprayed with potassium ferricyanide in aqueous ethylenediamine and dried. Of the other part of the paper the regions corresponding to the R_f-values of adrenaline, α-methyl noradrenaline, dihydroxy ephedrine and isoprenaline were eluted. The eluate of the adrenaline region was assayed fluorimetrically, the other eluates were assayed on the blood pressure of the pithed rat against standard doses of the individual amines. It can be seen from Table 2 that the medullary adrenaline was depleted by reserpine and showed a tendency to recover during the experimental period. The ratio adrenaline/noradrenaline was 2·6 after reserpine and two days of treatment with α-methyldopa but increased to about 8 after six days of treatment. A small amount of α-methyl noradrenaline was always detected. Apparently the amount was not related to the duration of treatment. In the eluates of the dihydroxy ephedrine regions of the chromatograms a depressor activity was found which was indistinguishable from the reference compound in tine course of response and dose-response relationship. Therefore the depressor activity was considered to be α-methyl adrenaline formed by N-methylation of α-methyl noradrenaline. The amount, expressed as *d,l*-dihydroxy ephedrine, varied considerably from animal to animal. However, there was a constant relationship to the amount of α-methyl noradrenaline found in the same adrenal. With one exception (the lowest figure of α-methyl adrenaline after five doses of α-methyldopa) approximately 10 per cent of both α-methylated amines consisted of α-methyl adrenaline. If one assumes that either the levo- or the dextrorotatory compound exerts the prominent depressor action and the enantiomorph has much less effect, the figures for α-methyl adrenaline in the table may have to be halved in order to represent the actual amounts of an optical active metabolite.

After rechromatography of an eluate the depressor activity was found exclusively in the dihydroxy ephedrine region. Eluates of paper strips adjacent to the dihydroxy ephedrine region (corresponding to the R_f-value of isoprenaline) did not contain a depressor substance. No depressor action was seen with eluates of the dihydroxy ephedrine or isoprenaline regions of chromatograms of adrenal extracts from two rabbits not treated with reserpine and α-methyldopa. The depressor action of the eluates, like that of dihydroxy ephedrine, was antagonized by pronethalol. Owing to the high concentration of adrenaline in the adrenal gland an adrenaline-like pressor activity was sometimes found in the eluates of the isoprenaline region. In one case this eluate was rechromatographed and then the pressor activity was only found in the adrenaline region. Presence of adrenaline contaminating the eluates containing α-methyl adrenaline could be revealed by repeating the assay after a dose of pronethalol. The depressor action of α-methyl adrenaline was abolished and the pressor activity of adrenaline intensified (Vanov and Vogt, 1963).

Apart from the theoretical reasons discussed above, there is evidence that the depressor metabolite of α-methyldopa has the same (*erythro*) configuration as dihydroxy ephedrine. This compound and (*threo*) dihydroxy pseudoephedrine have identical fluorescence characteristics and R_f-values in the chromatographic systems tried. However, the pseudoephedrine analogue exerts only $\frac{1}{20}$ to $\frac{1}{60}$ of the depressor activity of the ephedrine analogue. If the depressor activity were due to dihydroxy pseudoephedrine the figures in the last row of Table 2 had to be multiplied by a factor of 20–60. In this case the fluorescence reaction carried out with the $\frac{1}{8}$ aliquot of the adrenal extracts (see above) should have been positive for the dihydroxy ephedrine region in at least a few experiments. In no instance was there any fluorescence in this region indicating that the threshold of 0·1–0·2 μg of the reference compounds was not attained. On the other hand, distinct adrenaline and noradrenaline spots were always seen. A spot of α-methyl noradrenaline was detected only in the experiment with a total amount of 1·3 μg/adrenals. These results exclude the possibility that α-methyl adrenaline has the (*threo*) configuration and strongly suggest that it is configurated like (−) dihydroxy ephedrine (1R:2S).

Although the concentration of α-methyl adrenaline in organs after treatment of animals with α-methyldopa may be very low, a possible release may have functional consequences differing from those of noradrenaline release. Injection of 5–20 μg/kg dihydroxy ephedrine decreased the blood pressure of anaesthetized rats and rabbits by about 50 mm Hg. The possibility was investigated that α-methyl adrenaline incorporated into the amine store of the heart could be released by sympathetic stimulation and by drugs known to release catecholamines.

Rabbits anaesthetized with ether and pentobarbital sodium were given an infusion of 85 μg/kg×min dihydroxy ephedrine for 20 min. There was an initial fall of blood pressure and a secondary rise usually to a higher level than before the infusion. After termination of the infusion the blood pressure fell to subnormal levels. 20 min after the infusion was stopped the animal was bled. The plasma concentration of noradrenaline was 1·8±0·5 ng/ml

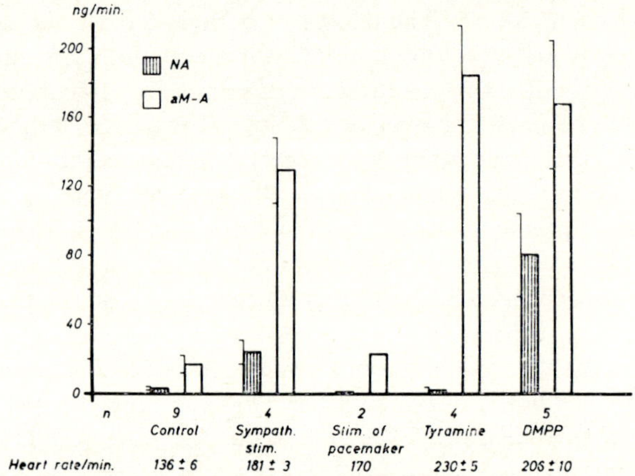

FIG. 2. Output of noradrenaline (NA) and *a*-methyladrenaline (*a*-MA) in ng/min from the isolated rabbit heart. Rabbits were infused with 85 μg/kg/min *a*-MA for 20 min and killed 20 min later. Tyramine (4–45 μg/ml.) and DMPP (9–20 μg/ml.) infused into aortic cannula for 4 min. n, number of estimations.

and that of dihydroxy ephedrine was 33·4±7·5 ng/ml. Results of bioassays agreed with the fluorimetric estimations ruling out the possibility of an unspecific fluorescence reaction. After the perfusion experiment the 9 hearts used in the study contained 0·51±0·10 μg/g noradrenaline and 0·80±0·09 μg/g dihydroxy ephedrine. These values were confirmed by paper chromatography in both solvent systems and by bioassay. Since the noradrenaline concentration of untreated rabbit hearts perfused for 2 hr was 1·4 μg/g, a partial noradrenaline depletion by the dihydroxy ephedrine taken up into the heart is obvious.

The control perfusates contained 2·6±0·7 ng/min of noradrenaline and 17·3±5·2 ng/min of dihydroxy ephedrine. During sympathetic stimulation the noradrenaline output rose to 24·2±7·0 and the dihydroxy ephedrine output to 129±19 ng/min. There was a positive inotropic action and a rise in heart rate from 136±6 to 181±3·4 per min. Increase in the heart rate by electrical stimulation of the pacemaker region with a frequency of 170/min did not result in a significant alteration of the amine output.

Lindmar and Muscholl (1961) have demonstrated that tyramine released small amounts of noradrenaline into the perfusate of the isolated rabbit heart. Under the same experimental conditions equieffective doses of DMPP released amounts of noradrenaline seven times higher than those released by tyramine. From the results shown by Fig. 2 it is obvious that after the infusion of dihydroxy ephedrine tyramine released mainly dihydroxy ephedrine. A significant release of noradrenaline (7 ng/min) occurred only in an experiment with the highest dose of tyramine administered (45 μg/ml). However, dihydroxy ephedrine was released at a maximum rate even by the smallest dose of tyramine (4 μg/ml).

Infusion of DMPP released substantial amounts of noradrenaline concomitantly with dihydroxy ephedrine. Therefore, DMPP simulates the effect of sympathetic stimulation much more than tyramine does.

The proportion of noradrenaline to dihydroxy ephedrine in the perfusates during release by sympathetic stimulation or by drugs was not conspicuously related to the proportion found in the hearts after termination of the perfusion experiment. This may be due to the possibility that dihydroxy ephedrine had not mixed completely with the endogenous noradrenaline and was released at a relatively higher rate than the more previously bound amine. In contrast, Muscholl and Maitre (1963) found that noradrenaline and α-methyl noradrenaline stored in the heart after administration of α-methyldopa to rabbits were released by sympathetic stimulation and DMPP in the same proportion as they were determined in the myocardium.

ACKNOWLEDGMENTS

This work was supported by the Deutsche Forschungsgemeinschaft. My thanks are due to Dr. Engelhardt for a sample of MHIP and to Dr. O. Thomä, C. H. Boehringer Sohn, Ingelheim, for synthesizing dihydroxy ephedrine and dihydroxy pseudoephedrine.

SUMMARY

The loss of noradrenaline (NA) from sympathetically innervated organs caused by drugs like reserpine and guanethidine is well known. Another depleting agent which recently has aroused interest is α-methyldopa. After its administration α-methylated amines are formed *in vivo* which can replace NA at the storage sites. In the heart, for instance, NA is partly replaced by a stoichiometric amount of α-methyl noradrenaline (α-MNA). Both amines are released by sympathetic stimulation. Like NA, α-MNA is removed from the circulation by an uptake process which terminates the biological action. Cocaine inhibits this uptake and consequently increases the actions of α-MNA.

α-MNA stored in the heart after administration of α-methyldopa to rabbits was found to be depleted by reserpine and guanethidine. However, α-MNA was more resistant to the action of both depleting agents than endogenous NA. Doses of reserpine and guanethidine which lowered but did not completely block the binding capacity of the amine stores caused a greater loss of NA than of α-MNA. Results obtained on perfused hearts indicate that monoamine oxidase activity is responsible for the preferential loss of NA.

Another metabolite of α-methyldopa is α-methyl adrenaline (α-MA). Small amounts of this amine were detected in adrenals of rabbits treated with reserpine followed by repeated doses of α-methyldopa. Similar to other catecholamines, α-MA is removed from the circulation by uptake into different organs. In rabbits, the heart NA was partly replaced by α-MA after an infusion of this agent. Hearts removed from rabbits given an infusion of α-MA released both NA and α-MA into the perfusion fluid during sympathetic stimulation. Similarly, the nicotinic drug DMPP released both NA and α-MA. In contrast, tyramine released only α-MA into the perfusate. Possible reasons for these results are discussed.

REFERENCES

AXELROD, J. (1964) The uptake and release of catecholamines and the effect of drugs, *Progress in Brain Research* **8**, 81–9.

AXELROD, J., G. HERTTING and L. POTTER (1962) Effect of drugs on the uptake and release of H^3-norepinephrine in the rat heart, *Nature, Lond.* **194**, 297.

BRODIE, B. B. and M. A. BEAVEN (1963) Neurochemical transducer systems, *Med. Exp.* (Basel) **8**, 320–51.

BROWN, G. L. and J. S. GILLESPIE (1957) The output of sympathetic transmitter from the spleen of the cat, *J. Physiol., Lond.* **138**, 81–102.

CAHN, R. S., C. K. INGOLD and V. PRELOG (1956). The specification of asymmetric configuration in organic chemistry, *Experientia* (Basel) **12**, 81–94.

CARLSSON, A. and M. LINDQVIST (1962) *In vivo* decarboxylation of α-methyldopa and α-methylmetatyrosine, *Acta Physiol. Scand.* **54**, 87–94.

CROUT, J. R., H. S. ALPERS, E. L. TATUM and P. A. SHORE (1964) Release of metaraminol from the heart by sympathetic nerve stimulation, *Science* **145**, 828–9.

DAY, M. D. and M. J. RAND (1964) Some observations on the pharmacology of α-methyldopa, *Brit. J. Pharmacol.* **22**, 72–86.

ELLMAN, G. L. (1958) Determination of epinephrine and related compounds on paper chromatograms, *Nature, Lond.* **181**, 768–9.

ENGELHARDT, A. (1965) Methode zur Auswertung von β-Adrenaloytika am isolierten Herzvorhof, *Arch. Exp. Path Pharmak*, **250**, 245–6.

EULER, U. S. VON (1956) *Noradrenaline.* Springfield, Ill., U.S.A., C. C. Thomas Publ., p. 107.

EULER, U. S. VON and I. ORWÉN (1955) Preparation of extracts of urine and organs for estimation of free and conjugated noradrenaline and adrenaline, *Acta Physiol. Scand.* **33**, Suppl. 118, 1–9.

FARRANT, J., J. A. HARVEY and J. N. PENNEFATHER (1964) The influence of phenoxybenzamine on the storage of noradrenaline in rat and cat tissues, *Brit. J. Pharmacol.* **22**, 104–12.

FAWAZ, G. (1963) Cardiovascular pharmacology, *Ann. Rev. Pharmacol.* **3**, 57–90.

FLECKENSTEIN, A. and H. BASS (1953) Zum Mechanismus der Wirkungsverstärkung und

Wirkungsabschwächung sympathomimetischer Amine durch Cocain und andere Pharmaka. 1. Mitteilung: Die Sensibilisierung der Katzen-Nickhaut für Sympathomimetika der Brenzkatechin-Reihe, *Arch. Exp. Path. Pharmak.* **220**, 143–56.

GOLDBERG, L. I., F. M. DA COSTA and M. OZAKI (1960) Actions of the decarboxylase inhibitor, α-methyl-3,4-dihydroxyphenylalanine, in the dog, *Nature, Lond.* **188**, 502–4.

GOODALL, McC. (1951) Studies of adrenaline and noradrenaline in mammalian heart and suprarenals, *Acta Physiol. Scand.* **24**, Suppl. 85.

GREEN, J. P. (1962) Binding of some biogenic amines in tissues, *Advances in Pharmacology* **1**, 349–422. Eds S. Garattini and P. A. Shore. New York, Academic Press.

HERTTING, G. and T. SCHIEFTHALER (1963) Beziehung zwischen Durchflussgrösse und Noradrenalin-Freisetzung bei Nervenreizung der isoliert durchströmten Katzenmilz, *Arch. Exp. Path. Pharmak.* **246**, 13–14.

HESS, S. M., R. H. CONNAMACHER, M. OZAKI and S. UDENFRIEND (1961) The effects of α-methyldopa and α-methylmetatyrosine on the metabolism of norepinephrine and serotonin *in vivo*, *J. Pharmacol.* **134**, 129–38.

HÖKFELT, B. (1951) Noradrenaline and adrenaline in mammalian tissues, *Acta Physiol. Scand.* **25**, Suppl. 92.

HUKOVIĆ, S. and E. MUSCHOLL, (1962) Die Noradrenalin-Abgabe aus dem isolierten Kaninchenherzen bei sympathischer Nervenreizung und ihre pharmakologische Beeinflussung, *Arch. Exp. Path. Pharmak.* **244**, 81–96.

IVERSEN, L. L. (1964) Inhibition of noradrenaline uptake by sympathomimetic amines, *J. Pharm. Pharmacol.* **16**, 465–7.

IVERSEN, L. L. (1965) Inhibition of noradrenaline uptake by drugs, *J. Pharm. Pharmacol.* **17**, 62–4.

KOPIN, I. J. (1964) Storage and metabolism of catecholamines: The role of monoamine oxidase, *Pharmacol. Rev.* **16**, 179–91.

KOPIN, I. J. and E. K. GORDON (1963) Origin of norepinephrine in the heart, *Nature, Lond.* **199**, 1289.

LINDMAR, R. and E. MUSCHOLL (1961) Die Wirkung von Cocain, Guanethidin, Reserpin, Hexamethonium, Tetracain und Psicain auf die Noradrenalin-Freisetzung aus dem Herzen, *Arch. Exp. Path. Pharmak.* **242**, 214–27.

LINDMAR, R. and E. MUSCHOLL (1964) Die Wirkung von Pharmaka auf die Elimination von Noradrenalin aus der Perfusionsflüssigkeit und die Noradrenalinaufnahme in das isolierte Herz, *Arch. Exp. Path. Pharmak.* **247**, 469–92.

LINDMAR, R. and E. MUSCHOLL (1965) Die Aufnahme von α-Methylnoradrenalin in das isolierte Kaninchenherz und seine Freisetzung durch Reserpin und Guanethidin *in vivo*, *Arch. Exp. Path. Pharmak.* **249**, 529–48.

MAITRE, L. and M. STAEHELIN (1963) Effect of α-methyl-DOPA on myocardial catecholamines, *Experientia* (Basel) **19**, 573–5.

Merck Index of Chemicals and Drugs (1960) 7th ed., Rahway, N. J., Merck & Co. Inc., p. 735.

MORAN, N. C. and M. E. PERKINS (1958) Adrenergic blockade of the mammalian heart by a dichloro analogue of isoproterenol, *J. Pharmacol.* **124**, 223–37.

MUSCHOLL, E. (1959) Die Konzentration von Noradrenalin und Adrenalin in den einzelnen Abschnitten des Herzens, *Arch. Exp. Path. Pharmak.* **237**, 350–64.

MUSCHOLL, E. (1961) Effect of cocaine and related drugs on the uptake of noradrenaline by heart and spleen, *Brit. J. Pharmacol.* **16**, 352–9.

MUSCHOLL, E. (1964) Drugs interfering with the storage and release of adrenergic transmitters, *Proc. 2nd International Pharmacological Meeting, Prague*, Vol. 3. *Pharmacology of Cholinergic and Adrenergic Transmission.* 291–302. Oxford, Pergamon Press.

MUSCHOLL, E. and L. MAITRE (1963) Release by sympathetic stimulation of α-methylnoradrenaline stored in the heart after administration of α-methyldopa, *Experientia* (Basel) **19**, 658–9.

MUSCHOLL, E. and M. VOGT (1958) The action of reserpine on the peripheral sympathetic nervous system, *J. Physiol., Lond.* **141**, 132–55.

PORTER, C. C., J. A. TOTARO and C. M. LEIBY (1961) Some biochemical effects of α-methyl-3,4-dihydroxyphenylalanine and related compounds in mice, *J. Pharmacol.* **134**, 139–45.

ROSELL, S., J. AXELROD and I. J. KOPIN (1964) Release of tritiated epinephrine following sympathetic nerve stimulation, *Nature, Lond.* **201**, 301.

SHORE, P. A. (1962) Release of serotonin and catecholamines by drugs, *Pharmacol. Rev.* **14**, 531–50.

SPECTOR, S., A. SJOERDSMA, P. ZALTZMAN-NIRENBERG, M. LEVITT and S. UDENFRIEND (1963) Norepinephrine synthesis from tyrosine-C^{14} in isolated perfused guinea pig heart, *Science* **139**, 1299.

STRÖMBLAD, B. C. R. (1961) Effects of insulin on adrenaline and noradrenaline content of some rat tissues, *Proc. Soc. Exp. Biol.* (N.Y.) **108**, 345–7.

THOENEN, H., A. HÜRLIMANN and W. HAEFELY (1964) Wirkungen von Phenoxybenzamin, Phentolamin und Azapetin auf adrenergische Synapsen der Katzenmilz, *Helv. Physiol. Pharmacol. Acta* **22**, 148–61.

VANOV, S. and M. VOGT (1963) Catecholamine-containing structures in the hypogastric nerves of the dog, *J. Physiol., Lond.* **168**, 939–44.

VAN ROSSUM, J. M. (1963) The relation between chemical structure and biological activity, *J. Pharm. Pharmacol.* **15**, 285–316.

VAN ROSSUM, J. M. and A. T. M. HURKMANS (1963) Reversal of the effect of α-methyldopa by monoamine oxidase inhibitors, *J. Pharm. Pharmacol.* **15**, 493–9.

WILLIAMS, B. J. and S. E. MAYER (1964) Interaction between norepinephrine and dichloroisoproterenol during adrenergic blockade of the isolated perfused rat heart, *J. Pharmacol.* **145**, 307–14.

ZAIMIS, E. (1964) Pharmacology of the autonomic nervous system, *Ann. Rev. Pharmacol.* **4**, 365–400.

DISCUSSION

ROSELL: Concerning false transmitters I would like to show a figure from a series of experiments by Axelrod, Kopin and me (*Nature* **201**, 301, 1964). We used H^3-adrenaline as a false transmitter in the perfused gracilis muscle from dogs. Following sympathetic nerve stimulation with physiological frequencies there was a pronounced increase in the venous outflow of H^3-adrenaline. In addition there was also a 50 per cent increase of the outflow of H^3-metanephrine. In similar experiments with H^3-noradrenaline we could demonstrate an increase in H^3-noradrenaline outflow following sympathetic nerve stimulation. However, no increase in the outflow of its metabolites could be demonstrated. Thus, there seems to be some difference in the inactivation of the released true transmitter as compared with the inactivation of the released false transmitter.

MUSCHOLL: In my opinion adrenaline is not necessarily a "false" transmitter. This term should be reserved for drugs or drug metabolites released from nerves. Adrenaline may be regarded as a transmitter which usually is not synthesized at the site of its release.

BROWN: I hope Dr. Muscholl and others will not devote too much time to proving that the original explanation of noradrenaline uptake that Gillespie and I gave was wrong. Gillespie himself has produced convincing evidence that it is the nerves and not the receptors of the effector cell that do this, and I have confessed our error on several occasions. I would, nevertheless, very much like to know what happens when the effector cell is exposed to the liberated transmitter.

MUSCHOLL: The amount of transmitter combining with the receptors is probably so small that present analytical methods cannot follow its fate.

CARLSSON: The easier release of α-MA compared with NA by, for example, tyramine might perhaps—as an alternative interpretation—be due to low affinity to binding sites in granules, in analogy with octopamine?

MUSCHOLL: The present experiments cannot decide this question. Preliminary experiments with α-methyl adrenaline taken up by the rat heart suggest that the loss from the heart occurs with a half time similar to that of noradrenaline newly taken up by the heart

REGULATION OF INCREASED ADRENOMEDULLARY SECRETION BY DRUGS

AIMO PEKKARINEN, KIMMO MANNINEN and MATTI MÄENPÄÄ

Department of Pharmacology, Turku University,
Turku, Finland

THE effect of drugs on the increased excretion of adrenaline, induced by insulin (4 I.U./kg i.m.) (Pitkänen, 1956), was followed in fasted rats. The rats were hydrated with 4 ml distilled water/100 g of body weight through stomach tube in Fluothane anesthesia, and the drugs were injected normally 15 min before the insulin injection. Urine samples were collected during 5 hr. Adrenaline and noradrenaline were determined by the automatic method (Pekkarinen).

The following groups of drugs were studied in Fluothane anesthesia. (I) CNS drugs (neuroleptic, psychostimulant and psychosedative drugs, analgesics and hypnotics). (II) Antihypertensive drugs. (III) Certain drugs of the autonomic nervous system. (IV) Some enzyme inhibitors for adrenaline and noradrenaline. The main purpose was to find drugs, which could effectively control and inhibit the increased secretion of adrenaline from the adrenal medulla during insulin shock. The most effective drugs were in the group of the CNS drugs the neuroleptic, psychostimulant and some psychosedative drugs (Tables 1, 2). The best structure for the adrenomedullary inhibition was the thiaxanthene group, observed in chlorprothixene, Sordinol, amitriptyline and nortriptyline. Effective also was the dibenzazepine nucleus of imipramine, desipramine and Insidon. Other effective neuroleptic drugs were chlorpromazine, thioridazine, protipendyl, prochlorperazine, dixyrazine, benactyzine, and in certain amount reserpine, pericyazide and oxypertine. The effective dose of the last two drugs also partially inhibited the urine flow. No clear inhibition was observed with phenotiazine neuroleptics with fluoratom (fluphenazine, fluanxol, trifluopromazine, trifluoperazine) nor haloperidol which have been given as a small dose, and with mepazine. All these effects are partially also depending upon the size of the dose. The studies on the dose relationship of these effects are continuing. It is important to remember that higher doses of certain drugs also can inhibit the urine flow, which also may contribute to their inhibiting action on the excretion of adrenaline in the urine. The inhibiting effect of many of these drugs on the excretion of adrenaline in the urine during insulin shock has been greater than the inhibiting effect of P-286, the specific adrenomedullary inhibitor (Table 3).

TABLE 1. EFFECT OF DRUGS ON THE URINARY EXCRETION OF ADRENALINE AND NORADRENALINE IN INSULIN SHOCK OF RATS (4 I.U./kg). URINE COLLECTED DURING 5 HR AFTER HYDRATION

		n	*A μg M±* *SEM*	*N μg M±* *SEM*	*Dose* *mg/kg*	*Urine* *volume* *ml*	*A* *p*
Controls							
Insulin	i.m.	166	2·71 ±0·16	0·30 ±0·02		24·6	
Distilled water	p.o.	28	0·15 ±0·05	0·14 ±0·04		17·8	(1)
A. *CNS Drugs*							
Chlorpromazine	i.m.	47	0·99 ±0·16	1·08 ±0·12	10	13·2	(1)
Chlorpromazine	i.m.	17	0·79 ±0·11	0·60 ±0·07	10	10·1	(1)
Chlorpromazine	i.m.	13	0·57 ±0·11	1·05 ±0·15	15	6·9	(1)
Chlorpromazine	i.m.	14	0·39 ±0·14	0·55 ±0·09	25	11·7	(1)
Chlorprothixene	i.m.	15	0·32 ±0·02	0·40 ±0·004	10	15·0	(1)
Imipramine	i.m.	71	1·55 ±0·20	0·24 ±0·03	10	23·0	(1)
Imipramine	i.m.	17	0·14 ±0·04	0·17 ±0·02	10	15·4	(1)
Imipramine	i.m.	15	1·30 ±0·30	0·93 ±0·13	20	21·9	(1)
Imipramine	i.m.	5	0·80 ±0·42	0·59 ±0·19	25	43·6	(1)
Amitriptyline	i.m.	18	1·01 ±0·26	0·17 ±0·03	5	15·8	
Chlordiazepoxide	i.m.	14	0·58 ±0·05	0·13 ±0·02	40	11·4	
Benactyzine	i.m.	14	1·66 ±0·19	0·29 ±0·04	2	20·8	(1)
Benactyzine	i.m.	14	1·13 ±0·13	0·17 ±0·07	4	14·0	(1)
Benactyzine	i.m.	15	3·07 ±0·20	0·15 ±0·03	20	18·7	
Mepazine	i.m.	15	3·41 ±0·55	0·02 ±0·01	25	18·2	(4)
Reserpine*	i.m.	33	1·82 ±0·30	0·20 ±0·06	1	22·8	(3)
Fenemal	i.m.	16	3·99 ±0·50	0	50	13·4	
B. *Antihypertensive Drugs*							
Tetrabenazine	i.p.	5	3·34 ±1·28	0·02 ±0·02	3	25·2	
Tetrabenazine	i.p.	12	2·09 ±0·11	0·12 ±0·07	40	15·8	(3)
Guanethidine†	i.m.	44	2·66 ±0·26	0·24 ±0·15	20	21·1	
α-Methyldopa‡	p.o.	19	2·57 ±0·33	1·42 ±0·17	25	21·1	
Bretylium§	p.o.	34	2·71 ±0·40	0·11 ±0·11	25	24·9	
Hexamethonium‖	i.m.	5	4·50 ±1·46	0·58 ±0·21	30	19·8	
Hexamethonium‖	i.m.	15	2·11 ±0·36	0·88 ±0·37	2×30	16·9	(4)
Hexamethonium‖	i.m.	14	0·87 ±0·22	0·67 ±0·10	3×30	16·9	(1)

(1) $p < 0·001$; (2) $p < 0·01$; (3) $p < 0·05$; (4) $p > 0·05$.

* 4 days' treatment: mean 1·75 μg A(17); drug 15 min before insulin, mean 1·89 μg (16).

† 4 days' treatment: mean 4·05 μg A(14); drugs 15 min before insulin, mean 1·96 μg (30).

‡ 4 days treatment.

§ 4 days' treatment: mean 2·42 μg A(20); drug 15 min before insulin, mean 3·12 μg A(14).

‖ 1–3 days' treatment.

In the group of psychosedatives only a partial inhibition has been observed with hydroxyzine, captodiamine, chlordiazepoxide and Valium (Table 2). Meprobamate was not effective, which might depend upon the small dose. In the group of hypnotics, e.g. paraldehyde, Heminevrin, glutethimide, chloralhydrate and ethinamate have been only partially effective. In the group of analgesics pethidine, ethylmorphine, methadone (Table 3) and Dromoran were only partially effective, while hydromorphone and morphine had no effects and the effect of ketobemidone was small (Table 3).

TABLE 2. EFFECT OF DRUGS ON THE URINARY EXCRETION OF ADRENALINE IN INSULIN SHOCK OF RATS (4 I.U./kg). URINE COLLECTED DURING 5HR AFTER HYDRATION

		n	*A μg*	*M±SEM*	*Dose mg/kg*	*Urine volume ml*	*A p*
Controls							
Insulin	i.m.	97	3·40	±0·14		20·0	
Distilled water	p.o.	8	0·09	±0·02		14·6	(1)
F. *Neuroleptic, psychostimulant and psychosedative drugs*							
1. Chlorprothixene	i.m.	25	0·22	±0·07	5	21·6	(1)
Chlorprothixene	i.m.	12	1·17	±0·28	10	10·4	(1)
1. Sordinol	i.m.	25	0·74	±0·11	10	8·4	(1)
Insidon	i.m.	25	0·88	±0·18	20	16·6	(1)
Thioridazine	i.m.	51	0·54	±0·09	30	13·5	(1)
Protipendyl	i.m.	6	0·77	±0·45	20	10·2	(1)
Prochlorperazine	i.m.	28	0·78	±0·15	5	15·6	(1)
Dixyrazine	i.m.	7	1·26	±0·11	2	11·9	(1)
Promazine	i.m.	28	1·36	±0·19	20	9·8	(1)
Pericyazine	i.m.	24	0·73	±0·21	5	3·9	(1)
Pericyazine	i.m.	16	1·40	±0·28	2	11·8	(1)
Fluphenazine	i.m.	19	2·46	±0·20	2	13·2	
Fluanxol	i.m.	10	2·73	±0·22	2	9·3	
Trifluoperazine	i.m.	51	2·99	±0·17	2	17·0	
Trifluopromazine	i.m.	27	3·29	±0·21	2	13·9	(4)
Oxypertine	i.m.	21	3·18	±0·22	2	9·0	(4)
Oxypertine	i.m.	5	0·56	±0·29	10	6·1	(3)
Haloperidol	i.m.	24·5	2·51	±0·15	0·5	14·7	(2)
2. Amitriptyline	i.m.	44	0·51	±0·13	5	16·9	(1)
Imipramine	i.m.	26	1·17	±0·21	10	16·1	(1)
Desipramine	i.m.	47	1·22	±0·13	10	19·9	(1)
Nortriptyline	i.m.	16	1·33	±0·20	10	13·7	
3. Hydroxyzine	i.m.	26	1·41	±0·27	25	22·6	(1)
Captodiamine	i.m.	27	1·74	±0·20	10	18·1	(1)
Chlordiazepoxide	i.m.	44	1·98	±0·29	20	17·8	(1)
Valium	i.m.	24	1·97	±0·13	5	15·2	(1)
Meprobamate	i.p.	26	3·34	±0·21	25	25·3	(4)

n = number of groups of 2 rats.

(1) $p < 0{\cdot}001$; (2) $p < 0{\cdot}01$; (3) $p < 0{\cdot}05$; (4) $p > 0{\cdot}05$.

The CNS regulation of adrenomedullary secretion or the control of the increased secretion of adrenaline in CNS are not known as well as the regulation of the mechanism of hypothalamic–pituitary–adrenocortical function. It would be important to know the site and location of the adrenomedullary pathways and their inhibition in the CNS. Insulin acts as the similar stimulator of the adrenomedullary secretion as the ACTH in the adrenocortical secretion. It would be important to find in CNS the active substances, which increase the adrenomedullary secretion.

TABLE 3. EFFECT OF DRUGS ON THE URINARY EXCRETION OF ADRENALINE IN INSULIN SHOCK OF RATS (4 I.U./kg). URINE COLLECTED DURING 5 HR AFTER HYDRATION

		n	*A μg M±* *SEM*		*Dose mg/kg*	*Urine volume ml*	*A p*
G. *Analgesics*							
Pethidine	i.m.	7	1·06	±0·28	25	13·9	(1)
Ethylmorphine	i.m.	7	1·45	±0·31	10	20·6	(1)
Methadone	i.m.	7	1·50	±0·22	4	15·5	(1)
Dromoran	i.m.	7	2·18	±0·23	1	12·8	(1)
Ketobemidone	i.m.	7	2·63	±0·35	2	21·8	
Hydromorphone	i.m.	7	3·55	±0·28	1	17·6	
Morphine	i.m.	14	5·26	±0·50	5	33·8	
Morphine	i.m.	12·5	3·70	±0·44	10	19·4	
H. *Hypnotics*							
Paraldehyde	i.p.	4	0·94	±0·10	300	7	(1)
Heminevrin	i.m.	6	1·48	±0·34	100	8·5	(1)
Glutethimide	i.m.	17	1·52	±0·24	50	7·0	(1)
Chloralhydrate	i.p.	7	1·75	±0·24	150	12·3	(1)
Ethinamate	i.m.	22	1·85	±0·25	40	16·9	(1)
Metaqualone	i.m.	7	3·53	±0·54	50	12·6	
I. P–286	i.m.	53·5	1·74	±0·16	5	21·1	(1)
Controls							
Insulin	i.m.	97	3·40	±0·14		20·0	
Distilled water	p.o.	8	0·09	±0·02		14·6	(1)

n = number of groups of 2 rats.

(1) $p < 0.001$.

Our findings cannot inform, whether the adrenomedullary secretion inhibiting drugs prevent the secretion of adrenaline from the adrenal medulla, have a cortisone-like effect on the blood glucose or inhibit the renal excretion of adrenaline. Our recent preliminary findings of the effects of these drugs on the blood glucose during insulin shock show that the most effective inhibitors do not inhibit the hypoglycemia during the insulin shock and do not have any cortisone-like effect on the blood glucose and also no clear potentiating or prolonging effect on the decrease of the blood glucose. In earlier studies it has been observed by several investigators, that many of the drugs, which have been effective inhibitors of the adrenomedullary secretion, are also the most effective inhibitors of radioactive adrenaline and noradrenaline uptake of tissues or granules. All the effective inhibitors of the adrenomedullary secretion of adrenaline in our studies have as a common feature an anxiolytic effect, although all anxiolytics have not been effective. Clinically the anxiolytic or sedative effect of neuroleptic, psychostimulant (thymoleptic), psychosedative, analgesic and hypnotic drugs is well known.

TABLE 4. EFFECT OF DRUGS ON THE URINARY EXCRETION OF ADRENALINE AND NORADRENALINE IN INSULIN SHOCK OF RATS (4 I.U./kg). URINE COLLECTED DURING 5 HR AFTER HYDRATION

		n	*A µg M±* *SEM*		*N µg M±* *SEM*		*Dose mg/kg*	*Urine volume ml*	*A p*
Controls									
Insulin	i.m.	166	2·71	±0·16	0·30	±0·02		24·6	
Distilled water	p.o.	28	0·15	±0·05	0·14	±0·04		17·8	(1)
Drugs									
C. Atropine	i.m.	29	4·37	±0·36	0·34	±0·07	1	20·2	
Trasentin	i.m.	15	3·27	±0·65	0·73	±0·14	20	21·7	
Scopolamine	i.m.	14	6·01	±0·83	—	—	1	18·5	
Scopolamine	i.m.	30	2·95	±0·40	0·31	±0·05	2	20·3	
Physostigmine	i.v.	15	3·29	±0·60	1·48	±0·24	200 µg	36·5	
Nicotine	i.m.	15	2·83	±0·31	0·20	±0·01	1	25·9	
Segontin	i.v.	5	0·13	±0·02	0·03	±0·02	5	18·8	
Segontin 4 days	i.p.	12	5·37	±0·67	0·02	±0·04	5	18·1	
Segontin	i.v.	20	1·43	±0·29	0·13	±0·02	10	13·7	(1)
Segontin	i.m.	28	1·34	±0·24	0·16	±0·02	10	10·9	(1)
D. Iproniazide	i.m.	21	3·50	±0·23	0·26	±0·06	100	23·0	
Isocarboxazide	p.o.	10	3·05	±0·67	0·81	±0·20	100	15·3	
Cocaine	i.m.	12	3·10	±0·57	0·44	±0·08	8	26·6	
Pyrogallol	i.v.	28	4·07	±0·39	0·38	±0·07	50	26·0	
Pyrogallol	i.v.	5	1·26	±0·49	0·12	±0·10	200	11·2	(3)
Quercetin	i.p.	9	3·53	±0·45	0·39	±0·10	2×150*	28·5	
Ascorbic acid	i.v.	15	3·85	±0·48	0·38	±0·08	200	23·9	
Thyroxine 4 days	p.o.	15	3·63	±0·89	0·40	±0·03	5 µg/d	34·9	
E. Iproniazide	i.m.						100		
Pyrogallol							50		
Ascorbic acid							200		
Thyroxine							5 µg		
Atropine							1		
Scopolamine							1		
Combined†		15	4·03	±0·10	0·40	±0·04		17·1	
Lidocain	i.m.	13	4·49	±0·37	0·15	±0·05	4	27·0	

n = number of groups of 2 rats.

(1) $p < 0{\cdot}001$; (3) $p < 0{\cdot}05$.

* 15 min before and 2·5 hr after insulin injection.

† 4 days' treatment.

Our method can also act as a pharmacological screening method for new drugs with adrenomedullary inhibiting action or in studying the increased excretion of adrenaline and noradrenaline and their regulation in certain normal and physiological conditions.

In the group of antihypertensive drugs no clear inhibiting effect was observed, except with high doses of hexamethonium and only a partial inhibition with reserpine (Table 1). Antihypertensive drugs, as the liberator of adrenaline and noradrenaline from granules, had also no increasing effect on the excretion of adrenaline in the insulin shock.

Amine oxidase inhibitors (iproniazide, isocarboxazide, cocaine), O-methyltransferase inhibitors (pyrogallol, quercetin) (Table 4) and ascorbic acid as antioxidant could not inhibit nor clearly increase the adrenomedullary secretion. The average excretion of adrenaline was then often above its control excretion in insulin shock. Thyroxine could not inhibit nor clearly increase the excretion of adrenaline. Lidocain in a small local anaesthetic dose could not inhibit the adrenaline excretion during insulin shock. It rather increased this excretion slightly.

Parasympatholytics or spasmolytics (atropine, scopolamine, Trasentin), physostigmine as anticholinesterase and nicotine as ganglionic stimulant could not diminish the adrenaline excretion (Table 4). Segontin in higher does decreased the adrenaline excretion.

REFERENCE

PITKÄNEN, E., (1956) *Acta Physiol. Scand.*, **38**, suppl. 129.

CALCIUM-DEPENDENT LINKS IN STIMULUS-SECRETION COUPLING IN THE ADRENAL MEDULLA AND NEUROHYPOPHYSIS

W. W. DOUGLAS

Department of Pharmacology, Albert Einstein College of Medicine, Yeshiva University, Bronx 61, New York, N.Y., U.S.A.

THE chromaffin cells of the adrenal medulla are known to be innervated by secretomotor nerves that are cholinergic. Acetylcholine is thus the immediate physiological stimulus that causes the chromaffin cells to release their hormones, adrenaline and noradrenaline, into the blood stream. Several years ago Dr. Rubin and I began experiments from which we hoped to learn something of the mechanism of action of acetylcholine on the chromaffin cell. The task we set ourselves was to find out, if we could, the sequence of events that acetylcholine sets in motion and that leads up to the extrusion of the catecholamines—a sequence of events we termed "stimulus-secretion coupling" (Fig. 1).

In casting around for an appropriate experimental approach to the problem we chose to be guided by what was already known of acetylcholine's action at other sites of cholinergic transmission that had been studied more carefully. There was strong evidence that acetylcholine accomplished its transmitter functions at sites such as the neuromuscular junction and the ganglionic synapse by some action restricted to the membrane of the post-synaptic cell; indeed, the elegant studies of del Castillo and Katz (1955), in which acetylcholine was applied by micropipette to the inside or outside of the motor end-plate, indicated that acetylcholine sensitivity was restricted to the outside of the membrane. Such a finding is, of course, consistent with the chemical nature of acetylcholine which is charged at physiological pH and would not be expected to penetrate cell membranes readily. Furthermore, there was also evidence that the effect of acetylcholine on the various cholinoceptive membranes that had then been studied was always to increase permeability to one or more of the common inorganic ions, the cations playing a major role. These considerations prompted us to enquire whether acetylcholine might exert its stimulant action on the chromaffin cell by the same general mechanism.

The results we obtained were in harmony with this possibility and led us to suggest that acetylcholine evoked secretion by promoting the inward movement of calcium across the cell membrane, the consequent appearance of an excess of calcium ions at some site in the cell providing the signal for catecholamine release (Douglas and Rubin, 1961). These results, and others that have subsequently strengthened the hypothesis, provide the main theme of this communication.

Later in the report, I shall briefly summarize a parallel series of experiments made on the neurohypophysis that suggests that Ca may play a similar role in the release of posterior pituitary hormones from the neurosecretory terminals (Douglas, 1963; Douglas and Poisner, 1964a, b). The posterior

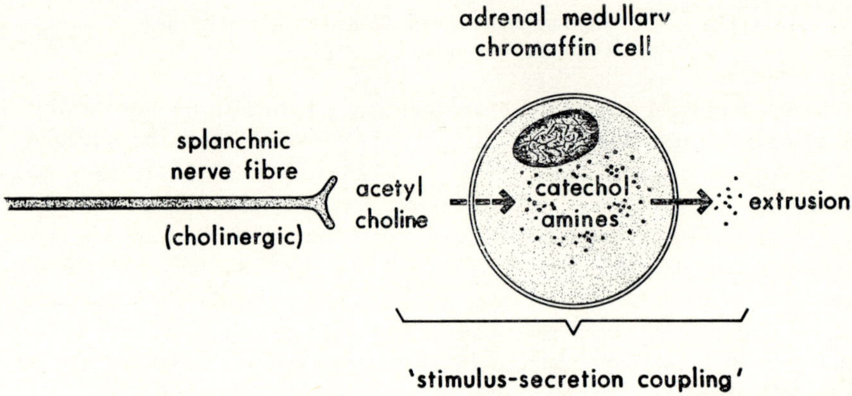

FIG. 1. A schema of the physiological events involved in nerve-evoked catecholamine release from a chromaffin cell in the adrenal medulla. The term "stimulus-secretion coupling" embraces all the events occurring between the application of the physiological secretogogue, acetylcholine, and the extrusion of the catecholamines.

pituitary hormones are, of course, polypeptides and not amines. But I believe the results are germane to the subject of the present symposium for they serve to emphasize the similarities in the mechanisms for release of quite dissimilar chemical substances from quite different types of secretory cells.

Finally, I should like to describe some recent evidence we have obtained (Douglas, Poisner and Rubin, 1965) that helps to characterize the catecholamine extrusion process in the medullary chromaffin cell.

METHODS

Adrenal gland. The experiments were performed on cats' adrenal glands that were acutely denervated and perfused with Locke's solution or some modified version of this solution. The glands were made to secrete at appropriate intervals by brief exposure to standard concentrations of acetylcholine dissolved in the control or test medium, and the effluent was collected from the adrenal vein for measurement of its catecholamine content by bioassay on the rabbit's aortic strip (Douglas and Rubin, 1961) or by the fluorimetric technique (Anton and Sayre, 1962). Typically the "resting" output of amines in these circumstances was low and comparable with that reported for cats' adrenal glands with normal circulation (Marley and Paton, 1961); while on exposure to acetylcholine (10^{-5} g/ml for 1 to 4 min), the rate of catecholamine output rose to 5 μg/min or more. The experimental arrangement is

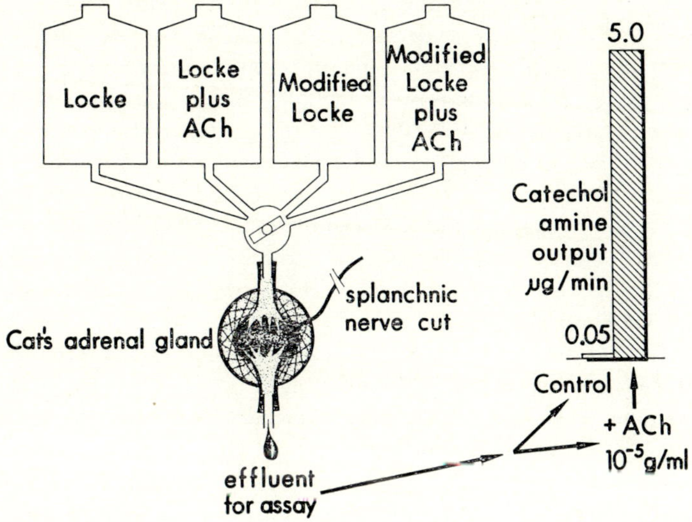

FIG. 2. A schematic representation of the experimental arrangement for perfusion of the cat's adrenal gland. The histogram in the right-hand side of the figure shows typical outputs of catecholamines from a gland perfused first with Locke's solution (control) and then for 2 min with Locke's solution with the addition of acetylcholine (10^{-5} g/ml.).

depicted in Fig. 2. AMP and other metabolites of ATP were measured by standard techniques (see Douglas, Poisner and Rubin, 1965).

Neurohypophysis. Rats' neurohypophyses were removed after decapitation. They were cut along the sagittal plane and were incubated *in vitro* in the appropriate medium: Locke's solution was the principal solution used. The glands were stimulated electrically (20 shocks/sec), or by increasing the K concentration 10-fold (from 5.6 to 56 mM) (the sodium concentration was lowered by a corresponding amount to maintain tonicity). For details of the method see Douglas and Poisner (1964a).

RESULTS

Adrenal gland. In the earliest experiments we tried the simple test of measuring the response to acetylcholine in Locke's solutions from which we had removed one or other of the cations, sodium, potassium or calcium. Acetylcholine continued to stimulate in the absence of sodium or potassium; indeed the responses were regularly enhanced in the absence of these ions. Obviously then, neither sodium nor potassium was needed for acetylcholine to exert its effect. In contrast, acetylcholine applied during perfusion with calcium-free Locke's solution had little or no effect, and calcium was evidently essential. Further study showed that variations of calcium concentration over a very wide range (0–18 mM), above as well as below the conventional value (2.2 mM), influenced the response to acetylcholine in graded fashion (Douglas and Rubin, 1961, 1963). Having in mind the concept of acetylcholine acting to facilitate ion flux across cell membranes, and being confronted with the then recent evidence of Bianchi and Shanes (1959) and others on calcium uptake in stimulated muscles, we hastened to examine the effect of acetylcholine on Ca^{45} uptake in the adrenal medulla. Exposure to ACh (10^{-5} g/ml) increased Ca^{45} uptake about 8-fold (Douglas and Poisner, 1962). Clearly now we had to ask ourselves if this effect of acetylcholine in promoting inward movement of calcium was sufficient to explain its stimulant effect on catecholamine secretion. To examine this possibility we set acetylcholine aside and tried other agents known or suspected to promote calcium influx. First among these was potassium, which had been found to increase Ca^{45} uptake into squid axons (Hodgkin and Keynes, 1957) and skeletal muscle (Bianchi and Shanes, 1959). In the presence of 56 mM potassium (10 times the conventional concentration) we found that Ca^{45} uptake in the medulla was enormously increased. This was encouraging since there was already evidence that excess potassium has a direct stimulant action on the medullary chromaffin cells (Vogt, 1952). And we soon found that potassium, like ACh, stimulated catecholamine secretion only when calcium was present in the external environment; and further, that the response to potassium became increasingly great as the concentration of calcium in the extracellular medium was increased (Douglas and Rubin, 1961, 1963). That potassium was a secretogogue was especially interesting, for it was difficult to imagine that its action was at any site other than the cell membrane: it is axiomatic that a stimulus must alter the environment of the thing to be stimulated, and presumably the cytoplasm of the chromaffin cell is already rich in potassium. The most probable explanation of the stimulant effect of excess potassium was therefore that it was the consequence of membrane depolarization resulting from a diminution the ratio of K inside to K outside the membrane. In any event, at this stage in our experiments we knew that two quite dissimilar secretogogues, acetylcholine and K, both promoted Ca influx in the medulla and required calcium for their action; and we strongly suspected that this calcium influx was the factor initiating secretion. This hypothesis was later strengthened by additional evidence that a variety of other adrenal secretogogues of widely divergent chemical structure all required calcium for their action: among the calcium-dependent secretogogues were various ACh congeners (Douglas and Rubin, 1961b; Douglas and Poisner, unpublished); the amines, 5-hydroxytryptamine and histamine; and the polypeptides, angiotensin and bradykinin (Poisner and Douglas, 1965).

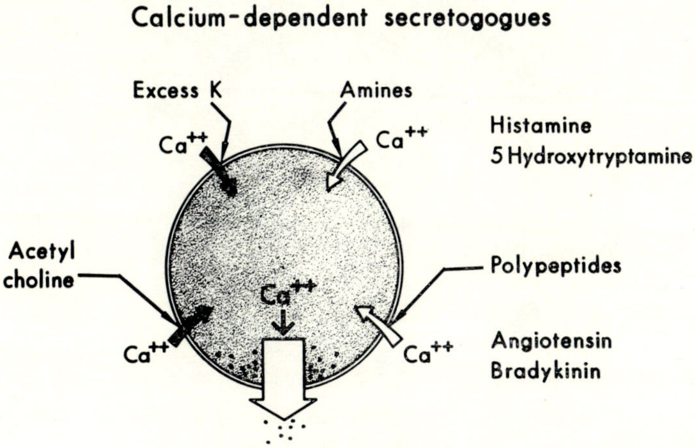

FIG. 3. The various adrenal medullary secretogogues requiring calcium for their effect. Ca influx in the medulla has been demonstrated with ACh and excess K. The other secretogogues are known to mobilize Ca in other tissues. All the secretogogues are supposed to act by promoting Ca influx into the chromaffin cell.

Although we did not examine the effect of these substances on calcium movement in the adrenal medulla they are known to promote Ca influx in muscles. There were thus grounds for supposing that Ca entry might be the common factor mediating the stimulant action of all these secretogogues (Fig. 3).

Obviously this hypothesis would be more satisfactory if we could show that calcium itself, in the absence of any of these secretogogues, is capable of stimulating catecholamine release under conditions in which it can penetrate the cell. One of the simplest and most widely known means of increasing membrane permeability in living cells of various kinds is, of course, to deprive them of calcium. It turned out that chromaffin cells perfused with calcium-free Locke's solution became exquisitely responsive to calcium when it was reintroduced to the medium: the reintroduction of calcium (in the conventional concentration, 2 mM) then always evoked the most intense secretion.

In contrast, a substantial rise in calcium concentration had little or no effect on secretion during perfusion with solutions already containing Ca, such as Locke's solution. Calcium deprivation thus "conditions" the cell in some way that leads to calcium becoming a secretogogue. I think this simple experiment is so valuable that I have chosen to depict the sequence of events

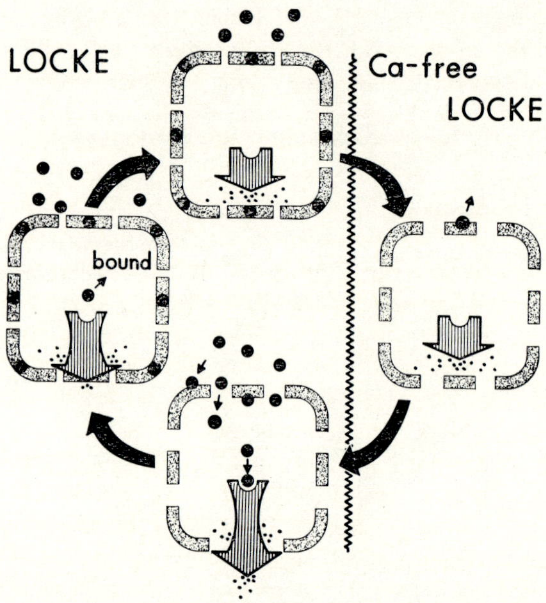

FIG. 4. A schematic portrayal of the events thought to underly the stimulant effect of Ca on catecholamine release when Ca is reintroduced to a chromaffin cell exposed for some minutes to Ca-free Locke's solution. The black circles represent Ca, either in the extracellular environment or the cell membrane (where it is responsible for maintaining the membrane relatively impermeable to Ca ions, as indicated by the small diameter of the membrane "pores"). The smaller black dots represent catecholamines. The Ca-deprived membrane becomes permeable to Ca. When Ca is reintroduced it enters the cell and activates the amine extrusion mechanism represented by the arrow. The stimulant effect is terminated as the membrane takes up Ca and becomes relatively impermeable to that ion, and as free Ca inside the cell is bound.

we imagine is occurring (Douglas and Rubin, 1961) in diagrammatic form (Fig. 4). We assume that perfusion with Ca-free Locke's solution leaches Ca off the cell membrane thereby increasing its permeability to various substances, including calcium, as indicated by the enlargement of the membrane "pores". When perfusion with calcium is restored, calcium enters the cell through the leaky membrane and activates the amine extrusion mechanism. This effect, although intense at first, wanes rapidly as the cell membrane takes up calcium and regains its normal state of relative impermeability to that ion—and also as the free calcium that has appeared in excess amount in the cell sap is bound or extruded. Inasmuch as we know that the level of ionized calcium within cells generally is very low, perhaps less than 1 per cent of that in the extracellular environment (e.g. Keynes and Lewis, 1956), it does not seem improbable that the sudden appearance of free calcium ions within the cell should provide the environmental change that is the essential attribute of a stimulus. This experiment involving "calcium-reintroduction" thus corroborates the calcium entry hypothesis of acetylcholine's action by demonstrating that calcium can evoke secretion in the absence of any of the familiar secretogogues.

Further support for the idea that the amine extrusion mechanism is something that is normally set in motion by calcium came from studies of the related alkaline-earth metal barium. The simple addition of barium during perfusion with Locke's solution proved itself sufficient to induce an intense secretion of catecholamines. The effect is apparently due to barium itself: it could be obtained during perfusion with Ca-free Locke's solution, Ca-free Locke's solution with the addition of EDTA, or a medium (isosmotic sucrose) devoid of sodium, potassium and calcium (Douglas and Rubin, 1964a, b). Barium is thus an exception to the rule that adrenal medullary secretogogues require calcium. But it is of the greatest interest that this exceptional secretogogue is an alkaline earth metal closely related to calcium. Studies on muscle have shown that barium penetrates more readily than does calcium, and the ease with which barium evokes catecholamine release may therefore be supposed to result from its ability to penetrate the chromaffin cell. Possibly, as suggested by Mullins (1961), barium passes into cells more easily because the singly hydrated barium ion fits pores in the membrane more closely than does the singly hydrated calcium ion. However, the hydrated Ba ion is of larger diameter than the hydrated Ca ion, and there is an interesting alternative explanation, namely that Ba modifies the cell membrane properties to facilitate its own entry. We were led to suspect such an action by experiments showing that the stimulant effect of barium on the adrenal gland could be inhibited by increasing the calcium concentration of the perfusion medium. Since our evidence indicated that calcium stimulated secretion whenever it was allowed to penetrate the chromaffin cell, its inhibitory effect during exposure to barium suggested that calcium was acting to impede barium entry. This led us to the view (Douglas and Rubin, 1964b) that possibly the basic mechanism of barium's action was to displace membrane calcium competitively, the Ba-membrane complex being more permeable than the Ca-membrane complex (Fig. 5). But whether or not this is the mechanism of barium entry—and there are still other possible explanations—the ability of barium to stimulate secretion without the aid of secretogogues such as ACh or K (and indeed, in the absence of any of the other common cations) strengthens the view that secretion is normally initiated by some mechanism responsive to the related alkaline earth metal, calcium. This mechanism, I might mention in passing, can also be activated by strontium, although strontium behaves more like calcium than like barium, and to obtain strong secretory responses with it requires some adjuvant treatment such as exposure to acetylcholine, excess potassium or Ca-free media, presumably to facilitate its inward movement. Various other divalent ions cannot substitute for Ca (Douglas and Rubin, 1964a).

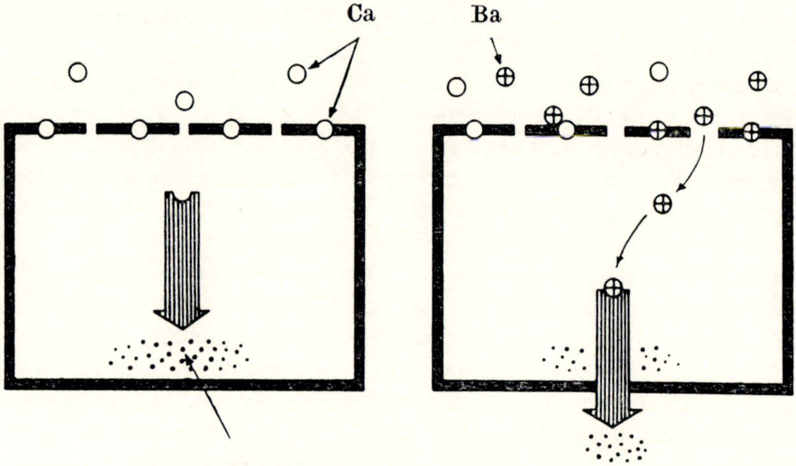

FIG. 5. A possible explanation of the stimulant effect of Ba on catecholamine secretion from the chromaffin cell. Ba is shown displacing membrane Ca competitively. The membrane that has taken up Ba is permeable to the alkaline earth metals which can enter and activate the catecholamine extrusion mechanism represented by the arrow. (From Douglas and Rubin, 1964b.)

Armed with this evidence that calcium and other alkaline earth metals are capable of acting as secretogogues in their own right, we may now return to the problem of the mechanism of action of acetylcholine. The hypothesis that we have proposed (Douglas and Rubin, 1961; Douglas, 1965) is depicted in Fig. 6. The cell at rest in the normal, calcium-containing, environment has calcium incorporated in its membrane and has a low permeability to extracellular calcium. The level of ionized calcium within the cell is very low and the calcium-activated secretory apparatus is dormant. When the cell is exposed to acetylcholine the permeability to calcium increases, calcium runs down its electrochemical gradient into the cell and the calcium ion concentration rises at some critical site. It must be emphasized that this site need not be with the cell sap, it could conceivably be the inner surface of the membrane itself. As a result of this rise in Ca concentration the secretory mechanism is set in motion and catecholamines are extruded. When the drug is removed there is recovery as the membrane regains its normal state of relative impermeability to calcium and as the excess calcium within the cell is bound or extruded.

If, at this point, we ask ourselves how acetylcholine brings about such a change in the membrane we are, of course, faced with one of the fundamental problems in physiology, how transmitter substances cause changes in cell membrane permeability. One possibility, admittedly naive, but clearly raised

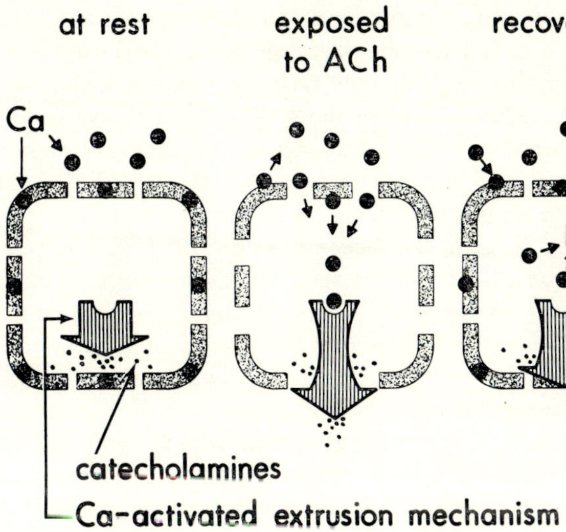

FIG. 6. The suggested mode of action of acetylcholine (ACh) on the adrenal medullary chromaffin cell. ACh is shown promoting influx of Ca (black circles) which then activates the amine extrusion mechanism. The possibility that ACh increases the permeability of the membrane by decalcifying it is indicated.

by some of the responses of the chromaffin cell I have already described, would be that the acetylcholine and the other secretogogues act by decalcifying the membrane. Our interpretations of the effects of calcium deprivation and of barium on the chromaffin cell both hint at such an action, and fragmentary support for the scheme was offered at an earlier meeting in Prague (Douglas, 1965).

The final piece of evidence supporting the view that calcium mediates the effects of medullary secretogogues such as acetylcholine, potassium and so on, came from study of the effects of magnesium. It is evident that if we are correct in supposing that influx of calcium initiates the secretory response, we should expect that a substance antagonizing the action of calcium in the medulla will also reduce the effectiveness of acetylcholine and the other secretogogues. On the general grounds of its antagonism towards calcium in a host of biological systems, magnesium was a natural choice for such an experiment; but before examining the effect of magnesium on the responses to acetylcholine and the like, we decided to seek direct evidence that it was indeed a calcium antagonist at the adrenal medulla. This we did simply by examining how it influenced the secretory response to calcium given after a period of calcium deprivation. Magnesium strongly inhibited this stimulant effect of calcium. With this evidence to hand, we went on to examine the effect of magnesium on the responses to acetylcholine and potassium. Magnesium again proved to be strongly inhibitory. (Douglas and Rubin, 1963.) Here then was evidence that responses to acetylcholine (and excess K) were inhibited by a substance demonstrated to be a calcium antagonist in the system. This finding nicely corroborated the "calcium-entry" hypothesis.

At this point I wish to broaden the scope of the discussion and draw attention to the striking parallels between the factors governing release of catecholamines from the adrenal medulla and those governing release of secretions from certain other cells.

Neurohypophysis. When it had become obvious that calcium played a key role in the extrusion of catecholamines from the medulla, it seemed to us worthwhile exploring the possibility that calcium might also be involved in the release of the posterior pituitary hormones: the neurosecretory cells of the hypothalamo–hypophyseal system, from which these hormones are released, share a common developmental origin with the chromaffin cells of the adrenal medulla; the morphological appearances of hormone storage in the two types of cell are quite similar; and, furthermore, Koelle and Geesey (1961) had suggested that acetylcholine might mediate release of posterior pituitary hormones from the neurosecretory terminals in much the same way as it mediates release of catecholamines from the chromaffin cell.

Now the neurohypophysis is a rather inaccessible gland, but fortunately it proved possible to extirpate it and work with it *in vitro*. The isolated neurohypophysis of the rat lent itself especially well to study of the problem of "stimulus–secretion coupling". This preparation, which consists essentially of the neurosecretory terminals of the hypothalamo–hypophyseal tract severed from their cell bodies, released posterior pituitary hormones to the incubation medium when it was stimulated electrically or when the potassium concentration of the incubation medium was raised. Presumably excess potassium mimics the effect of electrical stimulation by depolarizing the terminals. The isolated neurohypophyses, however, did not respond to acetylcholine or stable analogs of acetylcholine, and we were thus unable to obtain support for the suggested intervention of acetylcholine in the release process. It appeared simpler to suppose that the normal signal for vasopressin release is depolarization consequent upon the arrival of nerve impulses discharged down the hypothalamo-hypophyseal tract. The question was now: How does depolarization evoke the release of posterior pituitary hormones? The answers we obtained indicated that some calcium-dependent process was at work here, just as in the adrenal medulla. Thus we found (Douglas, 1963; Douglas and Poisner, 1964a, b):

(1) That excess K failed to evoke secretion of vasopressin when calcium was omitted from the incubation medium.

(2) That the secretory response to excess potassium increased with increasing calcium concentration of the medium over a wide range.

(3) That the secretory response to excess potassium was inhibited by magnesium (and that this effect could be overcome by adding more calcium; Douglas and Poisner, unpublished).

(4) That excess potassium increased the rate of Ca^{45} uptake in the neurohypophysis several-fold.

(5) That Ca acted as a secretogogue when introduced during exposure to excess K (and so also did Ba and Sr, but not Ni, Co, Zn or Mn. Douglas and Ishida, unpublished). And finally,

(6) That sodium was not required for the secretory response to excess potassium. On the contrary, the response to potassium was more than doubled in sodium-free media.

From these various pieces of evidence we concluded that the immediate stimulus for the release of the posterior pituitary hormones, as for the adrenal medullary hormones, was the entry of calcium ions into the cell. There was, of course, a difference between the two systems in the physiological mechanism for promoting calcium entry: at the chromaffin cell it was ACh; at the neurosecretory terminals it was depolarization consequent on the arrival of impulses.

A COMMON ROLE FOR CALCIUM IN "STIMULUS-SECRETION COUPLING"

At this stage of our work it seemed that the outlines of a broader scheme were emerging, for although the neurosecretory cells are classically considered as endocrine cells they clearly possess many of the morphological and electrophysiological characteristics of "conventional" neurones, and evidence had been available for a number of years that calcium played some critical role in the release of the neurohumour acetylcholine. Indeed, when Hodgkin and Keynes (1957) had demonstrated that activity in the squid giant axon was accompanied by an inward movement of calcium, they speculated that penetration of calcium into nerve endings might be one of the factors involved in releasing acetylcholine; and a similar proposal was made by Birks and MacIntosh (1957). Against this background, and with the strong evidence of calcium's involvement in secretion from chromaffin cells, Dr. Rubin and I suggested that there were obviously "grounds for supposing that calcium might be similarly involved in the secretion of noradrenaline on the arrival of impulses at sympathetic nerve endings" (Douglas and Rubin, 1961). The sympathetic nerve ending is, after all, a development homologue of the adrenal chromaffin cell. Subsequent events have lent substance to this supposition: Huković and Muscholl (1962), using perfused rabbit hearts, have reported that lowering the calcium concentration of the perfusion medium to one quarter of its conventional value approximately halved the output of noradrenaline in response to stimulation of the cardiac nerves; and indirect evidence has been obtained of stimulant effects of calcium and

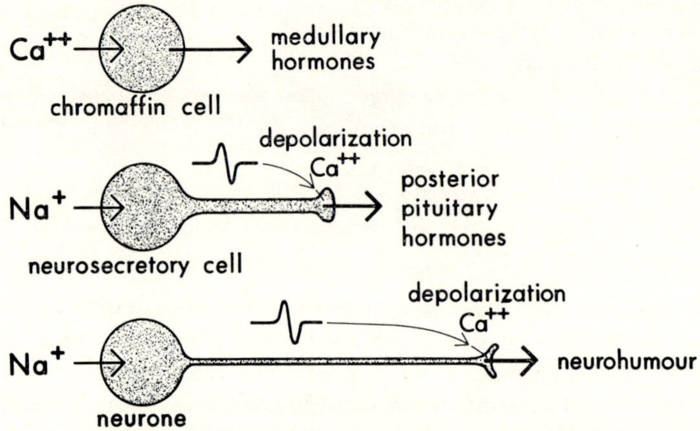

FIG. 7. A schema portraying the suggestion that Ca may play a common role in the release of neurohumours and neurohormones (see text).

inhibitory effects of magnesium on noradrenaline release from sympathetic fibers to the rabbit's ileum (Burn and Gibbons, 1964) and guinea pig vas deferens (Kuriyama, 1964).

From all this it looked as though a general principle was emerging. It was most tempting to assume that "calcium may have a general role in 'stimulus–secretion coupling' in neurones and endocrine glands of nervous origin" (Douglas, 1963) as portrayed in Fig. 7. But we had to admit that the evidence that calcium entry was responsible for the release of neurohumours and posterior pituitary hormones was less compelling than the evidence for such a role of calcium at the adrenal medulla. There was, for example, no evidence that calcium itself could act as a secretogogue. And about this time Birks reported evidence from work on the cholinergic neurones of the cat's sympathetic ganglion, that seemed to weaken this concept of a general role of calcium in the release process. Under the conditions of his experiment he found that the release of acetylcholine was critically dependent on sodium ions and he (Birks, 1963) was led to suggest that "sodium entry during the invasion of the nerve endings by an action potential facilitates ACh release by producing a local transient increase in sodium concentration at the inner surface of the terminal membrane". A similar view had been advanced by Fatt and Katz (1952) but was later abandoned by del Castillo and Katz (1956). This sodium entry hypothesis is clearly at odds with the scheme I and my colleagues had put forward to account for our findings on the adrenal chromaffin cell and neurohypophysis, in which inward movement of calcium was considered to be the key event. And certainly the finding that a fall in extracellular sodium depressed output of acetylcholine from the superior cervical ganglion was in vivid contrast with the effects of sodium deprivation on the adrenal medulla and neurohypophysis. But a recent communication from Edinburgh has sustained the view that the calcium entry hypothesis may be generally applicable to the release of neurohumours as well as neurones: Kelly (1965) has very recently reported that acetylcholine output from another set of cholinergic nerves, frog's motor neurones, is potentiated when extracellular sodium is lowered, just as is hormone output from the posterior pituitary and adrenal medulla; and he suggests, as did we (Douglas and Rubin, 1963; Douglas and Poisner, 1964a) that the effect may be due to competition between sodium and calcium ions. It seems unlikely that there should be a fundamental difference between the mechanisms for transmitter release in different cholinergic nerves and I restate the view (Douglas, 1965) that the effect of sodium deprivation in the perfused superior cervical ganglion is possibly connected not so much with the release process as with some "ancillary" function of sodium, such as the regulation of acetylcholine synthesis or its "mobilization" for release. It may be that these are especially important factors in the perfused superior cervical ganglion subjected to vigorous stimulation. At the frog neuromuscular junction the technique for measurement of ACh output is much more sensitive and the effects of Na deficiency can be studied at low frequencies of stimulation. It is pertinent to note that the results on the neurohypophysis (a preparation which may reasonably be compared with nerve) are not complicated by synthesis: the neurosecretory terminals act only as the storage and release depots and are unable to elaborate the hormones (Takabatake and Sachs, 1964).

Now the three cell types depicted in Fig. 7 share a common ancestry. But one should not conclude from this that the role of calcium we have postulated is limited to these cells: in fact there is some evidence (Douglas and Poisner, 1963) that calcium may be involved in a similar sort of way in the extrusion of protein from the submaxillary gland. This permits speculation that calcium may possibly be a common factor in "stimulus–secretion coupling" at a variety of sites. It is perhaps significant that in all the cells where we suspect there may be a calcium-activated secretory mechanism, the secretory product is one that is preformed and sequestered in membrane-limited vesicles. Perhaps the function of calcium is to participate in some common mechanism for the extrusion of the contents of such secretory vesicles (Douglas and Poisner, 1963).

Such a widespread participation of calcium in secretory mechanisms is in itself an interesting prospect, but a moment's reflection suggests that this may be part of a still more general involvement. From the beginning of our work, Dr. Rubin and I were struck by the many parallels between the responses of the secretory cells of the adrenal medulla and those of muscles, and we thought it appropriate to draw attention to this parallelism (Douglas and Rubin, 1961). Thus for each of the effects we could obtain on catecholamine secretion by altering the ionic environment of the chromaffin cell we found a corresponding effect reported on one muscle or another. It was, in part, because we were so impressed with this similarity between the two types of effector cell that we devised the term "stimulus–secretion coupling" a term patterned after that coined by Sandow (1952)—"excitation–contraction coupling"—for the corresponding events in muscle. With subsequent work the examples of similar behavior by secretory cells on the one hand and muscle on the other have multiplied, and we have continued to stress this parallelism (Douglas and Rubin, 1963; 1964a, b; Douglas and Poisner, 1964a, b). And Katz (1962), in his Croonian Lecture, has also drawn attention to the parallel requirement for calcium in the release of chemical transmitters and muscular contraction. Perhaps nature in her wisdom has found calcium entry, or "mobilization", a convenient way of initiating secretion or contraction in many cells; and it may further the efforts of those of us who are engaged in one or other of these fields to see our work within the framework of the more general concept of "stimulus–response coupling" (Fig. 8).

The nature of the extrusion process. Finally, I should like to consider with you the processes involved in catecholamine secretion from the adrenal medulla in which we believe calcium is playing a critical role. There is a plethora of theories concerning the mechanism of release of catecholamines, but facts are in less abundant supply. Morphologists have offered every shade of opinion. Thus it has been suggested repeatedly that apocrine—or even holocrine—secretion may be involved (see Franzen, 1964; Graumann, 1956; Bachmann, 1954). Other theories revolve around the membrane-limited chromaffin granules in which the bulk of the catecholamines is known to be stored: for example, Cramer (1928) claimed to have evidence that chromaffin granules were extruded whole into the blood stream and others since have reported similar findings (see Bachmann, 1954); and more recently De Robertis and Vaz Ferreira (1957) have suggested that the chromaffin granule may evacuate its contents by reverse pinocytosis ("exocytosis") or, alternatively, may give up its amines to the cell sap from where the amines diffuse to the cell exterior. But the evidence from electron microscopy is not persuasive for any of these possibilities (see Wetzstein, 1961). The last mentioned mechanism, however, has also been advocated on other grounds (e.g. Hillarp, Hökfelt and Nilson, 1954). A further complication is that two distinct classes of chromaffin granules have been described (Hillarp, 1960) one of which is rich in adenine nucleotides and the other of which possesses little or none. The fact that the second class of granules ("light" granules) gives up its amine more readily than the first when studied *in vitro* might suggest its participation in the normal release processes. However, matters are still further complicated by the existence of what appears to be a third, non-granular, pool of "free" amines in the cytoplasm (Hillarp, 1960). This too, is obviously something that the stimulus could tap. Indeed it is just

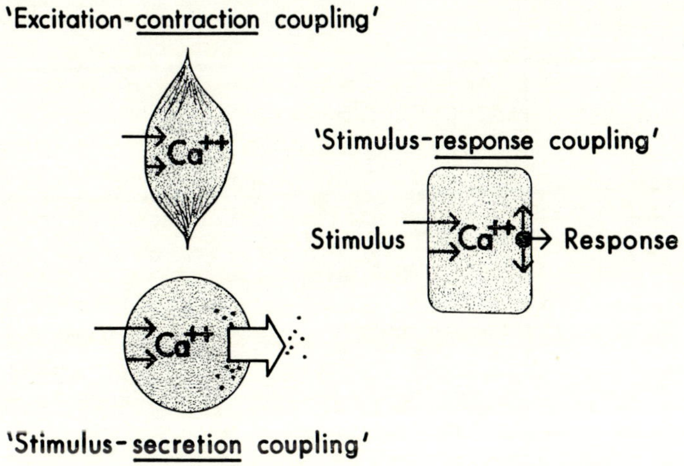

FIG. 8. A diagram to emphasize the similarity between the suggested role of Ca in "stimulus-secretion coupling" and its role in "excitation-contraction coupling", and suggesting that these may be but two facets of a more general involvement of Ca in what might be termed "stimulus-response coupling".

such a pool of "free" cytoplasmic amine which some authors (e.g. Schümann, 1961) suppose is released on stimulation of the splanchnic nerve.

My colleagues and I felt that it would help in deciding amongst these possibilities if information could be obtained on the substances other than catecholamines that are released during the secretory process. We argued that apocrine and holocrine secretion should result in protein appearing in the venous effluent—as should also the discharge of whole chromaffin granules into the blood vessels (protein accounts for about 12 per cent of the net weight of the chromaffin granules; Hillarp, 1960); and further, that the presence or absence of adenine nucleotides in the effluent might help to define the immediate "pool" of catecholamines involved in the secretory process.

Using Locke-perfused glands, stimulated with ACh and various other medullary secretogogues we have found that the discharge of catecholamines is accompanied by such small amounts of protein that we can rule out holocrine and apocrine secretion and also extrusion of whole granules into the blood stream. None of these can be contributing sufficiently to the acute secretory process (Douglas and Poisner, 1965). On the other hand, we have discovered that the discharge of catecholamines is accompanied by very large amounts of AMP (Douglas, Poisner and Rubin, 1965), and moreover that the amine and nucleotide effluxes are highly correlated. As shown in Fig. 9, the molar ratio catecholamines: AMP was about 6 : 1. In addition,

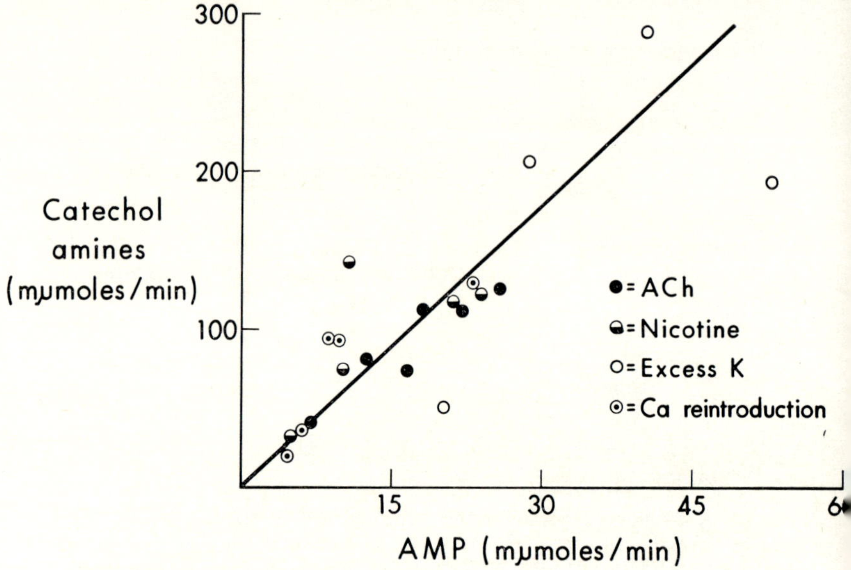

FIG. 9. The correlation between AMP and catecholamine efflux from cats' adrenal glands perfused with Locke's solution and stimulated with various medullary secretogogues. (From Douglas, Poisner and Rubin, 1965.)

we have found adenosine and some ATP and ADP. The sum of these adenine nucleotides and metabolites is such as to approximate the ratio of catecholamines to adenine nucleotides in the cat's "heavy" (nucleotide-rich) chromaffin granules which Hillarp and Thieme (1959) report to be about 3.7 : 1. This finding provides strong evidence that it is these "heavy" granules that are the immediate source of catecholamines during the acute secretory response and not any "free" pool of amines or nucleotide-poor "light" granules. The hypotheses on catecholamine extrusion that are consistent with the present findings and those that are not are depicted in Fig. 10.

Some comment is called for on the fact that in the venous effluent from secreting adrenal glands we found mainly AMP and adenosine, with much smaller amounts of ATP and ADP, whereas AMP accounts for only about 9 per cent of the adenosine phosphate in the cat's chromaffin granules with ADP constituting 22 per cent of the total and ATP 69 per cent (Hillarp and Thieme, 1959). At first sight this finding might suggest that some splitting of ATP and ADP may be playing a central part in the release process. This is an attractive possibility in view of calcium's action on secretion and its known stimulant effects on phosphatases (including ATPase) at other sites.

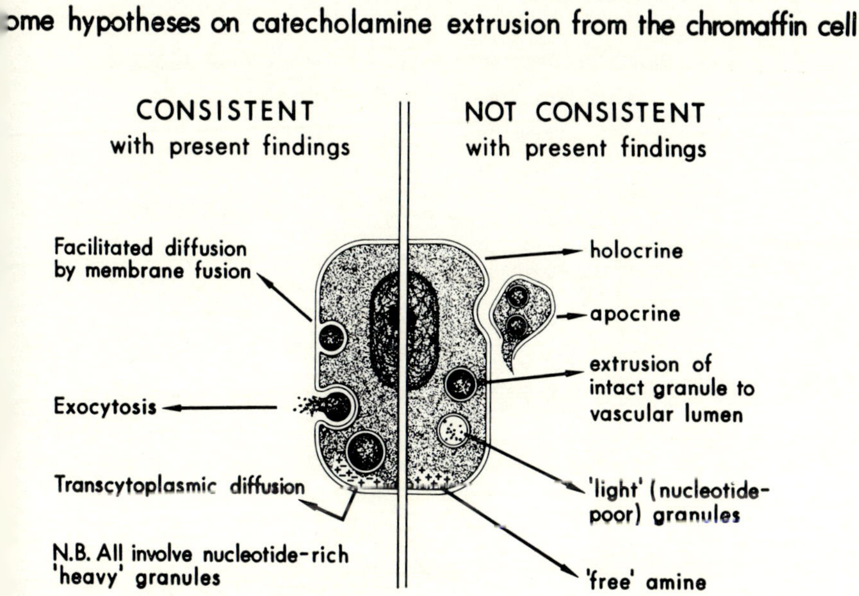

FIG. 10. The principal mechanisms of catecholamine extrusion that have been proposed by different authors (see text). They are divided into two groups: those that are not consistent with the present evidence on efflux of nucleotide and protein, and those that are. The "admissible" hypotheses (those consistent with the present findings) are depicted in greater detail in Fig. 11.

Iowever, we know from control experiments (Douglas, Poisner and Rubin, 965) in which we have injected ATP or ADP into the arterial supply to the drenal gland that these substances are rapidly metabolized to AMP and denosine (presumably by endothelial phosphatases) so that our findings are lso consistent with the possibility that the adenine nucleotides in the granule scape intact into the adrenal vessels.

Although the present results eliminate a number of postulated mechanisms f catecholamine secretion and indicate that the "heavy" granule is the source f amine drawn on during the secretory process, there still remains the roblem of how amine escapes from this granule. Some of the possibilities orth considering are depicted in the left half of Fig. 10 and, in more detail, in Fig. 11. First, we might suppose that the stimulus causes the granules somehow to discharge both catecholamines and adenine nucleotides to the cell sap from where they diffuse out of the cell across the plasma membrane. The fact that calcium can promote a parallel loss of catecholamines and adenine nucleotides from isolated chromaffin granules (Schümann and Philippu, 1963) seems at first sight to lend weight to this idea. However, magnesium has rather similar effects on isolated granules although, as we have seen, it is a powerful inhibitor on the whole cell so that the significance

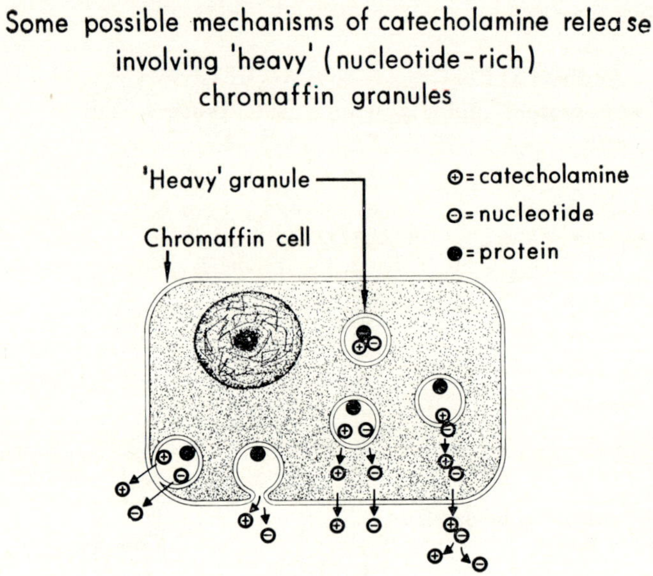

FIG. 11. A diagram of the adrenal chromaffin cell with "heavy" granules showing some possible mechanisms of secretion (see text).

of the effect of the alkaline earth metals on these isolated granules remains i[n] doubt. One difficulty with this concept that the granules give up amine an[d] adenine nucleotides to the cytoplasm is that these substances appear i[n] the venous effluent in a molar ratio approximating that found in the granule[s]. It would have to be assumed, therefore, that these different substance[s] traverse cytoplasm and cell membrane equally well, which seems rathe[r] unlikely. A way out of this difficulty might be to suppose that catecholamin[e] and nucleotide quit the granule and traverse the cytoplasm and cell mem[-] brane as a complex. This possibility has the additional virtue that the un[-] charged complex would probably penetrate the plasma membrane of the ce[ll] fairly easily. The rub here is that the catecholamine–nucleotide complex[es] that have been demonstrated experimentally are extremely weak and th[e] probability of their existing in cytoplasm is rather slender (Weiner, 1964

A simpler explanation for our observations would be that the chromaffin granule somehow discharges its content of catecholamines and adenine nucleotides directly to the cell exterior, for example, by a form of exocytosis (reverse pinocytosis). Such a mechanism has been suggested on the basis of certain electron micrographic evidence by De Robertis and Vaz Ferreira (1957), but others, e.g. Wetzstein (1961), have not found appearances consistent with this mechanism of catecholamine extrusion in their electron micrographs. Moreover, it is doubtful if the granule does evacuate all its contents as would be the case in true exocytosis, for there are several reasons for believing that the protein content of the granules may be largely retained. In the first place, analysis of adrenal glands subjected to prolonged stimulation showed that total granule protein was not as profoundly lowered as granule catecholamine and nucleotide (Carlsson, Hillarp and Hökfelt, 1957). And, second, electron micrographs (e.g. Wetzstein, 1961) of adrenal glands depleted of their amines show granules with persistent internal structure. While it is true that neither of these observations can distinguish between old granules that have lost amine and newly formed granules containing protein but as yet relatively little amine, such observations, and the present evidence that little protein appears in the venous effluent during the secretory response, are perhaps most simply explained by differential release of catecholamines and nucleotides without loss of granule protein. If this is what occurs during the secretory process, then clearly the hypothesis of reverse pinocytosis, as advanced by De Robertis and Vaz Ferreira (1957), would have to be modified in the way shown in Fig. 11 so that only some of the granule's components (the catecholamines and nucleotides) escape, while protein remains. An alternative possibility, however, that might explain retention of granular protein is depicted at the top left of Fig. 10 and at the extreme left of Fig. 11. In this scheme (here termed "facilitated diffusion by membrane fusion") the granule membrane and plasma (cell) membrane are supposed to fuse, and it is further supposed that the consequent rearrangement in the fused membrane provokes liberation and outward diffusion of catecholamine and nucleotide. But these conjectures raise more problems than they solve, and it must be conceded that while our results point squarely to the nucleotide-rich granule as the immediate source of amines drawn upon during evoked secretion, they still leave us confronted with a variety of possible mechanisms of release.

SUMMARY

In studies on cats' adrenal glands perfused with Locke's solution or variations of this medium, it has been found that acetylcholine, the physiological stimulus causing the chromaffin cells to release the catecholamines, is ineffective in the absence of calcium; that the response to acetylcholine increases with increasing calcium concentration over a wide range; that ACh increases the Ca^{45} uptake in the medulla; that the response to acetylcholine is inhibited by magnesium, a substance that is demonstrably a calcium antagonist in the adrenal medulla; that a variety of medullary secretogogues besides acetylcholine also require calcium, and that one of these, excess K, demonstrably increases the Ca^{45} uptake; and, finally, that calcium, strontium and barium are capable of evoking catecholamine secretion in appropriate circumstances in the absence of the familiar medullary secretogogues. From this it is concluded that influx of calcium into the chromaffin cell is responsible for initiating the hormone release process.

Studies on the isolated neurohypophysis of the rat, a preparation that contains the secretory terminals of the hypothalamo–hypophyseal tract from which the posterior pituitary hormones are liberated, have shown many parallels with the work on the adrenal medulla, although the immediate stimulus for the release of the posterior pituitary hormones is probably depolarization of the terminals consequent upon the arrival of impulses down the hypothalamo–hypophyseal tract and not ACh. Thus vasopressin was released when the isolated neurohypophysis was stimulated electrically or exposed to excess K but not when it was exposed to acetylcholine. However, the secretory response of the neurohypophysis to excess K was influenced by calcium and magnesium much as was the response to acetylcholine in the adrenal medulla, and potassium increased the Ca^{45} uptake. Again, influx of calcium is considered to be the event initiating hormone release.

In neither the adrenal gland nor the posterior pituitary gland, was sodium needed for the release of hormones; indeed, sodium had a depressant action at both sites, possibly due to competition with calcium.

The mechanisms of release of the medullary and posterior pituitary hormones are compared with those involved in the release of neurohumours; and attention is drawn to the remarkable parallelism between "stimulus–secretion coupling" in certain secretory cells and "excitation–contraction coupling" in muscles.

Finally, an account is given of experiments showing that the release of catecholamines from the adrenal medulla is accompanied by the efflux of large amounts of adenine nucleotides and metabolites, but by little protein. These findings are held to eliminate various suggested mechanisms of catecholamine extrusion and to narrow the choice to hypotheses involving release from the nucleotide-rich "heavy" chromaffin granules.

ACKNOWLEDGMENTS

The work was supported by Grants NB1093 and NB4006 from the National Institute of Neurological Diseases and Blindness, and Grant 2M6418 from the National Institute of Mental Health, United States Public Health Service

REFERENCES

ANTON, A. H. and D. F. SAYRE (1962) A study of the factors affecting the aluminum oxide-trihydroxyindole procedure for the analysis of catecholamines, *J. Pharmacol.* **138**, 360–75.

BACHMANN, R. (1954) Die Nebenniere. In: *Handbuch der Mikroskopischen Anatomie des Menschen*, Vol. VI, part 5, pp. 1–952.

BIANCHI, C. P. and A. M. SHANES (1959) Calcium influx in skeletal muscle at rest, during activity, and during potassium contracture, *J. Gen. Physiol.* **42**, 803–15.

BIRKS, R. I. (1963) The role of sodium ions in the metabolism of acetylcholine, *Canad. J. Biochem.* **41**, 2573–97.

BIRKS, R. I. and F. C. MACINTOSH (1957) Acetylcholine metabolism at nerve endings, *Brit. Med. Bull.* **13**, 157–61.

BURN, J. H. and W. R. GIBBONS (1964) The part played by calcium in determining the response to stimulation of sympathetic postganglionic fibres, *Brit. J. Pharmacol.* **22**, 540–8.

CARLSSON, A., N.-Å. HILLARP and B. HÖKFELT (1957) The concomitant release of adenosine triphosphate and catechol amines from the adrenal medulla, *J. Biol. Chem.* **227**, 243–52.

CRAMER, W. (1928) *Fever, Heat Regulation, Climate and the Thyroid-Adrenal Apparatus*, London: Longmans, Green & Co.

DEL CASTILLO, J. and B. KATZ (1955) On the localization of acetylcholine receptors, *J. Physiol., Lond.* **128**, 157–81.

DEL CASTILLO, J. and B. KATZ (1956) Biophysical aspects of neuromuscular transmission, *Progr. Biophys.* **6**, 121–70.

DE ROBERTIS, E. D. P., W. W. NOWINSKI and F. A. SAEZ (1960) *General Cytology*, 3rd ed., pp. 488–91. Philadelphia: W. B. Saunders Company.

DE ROBERTIS, E. D. P. and A. VAZ FERREIRA (1957) Electron microscope study of the excretion of catechol-containing droplets in the adrenal medulla, *Exp. Cell Res.* **12**, 568–74.

DOUGLAS, W. W. (1963) A possible mechanism of neurosecretion: release of vasopressin by depolarization and its dependence on calcium, *Nature, Lond.* **197**, 81–2.

DOUGLAS, W. W. (1965) Acetylcholine as a secretogogue: calcium-dependent links in "stimulus–secretion coupling" at the adrenal medulla and submaxillary gland. In: *Pharmacology of Cholinergic and Adrenergic Mechanisms*, vol. 3 of the *Proceedings of the Second International Pharmacological Meeting*, Prague, 1963. Pergamon Press, 1965; pp. 95–111.

DOUGLAS, W. W. and A. M. POISNER (1963) The influence of calcium on the secretory response of the submaxillary gland to acetylcholine or to noradrenaline, *J. Physiol., Lond.* **165**, 528–41.

DOUGLAS, W. W. and A. M. POISNER (1964a) Stimulus–secretion coupling in a neurosecretory organ: the role of calcium in the release of vasopressin from the neurohypophysis, *J. Physiol., Lond.* **172**, 1–18.

DOUGLAS, W. W. and A. M. POISNER (1964b) Calcium movement in the neurohypophysis of the rat and its relation to the release of vasopressin, *J. Physiol., Lond.* **172**, 19–30.

DOUGLAS, W. W. and A. M. POISNER (1965) Efflux of adenine nucleotides and their derivatives and of protein from adrenal glands during stimulation of the splanchnic nerve or exposure to acetylcholine, Abstract. 23rd Int. Congr. of Physiol. Sciences, Tokyo, September 1965.

DOUGLAS, W. W., A. M. POISNER and R. P. RUBIN (1965) Efflux of adenine nucleotides from perfused adrenal glands exposed to nicotine and other chromaffin cell stimulants, *J. Physiol., Lond.* **179**, 130–37.

DOUGLAS, W. W. and R. P. RUBIN (1961a) The role of calcium in the secretory response of the adrenal medulla to acetylcholine, *J. Physiol., Lond.* **159**, 40–57.

Douglas, W. W. and R. P. Rubin (1961b) Mechanism of nicotinic action at the adrenal medulla: calcium as a link in stimulus–secretion coupling, *Nature, Lond.* **192**, 1087–9.

Douglas, W. W. and R. P. Rubin (1963) The mechanism of catecholamine release from the adrenal medulla and the role of calcium in stimulus–secretion coupling, *J. Physiol., Lond.* **167**, 288–310.

Douglas, W. W. and R. P. Rubin (1964a) The effects of alkaline earths and other divalent cations on adrenal medullary secretion, *J. Physiol., Lond.* **175**, 231–41.

Douglas, W. W. and R. P. Rubin (1964b) Stimulant action of barium on the adrenal medulla, *Nature, Lond.* **203**, 305–7.

Euler, U. S. von and F. Lishajko (1962) Catecholamine release and uptake in isolated adrenergic nerve granules, *Acta Physiol. Scand.* **57**, 468–80.

Fatt, P. and B. Katz (1952) The effect of sodium ions on neuromuscular transmission, *J. Physiol., Lond.* **118**, 73–87.

Feldberg, W., B. Minz and H. Tsudzimura (1934) The mechanism of the nervous discharge of adrenaline, *J. Physiol., Lond.* **81**, 286–304.

Frankenhaeuser, B. and A. L. Hodgkin (1957) The action of calcium on the electrical properties of squid axons, *J. Physiol., Lond.* **137**, 218–44.

Franzen, D. (1964) Beiträge zur Morphologie und Chemohistologie des Nebennierenmarks des Goldhamsters (*Mesocricetus auratus*), *Anat. Anz.* **115**, 35–58.

Grauman, W. (1956) Beobachtungen über Bildung und Sekretion perjodreaktiver Stoffe im Nebennierenmark des Goldhamsters, *Z.f. Anat. u. Entwicklungsgesch.* **119**, 415–30.

Hillarp, N.-Å. (1960) Different pools of catecholamines stored in the adrenal medulla, *Acta Physiol. Scand.* **50**, 8–22.

Hillarp, N.-Å., B. Hökfelt and B. Nilson (1954) The cytology of the adrenal medullary cells with special reference to the storage and secretion of the sympathomimetic amines, *Acta Anat.* (Basel) **21**, 155–67.

Hillarp, N.-Å. and G. Thieme (1959) Nucleotides in the catecholamine granules of the adrenal medulla, *Acta Physiol. Scand.* **45**, 328–38.

Hodgkin, A. L. and R. D. Keynes (1957) Movements of labelled calcium in squid giant axons, *J. Physiol., Lond.* **138**, 253–81.

Huković, S. and E. Muscholl (1962) Die Noradrenalin-Abgabe aus dem isolierten Kaninchenherzen bei sympathischer Nervenreizung und ihre pharmakologische Beeinflussung, *Arch. Exp. Path. Pharmak.* **244**, 81–96.

Katz, B. (1962) The transmission of impulses from nerve to muscle and the subcellular unit of synaptic action, *Proc. Roy. Soc.* B. **155**, 455–77.

Kelly, J. S. (1965) Antagonism between Na^+ and Ca^{2+} at the neuromuscular junction, *Nature, Lond.* **205**, 296–7.

Keynes, R. D. and P. R. Lewis (1956) The intracellular calcium contents of some invertebrate nerve fibres, *J. Physiol., Lond.* **134**, 399–407.

Koelle, G. B. and C. N. Geesey (1961) Localization of acetylcholinesterase in the neurohypophysis and its functional implications, *Proc. Soc. Exp. Biol.* (N.Y.) **106**, 625–8.

Kuriyama, H. (1964) Effect of calcium and magnesium on neuromuscular transmission in the hypogastric nerve–vas deferens preparation of the guinea pig, *J. Physiol., Lond.* **175**, 211–30.

Marley, E. and W. D. M. Paton (1961) The output of sympathetic amines from the cat adrenal gland in response to splanchnic nerve activity, *J. Physiol., Lond.* **155**, 1–27.

Mullins, L. J. (1961) The macromolecular properties of excitable membranes. Current Problems in Electrobiology, *Ann. N.Y. Acad. Sci.* **94**, 390–404.

Poisner, A. M. and W. W. Douglas (1965) The requirements for calcium in adrenomedullary secretion evoked by histamine, serotonin, angiotensin and bradykinin, *Fed. Proc.* **24**, no. 2, part I, 488.

Sandow, A. (1952) Excitation–contraction coupling in muscular response, *Yale J. Biol. Med.* **25**, 176–201.

Schümann, H. J. (1961) Speicherung und Freisetzung der Brenzcatechinamine, *Symp. d.Dtsch. Ges.f. Endokrinologie* **8**, 23–32.

Schümann, H. J. and A. Philippu (1963) Zum Mechanismus der durch Calcium ur

Magnesium verursachten Freisetzung der Nebennierenmarkhormone, *Arch. Exp. Path. Pharmak.* **244**, 466–76.

TAKABATAKE, Y. and H. SACHS (1964) Vasopressin Biosynthesis III. *In vitro* synthesis, *Endocrinology* **75**, 934–42.

VOGT, M. (1952) The secretion of the denervated adrenal medulla of the cat, *Brit. J. Pharmacol.* **7**, 325–30.

WEINER, N. (1964) The catecholamines: Biosynthesis, storage and release, metabolism, and metabolic effects. In: *The Hormones*, vol. IV. N.Y., Academic Press, pp. 403–79.

WETZSTEIN, R. (1961) Die phäochromen Granula des Nebennierenmarks im elektronmikroskopischen Bild, *Symp. d.Dtsch. Ges.f. Endokrinologie* **8**, 33–41.

DISCUSSION

PEKKARINEN: Have you observed a correlation between adrenomedullary secretion in hypo- and hyperparathyroidism and low and high calcium concentrations in plasma?

DOUGLAS: We have no evidence bearing on this.

BRODIE: I am very impressed with Dr. Douglas's elegant dissertation. It brings out a very important point in biology—the properties of the nerve ending are more than the sum of the properties of its granules. This is important since the effects of drugs on the nerve endings are usually quite different than those on granules *in vitro*.

In a way, the results are reminiscent of blood clotting if one remembers the observations of Dr. Whittaker that under certain circumstances granules are sticky. Is it not possible that Ca^{++} in some way makes the granule "stick" to nerve ending to cause a fusion as a result of which it disgorges its contents.

DOUGLAS: I certainly am not at all persuaded that evidence on the behavior of granules *in vitro* can be carried over to explain the secretory responses of the whole cell. For example, Schümann and Philippu (*Arch. Exp. Path. Pharmak.* **244**, 466–76, 1963) have reported that Ca releases catecholamines from medullary granules but the concentration of calcium required to increase the rate of amine release was relatively high, greater than that normally present in plasma let alone cell sap; moreover, they obtained rather similar results with Mg, which we find has a powerful inhibitory effect on the whole cell. Dr. Rubin and I tended to favor (Douglas and Rubin 1961, 1963)—without much evidence I confess—a mechanism of release involving both granule and cell membrane in which Ca might participate. And, of course, adhesion of granule to membrane would be an early step in such a process—as I have indicated in Figs. 10 and 11.

STJÄRNE: Dr. Schümann and Dr. Philippu were unable to get a release of NA from the granule fraction isolated from the guinea pig heart by calcium addition, while calcium did accelerate the release from bovine adrenal medullary granules.

DOUGLAS: Yes, this too makes me question the relevance of the responses of the isolated adrenal chromaffin granules to Ca. One would like to believe that the common role of calcium I have described might have a similar explanation in the two cell types releasing the catecholamines.

WEINER: Another example of a Ca^{++} dependent secretory mechanism is that observed by Woodin and Wienecke (*Biochemical J.*, 1961–4). These workers have shown that leucocidin is selectively released from polymorphonuclear leucocytes and that this release is Ca^{++} and ATP dependent. Their results indicate that Ca^{++} is taken up in the depleted vesicles. We have observed (Borowitz, Fuwa, Weiner, *Nature*, 1965) that beef adrenal granules contain relatively large amounts of a Ca^{++} and that Ca^{++} is taken up into granules and concentrated therein, *in vitro*. Do you consider that Ca^{++} may exert its releasing effect by release of the amines from a storage complex subsequent to uptake into the granules?

DOUGLAS: The calcium-dependence of protein extrusion from the leucocyte poisoned with leucocidin is a fascinating observation—especially in view of the demonstrated importance of calcium in the release of protein from exocrine glands (Douglas and Poisner, 1962, *Nature* **196**, 379–80; Douglas and Poisner, 1963, *J. Physiol.* **165**, 528–41). As to the site of action of calcium: an action on the chromaffin granules is indeed one of the possibilities we have considered (Douglas and Rubin, 1961, *J. Physiol.* **159**, 40–57; *J. Physiol.* **167**, 288–310, 1963). It is attractive in view of the evidence of Schümann and Philippu that Ca can increase the rate of catecholamine loss from chromaffin granules but their evidence that magnesium also has this action casts doubt on the relevance of the Ca effect to the events occurring in the whole cell, as I pointed out in my reply to Dr. Brodie.

SHORE: You showed that acetylcholine promotes uptake of calcium and also that calcium releases catecholamines even in the presence of hexamethonium. What is the uptake of calcium in the presence of acetylcholine as well as hexamethonium?

DOUGLAS: It would be of interest, of course, to know whether hexamethonium's antagonistic action is accompanied by suppression of Ca influx. However, the influx experiments, as we have performed them, are rather tedious and we have not looked into this.

INTERACTION OF GUANETHIDINE AND BRETYLIUM WITH ADRENERGIC NEURONS

E. COSTA

Laboratory of Chemical Pharmacology
National Heart Institute
Bethesda, Maryland, U.S.A.

GUANETHIDINE and a number of basic organic compounds have many characteristics in common: they possess a two-carbon chain separating a ring system from a guanidine group; they decrease the norepinephrine (NE) content of peripheral tissues and they make adrenergic neurons unresponsive to nerve stimulation (Sheppard and Zimmerman, 1959; Cass, Kuntzman and Brodie, 1960; Maxwell, Plummer, Schneider, Povalski and Daniel, 1960; Boura, Copp, Green, Hodson, Ruffel, Sim, Walton and Grivsky, 1961; Costa, Kuntzman, Gessa and Brodie, 1962).

In contrast to guanethidine, bretylium does not decrease catecholamine of peripheral tissues, although it produces sympathetic blockade by interfering with the process by which nerve impulses invade sympathetic nerve terminals (Boura and Green, 1959; Costa, 1961). The similarity of the pharmacological effect of these two drugs has prompted several investigators to consider the action of guanethidine on NE storage process as distinct from its bretylium-like action (Zaimis, 1964). In fact, guanethidine blocks adrenergic transmission before it produces a measurable loss of NE stores, and the blockade disappears before the amine content is noticeably repleted (Cass and Spriggs, 1961). Additional support for the similarity of the mechanism of action of the two drugs is provided by the antagonism of amphetamine-like compounds against the sympatholytic effects of guanethidine and bretylium (Laurence and Rosenheim, 1960; Wilson and Long, 1960; Day and Rand, 1962; Matsumoto and Horita, 1962, and 1963). These findings led several investigators from our laboratory to study the nature of the guanethidine and bretylium binding sites to tissues innervated by sympathetic neurons (Costa, Chang and Brodie, 1964; Chang, Costa and Brodie, 1964; Schanker and Morrison, 1965; Chang, Costa and Brodie, 1965; Brodie, Chang and Costa, 1965).

These studies revealed that guanethidine is retained in tissues by two processes. One process localizes the drug nonspecifically and is not affected by amphetamine or reserpine; the other, suppressed by amphetamine, other sympathomimetics and reserpine, suggests a specific affinity of guanethidine for the storage sites of neuronal NE. The latter process is saturable and takes up three molecules of drug for each molecule of endogenous NE. Adrenergic blockade by guanethidine seems, therefore, related to the amount of drug that accumulates in sympathetic neurons. We have suggested that the presence of guanethidine causes a persistent change in the properties of adrenergic neurons leading to a reduction of the spike potential reaching the nerve endings. This may account for the depression of the NE output following nerve stimulation described after injection of guanethidine (Hertting, Axelrod and Patrick, 1962). According to this view, the persistent change induced by guanethidine in nerve terminals not only may reduce the size of the spike potential, but also cause an increase of permeability of nerve endings leading to a continuous release of NE. The rate of this release would be related to the amount of guanethidine taken up by the neurons. At various times after guanethidine, the degree of adrenergic blockade is not directly related to the NE level present in the store, but to the degree of persistent change of neuronal properties that causes the amine loss. Since bretylium does not release NE one can not propose that it causes a change of the properties of nerve endings similar to that caused by guanethidine. Thus, another mechanism must be invoked to explain the blockade of NE release by nerve impulses (Hertting, Axelrod and Patrick, 1962) and the antagonism of guanethidine effects, caused by bretylium (Costa, (1961).

MATERIAL AND METHODS

Male Sprague–Dawley rats, weighing about 200 g, were given the following drugs intravenously by the tail vein: guanethidine in pharmacological doses (8 mg/kg) and ^{3}H-guanethidine in tracer doses (20 μg/kg, S.A., 17.3 μc/μmole); ^{14}C-bretylium in pharmacological doses (6.2 mg/kg, S.A., 0.05 μc/μmole) and ^{14}C-bretylium in tracer doses (27 μg/kg, S.A., 49 μc/μmole). Other drugs were injected intraperitoneally. Doses and levels of drugs were calculated as the free base.

The animals were killed by cervical fracture 5–6 hr after the injection of guanethidine or bretylium. Tissues were immediately removed for assay of guanethidine, ^{3}H-guanethidine, and ^{14}C-bretylium by methods described elsewhere (Chang, Costa and Brodie, 1965; Brodie, Chang and Costa, 1965).

Tritium labeled guanethidine [2-^{3}H-(octahydro-1-azocinyl)-ethyl]-guanidine sulfate (17.3 μc/μmole) was donated by Dr. H. Sheppard of CIBA Pharmaceutical Company, Summit, New Jersey. The labeled bretylium (N-*o*-bromobenzyl-N-ethyl-N-methyl-N(^{14}C) methyl ammonium iodide) (4.9 μc/μmole) was donated by Dr. H. T. Openshaw, the Wellcome Research Laboratories, Beckenham, Kent, England, and diluted with non-labeled bretylium as required.

RESULTS

Localization of Guanethidine in Various Tissues

Reports from our laboratory demonstrated that in rat, after intravenous injections of guanethidine, the drug levels decline initially at a very fast rate, but after 2 hr the rate of disappearance becomes exponential and has a half-life of about 7 hr (Kuntzman, Costa, Gessa and Brodie, 1962). About half of the injected guanethidine has disappeared during the first 2 hr, but at this time, the residual drug is localized in tissues. Schanker and Morrison (1965), have studied the time course of guanethidine levels in various tissues

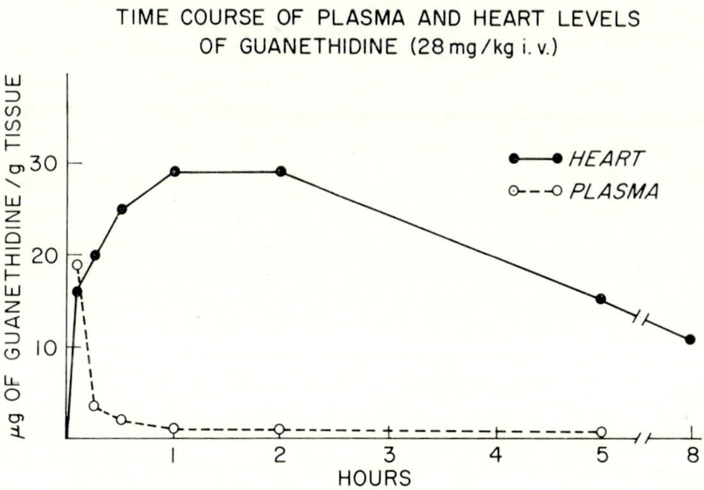

FIG. 1. Heart and plasma levels of guanethidine at various times after 28 mg/kv i.v. of guanethidine. Each point is the mean of four experiments.

after a dose of 28 mg/kg, i.v. Some of their data are reported in Fig. 1 to show that the plasma level declined rapidly during the first 30 min from a level of 19 μg/ml at 5 min, to a level of 1.7 μg/ml. In 2 hr the plasma level was about 0.5 μg/ml, which is approximately $\frac{1}{60}$ of the concentration present in heart.

The heart levels build up slowly and reach a plateau in about 1 hr, then they decline at a slow rate. Twenty-four hours later, the heart levels decline to about $\frac{1}{6}$ of the plateau level (not shown in Fig. 1). The fixation of the drug by heart differs in some respect from that by skeletal muscle: 2 hr after the injection of various doses of guanethidine (from 0.7 to 28 mg/kg) the skeletal muscle level of the drug, but not that of heart, is proportional to the dose injected (Chang, Costa and Brodie 1965). The lack of proportionality found in heart suggests that the process for uptake or storage had become saturated at the higher doses.

Time Course of the Effects of Guanethidine

Ptosis was elicited more rapidly by small doses of guanethidine than by large doses of the drug. The slow onset of action after a large dose arises from the antagonistic effect of the NE which is released by high doses of guanethidine in an active form onto receptor sites (Nash, Costa and Brodie, 1964). There was not an obvious relationship between adrenergic blockade and extent of NE depletion. Thus, sympathetic function was impaired in 15 min when the level of NE in heart had scarcely declined; this impairment had disappeared within 20 hr when the amine stores were still largely depleted.

Action of Reserpine and Amphetamine Pretreatment on the Effects of Guanethidine and Bretylium

In agreement with a previous report by Matsumoto and Horita (1963), we have found that amphetamine antagonized the adrenergic blockade and lowered the rate of NE depletion produced by 8 mg/kg of guanethidine. Pretreatment with amphetamine (0.75 mg/kg) blocked the heart uptake of a major fraction of small doses of guanethidine. However, the fraction of uptake blocked by amphetamine became progressively smaller as the dose of the drug was increased (Table 1). These results suggested that guanethidine

TABLE 1. EFFECT OF AMPHETAMINE ON THE UPTAKE OF GUANETHIDINE AND THE RELEASE OF NE FROM THE HEART

mg/kg of guanethidine	*% of NE depletion*		*μg/g of guanethidine in heart*	
	Controls	*Amphetamine pretreatment*	*Controls*	*Amphetamine pretreatment*
0.25	20	0	0.8	0.1
0.8	40	0	1.7	0.7
2.5	60	0	3.3	2.2
8.0	74	14	6.4	4.5
25·0	83	42	12.8	10.9

Rats were treated with amphetamine (0.75 mg/kg i.p.) 30 min before injecting various doses of guanethidine intravenously. Control rats were given guanethidine alone. Animals were killed 5 hr after injection of guanethidine.

was bound into tissues by two kinds of sites: specific sites that seem to be associated with the depletion of NE have a limited binding capacity and are blocked by amphetamine (Table 1); and unspecific sites, which are neither readily saturable nor affected by amphetamine (Table 1). Further evidence that guanethidine is retained in tissues by more than one kind of binding is provided by the data in Table 1.

After pretreatment with amphetamine, guanethidine in a dose of 8 mg/kg, depleted heart NE by only 14 per cent despite a guanethidine level of 4.5 μ/gg (Table 1); in rats not treated with amphetamine, this concentration is more than sufficient to deplete the catecholamine by 60 per cent or more. This suggests that the guanethidine retained by the heart after treatment with amphetamine was bound mainly by unspecific sites and that amphetamine had specifically prevented the retention of guanethidine by sites related to the storage of NE.

In other experiments (Chang, Costa and Brodie, 1965) we have found that amphetamine administered 3 or 8 hr after guanethidine, released about 30 per cent of the drug and almost immediately counteracted the adrenergic blockade caused by guanethidine, although no change in the NE level occurred. These results show that the reversal of the adrenergic blockade caused by amphetamine is closely related to the release of guanethidine from specific binding sites, but is not directly related to the level of NE.

The specific uptake of guanethidine was compared in heart and spleen, tissues which contain NE in large amounts, and in skeletal muscle, which contains only trace amounts of the amine (Table 2). After pretreatment

TABLE 2. EFFECT OF AMPHETAMINE ON UPTAKE OF GUANETHIDINE BY HEART, SPLEEN AND SKELETAL MUSCLE OF RATS

	Guanethidine level ($\mu g/g \pm S.E.$)		
	Heart	*Spleen*	*Skeletal muscle*
Controls	6.5±0.2 (5)	2.9±0.1 (3)	3.1±0.2 (4)
Pretreated with amphetamine (7.5 mg/kg)	3.3±0.1* (5)	1.4±0.1* (3)	2.7±0.2 (5)
% blockade of uptake	49	50	13

* $P < 0{\cdot}01$.

Rats were pretreated with amphetamine (7·5 mg/kg i.p.) and 30 min later given guanethidine (8 mg/kg i.v.). Control rats were killed 5 hr after injection of guanethidine. Figure in parentheses indicate number of experiments. In each experiment tissues of 3 rats were pooled.

with 7.5 mg/kg of amphetamine, the uptake of guanethidine by heart and spleen was markedly lowered, but little or no effect was observed on the uptake by skeletal muscle.

In other experiments (Chang, Costa and Brodie, 1965), it was found that the effect of amphetamine on the uptake of guanethidine was dose dependent. The larger the dose of amphetamine, the more the uptake of guanethidine was blocked, until a maximal displacement, about 3 μg/g was achieved by amphetamine in doses of 3 to 6 mg/kg. Similarly, the larger the dose of amphetamine, the more the loss of NE was prevented, and was entirely blocked by doses of 3 to 6 mg/kg. Thus, the release of NE and the uptake of guanethidine were blocked in an almost parallel fashion. Additional experiments with a larger dose of guanethidine (16 mg/kg) have shown that the uptake of drug was also suppressed by about 3 μg/g by a pretreatment with doses of 3 to 6 mg/kg of amphetamine. The uptake of guanethidine by specific sites was found to be relatively constant for a period of 5 hr after drug administration. During the first hour, however, the blockade of the retention of guanethidine

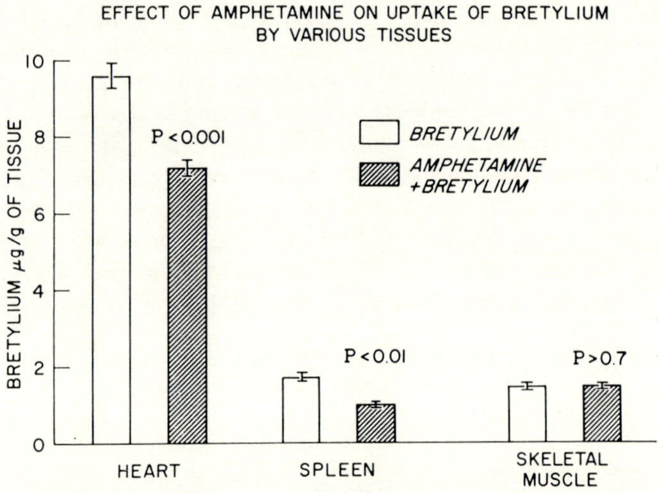

FIG. 2. Effect of amphetamine on uptake of bretylium by various tissues. Rats were treated with amphetamine and 30 min later were given ^{14}C-bretylium (7 mg/kg, i.v. S.A., 0.05 μc/μmole). Controls were given bretylium only. Animals were killed 6 hr after injection of bretylium. Each bar represents the mean $\pm$ S.E. of groups of at least four animals.

was not measured because, as shown in Fig. 1, the tissue level of the drug had not yet equilibrated with the rapidly changing plasma levels.

These results indicate that both the specific uptake of guanethidine and the loss of NE after single doses of guanethidine, could be manipulated by pretreatment of the animals with various doses of amphetamine. On plotting the specific uptake of guanethidine against the depletion of heart NE, a proportionality was found, indicating that the amount of guanethidine taken up by these sites was roughly proportional to the rate of NE depletion as defined by the per cent of amine released in 5 hr. Since amphetamine also prevents or reverses the sympatholytic action of bretylium, it was of interest to determine whether it would affect the uptake of bretylium. As shown in Fig. 2, amphetamine (0.8 mg/kg) reduced by about 30 per cent the uptake of bretylium by rat heart and spleen, but did not affect the uptake by skeletal muscle. At first glance, these results support the view that bretylium and guanethidine are taken up by similar sites in nerve endings, and produce adrenergic blockade by a similar mechanism.

Previously we demonstrated that reserpine blocks the uptake of guanethidine, and releases it from sympathetically innervated tissues (Chang, Costa and Brodie, 1964; Chang, Costa and Brodie, 1965). As shown in Fig. 3, pretreatment of rats with reserpine had no effect on the uptake of bretylium

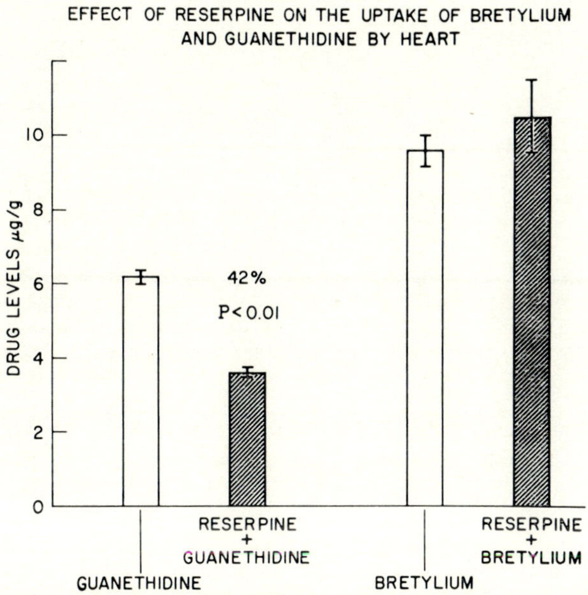

FIG. 3. Effect of reserpine on the uptake of bretylium and guanethidine by heart. Rats were treated with reserpine (5 mg/kg, i.p.) and 24 hr later with guanethidine (8 mg/kg i.v.) or ^{14}C-bretylium (7 mg/kg, i.v. S.A., 0.05 $\mu c/\mu$mole). Controls were given guanethidine or bretylium only. Animals were killed 5 hr after guanethidine and 6 hr after bretylium. Each bar represents the mean $\pm$ S.E. of groups of at least four animals.

by heart, though reserpine blocked the uptake of guanethidine by a degree comparable to that caused by maximal doses of amphetamine (Table 2). In previous experiments, we had in fact shown that no additional inhibition of guanethidine uptake was produced by the subsequent administration of amphetamine in reserpine-treated animals (Chang, Costa and Brodie, 1965). Figure 4 shows that reserpine did not affect the uptake of tracer doses of ^{14}C-bretylium though it decreased by 78 per cent the uptake of tracer doses of ^{3}H-guanethidine. These results suggest that guanethidine, but not bretylium, is taken up by the same mechanism that is responsible for the storage of noradrenaline. Further evidence that the two drugs are accumulated in

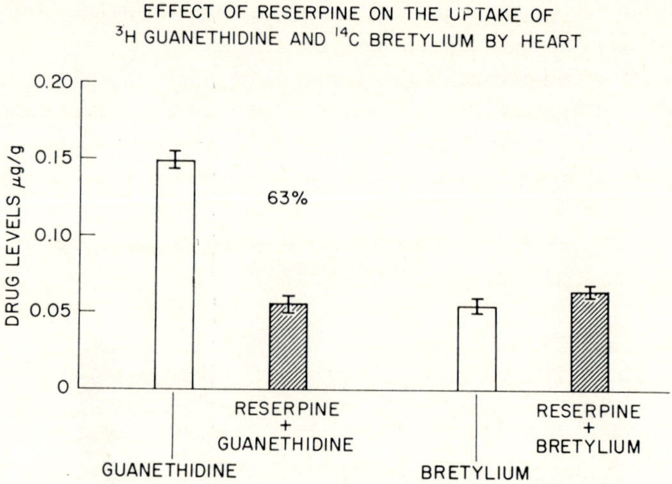

FIG. 4. Effect of reserpine on the uptake of ${}^{3}H$-guanethidine and ${}^{14}C$-bretylium by heart. Rats were treated with reserpine (1 mg/kg, i.p.) and 3 hr later with ${}^{3}H$-guanethidine (20 μg/kg, i.v. S.A., 17.3 μc/μmole) or ${}^{14}C$-bretylium (27 μg/kg, i.v. S.A., 4.8 μc/μmole). Controls were given ${}^{3}H$-guanethidine or ${}^{14}C$-bretylium only. Animals were killed 5 hr after the injection of guanethidine or bretylium Each bar represents the mean ± S.E. of groups of at least 4 animals.

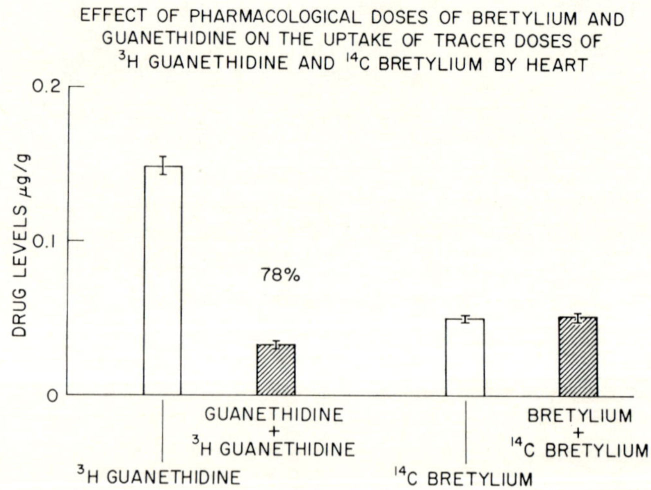

FIG. 5. Effect of pharmacological doses of bretylium and guanethidine on the uptake of tracer doses of ${}^{3}H$-guanethidine and ${}^{14}C$-bretylium by heart. Rats were treated with bretylium (6.2 mg/kg, i.v.) or guanethidine (8 mg/kg, i.p.) and 30 min later with ${}^{3}H$-guanethidine (20 μg/kg, i.v. S.A., 17.3 μc/μmole) or ${}^{14}C$-bretylium (27 μg/kg, i.v. S.A., 4.8 μc/μmole). Controls were given ${}^{3}H$-guanethidine or ${}^{14}C$-bretylium only. Animals were killed 5 hr after the injection of guanethidine or bretylium. Each bar represents the mean ± S.E. of groups of at least 4 animals.

different sites was obtained by comparing the effects of large doses of unlabeled guanethidine or bretylium on the uptake of tracer doses of the labeled drugs by rat heart. Figure 5 shows that pretreatment with a large dose of guanethidine (8 mg/kg) markedly lowered the uptake of a tracer dose of ^{3}H-guanethidine (20 μg/kg i.v.). On the other hand, a large dose of bretylium (6.2 mg/kg) failed to lower the uptake of tracer doses of ^{14}C-bretylium (27 μg/kg).

A previous paper has reported that bretylium counteracts the release of NE by guanethidine (Kuntzman, Costa, Gessa and Brodie, 1962). We considered the possibility that bretylium elicits this effect by preventing the uptake of guanethidine by specific sites. Bretylium blocked the uptake of ^{3}H-guanethidine given in tracer doses (Brodie, Chang and Costa, 1965), but

TABLE 3. ACTION OF BRETYLIUM ON THE UPTAKE OF GUANETHIDINE AND THE RELEASE OF NE FROM THE HEART OF RATS

Bretylium mg/kg i.p.	*Heart NE content % of controls*	*Guanethidine level in heart (µg/g) mean ± S.E.*
none (4)	29	5.6 ± 0.1
0.6 (4)	70	5.0 ± 0.2
3.1 (4)	84	5.4 ± 0.6

Rats were pretreated with various doses of bretylium and 30 min later given guanethidine (8 mg/kg i.v.). Control rats were given only guanethidine. Figures in parentheses indicate number of animals. Heart NE content of normal rats was 1.11 ± 0.14 μg/g S.E. ($n = 24$.)

it did not block the uptake of guanethidine given in doses large enough to produce adrenergic blockade (Table 3). Although bretylium failed to prevent the uptake of guanethidine, it still inhibited the release of NE elicited by guanethidine.

Relationship between the Uptake of Bretylium and Guanethidine with the Process that Stores Neuronal NE

If guanethidine or bretylium are taken up by the mechanism responsible for the storage of NE, catecholamines should interfere with their uptake. Pretreatment with NE or epinephrine (E) (2.5 mg/kg s.c.) prevented the uptake of tracer doses of ^{3}H-guanethidine. Isoprenaline (ISOP) was less effective and normetepinephrine (N MET) had no effect in blocking guanethidine uptake. The catecholamines did not affect the uptake of ^{14}C-bretylium (Fig. 6). We have reported elsewhere (Brodie, Chang and Costa, 1965) that the binding of tracer doses of ^{3}H-guanethidine was blocked by 70 per cent or more by amphetamine, tyramine, ephedrine and metaraminol. As reported

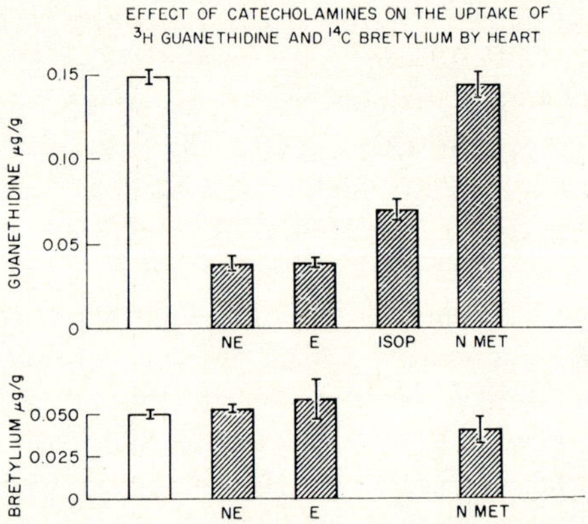

FIG. 6. Effect of catecholamines on the uptake of ^{3}H-guanethidine and ^{14}C-bretylium by heart. Rats were pretreated with various drugs and 30 min later given ^{3}H-guanethidine (20 μg/kg, i.v. S.A., 17.3 μc/μmole) or ^{14}C-bretylium (27 μg/kg, i.v. S.A., 4.8 μc/μmole). Control rats were given ^{3}H-guanethidine or ^{14}C-bretylium only. Animals were killed 5 hr after the injection of guanethidine or bretylium. Each bar represents mean ± S.E. of groups of at least 4 animals.

previously (Chang, Costa, Brodie, 1965) amphetamine and ephedrine antagonized the binding of guanethidine given in pharmacological doses; this has not yet been established for tyramine and metaraminol.

DISCUSSION

The data reported support the view that guanethidine is retained in tissues by two types of binding: the first is an unspecific adsorption to tissues components; the second involves a specific affinity for NE storage compartment. The two types of binding can be distinguished by the action of amphetamine-like drugs which selectively displace guanethidine from the specific sites.

The view that guanethidine is stored in adrenergic neurons is supported by the observation that reserpine reduces the uptake of guanethidine in tissues such as heart and spleen, but it does not affect the uptake by skeletal muscle which contains only small amounts of NE. Moreover, reserpine releases ^{3}H-guanethidine from heart, but not from skeletal muscle (Chang, Costa and Brodie, 1965).

The specific sites in heart are saturable, taking up a maximum of about three molecules of the drug for each molecule of endogenous NE. In contrast, the unspecific sites are not saturated at levels of drug as high as 15 μg/g. This, and the release of guanethidine taken up in specific sites by reserpine, suggest that a specialized transport process might function in the accumulation of the drug by sympathetic neurons. Moreover, results from our laboratory (Chang, Costa and Brodie, 1965) imply that a measurable response such as the loss of NE from neuronal stores in 5 hr, is related to the number of guanethidine molecules initially taken up by the sympathetic nerve endings. Considerable evidence (Chang, Costa and Brodie, 1965), also suggests that the adrenergic blockade is related to the amount of guanethidine that accumulates in sympathetic neurons. Pretreatment with amphetamine prevents both the adrenergic blockade and the specific uptake of guanethidine. Moreover, amphetamine given after guanethidine displaces this drug from sympathetic nerve endings and reverses the neuronal blockade. Thus, the amount of guanethidine in sympathetic neurons, the degree of neuronal blockade, and the rate of NE depletion are strictly related. However, this is not causal relationship, for the NE levels are not appreciably reduced shortly after giving the drug at a time when sympathetic blockade is complete. Moreover, the catecholamines are still depleted 20 hr after 8 mg/kg of guanethidine, when the blockade has disappeared.

Since the neuronal uptake of guanethidine after single doses of drug is correlated with the rate of NE loss as well as with the degree of neuronal blockade, both effects could be the consequence of a single action of the drug. In the conventional description of the physiological events that cause release of the transmitter from nerve endings, nerve impulses are said to invade the nerve terminals, depolarize them and, as a consequence, cause an enormous acceleration in the release of the transmitter (Eccles, 1964). A number of workers have emphasized that the role of the mechanism that takes up NE at adrenergic neurons is to recapture into neurons, after the membrane is repolarized, the major part of the NE discharged from nerve endings during depolarization (Hertting and Axelrod, 1961; Huković and Muscholl, 1962; Boullin, Costa and Brodie, 1964). In support of this view, Malmfors (1964) has shown that during nerve stimulation, the inhibition of NE recapture by imipramine leads to a partial NE depletion of nerve terminals.

If guanethidine were to produce persistent change of presynaptic terminals similar to a persistent depolarization, the NE released onto receptors could not be returned to storage; hence, there would be a massive loss of the amine. In this case, adrenergic blockade and NE loss would result from the same primary action—persistent change of the basic properties of adrenergic terminals. Moreover, the adrenergic blockade and NE loss would not be related to the level of tissue NE, but to the action that produces the depletion. A crucial test of this hypothesis was the demonstration that a large proportion of labeled NE released by guanethidine appears in the blood stream unchanged (Nash, Costa, Brodie, 1964). Moreover, in accord with the hypothesis, pretreatment of rats with bretylium, which prevents depolarization of sympathetic nerve endings, antagonizes the depletion of NE by guanethidine (Kuntzman, Costa, Gessa and Brodie, 1962).

Moreover, a number of differences can be found between the pharmacological effects of guanethidine and the effects of indirect sympathomimetics:

(1) The rate of NE release elicited by guanethidine is slower than that of indirect sympathomimetic. (2) The blockade of sympathetic nerves by guanethidine precedes its effects on NE depletion; indirect sympathomimetics do not block the responses to stimulation of sympathetic nerves. (3) The depletion of NE by guanethidine is blocked by pargyline; the effects of tyramine on endogenous NE are not altered by pargyline. Independently from any speculation on the mode of action of guanethidine at sympathetic nerve endings, our results demonstrate that a selective localization of guanethidine is an integral part of the blocking mechanism. A report by Lindmar, Bisson and Muscholl (1964), shows that pretreatment of rats with reserpine does not prevent the uptake of guanethidine by heart. These authors measured the uptake of guanethidine by heart 30 min after the administration of 15 mg/kg of the drug. After such a large dose, most of the uptake is due to nonspecific binding. Furthermore, as shown in Fig. 1, in 30 min, the tissue level of guanethidine is still undergoing rapid change and has not yet equilibrated with that of plasma.

Evidence for the view that guanethidine and bretylium act by different mechanisms is based on the structure activity relationship of a number of compounds that produce sympathetic blockade (Costa, Kuntzman, Gessa and Brodie, 1962). Bretylium and certain benzylguanidines, possessing a single carbon separating a ring system from a strongly basic side chain nitrogen, were shown to prevent the physiological release of NE, and to prevent the depletion of NE by guanethidine. In fact, drugs potent in blocking the release of NE by guanethidine are also potent in producing adrenergic blockade. Despite this evidence, a number of workers still believe that guanethidine and bretylium act by a similar mechanism (Zaimis, 1964). However, if bretylium and guanethidine act by a similar mechanism, they should be taken up into tissues by similar processes. Our studies indicate that bretylium and guanethidine are incorporated into different sites. For example, reserpine blocks the uptake of guanethidine by adrenergic neurons in rat heart, and releases the drug from its sites of uptake (Chang, Costa and Brodie, 1964) but has no effect on the uptake of bretylium. Furthermore, large doses of E and NE inhibit the uptake of tracer doses of ^{3}H-guanethidine, but do not affect the uptake of ^{14}C-bretylium. These results suggest that there is a common mechanism for the uptake and storage of catecholamines and guanethidine that is not shared by bretylium. Perhaps the most definitive evidence that bretylium and guanethidine are retained by different sites is provided by the studies showing that pretreatment with large doses of guanethidine antagonizes the uptake of tracer doses of ^{3}H-guanethidine by the heart. In contrast, the uptake of ^{14}C-bretylium is unaffected by pretreatment with large doses of bretylium. These results indicate that the sites of ^{3}H-guanethidine fixation by heart are readily saturable, whereas those of bretylium show no sign of saturation.

Amphetamine blocks the sympatholytic action of bretylium and interferes to some extent with its uptake by heart: it is not yet apparent whether the lowering of bretylium binding is associated directly with the loss of the sympatholytic action. Since only amphetamine but not the other sympathomimetics block the uptake of bretylium, it is pertinent to keep in mind the high liposolubility of this drug which may be the cause of the antagonism of bretylium uptake.

In conclusion it may be inferred from these results that bretylium does not prevent the release of NE through displacement of guanethidine from nerve endings, but by antagonizing the action of guanethidine within the neuron itself.

SUMMARY

Guanethidine is localized in rat heart and spleen by unspecific binding sites and by sites specifically concerned with the storage of NE. The two types of binding may be distinguished and the specific uptake measured from the amount of guanethidine displaced by amphetamine and reserpine. The specific sites in heart are readily saturable, and on increasing the dose of guanethidine take up a maximum of about three molecules of the drug for each molecule of endogenous NE. Blockade of adrenergic neurons and loss of NE from nerve endings are a consequence of the same primary action of guanethidine. The uptake of bretylium by adrenergic neurons differs from that of guanethidine: (*a*) Bretylium uptake is not affected by reserpine; (*b*) NE and E block the uptake of small doses of ^{3}H-guanethidine, but not that of ^{14}C-bretylium; (*c*) pretreatment with large doses of guanethidine antagonizes the uptake of small doses of ^{3}H-guanethidine; in contrast, the uptake of similar doses of ^{14}C-bretylium is unaffected by pretreatment with large doses of bretylium; (*d*) pretreatment with bretylium does not prevent the uptake of pharmacological doses of guanethidine, though it prevents guanethidine from depleting heart NE stores.

These results indicate that the sites of ^{3}H-guanethidine uptake in the heart are readily saturable while those of bretylium are not. Bretylium does not prevent the depletion of NE by guanethidine because it interferes with its uptake by heart, but because it antagonizes the action of guanethidine within the neuron itself.

REFERENCES

BOULLIN, D. J., E. COSTA and B. B. BRODIE (1964) The effect of dibenzyline on the release and re-uptake of norepinephrine (NE) by the isolated cat colon after nerve stimulation, *Pharmacologist* **6**, 196.

BOURA, A. L. A., F. C. COPP, A. F. GREEN, H. F. HODSON, G. E. RUFFEL, M. F. SIM, E. WALTON and E. M. GRIVSKY (1961) Adrenergic neurone-blocking agents related to choline, 2,6-xylyl ether bromide (TM10), bretylium and guanethidine, *Nature, Lond.* **191**, 1312–13.

BOURA, A. L. A. and A. F. GREEN (1959) The actions of bretylium: adrenergic neuron-blocking and other effects, *Brit. J. Pharmacol.* **14**, 536–48.

BRODIE, B. B., C. C. CHANG and E. COSTA (1965) On the mechanism of action of guanethidine and bretylium, *Brit. J. Pharmacol.* **25**, 171–8.

CASS, R., R. KUNTZMAN and B. B. BRODIE (1960) Norepinephrine depletion as a possible mechanism of action of guanethidine (SU–5864), a new hypotensive agent, *Proc. Soc. Exp. Biol.* (N.Y.) **103**, 871–2.

CASS, R. and T. L. B. SPRIGGS (1961) Tissue amine levels and sympathetic blockade after guanethidine and bretylium, *Brit. J. Pharmacol.* **17**, 442–50.

CHANG, C. C., E. COSTA and B. B. BRODIE (1964) Reserpine-induced release of drugs from sympathetic nerve endings, *Life Sci.* **3**, 839–44.

CHANG, C. C., E. COSTA and B. B. BRODIE (1965) Interaction of guanethidine with adrenergic neurons, *J. Pharmacol.* **47**, 303–12.

COSTA, E. (1961) Recent concepts of the role of peripheral versus central action of catecholamines in blood pressure regulation. In: *Hypertension, Recent Advances.* Ed: Brest, A. N. and Moyer, J. H. Philadelphia, Lea and Febiger, pp. 348–53.

COSTA, E., C. C. CHANG and B. B. BRODIE (1964) Interaction of guanethidine with adrenergic neurons at a molecular level, *Pharmacologist* **6**, 174.

COSTA, E., R. KUNTZMAN, G. L. GESSA and B. B. BRODIE (1962) Structural requirements for bretylium and guanethidine-like activity in a series of guanidine derivatives, *Life Sci.* **1**, 75–80.

DAY, M. D. and M. J. RAND (1962) Antagonism of guanethidine by dexamphetamine and other related sympathomimetic amines, *J. Pharm. Pharmacol.* **14**, 541–9.

ECCLES, J. C. (1964) *The Physiology of Synapses.* New York, Academic Press, Inc.

HERTTING, G. and J. AXELROD (1961) Fate of tritiated noradrenaline at the sympathetic nerve-endings, *Nature, Lond.* **192**, 172–3.

HERTTING, G., J. AXELROD and R. W. PATRICK (1962) Actions of bretylium and guanethidine on the uptake and release of ^{3}H-noradrenaline, *Brit. J. Pharmacol.* **18**, 161–6.

HUKOVIĆ, S. and E. MUSCHOLL (1962) Die Noradrenalin-Abgabe aus dem isolierten Kaninchenherzen bei sympathischer Nervenreizung und ihre pharmakologische Beeinflüssung, *Arch. Exp. Path. Pharmak.* **246**, 81–96.

KUNTZMAN, R., E. COSTA, G. L. GESSA and B. B. BRODIE (1962) Reserpine and guanethidine action on peripheral store of catecholamines, *Life Sci.* **1**, 65–74.

LAURENCE, D. R. and M. L. ROSENHEIM (1960) Clinical effects of drugs which prevent the release of adrenergic transmitter. In: *CIBA Symposium on Adrenergic Mechanisms.* Ed: Vane, J. F., Wolstenholme, G. E. W. and O'Connor, M., London, Churchill, pp. 201–8.

LINDMAR, R., G. M. BISSON and E. MUSCHOPP (1964) Aufnahme von Guanethidin in Gewebe nach sympathischer Denervierung und Vorbehandlung mit Reserpin, *Arch. Exp. Path. Pharmak.* **247**, 303–4.

MALMFORS, T. (1964) Release and depletion of the transmitter in adrenergic terminals produced by nerve impulses after the inhibition of noradrenaline synthesis or re-absorption, *Life Sci.* **3**, 1397–1402.

MATSUMOTO, C. and A. HORITA (1962) The antagonism of bretylium by sympathomimetic amines, *Nature, Lond.* **195**, 1212–13.

MATSUMOTO, C. and A. HOBITA (1963) Studies on the antagonism of guanethidine by methamphetamine, *Biochem. Pharmacol.* **12**, 295–7.

MAXWELL, R. A., A. J. PLUMMER, F. SCHNEIDER, H. POVALSKI and A. J. DANIEL (1960) Pharmacology of [2-(octahydro-1-azocinyl)-ethyl] guanidine sulfate (SU-5864), *J. Pharmacol.* **128**, 22–9.

NASH, C. W., E. COSTA and B. B. BRODIE (1964) The actions of reserpine, guanethidine and metaraminol on cardiac catecholamine stores, *Life Sci.* **3**, 441–9.

SCHANKER, L. S. and A. S. MORRISON (1965) Physiological disposition of guanethidine in the rat and its uptake by heart slices, *Int. J. Neuropharmacol.* **4**, 27–39.

SHEPPARD, H. and J. ZIMMERMAN (1959) Effect of guanethidine (SU-5864) on tissue catecholamines, *Pharmacologist* **1**, 69.

WILSON, R. and C. LONG (1960) Action of bretylium antagonized by amphetamine, *Lancet ii*, 262.
ZAIMIS, E. (1964) Pharmacology of the autonomic nervous system, *Ann. Rev. Pharmacol.* **4**, 365–400.

DISCUSSION

CARLSSON: Did you try desmethylimipramine as a blocker of guanethidine uptake? If so, is it more or less potent in this respect than amphetamine?

COSTA: Desmethylimipramine is less potent than ampethamine in blocking guanethidine uptake.

KOPIN: Is guanethidine released by nerve stimulation? How does this occur if there is depolarization?

COSTA: Dr. Boullin in our laboratory released ^{3}H-guanethidine by nerve stimulation. In this condition the amount of drug (20 μg i.v. of guanethidine) taken up was not sufficient to cause an adrenergic blockade. We do not have any experience with pharmacological doses of guanethidine.

THE ACTIONS OF RESERPINE ON THE BIOSYNTHESIS AND STORAGE OF CATECHOLAMINES

N. WEINER and C. O. RUTLEDGE

Department of Pharmacology, Harvard Medical School
Boston, Massachusetts, U.S.A.

RESERPINE is well known to exert pronounced effects on indole and catecholamine metabolism. Among its principal effects on the metabolism of catecholamines are: (1) Reserpine is able to produce severe depletion of the catecholamine content of all the tissues of the body (Carlsson, Rosengren, Bertler and Nilsson, 1956; Brodie, Olin, Kuntzman and Shore, 1957); (2) This depletion can be produced with small doses of reserpine and the depletion may be very prolonged. It is not accompanied by marked sympathomimetic effects, however, and the release of noradrenaline is difficult to demonstrate (Paasonen and Krayer, 1958; Kopin and Gordon, 1963). The adrenal gland is less susceptible to the depleting action of reserpine than are other tissues (Carlsson, Rosengren, Bertler and Nilsson, 1956). (3) Reserpine is able to block the uptake of catecholamines into tissues, as well as accelerate the release of bound H^3-noradrenaline (Axelrod, Whitby and Hertting, 1961). (4) In low doses, reserpine blocks the spontaneous release of catecholamines from particulate fractions of nerve homogenates and from adrenal granules (Euler and Lishajko, 1961); (5) The uptake of dopamine and noradrenaline into chromaffin granules is blocked by reserpine. As a consequence of this inhibition of dopamine uptake, the conversion of dopamine to noradrenaline, which occurs within the granules, is also inhibited (Kirshner, 1962; Kirshner, Rorie and Kamin, 1963). All of these actions of reserpine have been explained in accordance with our present concept of the storage of catecholamines within either vesicles or granules which spontaneously lose and actively take up catecholamines. Reserpine presumably blocks this uptake and the free amine is oxidatively deaminated by monoamine oxidase and released as inactive metabolites (Axelrod, 1964). Perhaps it would be more correct to admit the converse; i.e., that the present model of the nerve terminal has been evolved to explain the actions of reserpine and those of other agents which affect catecholamine metabolism.

Shortly after the depleting effect of reserpine was elucidated and the effect of reserpine on both innervated and denervated tissues was demonstrated (Kroneberg and Schümann, 1957; Mirkin, 1958; Weiner, Perkins and

Sidman, 1962), we initiated studies on the adrenal slice, anticipating that an *in vitro* model of this type would help to elucidate further the actions of this agent. It soon became obvious, however, that reserpine exerted no detectable depleting effect on the adrenal slice, even with incubations of two hours duration in the presence of high concentrations of the drug (Draskóczy, Burack and Weiner, 1963). Since we observed a rapid spontaneous release of catecholamines from control adrenal slices, the negative results indicated that there might be an impaired amine uptake into binding sites in the incubated slices which could not be further affected by reserpine. Alternatively, the absence of an action of reserpine in the incubated slice might be due to a dual effect of the agent. The drug may have been blocking the spontaneous loss of catecholamines from the granules (Euler and Lishajko, 1961), which is ordinarily so prominent in the incubated slice, in addition to inhibiting uptake of the released amines.

In view of Kirshner's studies on the effect of reserpine on noradrenaline synthesis, further studies on the actions of reserpine have been performed both in the adrenal slice and in the perfused rabbit heart, using labeled precursors of catecholamines. The results demonstrate a marked inhibitory effect of reserpine on the formation of noradrenaline from labeled precursor amino acids and suggest that this effect of reserpine may play an important role in the catecholamine depleting effect of this drug.

MATERIALS AND METHODS

Adrenal slices. Cow adrenal glands were obtained from a local abattoir, immersed in ice and brought to the laboratory as quickly as possible, usually within 2 hr. The medullae were dissected out and placed in cold sucrose. The tissue was sliced free hand, using a frosted glass slide and a Stadie–Riggs microtome blade. Approximately 200 mg of tissue (6–8 slices) were placed in each flask containing modified Krebs–Ringer phosphate buffer, and glucose (Draskóczy, Burack and Weiner, 1963). Each flask also contained: ATP (5×10^{-4} M); reduced glutathione (1×10^{-3} M); AMP (6×10^{-4} M); and sodium fumarate (2×10^{-2} M). Reserpine was dissolved in a small amount of acetic acid, diluted in distilled water, and used at a final concentration of 6×10^{-5} M. An equivalent amount of acetic acid was added to the control flasks.

Flasks were preincubated at 37°C for 15 min and then either L-3,5-H^3-tyrosine (New England Nuclear Corporation; 5.6 C/mM), 0.1 mC, or L-3,4-dihydroxyphenylalanine (dopa) (2,5,6-H^3) (Nuclear Chicago Corporation; 5.6 C/mM), 0.1 mC, was added and the flasks were incubated further for either 30 or 60 min. Zero-time flasks were processed immediately after addition of the labeled amino acid. At the end of the incubation, the flasks were cooled to 0°C, slices were removed, rinsed and homogenized in 20 volumes of cold 0.32 M sucrose. An aliquot was removed and precipitated with 10 per cent trichloroacetic acid (TCA), and the remainder was subjected to differential centrifugation. Gradient density centrifugation was also performed on occasion, but it was found that almost all the labeled amines were in the chromaffin granule layer, and the routine use of this additional separation procedure did not seem warranted. The fractions were precipitated with 10 per cent TCA, centrifuged and the pellets were washed twice with 5 per cent TCA. The supernatants were combined and saved for chromatography on Dowex-50 to separate the amines.

Perfused hearts. Male albino rabbits weighing between 1.5 and 2.5 kg were injected with reserpine and at varying times thereafter were killed by a blow on the head. The animals were rapidly exsanguinated and the heart was carefully dissected out. The aorta was cannulated and the heart rinsed free of blood. The isolated heart was perfused with Krebs–Henseleit solution, equilibrated with 95 per cent O_2–5 per cent CO_2, at a pressure of 100 cm water for 5 min, after which the perfusion was continued with 400 ml of an identical solution containing 5×10^{-5} M L-tyrosine and 100 μC of 3,5-H^3-L-tyrosine (Spector, Sjoerdsma, Zaltzman-Nirenberg, Levitt and Udenfriend, 1963). The perfusate was returned to the reservoir by means of a peristaltic pump (average perfusion rate = 75–80 ml/min). After either 30 or 60 min of perfusion, the hearts were removed, frozen in liquid N_2, pulverized with a steel homogenizer, and the tissue was precipitated with 10 per cent TCA while still frozen. The acidified tissue was further homogenized in a glass homogenizer, centrifuged and the pellet was washed twice with 5 per cent TCA. The supernatants were combined and saved for assay of noradrenaline and for chromatography on Dowex-50 to separate the amines.

For studies of H^3-noradrenaline uptake, after 5 minutes perfusion with Krebs–Henseleit solution, the hearts were perfused for 10 min with an identical solution containing HDL-7-H^3-noradrenaline (New England Nuclear Corporation; 4.7 C/mM), 0.1 μC, and isotope-free DL-noradrenaline to make the final concentration of noradrenaline equal 1.25 μg/l (Kopin, Hertting and Gordon, 1962). After this, the hearts were again perfused with noradrenaline-free Krebs–Henseleit buffer for 2 min. In these experiments, the perfusate was not returned to the reservoir. The tissues were processed as described above, except that the catecholamines were separated from O-methylated derivatives on alumina prior to the separation of the catecholamines on Dowex-50. It was found that the radioactive material which was not adsorbed to alumina amounted to only 10–20 per cent of the total radioactivity found in the heart.

Separation and assay of catecholamines. The acid extracts were neutralized to pH 6.5 and adsorbed onto Dowex-50 Na^+ columns (5 mm × 35 mm). (After washing with (*a*) phosphate buffer, pH 6.5, (*b*) EDTA and (*c*) H_2O Häggendal, 1963) the noradrenaline (and adrenaline) was eluted with 1 N,

HCl and dopamine was subsequently eluted with 2 N HCl. The eluate was collected in either 2 or 4 ml aliquots in scintillation vials, and the HCl was subsequently removed by vacuum evaporation over H_2SO_4 and KOH flakes. To the dried residue, 2 ml H_2O was added, followed by 14 ml of scintillation fluid (Weiner and Trendelenburg, 1962) and the samples were counted in a Packard tricarb liquid scintillation counter (efficiency about 9 per cent). Samples were corrected for quenching by the addition of internal standards.

Prior to Dowex-50 chromatography, an aliquot of the heart extracts was adsorbed on alumina, eluted with acetic acid, and assayed for noradrenaline by the trihydroxyindole procedure (modified from Weiner, Draskóczy and Burack, 1962).

In several experiments, the perfusate was not reperfused, but was immediately collected in acid in an ice bath. The catecholamines of the perfusate were adsorbed on alumina and eluted with acetic acid. The eluate was then adsorbed on Dowex-50 and the amines were separated as described. In control experiments, the content of H^3-noradrenaline and H^3-dopamine in the perfusate was found to be less than 10 per cent of that in the heart. After reserpine, the content of these labeled amines in the perfusate was not different from that found in perfusates from control hearts.

RESULTS

Adrenal slices. H^3-tyrosine is rapidly converted to dopamine, noradrenaline (and adrenaline) in cow adrenal slices (Figs. 1 and 2). About 40 per cent of the total dopamine formed is found in the incubation medium. Approximately 30 per cent of the β-OH amines formed is found in the incubation fluid. This large loss of newly formed amines can only be partly explained by deterioration of the slices, since a much smaller fraction of the total amines of the system are lost from the tissue. It appears that newly formed amines are more easily lost from the tissue slice than endogenous amines, since the specific activity of the β-OH amines in the medium exceeds that of the slice.

Similar results were obtained when H^3-dopa was used as the amino acid precursor, except that the amount of H^3-dopamine found in the medium was much greater than that found with H^3-tyrosine. Control studies have shown that the H^3-dopamine in the medium can be accounted for in part by the loss of dopa decarboxylase into the medium and formation of dopamine from dopa outside the slice. In contrast, H^3-tyrosine apparently must first be converted to dopa in a particulate fraction of the tissue (Nagatsu, Levitt and Udenfriend, 1964), and its subsequent conversion to dopamine and noradrenaline probably occurs largely or exclusively within the cells of the tissue slice.

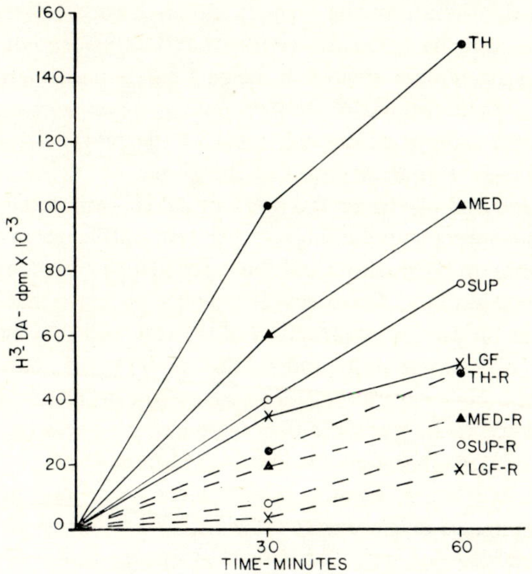

FIG. 1. Formation of H^3-dopamine from H^3-tyrosine in cow adrenal medulla slices, and the effect of reserpine (R). TH, total homogenate (●); LGF, large granule fraction (mitochondria and granules) (×); Sup, cell supernatant (○); Med, incubation medium (▲). Solid lines represent incorporation into fractions in absence of reserpine. Broken lines represent incorporation in the presence of reserpine (6×10^{-5} M). Each point is the mean of two experiments.

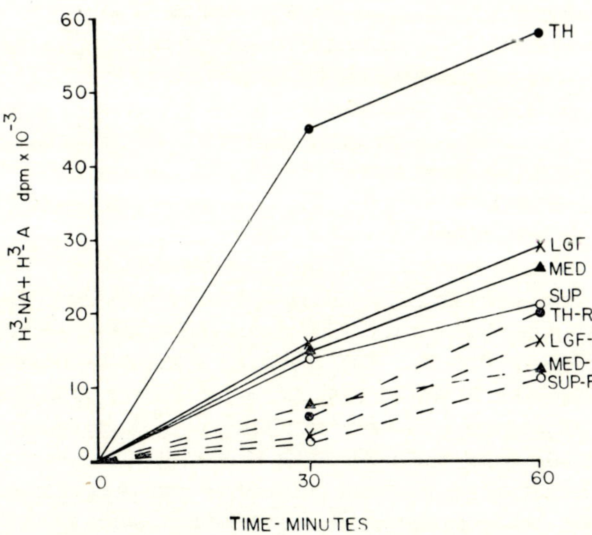

FIG. 2. Formation of H^3-noradrenaline plus H^3-adrenaline in cow adrenal medulla slices and the effect of reserpine (R). See legend of Fig. 1.

A considerable fraction of the formed amines is found in the cell supernatant. This also appears to be the result of cell deterioration, since most of the endogenous amines are located in the particulate fraction in nonincubated adrenal tissue. Furthermore, if the precursor is given *in vivo* and the glands are removed and rapidly processed, most of the newly formed amines are present in the large granule fraction of the gland.

In the presence of reserpine the amount of H^3-dopamine and H^3-β-OH amines is reduced approximately 60–75 per cent. All fractions of the adrenal cell, as well as that in the medium, exhibit a proportionately reduced content of the newly formed amines. This provides further evidence that the noradrenaline is formed in the particulate fraction of the cell, and that the newly formed amines in the cell supernatant and in the medium originated in the large granule fraction. No inhibition of formation of the β-OH amines by reserpine would be expected if the noradrenaline were being formed in non-particulate regions of the system (Kirshner, Rorie and Kamin, 1963).

Perfused hearts. Reserpine was administered to rabbits in doses ranging from 1 to 5000 μg/kg, and the animals were killed 1 or 24 hours later. The hearts were removed and perfused with H^3-tyrosine and after perfusion they were analyzed for their content of noradrenaline, H^3-noradrenaline and H^3-dopamine. It can be seen from Fig. 3 that the inhibition of formation of H^3-noradrenaline in the heart is more pronounced 1 hr after the injection of reserpine than 24 hr after the administration of the same amount of drug. In contrast, noradrenaline content of the heart is reduced to a greater extent at the later time. Between 1 and 24 hr, the dose-response curve for noradrenaline content of the heart shifts to the left, whereas that for the formation of H^3-noradrenaline in the heart is displaced to the right. For example, at 1 hr afteı 10 μg/kg reserpine, the H^3-noradrenaline content of the isolated heart is considerably less than that in hearts from animals given the same dose of reserpine 24 hr before; while the noradrenaline content is lower in those hearts removed from animals 24 hr after the injection of reserpine. At both 1 and 24 hr after 100 μg/kg reserpine, no H^3-noradrenaline is detectable in the perfused hearts.

The H^3-dopamine content is markedly reduced in hearts from animals given reserpine in doses of 10 or 100 μg/kg 1 hr before killing (Fig. 4). After 24 hr the H^3-dopamine content is also reduced in animals given reserpine in doses exceeding 10 μg/kg. It is of interest that the maximum inhibition of the H^3-dopamine content in the heart which we were able to achieve did not exceed 70 per cent. The dose-response curve is flat between doses of reserpine of 500 μg/kg and 5000 μg/kg. As with noradrenaline, the H^3-dopamine content of the perfused heart is reduced considerably more 1 hr after reserpine than when the same dose of the alkaloid is given 24 hr before the challenge with H^3-tyrosine is made (Fig. 4).

In order to more fully evaluate the actions of reserpine on noradrenaline

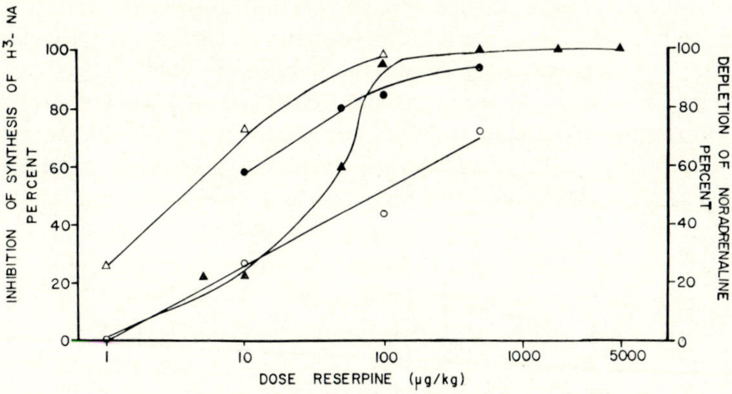

FIG. 3. Effect of various doses of reserpine on the formation of H^3-NA and the content of NA in the isolated perfused rabbit heart. ○——○, NA content of heart 1 hr after reserpine; ●——●, NA content of heart 24 hr after reserpine; △——△, H^3-NA content of the heart 1 hr after reserpine; ▲——▲, H^3-NA content of the heart 24 hr after reserpine. Results are expressed relative to controls, either as per cent inhibition of formation of H^3-NA or as per cent depletion of NA content. Reserpine was given either intravenously (1 hr) or intraperitoneally (24 hr).

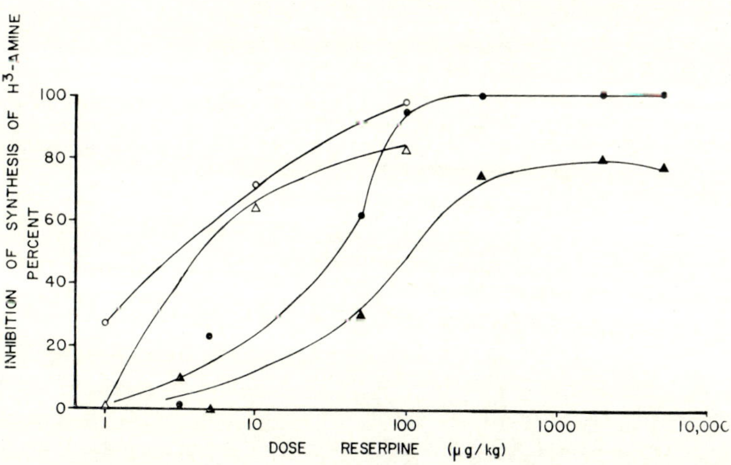

FIG. 4. Effect of various doses of reserpine on the formation of H^3-dopamine and H^3-NA in the isolated perfused rabbit heart. ○——○, H^3-NA content in the heart 1 hr after reserpine; ●——●, H^3-NA content of the heart 24 hr after reserpine; △——△, H^3-Dopamine content of heart 1 hr after reserpine; ▲——▲, H^3-Dopamine content of heart 24 hr after reserpine. Results are expressed as per cent inhibition of synthesis of H^3-amine, compared with control hearts. Reserpine was given either intravenously (1 hr) or intraperitoneally (24 hr).

and dopamine synthesis in the isolated heart, we initiated a series of experiments on the effects of reserpine on catecholamine formation, noradrenaline content and noradrenaline uptake at varying times after intravenous reserpine (Fig. 5). The dose of 10 μg/kg was chosen since definite effects on H^3-catecholamine content in the heart were observed with this relatively low dose. Reserpine was found to have a marked, early effect on the amount of H^3-dopamine and H^3-noradrenaline formed from tyrosine in the isolated heart. There was 60–70 per cent reduction in the amounts of these labeled

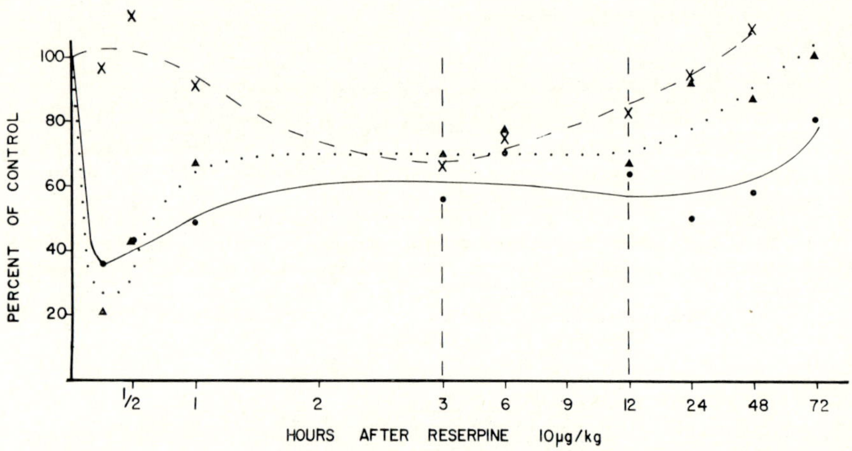

FIG. 5. The formation of H^3-dopamine and H^3-NA from H^3-tyrosine and the uptake of H^3-NA in the isolated perfused rabbit heart at various times after the intravenous administration of 10 μg/kg reserpine. ●——●, Newly formed H^3-NA in the heart; ▲——▲, Newly formed H^3-dopamine in the heart; ×——×, Uptake of H^3-NA in the heart. Results are expressed as the per cent of mean control values. Each point is the mean value of two or more experiments.

amines between 15 and 60 min after reserpine administration. After 1 hr the H^3-dopamine levels in the heart progressively rose to normal levels during the subsequent 24 hr, but the H^3-noradrenaline levels in the heart reached a plateau at about 6 hr, which was approximately 70 per cent of that found in control hearts, and remained at this level for at least 48–72 hr. Noradrenaline content of the heart fell slightly in the first 30 min after reserpine and eventually stabilized at about 70 per cent of control values for at least 72 hr. The ability of the isolated heart to take up exogenous H^3-noradrenaline was not impaired for at least an hour after the injection of reserpine. Impairment gradually progressed to a maximum of about 70 per cent of the control value at 6 hr, and remained at this level until 24 hr after reserpine, after which the uptake began to return toward control levels (Fig. 5).

DISCUSSION

The results on adrenal slices and the isolated perfused heart indicate that the effects of reserpine on biosynthesis of catecholamines may be a prominent aspect of the depleting action of this alkaloid. Since free noradrenaline appears to be quite stable in the presence of adrenal slices (Drasköczy, Burack and Weiner, 1963) the reduction in the content of H^3-noradrenaline in adrenal slices cannot be ascribed to increased breakdown of the newly formed amine unless one presumes that newly formed noradrenaline is selectively destroyed in the presence of reserpine. Studies are now under way to determine whether or not this is true. It appears more likely that noradrenaline synthesis is inhibited in these systems and that this effect is due to impaired uptake of dopamine into granules and the consequent inhibition of the formation of noradrenaline which is presumed to occur within this particulate. The reduced formation of dopamine appears to be only apparent and not real, and can be most reasonably attributed to increased breakdown of this amine by monoamine oxidase (MAO) as a result of impaired uptake and protection of the amine within the binding sites. Dopamine is known to be very rapidly deaminated by this enzyme (Weiner, 1960).

In the heart, inhibition of synthesis comes on very early, within 15–30 min after reserpine administration. This action of reserpine precedes any detectable impairment of binding of H^3-noradrenaline which is added to the perfusate, and suggests either that: (*a*) reserpine exerts an effect on synthesis which is independent of its action on amine uptake; or that (*b*) endogenously formed noradrenaline is synthesized and stored in a compartment different from that which takes up exogenous noradrenaline; or that (*c*) dopamine uptake is more sensitive to the action of reserpine than is noradrenaline uptake. Although neither of the first two hypotheses can be excluded at present, if one assumes that levels of free and bound dopamine and noradrenaline are determined by: (*a*) rate of biosynthesis; (*b*) rate and efficiency of uptake into and retention in protective binding sites and (*c*) rate of oxidative deamination, the actions of reserpine can be best explained by the last hypothesis. Thus, reserpine blocks uptake of dopamine into granules and, as a consequence, blocks the synthesis of noradrenaline. Large doses will completely block this uptake and hence completely block noradrenaline synthesis. In contrast, dopamine synthesis is not blocked, and the equilibrium levels of new dopamine will be determined solely by the relative rates of dopamine formation and breakdown. It appears that this equilibrium is achieved when H^3-dopamine levels are about 30 per cent of those found when uptake, binding and conversion to noradrenaline are unimpaired.

The observation that noradrenaline and "apparent" dopamine synthesis may be severely inhibited when noradrenaline uptake is unimpaired suggests that noradrenaline uptake is less sensitive to the action of reserpine than is dopamine uptake. Again, since noradrenaline is a much poorer substrate for MAO than dopamine, this differential susceptibility to reserpine action may be the result of the more rapid destruction of free dopamine. If the rates of both noradrenaline and dopamine uptake are reduced proportionately by a competitive action of reserpine at the sites of uptake, but the deamination of dopamine is increased to a much greater degree as a result (because it is a better MAO substrate), then the action of reserpine either in low doses or

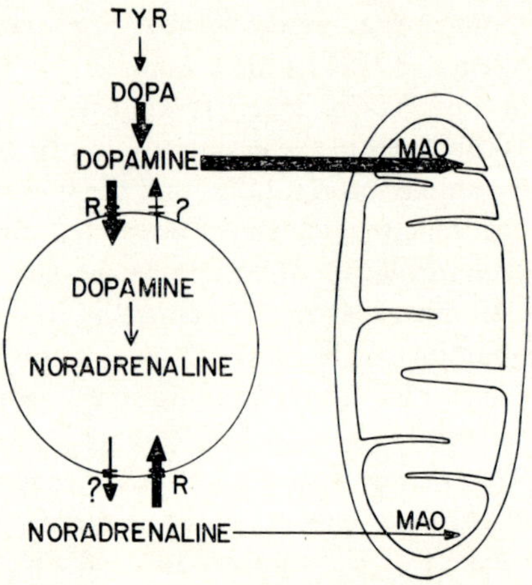

FIG. 6. Diagrammatic representation of the effect of reserpine on catecholamine metabolism at the nerve terminal and in the chromaffin cell. Width of arrows are intended to crudely reflect the relative rates of the various transformations. Site of reserpine action is indicated by a pair of cross-bars on the arrows. Primary action is presumed to be the block of amine uptake, although the drug may also act to prevent loss from the sites of storage. See text for details.

before maximum effect is attained may be evident only with regard to dopamine synthesis, dopamine uptake, and noradrenaline synthesis (secondarily), but not with respect to noradrenaline uptake (Fig. 6). This action of reserpine can also explain the ability of MAO inhibitors to block the depleting effect of reserpine. By inhibiting MAO, these inhibitors prevent the rapid destruction of dopamine and the amine may then more effectively compete with reserpine for the uptake sites into the storage particles where it can be converted to noradrenaline.

SUMMARY

Employing H^3-tyrosine as precursor, active formation of dopamine and noradrenaline in cow adrenal medulla slices and in isolated, perfused rabbit hearts has been demonstrated. Reserpine (6×10^{-5} M) is able to markedly inhibit the formation of dopamine and noradrenaline in adrenal slices. Administration of reserpine (10 μg/kg intravenously) to rabbits results in marked inhibition of formation of H^3-dopamine and H^3-noradrenaline in the heart, when the heart is isolated and perfused 15–30 min after the injection. Impairment of H^3-dopamine formation gradually subsides over 24 hr, but noradrenaline formation remains moderately depressed for at least 72 hr. Uptake of H^3-noradrenaline by the heart is not impaired until at least 1 hr after the injection of reserpine, and is only moderately depressed 3–48 hr after this dose. Higher doses of reserpine produce proportionately greater impairment of H^3-dopamine and H^3-noradrenaline formation in the heart and of H^3-noradrenaline uptake, and deplete the content of endogenous noradrenaline in a dose-related manner.

The results indicate that reserpine blocks the uptake of dopamine into the storage particles and consequently blocks the formation of noradrenaline which ordinarily takes place in these particulates. The newly-formed dopamine, unprotected by binding, is rapidly deaminated by monoamine oxidase. The precise quantitative importance of this action of reserpine in achieving tissue catecholamine depletion remains to be assessed.

ACKNOWLEDGMENTS

This work was supported by grants NB-04663 and NB-02947, training grant 5 T1-GM-95 and research career program award K3 GM-4954 of the U.S. Public Health Service and by USAF SAM contract 41(657)-414.

We are indebted to Dr. Gordon L. Coppoc for valuable assistance during a portion of this investigation.

REFERENCES

AXELROD, J. (1964) The uptake and release of catecholamines and the effect of drugs. In: *Progress in Brain Research*, vol. 8, *Biogenic Amines* (H. E. Himwich and W. A. Himwich, eds.), Amsterdam, Elsevier Publ. Co., pp. 81–9.

AXELROD, J., L. G. WHITBY and G. HERTTING (1961) Effect of psychotropic drugs on the uptake of H^3-norepinephrine by tissues, *Science* **133**, 383–4.

BRODIE, B. B., J. S. OLIN, R. G. KUNTZMAN and P. A. SHORE (1957) Possible interrelationships between release of brain norepinephrine and serotonin by reserpine, *Science* **125**, 1293–4.

CARLSSON, A., E. ROSENGREN, Å. BERTLER and J. NILSSON (1956) Effect of reserpine on the metabolism of catechol amines. In: *Psychotropic Drugs* (S. Garattini and V. Ghetti, eds.), Amsterdam, Elsevier Publ. Co., pp. 363–72.

DRASKÓCZY, P. R., W. R. BURACK and N. WEINER (1963) Catecholamine release from adrenal slices and rate of respiration, *J. Pharmacol.* **141**, 58–64.

EULER, U. S. VON and F. LISHAJKO (1961) Effect of reserpine on the release of catecholamines from isolated nerve and chromaffin cell granules, *Acta Physiol. Scand.* **52**, 137–45.

HÄGGENDAL, J. (1963) An improved method for fluorimetric determination of small amounts of adrenaline and noradrenaline in plasma and tissues, *Acta Physiol. Scand.* **59**, 242–54.

KIRSHNER, N. (1962) Uptake of catecholamines by a particulate fraction of the adrenal medulla, *J. Biol. Chem.* **237**, 2311–17.

KIRSHNER, N., M. RORIE and D. L. KAMIN (1963) Inhibition of dopamine uptake *in vitro* by reserpine administered *in vivo*, *J. Pharmacol.* **141**, 285–9.

KOPIN, I. J. and E. K. GORDON (1963) Metabolism of administered and drug-released norepinephrine-7-H^3 in the rat, *J. Pharmacol.* **140**, 207–16.

KOPIN, I. J., G. HERTTING and E. K. GORDON (1962) Fate of norepinephrine-H^3 in the isolated perfused rat heart, *J. Pharmacol.* **138**, 34–40.

KRONEBERG, G. and H. J. SCHÜMANN (1957) Die Wirkung des Reserpins auf den Hormongehalt des Nebennierenmarks, *Arch. Exp. Path. Pharmak.* **231**, 349–60.

MIRKIN, B. L. (1958) Catechol amine depletion in the rat's denervated adrenal gland following chronic administration of reserpine, *Nature, Lond.* **182**, 113–14.

NAGATSU, T., M. LEVITT and S. UDENFRIEND (1964) Tyrosine hydroxylase. The initial step in norepinephrine biosynthesis, *J. Biol. Chem.* **239**, 2910–17.

PAASONEN, M. K. and O. KRAYER (1958) The release of norepinephrine from the mammalian heart by reserpine, *J. Pharmacol.* **126**, 153–60.

SPECTOR, S., A. SJOERDSMA, P. ZALTZMAN-NIRENBERG, M. LEVITT and S. UDENFRIEND (1963) Norepinephrine synthesis from tyrosine-C^{14} in isolated perfused guinea pig heart, *Science* **139**, 1299–1301.

WEINER, N. (1960) Substrate specificity of brain amine oxidase of several mammals, *Arch. Biochem.* **91**, 182–8.

WEINER, N., P. R. DRASKÓCZY and W. R. BURACK (1962) The ability of tyramine to liberate catecholamines *in vivo*, *J. Pharmacol.* **137**, 47–55.

WEINER, N., M. PERKINS and R. L. SIDMAN (1962) Effect of reserpine on noradrenaline content of innervated and denervated brown adipose tissue of the rat, *Nature, Lond.* **193**, 137–8.

WEINER, N. and U. TRENDELENBURG (1962) The effect of pretreatment with reserpine and with cocaine on the uptake of tyramine-2-C^{14} and *dl*-epinephrine-2-C^{14} into heart and spleen, *J. Pharmacol.* **137**, 56–61.

ON THE MECHANISM OF NOREPINEPHRINE DEPLETION BY RESERPINE, METARAMINOL AND RELATED COMPOUNDS AND ANTAGONISM BY MONOAMINE OXIDASE INHIBITION

P. A. Shore, H. S. Alpers and D. Busfield*

Department of Pharmacology
University of Texas Southwestern Medical School
Dallas, Texas, U.S.A.

The mechanism whereby α-methyldopa (α-MD) and α-methyl-*m*-tyrosine (α-MMT) lower the norepinephrine (NE) content of various tissues lies in the conversion of these amino acids to amine products which are the active NE releasers. Thus, inhibition of the decarboxylation of α-MD or α-MMT also inhibits their ability to lower NE levels, while administration of amine products of the amino acids leads to NE depletion, and amine products persist in tissues throughout NE depletion (Carlsson and Lindqvist, 1962; Udenfriend and Zaltzman-Nirenberg, 1962; Gessa, Costa, Kuntzman and Brodie, 1962; Porter, Totaro and Leiby, 1961).

We have recently presented evidence that *l*-metaraminol is the active releasing agent when α-MMT or its decarboxylation product, α-methyl-*m*-tyramine, is administered (Shore, Busfield and Alpers, 1964). In these studies it was found that *l*-metaraminol, but not *d*-metaraminol or α-methyl-*m*-tyramine, is taken up and bound for prolonged periods by tissues rich in adrenergic innervation. The *l*-metaraminol thus bound behaves as a false sympathetic transmitter in that it replaces norepinephrine on a mole-for-mole basis; the drug is released by compounds known to release NE, and its uptake is inhibited by sympathetic denervation (immunosympathectomy) or by drugs known to inhibit NE uptake. Metaraminol is also released by NE administration. In other experiments metaraminol was found to be released from heart by sympathetic nerve stimulation (Crout, Alpers, Tatum and Shore, 1964). It was thus concluded that metaraminol releases NE by competitive displacement from the major NE storage sites and that the drug then acts as a false sympathetic transmitter.

Reserpine, the first known NE-depleting drug, has been thought to act by an entirely different mechanism, mainly because measurable reserpine levels disappear from tissues long before NE levels return to normal. Theories on the mechanism of action of reserpine have centered on the possible

* Present address: Glaxo Laboratories, Greenford, Middlesex, England.

inhibition of a transport system which serves to maintain NE stores. These theories have been discussed in recent reviews (Shore, 1962; Kopin, 1964; Carlsson, 1964).

The present paper describes evidence which further emphasizes the competitive nature of metaraminol-induced NE lowering, and also demonstrates that certain similarities are evident in the action of reserpine. It is suggested that reserpine competes with NE at a site of action which may be the NE carrier, while metaraminol competes with a larger NE pool.

METHODS AND MATERIALS

Female Sprague–Dawley rats, 150–180 g, were used in all experiments and were killed by chloroform inhalation. Hearts were removed rapidly and trimmed free of fat and atria. Metaraminol and NE were measured fluorimetrically as described previously (Shore and Alpers, 1964; Shore and Olin, 1958). Unless otherwise specified, metaraminol signifies the levo isomer.

Other experimental details are described under Results.

RESULTS

Effect of Monoamine Oxidase Blockade on the NE-lowering Action of Metaraminol and on Metaraminol Uptake

Various doses of metaraminol were injected intraperitoneally into normal rats or into rats treated 18 hr previously with the monoamine oxidase (MAO) inhibitor pheniprazine (5 mg/kg i.p.). Three hours later the animals were killed and heart NE content assayed. As shown in Fig. 1, the dose-response

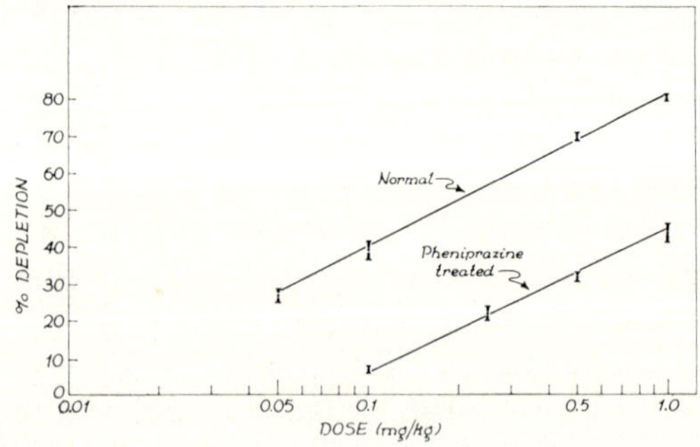

FIG. 1. Various doses of metaraminol were administered intraperitoneally to normal rats and to rats given pheniprazine (5 mg/kg i.p.) 18 hr previously. Animals were killed 3 hr later and NE content of hearts measured. Markers denote mean percent depletion and standard errors.

curve of metaraminol-induced heart NE lowering was shifted to the right in a parallel fashion by pretreatment with pheniprazine.

In animals receiving 0.1 or 0.5 mg/kg metaraminol, the concentration of this drug in heart was also measured. As shown in Table 1, the amount of

TABLE 1. STOICHIOMETRY OF NE RELEASE AND METARAMINOL UPTAKE IN HEARTS OF NORMAL AND PHENIPRAZINE-TREATED RATS

	Dose Metaraminol (mg/kg)	*NE absent ($m\mu$ mol/g)*	*Metaraminol present ($m\mu$ mol/g)*
Normal	0.1	3.1 (2.3–3.8)	3.0 (2.8–3.1)
7.8 (6.4–8.8)	0.5	5.4 (5.2–5.7)	5.5 (4.8–5.8)
Pheniprazine-treated	0.1	1.0 (0.7–1.3)	1.3 (1.2–1.5)
12.2 (9.3–14.1)	0.5	4.0 (3.4–4.9)	3.6 (3.3–4.1)

Metaraminol (0.1 or 0.5mg/kg i.p.) was given to normal rats or to rats treated with pheniprazine (5 mg/kg i.p.) 18 hr previously. Animals were killed 3hr after metaraminol and hearts analyzed for NE and metaraminol. Figures denote means and ranges of values. Column showing NE absent was derived by subtracting NE concentration in metaraminol-treated rats from mean NE concentration in appropriate control population, shown in leftmost column.

metaraminol taken up closely approximated the amount of NE displaced n both normal and pheniprazine-treated rats, i.e. blockade of MAO inhibited both NE lowering and metaraminol uptake in a parallel fashion.

Effect of Pretreatment with α-Methyldopa on Uptake of Metaraminol

Rats were given *dl*-α-MD (400 mg/kg i.p.), and 18 hr later these and normal rats were given metaraminol (0.1 mg/kg i.p.). Three hours later the animals

TABLE 2. EFFECT OF α-METHYLDOPA PRETREATMENT ON UPTAKE OF METARAMINOL BY RAT HEART

Condition	*Metaraminol conc. $\mu g/g \pm SE$*	
Normal	0.54 ± 0.02	$P < 0.01$
α-Methyldopa treated	0.30 ± 0.01	

Metaraminol (0.1 mg/kg i.p.) was given to normal rats or to rats treated 18 hr previously with α-methyldopa (400 mg/kg i.p.). Animals were killed 3 hr later and hearts analyzed for metaraminol.

were killed and heart metaraminol content measured. As shown in Table 2, there was a greatly lessened uptake of metaraminol in rats pretreated with α-MD.

Effect of MAO Blockade on NE-lowering Action of Reserpine

Various doses of reserpine were injected intravenously into normal rats and into rats treated 18 hr previously with pheniprazine (5 mg/kg i.p.). Three hours later the rats were killed and heart NE content measured. As depicted in Fig. 2, the dose-response curve of reserpine-induced NE lowering was shifted greatly to the right, essentially in a parallel fashion. In order to insure

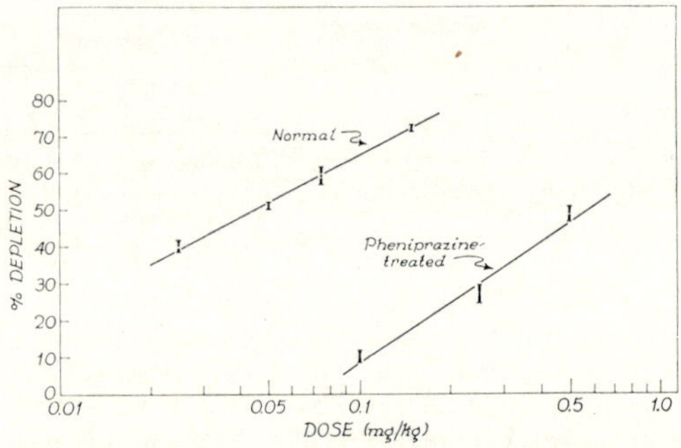

FIG. 2. Various doses of reserpine were administered intravenously to normal rats and to rats given pheniprazine (5 mg/kg i.p.) 18 hr previously. Animals were killed 3 hr later and NE content of hearts measured. Markers denote mean percent depletion and standard errors.

that the ability of larger doses of reserpine to overcome the effect of MAO inhibition was not due to a central effect of reserpine, other experiments were carried out with 0.5 mg/kg reserpine, but using MAO-blocked rats also given the ganglionic blocking drug, chlorisondamine (5 mg/kg i.p.), 15 min before reserpine. It was found that ganglionic blockade had no effect on the ability of this higher dose of reserpine to overcome the effect of MAO inhibition on NE release.

Effect of MAO Blockade on the Metaraminol-lowering Action of Reserpine

Rats were given metaraminol (1 mg/kg i.p.). Four hours later some were treated with pheniprazine (5 mg/kg i.p.). Eighteen hours later the animals were given various doses of reserpine intravenously. Three hours later the animals were killed and hearts analyzed for metaraminol. It was found (Fig. 3) that MAO inhibition had no effect on reserpine-induced metaraminol lowering.

Inhibition by NE Pretreatment of Reserpine-induced NE Lowering

NE (1 mg/kg s.c.) was administered to rats, and 15 min later some were given reserpine, 25 or 50 μg/kg intravenously. The animals were killed 24 h

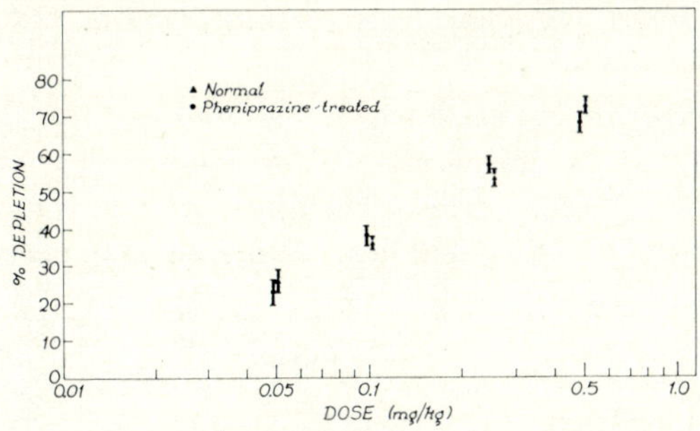

FIG. 3. Metaraminol (1 mg/kg i.p.) was given to rats. Four hours later some rats were given pheniprazine (5 mg/kg i.p.). Eighteen hours later various doses of reserpine were given intravenously, and 3 hr after reserpine, animals were killed and metaraminol content of hearts measured. Markers denote mean percent depletion and standard errors.

later and hearts analyzed for NE. In other experiments rats were given the same dose of NE 4 hr after reserpine (25 μg/kg i.v.) and were killed 24 hr after reserpine administration.

It was found (Table 3) that in rats given NE alone, heart NE levels had returned to normal 24 hr later. Reserpine greatly lowered heart NE levels, but in animals given NE shortly before reserpine, heart NE levels 24 hr later

TABLE 3. INHIBITION OF RESERPINE-INDUCED NE RELEASE FROM RAT HEART BY NE PRETREATMENT

Treatment	*Heart NE* ($\mu g/g \pm SD$)	
None	1.04 ± 0.13	
NE (1mg/kg s.c.)	1.03 ± 0.22	
Reserpine (50 μg/kg i.v.)	0.25 ± 0.02	$P < 0.01$
NE 15min before reserpine	0.40 ± 0.08	
Reserpine (25μg/kg i.v.)	0.37 ± 0.03	
NE 15 min before reserpine	0.65 ± 0.10	$P < 0.001$
Reserpine 4 hr before NE	0.32 ± 0.03	

Rats killed 24 hr after reserpine or, in the case of receiving NE alone, 24 hr after NE.

were significantly higher than in animals given reserpine alone. Furthermore, at this time a lesser degree of ptosis was evident in the NE-pretreated rats. In animals given NE 4 hr after reserpine, NE levels 24 hr later were no different from those in animals receiving reserpine alone.

Comparison of Heart NE-lowering Activity of Various Substituted α-Alkyl Phenylethylamines

Preliminary experiments established that maximal NE lowering in rat heart was produced by metaraminol 3 hr after intraperitoneal administration of 1 mg/kg. Various drugs structurally related to metaraminol were given intraperitoneally to rats, and heart NE levels were measured 3 hr later.

DRUG	Dose mg/kg	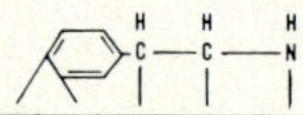					% NE Decrease
		p	m	β	α	N	
dl amphetamine	2	H	H	H	CH_3	H	0 (< 10%)
*l*p-OH amphetamine	1·5	OH	H	H	CH_3	H	0 (< 10%)
*d*p-OH amphetamine	1·5	OH	H	H	CH_3	H	0 (< 10%)
*dl*m-OH amphetamine	2	H	OH	H	CH_3	H	0 (< 10%)
*dl*m-OH amphetamine	10	H	OH	H	CH_3	H	48 (40-56) *
dl norephedrine	2	H	H	OH	CH_3	H	0 (< 10%)
*dl*p-OH norephedrine	2	OH	H	OH	CH_3	H	47 (44-49)
*l*m-OH norephedrine	1	H	OH	OH	CH_3	H	81 (77-83)
*d*m-OH norephedrine	1	H	OH	OH	CH_3	H	0 (< 10%)
p-OH ephedrine	2	OH	H	OH	CH_3	CH_3	45 (38-55) †
dl-1-(3-OH phenyl)-2-aminobutanol	2-10	H	OH	OH	CH_3CH_2	H	0 (< 10%)
dl-1-(3-methoxyphenyl) 2-aminopropanol	2	H	OCH_3	OH	CH_3	H	72 (65-77) ‡

* Inhibited by benzyloxyamine

† Optical characteristics unknown. Not inhibited by SKF 525-A

‡ Inhibited by SKF 525-A

FIG. 4. Effect of various substituted α-alkyl phenylethylamines on rat heart NE content. The various drugs were given intraperitoneally to rats in the doses indicated. Three hours later animals were killed and heart NE content measured. Benzyloxyamine (200 mg/kg i.p.), when used, was given 30 min before test drug. SKF 525-A (20 mg/kg i.p.), when used, was given 30 min before test drug.

Of the drugs tested (Fig. 4), only those possessing the combination of a phenolic group, a β-hydroxyl group, and an α-alkyl group no larger than methyl significantly lowered heart NE levels at the dose level of 1 mg/kg (or 2 mg/kg in the case of racemic mixtures). Furthermore, only the levorotatory form (presumably reflecting configuration of the β carbon, see Shore, Busfield and Alpers, 1964) showed activity. NE levels were lowered by a high dose (10 mg/kg) of α-methyl-*m*-tyramine, but this lowering was inhibited by prior treatment of rats with benzyloxyamine, an inhibitor of dopamine β-oxidase (Nikodijevic, Creveling and Udenfriend, 1963). NE levels were also lowered by the *m*-methoxy analog of metaraminol (1-[3-methoxyphenyl]-2-aminopropanol), but it was found that this compound is converted in the rat to metaraminol. Prior treatment of rats with SKF 525-A, a compound known to inhibit various routes of drug metabolism, greatly inhibited the NE-lowering action of this compound, but not that of metaraminol.

In other studies, hearts were examined for the presence of *l*-metaraminol, *d*-metaraminol, α-methyl-*m*-tyramine, or the α-ethyl analog of metaraminol (1-[3-hydroxyphenyl] 2-aminobutanol) 3 hr after injection of these drugs. Of these four compounds, only *l*-metaraminol persisted in measurable levels.

DISCUSSION

The effect of monoamine oxidase inhibition in shifting the metaraminol-NE depletion dose-response curve as well as metaraminol uptake to the right in a parallel fashion (Fig. 1, Table 1) suggests that a competitive mechanism is operant. It does not seem likely that metaraminol is competing here with pheniprazine *per se*, as the MAO inhibitor is an "irreversible" MAO inhibitor; and as it was given 18 hr before the experiment, little free pheniprazine would remain in tissues (Horita and Chinn, 1964). The alternative conclusion is that competition exists between metaraminol and NE which cannot now be metabolized by MAO and thus is free to compete with metaraminol for a common site.

If this concept is true, then upon substitution of NE by an amine not metabolized by MAO, a subsequent injection of metaraminol should result in a lessened uptake of metaraminol even in the absence of MAO inhibition. As shown in Table 2, pretreatment of rats with α-methyldopa, so as to replace some NE stores by α-methylnorepinephrine (Carlsson and Lindqvist, 1962; Muscholl and Maitre, 1963), resulted in considerably lessened uptake of metaraminol after its injection.

MAO blockade also shifted the reserpine-induced NE depletion dose-response curve to the right (Fig. 2), again suggesting a competitive, surmountable inhibition, and again it follows that reserpine must be competing with norepinephrine for a common site. To gain further evidence that stabilized NE, and not the MAO inhibitor *per se* competes with reserpine, it was shown (Fig. 3) that the ability of reserpine to release metaraminol (not a MAO substrate) is not affected by pretreatment with a MAO inhibitor.

A consequence of the hypothesis that NE and reserpine compete for some common site should be that the presence of a high concentration of NE during the brief sojourn of circulating reserpine (Hess, Shore and Brodie, 1956) should inhibit the action of reserpine by successfully competing for its site of action. As described in Table 3, injection of a large dose of NE inhibited the long-term action of a subsequently injected dose of reserpine.

Other evidence that reserpine and NE compete for a common site comes from experiments *in vitro* which show that increasing the concentration of catecholamine partially reversed the inhibitory effect of reserpine on C^{14} catecholamine uptake by isolated adrenal granules (Jonasson, Rosengren and Waldeck, 1964).

Findings presented elsewhere and in this paper make it apparent that metaraminol substitutes mole-for-mole for NE. This stoichiometry appears to break down when large doses of metaraminol are given (Udenfriend and Zaltzman-Nirenberg, 1964), but in this context the findings of Jasmin and Bajusz (1964), who demonstrated ischemic cardiac and kidney lesions in animals treated with large doses of metaraminol, may be pertinent. The general conclusion, then, seems valid, that a stoichiometric replacement is operant in the case of metaraminol.

No such stoichiometry exists for reserpine, which has been shown to rapidly leave tissues after injection, so that at times of maximal NE depletion, little or no reserpine can be detected in tissues (Hess, Shore and Brodie, 1956). Certainly no over-all stoichiometry exists in the case of reserpine.

Other experiments, however, using tritiated reserpine have revealed that reserpine persists in tissues in very low concentrations (0.007–0.008 μg/g in brain) for at least 48 hr after injection (Plummer, Sheppard and Schulert, 1957). In the light of the present findings it seems possible that this small quantity of reserpine may have displaced NE from a small NE store. If true, this obviously represents a different pool from that displaced by metaraminol, and as the minute amounts of bound reserpine are associated with considerable NE depletion and inhibition of NE uptake, it seems likely that reserpine binds specifically to the NE carrier, thereby leading to depletion. Once reserpine is attached, it may be difficult to dislodge, thus maintaining NE depletion.

The action of a MAO inhibitor in inhibiting reserpine-induced NE depletion thus seems clear. MAO inhibition allows stabilized NE in the region of the carrier to compete with the drug for the site of binding.

As it seems clear that NE depletion and metaraminol uptake go hand-in-hand, some insight into structural requirements of a false transmitter may be obtained from a study of the NE-depleting action of compounds structurally related to metaraminol. As described in Fig. 4, specificity is considerable. Thus, for marked displacement there must be at least one phenolic hydroxyl group, a β-hydroxyl group, and an α-alkyl group no larger than methyl. The optical configuration, presumably of the β carbon, must be levorotatory. Results suggestive of these requirements have also been obtained by other experiments (Fischer, Musacchio, Kopin and Axelrod, 1964) showing that octopamine, but not tyramine, is bound persistently.

SUMMARY

Administration of *l*-metaraminol or its precursors, α-methyl-*m*-tyramine or α-methyl-*m*-tyrosine, leads to a stoichiometric substitution of tissue norepinephrine (NE) stores by *l*-metaraminol for prolonged periods of time. The drug is taken up specifically by NE storage depots where it acts as a false transmitter. Thus, *l*-metaraminol is released from heart by nerve stimulation or by drugs known to release norepinephrine (reserpine, guanethidine, tyramine), and uptake of the amine is inhibited by drugs known to inhibit NE uptake (reserpine, guanethidine, imipramine) or by immunosympathectomy. *l*-Metaraminol is also released by NE. These findings suggest competition between these amines. In the present paper, a competitive mechanism is demonstrated by the finding that inhibition of monoamine oxidase (MAO) activity inhibits both uptake of *l*-metaraminol and NE depletion, the dose-response curve of metaraminol-induced NE depletion and metaraminol uptake being shifted to the right in a parallel fashion. Uptake of metaraminol is also inhibited by prior treatment with α-methyldopa.

Similar experiments with reserpine show that the reserpine-induced NE depletion dose-response curve is also shifted to the right by MAO inhibition. The MAO-inhibiting drug does not *per se* inhibit reserpine action as release of metaraminol (not a MAO substrate) by reserpine is not inhibited by MAO inhibition. Pretreatment of animals with high doses of NE inhibits the action of reserpine *in vivo*. As no over-all stoichiometry is evident in the case of reserpine, the latter drug must be competitive with NE at a smaller store, perhaps at the site of a specific NE carrier. MAO inhibitors thus act by allowing endogenous NE to compete with the depleting drugs for their sites of action.

Studies with compounds related structurally to metaraminol reveal considerable specificity in structural requirements for displacement of NE.

ACKNOWLEDGMENTS

We wish to express our thanks to Merck Sharp & Dohme Research Laboratories, Smith Kline & French Laboratories, Sterling-Winthrop Research Institute, Wellcome Research Laboratories, and Lilly Research Laboratories for their kind gifts of drugs used in this study.

This work was supported by U.S. Public Health Service Grant No. MH-05831.

REFERENCES

CARLSSON, A. (1964) Drugs which block the storage of 5-hydroxytryptamine and related amines. In: *Handbuch der Experimentellen Pharmakologie.* In press. *Ergänzungwerk*, vol. XIX: Indole alkylamines and related compounds.

CARLSSON, A. and M. LINDQVIST (1962) *In vivo* decarboxylation of α-methyldopa and α-methyl-meta-tyrosine, *Acta Physiol. Scand.* **54**, 87–94.

CROUT, J. R., H. S. ALPERS, E. L. TATUM and P. A. SHORE (1964) Release of metaraminol (Aramine) from the heart by sympathetic nerve stimulation, *Science* **145**, 828–9.

FISCHER, J. E., J. MUSACCHIO, I. J. KOPIN and J. AXELROD (1964) Effects of denervation on the uptake and β-hydroxylation of tyramine in the rat salivary gland, *Life Sci.* **3**, 413–19.

GESSA, G. L., E. COSTA, R. KUNTZMAN and B. B. BRODIE (1962) On the mechanism of norepinephrine release by α-methyl-meta-tyrosine, *Life Sci.* **1**, 353–60.

HESS, S. M., P. A. SHORE and B. B. BRODIE (1956) Persistence of reserpine action after the disappearance of drug from brain, *J. Pharmacol.* **118**, 84–9.

HORITA, A. and C. CHINN (1964) An analysis of the interaction of reversible and irreversible monoamine oxidase inhibitors, *Biochem. Pharmacol.* **13**, 371–8.

JASMIN, G. and E. BAJUSZ (1964) Étude pathogénique des lésions cardiaques rénales et musculaires provoquées chez le rat par la méthoxamine et le métaraminol, *Rev. Can. Biol.* **23**, 67–76.

JONASSON, J., E. ROSENGREN and B. WALDECK (1964) Effects of some pharmacologically active amines on the uptake of arylalkylamines by adrenal medullary granules, *Acta Physiol. Scand.* **60**, 136–40.

KOPIN, I. J. (1964) Storage and metabolism of catecholamines: The role of monoamine oxidase, *Pharmacol. Rev.* **16**, 179–92.

MUSCHOLL, E. and L. MAITRE (1963) Release by sympathetic stimulation of α-methyl-noradrenaline stored in the heart after administration of α-methyldopa, *Experientia* (Basel) **19**, 658–9.

NIKODIJEVIC, B., C. R. CREVELING and S. UDENFRIEND (1963) Inhibition of dopamine β-oxidase *in vivo* by benzyloxyamine and benzylhydrazine analogs, *J. Pharmacol.* **140**, 224–8.

PLUMMER, A. J., H. SHEPPARD and A. R. SCHULERT (1957) The metabolism of reserpine. In: *Psychotropic Drugs*, Garattini, S. and Ghetti, V., Eds., Amsterdam, Elsevier Publ. Co., pp. 350–62.

PORTER, C. C., J. A. TOTARO and C. M. LEIBY (1961) Some biochemical effects of α-methyl-3,4-dihydroxyphenylalanine and related compounds in mice, *J. Pharmacol.* **134**, 139–45.

SHORE, P. A. (1962) Release of serotonin and catecholamines by drugs, *Pharmacol. Rev.* **14**, 531–50.

SHORE, P. A. and H. S. ALPERS (1964) Fluorimetric estimation of metaraminol and related compounds, *Life Sci.* **3**, 551–4.

SHORE, P. A., D. BUSFIELD and H. S. ALPERS (1964) Binding and release of metaraminol: mechanism of norepinephrine depletion by α-methyl-*m*-tyrosine and related agents, *J. Pharmacol.* **146**, 194–9.

SHORE, P. A. and J. S. OLIN (1958) Identification and chemical assay of norepinephrine in brain and other tissues, *J. Pharmacol.* **122**, 295–300.

UDENFRIEND, S. and P. ZALTZMAN-NIRENBERG (1962) On the mechanism of the nor-epinephrine release produced by α-methyl-meta-tyrosine, *J. Pharmacol.* **138**, 194–9.

UDENFRIEND, S. and P. ZALTZMAN-NIRENBERG (1964) On the mechanism of norepinephrine depletion by aramine, *Life Sci.* **3**, 695–702.

DISCUSSION

ALMGREN: We have studied the uptake and disappearance of H^3-metaraminol in the *dl*-form with the rat salivary gland intact and after pre- and postganglonic sympathectomy after an intravenous injection of 0·02 mg/kg body weight. One can clearly see that the sympathetic nerves take up metaraminol and also that the level of metaraminol in the preganglionically sympathectomized gland shows a much slower decrease than in the intact one, indicating that this amine is released by physiological activity and that this is a rather important factor for its disappearance.

SHORE: This, then, is further evidence that metaraminol in sympathetic nerves acts as a false transmitter.

UDENFRIEND: In the guinea pig heart, there is not a stoichiometric relationship between metaraminol and noradrenaline. Furthermore, Dr. Spector has shown that noradrenaline does not release the metaraminol taken up by the heart.

SHORE: Using the sensitive fluorimetric method for metaraminol, we find that stoichiometry exists in guinea pig heart as well as in tissues of other species (rat heart, rabbit brain, etc.). I think the difference you observe is methodologic. Furthermore, very high doses of metaraminol were injected in your experiments. It has been shown that high doses of metaraminol cause myocardial lesions. Thus the high doses introduce a complicating factor.

BRODIE: I find myself confused by the terminology. If a MAO inhibitor slows the rate at which reserpine releases NA from heart but not the final extent of depletion, should we call this a blocking effect? Similarly small doses of metaraminol do not deplete NA in 2 hr, but might they not do so in, say, 10 hr? Should not the stoichiometric aspects of small doses of aramine in heart be referred to maximal depletion over many hours rather than to a short time period?

SHORE: I think the terms "inhibition" and "antagonism" are better than the term "blocking". In our experiments we used small doses of reserpine as well as metaraminol. As far as stoichiometry goes, we have demonstrated stoichiometry with various doses of metaraminol as well as at various times at any single dose.

NOTE ADDED IN PROOF

Recent studies by L. Maitre (personal communication) have shown that after large doses of metaraminol, considerable amounts of α-methylnorepinephrine appear in guinea-pig heart, thus explaining the reported lack of stoichiometry in this species after treatment with high doses of the depleting agent. Preliminary studies by Maitre indicate that a similar phenomenon does not occur in the rat.

PHYSIOLOGICAL AND PHARMACOLOGICAL RELEASE OF MONOAMINES IN THE CENTRAL NERVOUS SYSTEM

ARVID CARLSSON

Department of Pharmacology, University of Göteborg, Sweden

RELEASE is a rather vaguely defined term, as it is currently used in the monoamine field. Physiologically, at least three kinds of release of monoamines at nerve endings can be distinguished (Fig. 1):

(A) release by the nerve impulse, which appears to act directly or indirectly on a small trasmitter pool in the storage particles (granules) (cf. Carlsson, 1964, 1965) (arrow no. 7);
(B) spontaneous leakage through the axon membrane (arrow no. 5);
(C) spontaneous loss from the granules into the axoplasm (arrow no. 6).

The absolute rates of these three processes are not easy to measure exactly, because they are partly balanced by opposite processes:

(D) pumping through the axon membrane back into the axoplasm (arrow no. 2);
(E) incorporation into the complex of the storage granules (arrow no. 3).

These two powerful uptake mechanisms are quite independent, as shown by selective blockade. The first mechanism (D) is selectively blocked by a number of antidepressive agents of imipramine type, the second (E) by compounds such as reserpine, tetrabenazine and prenylamine, generally known as Segontin®. Blockade of the second mechanism results in net loss—or "release"—of transmitter; under certain conditions this is true also of the first mechanism. In addition to the true blocking agents, we have to consider compounds which serve as "substrates" for the uptake mechanisms and may thus compete with the endogenous amines for uptake mechanisms and binding sites. As might be expected, most of these compounds are chemically closely related to the endogenous amines. However, the structural specificity of the uptake mechanisms is not very high.

Other important processes which influence the release of monoamines are of course their formation and breakdown.

In this paper these various processes and the consequences of their blockade will be discussed, with particular reference to the central nervous system.

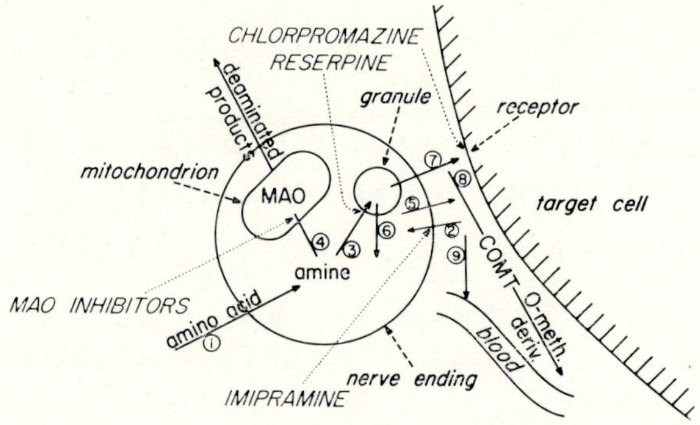

FIG. 1. Hypothetical model of a monoamine transmission unit, depicting the chief sites of attack of 4 major classes of psychotropic drugs. For explanation, see text.

THE STORAGE GRANULES

As suggested several years ago (Carlsson *et al.*, 1957), the monoamine release, or block of uptake, induced by reserpine is largely an intracellular event, oxidative deamination taking place before the amine released from the store escapes from the cell. Subsequent experiments demonstrated the direct action of reserpine on storage granules *in vivo* (Bertler, Hillarp and Rosengren, 1960, 1961) and *in vitro* (Kirshner, 1962a and b; Carlsson, Hillarp and Waldeck, 1962, 1963). Owing to the different distribution of monoamine oxidase and catechol-O-methyl transferase (Carlsson, Lindqvist and Magnusson, 1960; Carlsson, 1960; Carlsson and Hillarp, 1962), the catecholamines lost from the granules are preferentially undergoing oxidative deamination (Andén, Roos and Werdinius, 1964) with diminished formation of O-methylated amines (Häggendal, 1963). Additional evidence has been supplied by Kopin and his coworkers (see his paper presented at this Symposium).

In the initial stage the block induced by reserpine is competitive (later it becomes irreversible) and may thus be prevented by high amine concentrations induced by monoamine oxidase inhibition (Carlsson, Hillarp and Waldeck, 1963; Jonason, Rosengren and Waldeck, 1964) or by other agents with the same point of attack, such as tetrabenazine (Quinn, Shore and Brodie, 1959) and, according to recent observations in our laboratory, prenylamine (Segontin®) (Fig. 2). This drug is a potent blocker of amine uptake by storage granules (Carlsson, Hillarp and Waldeck, 1963) and should perhaps chiefly be looked upon as a reversibly acting reserpine-like drug, although it has certain other effects in addition. The most interesting thing about prenylamine is that its structure is entirely different from those of reserpine and tetrabenazine. It thus represents a new type of monoamine-storage blocking agent.

Although the release induced—or rather unmasked—by these agents is relatively slow, a certain fraction seems to be released quite rapidly. This is apparent from our earlier data (Carlsson, Rosengren, Bertler and Nilsson, 1957) and from more recent data of Stjärne (1964). Labelled noradrenaline injected before the blocking agent seems to behave essentially like endogenous noradrenaline. Interestingly enough, if the experiment is done with labelled

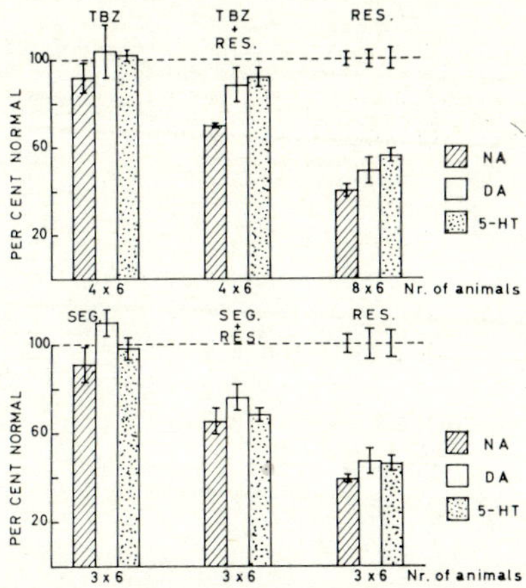

FIG. 2. Protective effect of tetrabenazine (TBZ; 100 mg/kg i.p. 30 min before reserpine, RES) and prenylamine (SEG; 10+10 mg/kg i.v. 30 and 5 min before reserpine) on monoamine depletion of mouse brain by reserpine (1 mg/kg i.p.). The animals were killed 23·5 hr after reserpine. Each monoamine analysis was performed on 6 pooled brains. Values are means ± standard errors (Carlsson and Lindqvist).

octopamine (formed from injected labelled tyramine) instead of noradrenaline, practically all of this amine is quickly released, as if it were able to enter only the quickly-released fraction (Fig. 3). This indicates a relatively low affinity of octopamine for the binding sites of the storage complex.

This brings us to the transmitter analogues. These compounds first utilize the uptake mechanisms of the axon membrane and storage granules, respectively, and then occupy the binding sites of the storage complex in the granules. They thus compete with the physiological monoamines. The most successful competitors must be assumed to be those with the highest affinities for uptake mechanisms and binding sites and, at the same time, highest metabolic stability. So far metaraminol seems to be unbeaten in this respect. Yet it is interesting to note that the release induced by metaraminol does not seem to be quicker than the release unmasked by a pure storage blocking agent like reserpine (Fig. 3). Similar observations have been made by Andén (1964), comparing reserpine with (+)- and (−)-adrenaline as displacing agents. This suggests that the maximum rate of "release" from granules that can be attained is equal to the spontaneous rate of release. Normally this is balanced by uptake, but it is unmasked if uptake is blocked.

The behaviour of metaraminol and the closely related α-methyl noradrenaline in the central nervous system can be studied after administration of their precursors α-methyl metatyrosine and α-methyl DOPA, respectively, which unlike the amines are able to penetrate the blood-brain barrier (Carlsson and Lindqvist, 1962). This barrier, in conjunction with high affinity

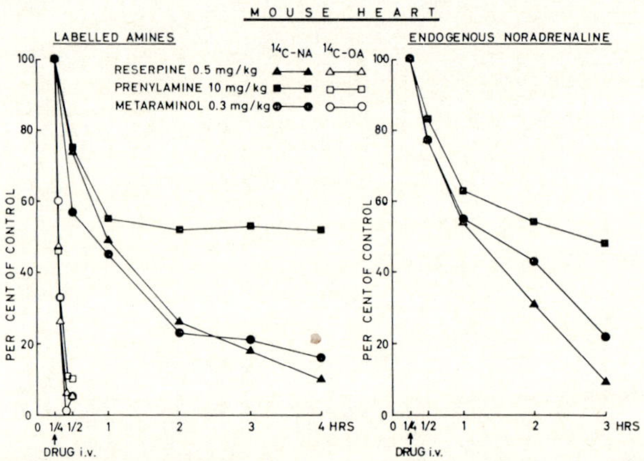

FIG. 3. Effect of reserpine, prenylamine and metaraminol on labelled and endogenous noradrenaline (NA) and on labelled octopamine (OA) (formed *in vivo* from injected ^{14}C-tyramine) in mouse heart. All injections intravenous. Labelled amines were injected at zero time (Carlsson and Waldeck). *Note.* The deviation of the prenylamine curves was probably due to rapid elimination of the drug, as supported by re-injection experiments.

for storage sites and high metabolic stability, is probably responsible for the marked resistance of these amines against the depleting action of reserpine, which—in the case of α-methyl noradrenaline—can be utilized for histochemical detection of the β-hydroxylase-containing noradrenaline neurons in the brain (Carlsson, Dahlström, Fuxe and Hillarp, 1965; cf. Muscholl, this Symposium).

Superimposed on the spontaneous leakage from granules is the release induced by the nerve impulses. In the peripheral adrenergic nerves, the spontaneous impulse flow thus causes a certain enhancement of reserpine-induced loss of transmitter (for references see Carlsson, 1965; Sedvall and Thorson, 1965). Electrical stimulation at high frequency can accelerate the release considerably (Malmfors, 1965). In the adrenal medulla, where spontaneous leakage from the granules appears to be slow, the influence of nerve activity is pronounced. No data on the behaviour of central neurons in this respect are available.

As has been shown by Drs. Muscholl, Shore (see their papers presented at this Symposium) and others, various transmitter analogues taken up by granules may be released by high-frequency stimulation of the nerves. Also the physiological impulse flow results in release of, for example, metaraminol (Almgren, this Symposium).

THE CELL MEMBRANE

The dual uptake mechanism for the adrenergic transmitter, first postulated two years ago (Carlsson, Hillarp and Waldeck, 1963), is now well established. It has been demonstrated both histochemically (Hamberger, Malmfors, Norberg and Sachs, 1964; Hillarp and Malmfors, 1964) and biochemically (Carlsson and Waldeck, 1965a). The histochemical evidence is that even after blockade of the granular storage mechanism by a large dose of reserpine, the adrenergic nerves are able to take up and concentrate α-methylnoradrenaline and—provided that monoamine oxidase has been inhibited beforehand—noradrenaline. The distribution of the amine thus taken up differs from the normal distribution, which is closely correlated to the distribution of storage granules, and indicates an extragranular localization. The uptake is blocked by cocaine (Hillarp and Malmfors, 1964) and compounds of the imipramine type (Malmfors, 1965). The biochemical evidence is analogous: even after a large dose of reserpine labelled noradrenaline (after monoamine oxidase inhibition) and metaraminol are still taken up—though retained less efficiently than normally—by adrenergically innervated tissues (Fig. 4). The uptake is blocked by agents of the imipramine group.

There can be little doubt that the imipramine-sensitive uptake mechanism is located in the cell membrane. It should be noted that blockade of the two uptake mechanisms results in opposite functional effects: blockade of granular storage by, for example, reserpine leads to disruption, blockade of the membrane pump by, for example, desmethylimipramine leads to sensitization of the transmission mechanism.

The question now arises: how rapid is the spontaneous leakage of transmitter through the cell membrane? It is probably very slow, since even practically complete blockade of the membrane "pump" by, for example, desmethylimipramine does not result in any appreciable decrease in transmitter levels or in a sympathomimetic response. This may be due to impermeability of the cell membrane or to the absence of transmitter in any significant concentration in the axoplasm under normal conditions. Probably the latter alternative is true. There is good evidence that normally a hard competition for transmitter occurs between the storage granules and the monoamine oxidase: only if both competitors are put out of function can a significant accumulation of the transmitter or one of its analogues occur in the axoplasm. If such accumulation has been brought about and then the membrane pump is blocked by desmethylimipramine, a very rapid release of axoplasmic amine takes place (Fig. 5). It is interesting to note that this release is about as rapid as that induced by tyramine and metaraminol under the same conditions. In other words, it appears that these releasing agents act by utilizing and thus blocking the membrane pump, thereby unmasking

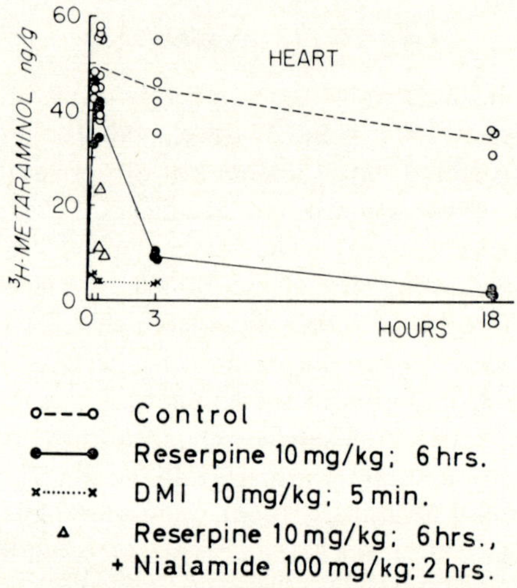

FIG. 4. ^{3}H-metaraminol levels in mouse heart at various intervals following i.v. injection of the labelled amine (at zero time). Injections of reserpine and nialamide were intraperitoneal, desmethylimipramine (DMI) intravenous. The drugs were given before the labelled amine at the intervals indicated (Carlsson and Waldeck, 1965a).

the spontaneous leakage of amine through the cell membrane. The situation thus seems to be analogous to that earlier descıibed for the storage granules.

The membrane pump is of relatively low structural specificity and may thus be utilized even by amines which are not closely related to the physiological transmitter, for example, guanethidine, whose presynaptic effects are blocked by, for example, desmethylimipramine. On the other hand, the structural requirements among the membrane-pump blocking agents of imipramine type are rather strict (cf. Carlsson and Waldeck, 1965b). Lipid-soluble amines such as reserpine and prenylamine need not utilize the membrane pump for penetration into the nerves.

Does the membrane pump operate in central monoamine neurons? If so, is it blocked by the antidepressive agents of imipramine type? The answer to both questions is very probably yes. Again, histochemical and biochemical evidence support each other. Histochemically, uptake of monoamines by specific nerve terminals of cortical slices has been detected *in vitro* (Hamberger and Masuoka, 1965). This uptake is blocked by desmethylimipramine and cocaine. Remarkably enough, also the pericytes around the capillaries of cortical slices take up considerable amounts of amines, and also this uptake is blocked by desmethylimipramine. This calls for caution in the biochemical approach to the problem.

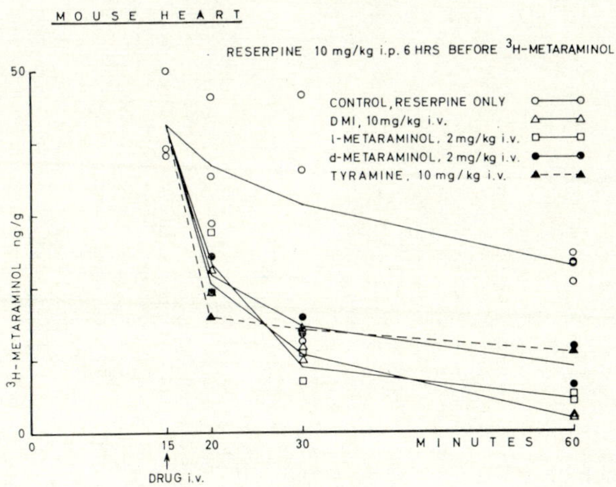

FIG. 5. Effect of desmethylimipramine (DMI), *l*- and *d*-metaraminol and tyramine on ^{3}H-metaraminol levels in heart of reserpine-pretreated mice. The labelled amine was given at zero time and the drugs 15 min later (at arrow) (Carlsson and Waldeck).

Certain parts of central neurons are located outside the blood–brain barrier. Histochemically such parts have been found to take up and concentrate circulating monoamines even after treatment with a large dose of reserpine (Fuxe and Hillarp, 1964).

There is some further evidence, histochemical as well as biochemical, that membrane-pumps operate *in vivo* and are blocked by desmethylimipramine also in the central monoamine neurons (Carlsson, Dahlström, Fuxe and Lindqvist, 1965). Histochemically, it has been found that if DOPA is injected into reserpine-nialamide treated rats, green fluorescence typical of catecholamines appears in noradrenaline neurons, whose terminals show up distinctly (Fig. 6). This appearance of fluorescence is prevented by desmethylimipramine. Biochemically analogous observations were made (Fig. 7). The accumulation of noradrenaline in the brains of rats treated in the same way was largely prevented by desmethylimipramine. Blockade of the membrane pump should result in diminished synthesis of noradrenaline, since the β-hydroxylase is probably present in the storage granules of the noradrenaline neurons (cf. Kirshner, this Symposium). It therefore should be offered less

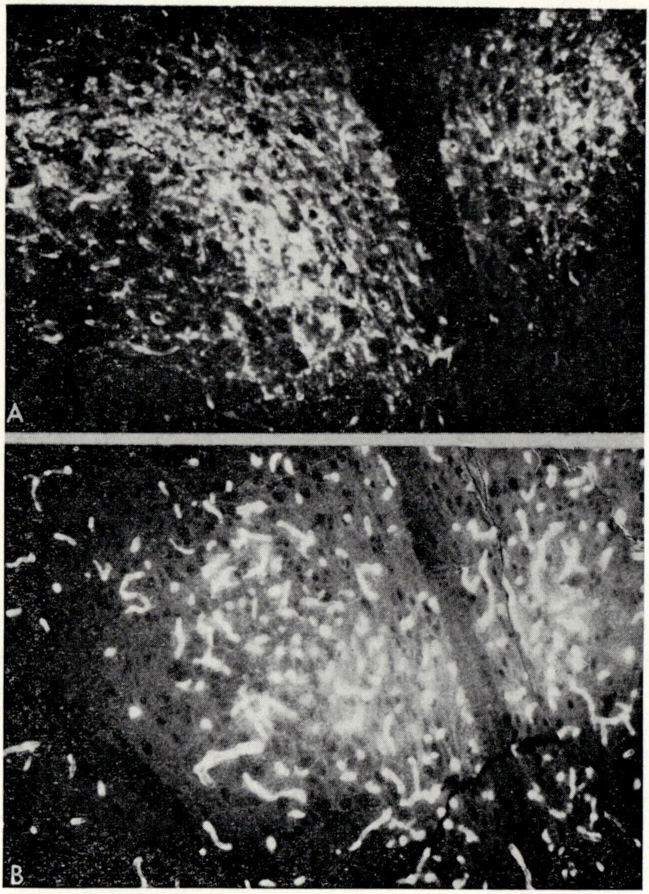

FIG. 6. Histochemical preparation from the paraventricular nucleus of the rat, showing greenish-yellow fluorescence typical of catecholamines. *Top.* Control rat treated with reserpine 10 mg/kg i.p. 22 hr, nialamide 50 mg/kg i.p. 4 hr and L-dopa 50 mg/kg s.c. 0·5 hr before sacrifice. *Bottom.* Rat treated in the same manner except that desmethylimipramine (25 mg/kg i.p.) was injected 15 min before the L-dopa. *Note.* In the control rat nerve terminals normally containing noradrenaline are strongly fluorescent. Fluorescence is seen also in capillaries and pericytes. In the desmethylimipramine treated rat no nerve terminals are visible, but only the capillaries and pericytes. In addition a diffuse background fluorescence is seen in the paraventricular nucleus. (Carlsson, Dahlström, Fuxe and Lindqvist, 1965.) Magnification (both figures) $\times 100$.

substrate, i.e. dopamine, after blockade of the pump and subsequent loss of dopamine from the neurons. Since dopamine hardly occurs free in the axoplasm normally, desmethylimipramine is not very active in reducing noradrenaline levels in normal tissues.

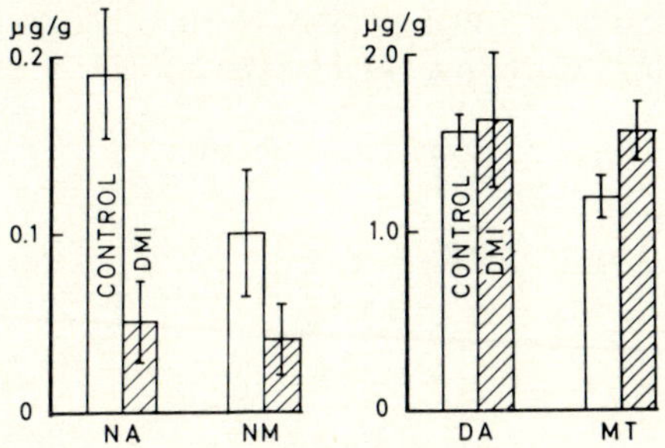

FIG. 7. Noradrenaline (NA), dopamine (DA), and their O-methylated metabolites, normetanephrine (NM) and 3-methoxytyramine (MT), respectively, in rat brain following drug treatments as described in legend of Fig. 6. (Carlsson, Dahlström, Fuxe and Lindqvist, 1965.)

These findings, when considered together, leave little doubt that membrane pumps operate also in central monoamine neurons. Blockade of these pumps is probably responsible for some of the peripheral actions of the imipramine group, such as noradrenaline potentiation and enhanced responses to sympathetic stimulation, and for central actions, such as antireserpine effect and DOPA potentiation (Everett, Wiegand and Rinaldi, 1963).

THE MONOAMINE OXIDASE (MAO)

The central stimulation caused by a relatively "pure" MAO inhibitor like nialamide is probably mediated by an accumulation of monoamines in the brain: when this accumulation is prevented by one of the dopacetamide derivatives that inhibit monoamine synthesis, excitation fails to appear (Carlsson, Corrodi and Waldeck, 1963).

However, many puzzling effects of MAO inhibitors have been described. It seems unquestionable that, among other things, they may cause release of monoamines. Five years ago we described a central noradrenaline-releasing effect of tranylcypromine (Carlsson, Lindqvist and Magnusson, 1960). This effect was very rapid in onset. We have now observed a more slowly developing release by other MAO inhibitors. This has been found in two independent experimental series.

First we have observed that nialamide, given to mice, may cause release (Fig. 8)—or blockade of uptake (Fig. 4)—of labelled metaraminol in the heart. The effect develops gradually. It does not seem to be due simply to displacement by accumulation of noradrenaline, because the phenomenon may be observed also when this accumulation is slight. We are rather inclined to believe that accumulation of other monoamines such as octopamine may be responsible for the effect. In similar manner labelled noradrenaline seems to be released, or prevented from being taken up.

In the other series, which was performed in collaboration with Hillarp's group in Stockholm (Carlsson, Dahlström, Fuxe and Lindqvist, 1965),

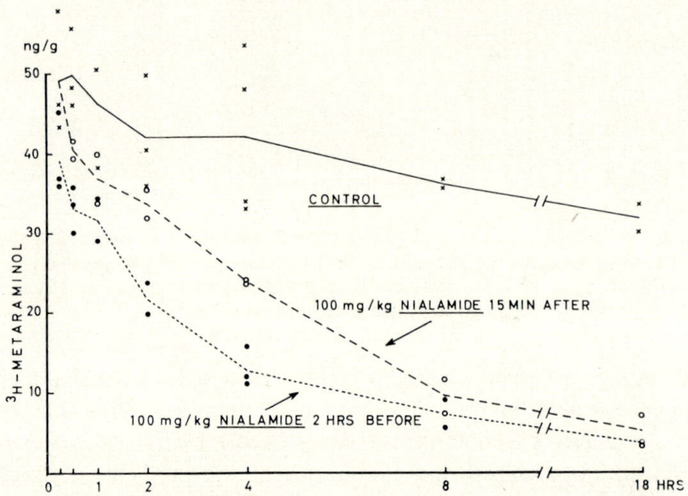

FIG. 8. ^{3}H-metaraminol in mouse heart at various intervals following i.v. injection of the labelled amine (at zero time). *Note.* Nialamide (100 mg/kg. i.p.), injected either before or after the labelled amine, causes a reduction in amine levels. (Carlsson and Waldeck.)

pargyline was injected into rats in a dose of 400 mg/kg i.p. It was found that central excitation and other symptoms developed much faster and became much more severe if the animals were kept in a warm environment. We therefore compared animals kept at room temperature (22°–23°C) and at 29°C. The changes observed in the brains were correlated to the behavioural effects. Extremely severe histochemical changes were seen in the brains of animals who were passing into a syndrome of convulsions and exhaustion, i.e. release of monoamines from the nerve terminals (and from the cell bodies) into the surrounding space; distinct nerve terminals could no more be seen (Fig. 9). No such changes were observed in the animals given the same dose of pargyline but kept at room temperature.

The corresponding biochemical findings (Fig. 10) revealed that the high-temperature animals had lower—and actually subnormal—noradrenaline levels and higher levels of O-methylated metabolites than their room-temperature controls. Evidently the increased release of catecholamines from

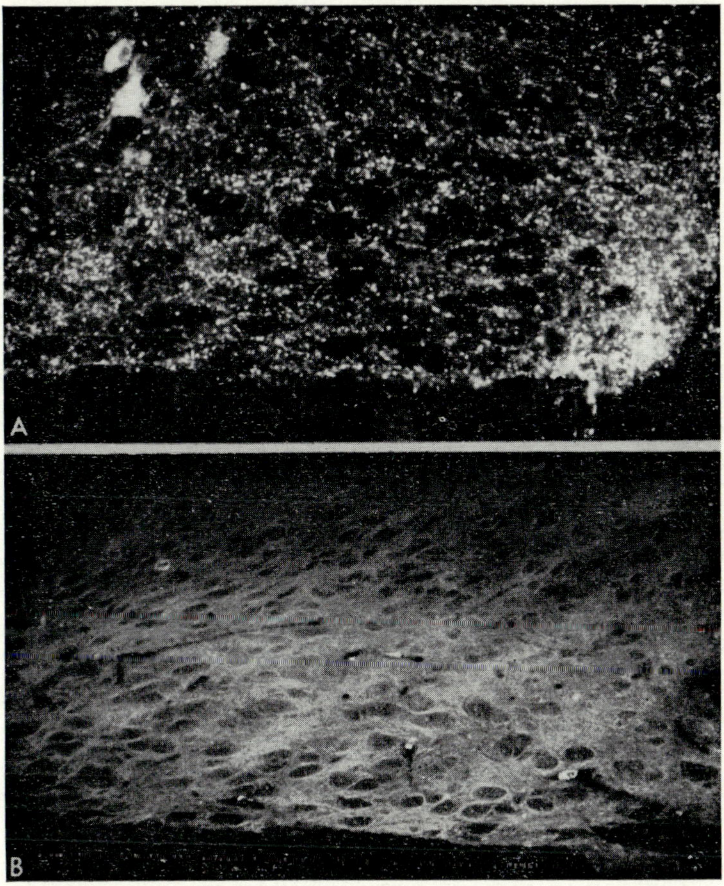

FIG. 9. Histochemical pictures showing monoamine fluorescence in autonomic efferent vagus nucleus of the rat. *Top.* Pargyline (400 mg/kg i.p.) 3 hr before sacrifice. Room temperature (22–23°C). *Bottom.* Same treatment. Ambient temperature 29°C. *Note.* Distinct noradrenaline-containing nerve terminals surrounding non-fluorescent cell bodies can be seen only in the animal kept at room temperature. In the animal kept at 29°C fluorescent material occurs largely outside the nerve terminals. (Carlsson, Dahlström, Fuxe and Lindqvist, 1965.) Magnification (both figures) ×180.

the neurons had resulted in an increased rate of O-methylation, in agreement with earlier findings (Carlsson, Lindqvist and Magnusson, 1960). A similar release was seen—though much less pronounced—after treatment with nialamide and tranylcypromine (histochemical data only).

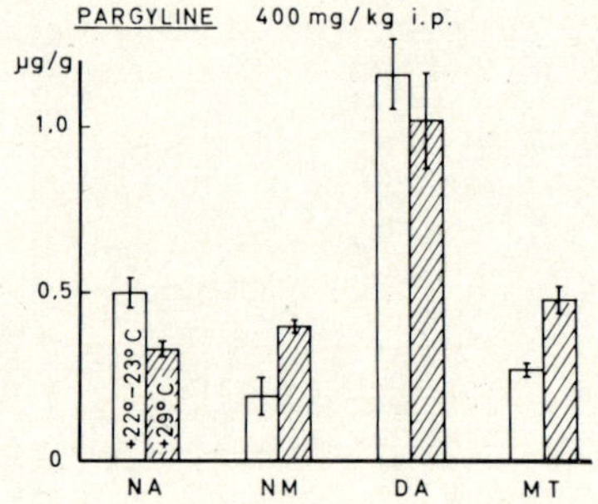

FIG. 10. Noradrenaline (NA), dopamine (DA) and their O-methylated metabolites, normetanephrine (NM) and 3-methoxytyramine (MT) in the brains of rats treated as described in legend of Fig. 9. (Carlsson, Dahlström, Fuxe and Lindqvist, 1965.)

Whatever may be the explanation of the phenomenon, the data show that release phenomena in the brain may be revealed by histochemical as well as biochemical techniques, and preferably by a combination of the two.

Experiments are in progress to elucidate the mechanism involved. Perhaps the 5-HT accumulating and leaking out from neighbour nerve terminals is taken up by catecholamine nerve terminals, causing release of the transmitter.

OTHER MECHANISMS

Chlorpromazine and haloperidol have been found to exert a profound and specific influence on catecholamine metabolism in brain. They enhance already in low dosage the accumulation of O-methylated catecholamine metabolism following MAO inhibition by nialamide (Carlsson and Lindqvist, 1963) and increase the levels of the deaminated dopamine metabolites 3,4-dihydroxyphenylacetic acid and homovanillic acid (Andén, Roos and Werdinius, 1964). The amine levels are not influenced. The mechanism of this effect has not been elucidated. Apparently a good correlation between accumulation of the acids and clinical potency exists. Thus the effect appears to be closely linked to this type of neuroleptic activity. We have earlier suggested that the effect represents some kind of compensatory activation, secondary to receptor blockade. In other words, some kind of feed-back mechanism appears to be responsible for the phenomenon.

Of considerable interest in this respect is the observation that the level of 5-hydroxyindoleacetic acid in brain was uninfluenced by these drugs. This suggests that the catecholamines rather than 5-HT are involved in their actions.

USE OF INHIBITORS OF MONOAMINE SYNTHESIS FOR RECORDING THE ACTIVITY OF MONOAMINE NEURONS

Malmfors (1964, 1965) has shown that electrical stimulation of the cervical sympathetic leads to rapid depletion of the adrenergic transmitter store, provided that the transmitter synthesis has been inhibited by one of the new potent inhibitors of the first step in monoamine synthesis, i.e. one of the dopacetamides (Carlsson, Corrodi and Waldeck, 1963) or α-methyl tyrosine (Udenfriend, this Symposium). This principle appears to be applicable to central monoamine neurons. Thus it has been shown that different kinds of stress in conjunction with inhibition of monoamine synthesis lead to depletion of the transmitter store of the dopamine neurons terminating in the primary capillary plexus of the pituitary portal system (Fuxe, this Symposium). In future analysis of the activity of monoamine neurons this principle will probably prove very useful.

CONCLUSION

Table 1 summarizes the mechanisms of "internal" and "external release" induced by drugs, according to the present concept. "Internal release"

TABLE 1. MECHANISMS OF "INTERNAL" AND "EXTERNAL RELEASE" BY DRUGS

	Type of Agent	Mechanism
"Internal"	Reserpine Tetrabenazine Prenylamine	Block of granule storage
"External"	Tyramine Other analogues Guanethidine MAOI+Reserpine DMI+Reserpine	Block of granule storage + Block of membrane pump

appears to be caused by blockade of granule storage, e.g. by reserpine, tetrabenazine and prenylamine. "External release" appears to require the combined blockade of granule storage and membrane pump, e.g. by tyramine and other analogues, guanethidine, a monoamine oxidase inhibitor alone or—with more striking effects—followed by reserpine, or desmethylimipramine followed by reserpine.

ACKNOWLEDGMENTS

The work described in this review has been sponsored by the Air Force Office of Scientific Research under Grant AF EOAR 64-30 through the European Office of Aerospace Research (OAR), United States Air Force, from a Public Health Service Research Grant (NB 04359-02) from the National Institute of Neurological Diseases and Blindness, the Swedish State Medical Research Council and the Foundation of Knut and Alice Wallenberg.

REFERENCES

ANDÉN, N.-E. (1964) Uptake and release of dextro- and laevo-adrenaline in noradrenergic stores, *Acta Pharmacol.* (Kbh.) **21**, 59–75.

ANDÉN, N.-E., B.-E. ROOS and B. WERDINIUS (1964) Effects of chlorpromazine, haloperidol and reserpine on the levels of phenolic acids in rabbit corpus striatum, *Life Sci.* **3**, 149–58.

BERTLER, Å., N.-Å. HILLARP and E. ROSENGREN (1960) Some observations on the synthesis and storage of catecholamines in the adrenaline cells of the suprarenal medulla, *Acta Physiol. Scand.* **50**, 124–31.

BERTLER, Å., N.-Å. HILLARP and E. ROSENGREN (1961) Effect of reserpine on the storage of new-formed catecholamines in the adrenal medulla, *Acta Physiol. Scand.* **52**, 44–8.

CARLSSON, A. (1960) Discussion remark. In: *Ciba Symposium on Adrenergic Mechanisms.* Eds.: J. R. Vane, G. E. W. Wolstenholme and M. O'Connor, London, J. &. A. Churchill, pp. 432–9.

CARLSSON, A. (1964) Functional significance of drug-induced changes in brain monoamine levels, *Progress in Brain Research* **8**, 9–27.

CARLSSON, A. (1965) Drugs which block the storage of 5-hydroxytryptamine and related amines. In: *Handbuch der Exp. Pharmak. Ergänzungswerk.* Ed.: V. Erspamer, Springer Verlag, Berlin-Göttingen-Heidelberg.

CARLSSON, A., H. CORRODI and B. WALDECK (1963) α-Substituierte Dopacetamide als Hemmer der Catechol-O-methyl-transferase und der enzymatischen Hydroxylierung aromatischer Aminosäuren. In der Catecholaminmetabolismus eingreifende Substanzen. 2. *Helv. Chim. Acta* **46**, 2270–85.

CARLSSON, A., A. DAHLSTRÖM, K. FUXE and N.-Å. HILLARP (1965) Failure of reserpine to deplete noradrenaline neurons of α-methylnoradrenaline formed from α-methyl DOPA, *Acta Pharmacol.* (Kbh.) **22**, 270–6.

CARLSSON, A., A. DAHLSTRÖM, K. FUXE and M. LINDQVIST (1965) Histochemical and biochemical detection of monoamine release from brain neurons, *Life Sci.* **4**, 809–16.

CARLSSON, A. and N.-Å. HILLARP (1962) Formation of phenolic acids in brain after administration of 3,4-dihydroxyphenylalanine, *Acta Physiol. Scand.* **55**, 95–100.

CARLSSON, A., N.-Å. HILLARP and B. WALDECK (1962) A Mg^{++}-ATP dependent storage mechanism in the amine granules of the adrenal medulla, *Med. Exp.* (Basel) **6**, 47–53.

CARLSSON, A., N.-Å. HILLARP and B. WALDECK (1963) Analysis of the M^{++}-ATP dependent storage mechanism in the amine granules of the adrenal medulla, *Acta Physiol. Scand.* **59**, Suppl. 215, 1–38.

CARLSSON, A. and M. LINDQVIST (1962) *In vivo* decarboxylation of α-methyl DOPA and α-methyl metatyrosine, *Acta Physiol. Scand.* **54**, 87–94.

CARLSSON, A. and M. LINDQVIST (1963) Effect of chlorpromazine or haloperidol on formation of 3-methoxytyramine and normetanephrine in mouse brain, *Acta Pharmacol.* (Kbh.) **20**, 140–4.

CARLSSON, A., M. LINDQVIST and T. MAGNUSSON (1960) On the biochemistry and possible functions of dopamine and noradrenaline in brain. In: *Ciba Symposium on Adrenergic Mechanisms*. Eds.: J. R. Vane, G. E. W. Wolstenholme and M. O'Connor, London, J. & A. Churchill, pp. 432–9.

CARLSSON, A., E. ROSENGREN, Å. BERTLER and J. NILSSON (1957) Effect of reserpine on the metabolism of catecholamines. In: *Psychotropic Drugs*. Eds.: S. Garattini and V. Ghetti, Amsterdam, Elsevier Publ. Co., pp. 363–72.

CARLSSON, A. and B. WALDECK (1965a) Mechanism of amine transport in the cell membranes of the adrenergic nerves, *Acta Pharmacol.* (Kbh.) **22**, 293–300.

CARLSSON, A. and B. WALDECK (1965b) Inhibition of ^{3}H-metaraminol uptake by antidepressive and related agents, *J. Pharm. Pharmacol.* **17**, 243–4.

EVERETT, G. M., R. G. WIEGAND and F. U. RINALDI (1963) Pharmacologic studies of some nonhydrazine MAO inhibitors, *Ann. N.Y. Acad. Sci.* **107**, 1068–80.

FUXE, K. and N.-Å. HILLARP (1964) Uptake of L-dopa and noradrenaline by central catecholamine neurons, *Life Sci.* **3**, 1403–6.

HAMBERGER, B., T. MALMFORS, K.-A. NORBERG and CH. SACHS (1964) Uptake and accumulation of catecholamines in peripheral adrenergic neurons of reserpinized animals, studied with a histochemical method, *Biochem. Pharmacol.* **13**, 841–4.

HAMBERGER, B. and D. MASUOKA (1965) Localization of catecholamine uptake in rat brain slices, *Acta Pharmacol.* (Kbh.) **22**, 363–8.

HILLARP, N.-Å. and T. MALMFORS (1964) Reserpine and cocaine blocking of uptake and storage mechanisms in adrenergic nerves, *Life Sci.* **3**, 703–8.

HÄGGENDAL, J. (1963) The presence of 3-O-methylated noradrenaline (normetanephrine) in normal brain tissue, *Acta Physiol. Scand.* **59**, 261–8.

JONASON, J., E. ROSENGREN and B. WALDECK (1964) Effects of some pharmacologically active amines on the uptake of arylalkylamines by adrenal medullary granules, *Acta Physiol. Scand.* **60**, 136–40.

KIRSHNER, N. (1962a) Uptake of catecholamines by a particulate fraction of the adrenal medulla, *Science* **135**, 107–8.

KIRSHNER, N. (1962b) Uptake of catecholamines by a particulate fraction of the adrenal medulla, *J. Biol. Chem.* **237**, 2311–17.

MALMFORS, T. (1964) Release and depletion of the transmitter in adrenergic terminals produced by nerve impulses after the inhibition of noradrenaline synthesis or reabsorption, *Life Sci.* **3**, 1397–1402.

MALMFORS, T. (1965) Studies on adrenergic nerves, *Acta Physiol. Scand.* **64**, Suppl. 248.

QUINN, G. P., P. A. SHORE and B. B. BRODIE (1959) Biochemical and pharmacological studies of Ro 1–9569 (tetrabenazine), a non-indole tranquilizing agent with reserpine-like effects, *J. Pharmacol.* **127**, 103–9.

SEDVALL, G. and J. THORSON (1965) Adrenergic transmission at vasoconstrictor nerve terminals partially depleted of noradrenaline, *Acta Physiol. Scand.* **64**, 251–8.

STJÄRNE, L. (1964) Studies of catecholamine uptake, storage and release mechanisms, *Acta Physiol. Scand.* **62**, Suppl. 228, 1–97.

DISCUSSION

AXELROD: In view of the large amounts of extragranular noradrenaline do you believe it is free or bound?

CARLSSON: The histochemical evidence does not support the view that a large pool of extragranular noradrenaline exits in the adrenergic nerve endings normally.

KOPIN: How does your hypothesis explain selective release of unlabelled norepinephrine by tyramine and that reserpine releases both at equal rates, and why does reserpine inhibit spontaneous norepinephrine release (*in vitro*)?

CARLSSON: As to the first question, time did not permit me to go into the question of different pools of stored norepinephrine, the existence of which I think we agree upon. This of course may explain the kind of differences you mentioned. As to the second question, the inhibition of spontaneous release by reserpine has, to my knowledge, been demonstrated only *in vitro*, under conditions yielding a very high rate of spontaneous release. There is no indication that similar phenomenon occurs *in vivo*.

EULER: How do you suggest that a nerve stimulus causes a primary release of amine from the granules?

CARLSSON: I think that we have to postulate that the membrane potential acts, directly or indirectly on the granules to cause release from the granules because it has been found histochemically that extragranular amines that have been accumulated by means of experimental procedures can not be released by nerve stimulation.

RELEASE OF MONOAMINES DURING ELECTRICAL STIMULATION OF SELECTED BRAIN AREAS IN CATS

LARS M. GUNNE and TOMMY LEWANDER

ELECTRICAL stimulation of certain brain regions, belonging to the limbic and hypothalamic systems, is known to produce rage attacks in unrestrained cats with chronically implanted electrodes. Initially on stimulation the animals are alerted, then they become aggressive, hissing, clawing and finally attacking. At the same time there are signs of autonomic activity: salivation, dilated pupils and pilo-erection.

Two brain regions where such responses can be elicited were studied, namely the amygdaloid nucleus of the temporal lobe and the lateral hypothalamus. If the stimulating electrode is placed within the amygdala the resulting rage attacks are poorly directed and will appear only after 1–2 min delay, while stimulation of the lateral hypothalamus produces directed aggressive behaviour, starting rapidly on each stimulation. When offered a rat some of the hypothalamus stimulated cats bite its neck and kill it within a few seconds after the start of stimulation. Occasionally these cats may even attack the experimenter (cf. Egger and Flynn, 1963).

In a preliminary study (Gunne and Reis, 1963) electrical stimulation of the amygdaloid nuclei was found to result in a reduction of brain noradrenaline. The stimuli had to be repeated until 16 to 24 rage attacks had been produced within 2–3 hr. Otherwise no definite effect was noted on the brain catecholamines, probably due to resynthesis and other compensatory mechanisms.

Histochemically there was a disappearance of fluorescence within the catecholamine terminals of practically all regions of the forebrain (Fuxe and Gunne, 1964). The effect was more pronounced if the cats had been pretreated with an inhibitor of the catecholamine synthesis (H22/54, Hässle Ltd).

According to fluorimetric estimations the depletion appeared to be confined to brain noradrenaline and did not involve serotonin or dopamine. Together with the reduction of brain noradrenaline there was also a marked depression of the adrenal gland content of catecholamines.

Chromatographic separation of the catecholamines of single cat brains was undertaken, essentially according to Häggendal (1962). It was found that stimulation of the amygdala resulted in a lower noradrenaline peak, while normetanephrine had increased compared with the control, indicating an increased breakdown of noradrenaline. Dopamine and its methoxylated product 3-methoxytyramine appeared to remain unaffected by amygdaloid stimulation.

REFERENCES

Egger, M. D. and J. P. Flynn (1963) Effects of electrical stimulation of the amygdala on hypothalamically elicited attack behavior in cats, *J. Neurophys.* **26**, 705–20.

Fuxe, K. and L. M. Gunne (1964) Depletion of the amine stores in brain catecholamine terminals on amygdaloid stimulation, *Acta Physiol. Scand.* **62**, 493–4.

Gunne, L. M. and D. J. Reis (1963) Changes in brain catecholamines associated with electrical stimulation of amygdaloid nucleus, *Life Sci.* **2**, 804–9.

Häggendal, J. (1962) On the use of strong exchange resins for determinations of small amounts of catecholamines, *Scand. J. Clin. & Lab. Invest.* **14**, 537–44.

THE RELEASE OF BIOGENIC AMINES BY BACTERIAL TOXINS

H. Rašková, Z. Jiřička, K. Mašek, R. Smetana,
J. Švihovec and J. Vaněček

Institute of Pharmacology, Czechoslovak Academy of Sciences and
Charles University, Prague

Within the last years endotoxins have become more and more the interest of pharmacological research. The release of vasoactive biogenic amines is considered in the explanation of their mechanism of action (Hinshaw, 1964; Rašková and Vaněček, 1964). In general, however, and especially with other toxins, too little attention is paid to the role of biogenic amines in the various patterns of intoxication by different bacterial toxins.

We should like to demonstrate on the basis of our own experiences that there exist common and specific mechanisms of release of biogenic amines for different toxins. This can be demonstrated by results concerning the action of purified endotoxin of *S. marcescens**, *Staphylococcus* α†, and *Shigella shigae* toxin‡. The last two products are exotoxins; *Staphylococcus* toxin is characterized by a hemolytic activity.

MATERIAL AND METHODS

The changes in catecholamine contents of rabbit adrenals were assayed 3 hr after the administration of 0·1 μg of endotoxin.

Vanes technique (1964), superfusing isolated organs by the blood of a cat or rabbit, to whom the toxin is administered intravenously, was used. The blood pressure was recorded. By means of an extracorporal circulation isolated organs were superfused (isolated stomach strip, duodenum of the rat and guinea pig ileum).

With *Shigella shigae* toxin changes of catecholamine contents were determined in brain, heart and adrenals of mice. Moreover, changes in 5-HT content were followed in brain. The toxin was injected also in the lateral ventricle of cats and behavioural changes were followed. After the administration of *Staphylococcus* α toxin the changes in catecholamines were determined

* We are very much indebted to Dr. Alois Nowotny from Temple University, Philadelphia, for the generous gift of this toxin.

† The product was generously supplied by Dr. Součková from the Institute for Sera and Vaccines in Prague.

‡ The toxin was generously supplied by Dr. V. Matějovská from the Institute of Epidemiology and Microbiology in Prague.

in the adrenals, brain and heart of mice and in the adrenals of rabbits. The above-described perfusion technique of Vane was also used. Furthermore the ventricles were perfused using the technique of Bhattacharya and Feldberg (1957). The perfusate was assayed on the isolated stomach strip according to the method of Vane (1957).

For the catecholamine assay the technique of Bertler, Carlsson and Rosengren (1958) was used, a fluorimetric device designed by one of us (Smetana, 1962) being used for the actual determination. 5-HT was determined biologically, using the extraction of Amin, Crawford and Gaddum (1954). Significant relative changes could be detected, although the extraction is only partial.

RESULTS

On Fig. 1 the changes in catecholamine content of the rabbit adrenals 3 hr after the injection of the very small doses of endotoxin of 0·1 μg per kilo are shown. There is a significant decrease of catecholamines in the rabbit adrenals representing an average 150 μg of adrenaline and 20 μg of noradrenaline. It is very interesting that a mixture of 150 μg of adrenaline and

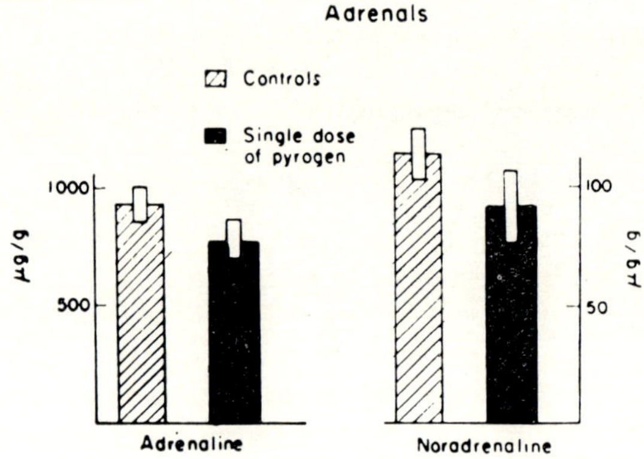

FIG. 1. Depletion of catecholamines from the rabbit adrenal 3 hr after intravenous administration of 0·1 μg of endotoxin.

20 μg of noradrenaline provokes in the rabbit a two-peak elevation of temperature, which corresponds roughly to the course of fever elicited with 0·1 μg of endotoxin (Fig. 2), whereas both components given separately act differently, adrenaline with a peak soon after administration, noradrenaline after a latency period.

On Fig. 3 the result of an autoperfusion experiment is demonstrated. After 1 mg of endotoxin per kilo there is no noticeable change of the blood

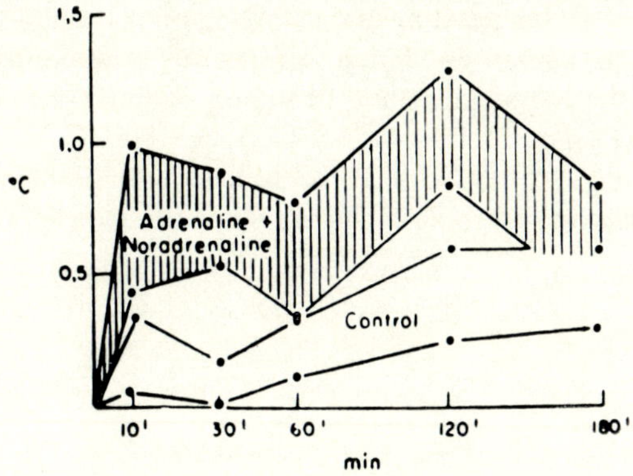

FIG. 2. Fever response after intravenous injection of a mixture of 150 μg adrenaline and 20 μg of noradrenaline to rabbits.

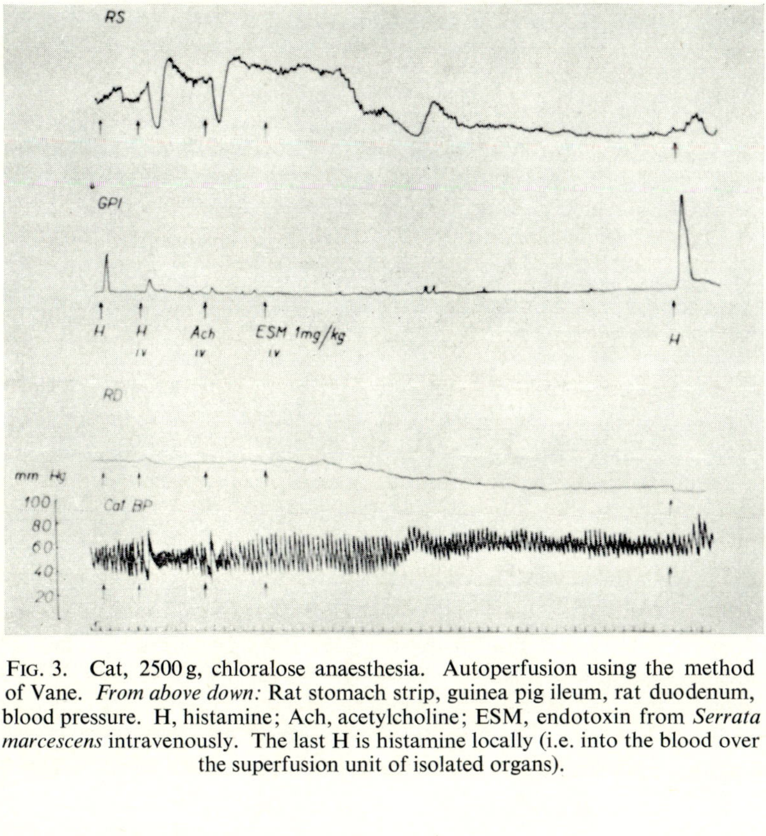

FIG. 3. Cat, 2500 g, chloralose anaesthesia. Autoperfusion using the method of Vane. *From above down:* Rat stomach strip, guinea pig ileum, rat duodenum, blood pressure. H, histamine; Ach, acetylcholine; ESM, endotoxin from *Serrata marcescens* intravenously. The last H is histamine locally (i.e. into the blood over the superfusion unit of isolated organs).

pressure. A small contraction of the guinea pig ileum speaks for possible histamine release, the long-lasting and changing relaxation of the rat stomach strip and a relaxation of duodenum speaks for a liberation of catecholamines, though the long-lasting and changing pattern on the stomach strip needs further analysis.

For *Shigella shigae* toxin, as may be seen from Fig. 4, there is a highly significant decrease in catecholamine content of the mouse brain following a

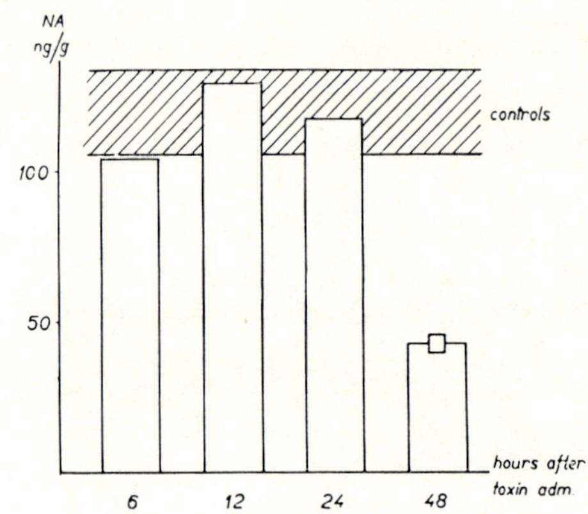

FIG. 4. Depletion of noradrenaline from mouse brain after the administration of *Shigella shigae* toxin.

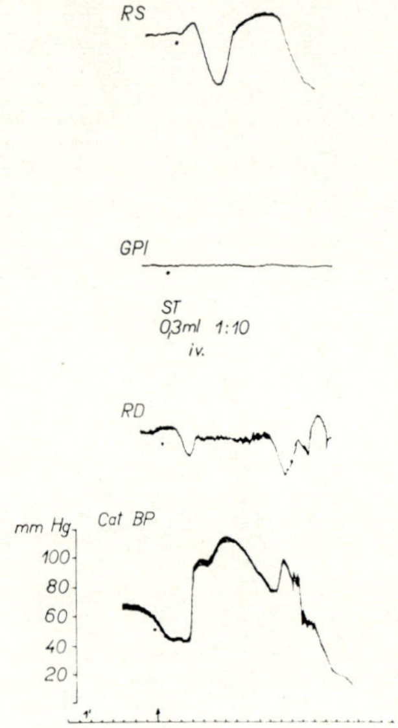

FIG. 5. Cat 1800 g, chloralose anaesthesia. Autoperfusion using the method of Vane. From above down: Rat stomach strip, guinea-pig ileum, rat duodenum, blood pressure. ST, *Staphylococcus*-α toxin intravenously.

48 hr latency period after toxin administration. The heart content of catecholamine was not changed, but a significant adrenaline and noradrenaline decrease in the adrenals was always found. Behavioural changes of cat reminded us of those seen by Gaddum and Vogt (1956) after intraventricular injection of 5-HT. As a significant decrease of 5-HT was found in the mouse brain 48 hr after the administration of *Shigella* toxin, this is an almost significant correlation.

Staphylococcus α toxin differs from the previous ones by its very specific action on smooth muscle, which develops contracture after a latency period. This is often irreversible and after a transient increased sensitivity to catecholamines the isolated organs or blood vessels become completely unresponsive.

Whereas endotoxin is either without any initial influence on blood pressure or causes hypotension, *Staphylococcus* α toxin produces a marked hypertension in different species. This can be seen from Fig. 5, which reproduces the effect of an intravenous injection of toxin using the above-described superfusion of isolated organs with the blood of the cat.

In this case there is no reaction on the guinea pig ileum, whereas the typical relaxation of the rat stomach strip and the rat duodenum speak for the liberation of catecholamines. When the cat brain is perfused and *Staphylococcus* α toxin injected intraventricularly, the rat stomach is contracted by the perfusate. This action is antagonizable by serotonin antagonists. There is marked catecholamine depletion in the adrenals of mice 3 hr after *Staphylococcus* α toxin administration and at 30 min after the toxin administration the depletion of catecholamines is already evident.

DISCUSSION

It could be expected that the administration of toxins produces a release of catecholamines from the adrenals. This is indeed so with all the three investigated toxins. It depends on the rapidity of toxin action whether the changes occur rapidly as with *Staphylococcus* α toxin, or with a considerable delay like with *Shigella shigae* toxin. From our preliminary autoperfusion experiments in cats and rabbits it seems that at least after the administration of either endotoxin or *Staphylococcus* α toxin, although they differ very much in their general action, catecholamines are released almost instantly. The relaxation of the rat stomach strip and of the rat duodenum speaks for this. After endotoxin a contraction of guinea pig ileum points to a secondary catecholamine release produced by histamine.

The difference between the action of endotoxin and *Staphylococcus* α toxin on blood pressure probably is due to the basic difference that *Staphylococcus* α toxin as demonstrated by Švihovec, Mašek and Jiřička (1965) has a direct contracting activity on vessel (and other) smooth muscle for which the release of biogenic amines is only an adjuvant factor, whereas for the effect of endotoxins this is one of the main factors.

On the other hand *Shigella shigae* toxin has a more reserpine-like activity, at least in the brain. This toxin decreases the content of catecholamines and 5-HT in the brain after a considerable latency period. The analogy of 5-HT and *Shigella shigae* toxin administration directly into the brain in behavioural changes speaks for the possibility that for *Shigella shigae* toxin effects, the liberation of 5-HT in the brain might be of importance.

Certainly the described examples of liberation of biogenic amines are not the whole story. Toxins are complex substances which certainly liberate a considerable number of biologically active substances. Their effects can be only understood when an attempt is made to study not only the release and action of one substance but the complex of effects produced by the integration of various partial actions. The necessity to have adrenaline and noradrenaline in a mixture corresponding to the amount of liberated catecholamine to imitate endotoxin fever can serve as one small example.

SUMMARY

Endotoxin, *Shigella shigae* toxin and *Staphylococcus* α toxin liberate catecholamines from the adrenals. *Shigella shigae* toxin decreases, after a considerable latency period, the content of the brain in catecholamines and 5-HT. In autoperfusion experiments it has been shown that *Staphylococcus* α toxin liberates catecholamines, endotoxin probably only through histamine liberation. As toxins liberate a number of substances or lead to increased formation of some (Schayer, 1964) a complex approach should be followed in investigating the release of biogenic amines through bacterial toxins.

REFERENCES

AMIN, A. H., T. B. B. CRAWFORD and J. H. GADDUM (1954) The distribution of substance P and 5-hydroxytryptamine in the central nervous system of dog, *J. Physiol.* **126**, 595–618.

BERTLER, Å., A. CARLSSON and E. ROSENGREN (1958) A method for the fluorimetric determination of adrenaline and noradrenaline in tissue, *Acta Physiol. Scand.* **44**, 273–92.

BHATTACHARYA, B. K. and W. FELDBERG (1958) Perfusion of cerebral ventricles: effects of drugs on outflow from the cisterna and aqueduct, *Brit. J. Pharmacol.* **13**, 156–62.

BROWN, D. A., M. C. CASEWELL and J. P. QUILLIAM (1965) Observations on the spasmogenic action of the alphatoxin of *Staphylococcus Pyogenes* on isolated intestinal muscle, *Recent Advances in the Pharmacology of Toxins*, Pergamon Press, pp. 207–11.

GADDUM, J. H. and M. VOGT (1956) Some central actions of 5-hydroxytryptamine and various antagonists, *Brit. J. Pharmacol.* **11**, 175–9.

HINSHAW, B. L. (1964) *Bacterial Endotoxins*, Institute of Microbiology, Rutgers, The State University, New Brunswick, N.J., pp. 118.

RAŠKOVÁ, H. and J. VANĚČEK (1964) Pharmacology of some toxins, *Pharmacol. Rev.* **16**, 1–45.

SCHAYER, R. W. (1964) *Relationship of Induced Synthesis of Histamine to Some Biological Actions of Endotoxins in Bacterial Endotoxins*, Institute of Microbiology, Rutgers, The State University, New Brunswick, N.J., pp. 182–6.

SMETANA, R. (1962) A simple device for the determination of catecholamines, *Int. J. Neuropharmacol.* **1**, 189–92.

ŠVIHOVEC, J., K. MAŠEK and Z. JIŘIČKA (1965) *Recent Advances in Pharmacology of Toxins*, Pergamon Press, pp. 193–7.

THE PHYSIOLOGICAL RELEASE AND ACTION OF MONOAMINES IN THE SPINAL CORD

R. W. RYALL

THERE is evidence that noradrenaline and 5-hydroxytryptamine may be transmitter substances liberated from descending pathways in the spinal cord. This evidence derives mainly from the fluorescence studies of Hillarp and his colleagues and it receives further support from the pharmacological studies carried out in Göteborg by Lundberg *et al.* They have shown that the effects obtained from stimulation of flexor reflex afferents (FRA) in the spinal cat treated with dihydroxyphenylalanine (DOPA) or 5-hydroxytryptophan are similar to those obtained in the decerebrate animal which had not been treated with these amine precursors. Similar effects were obtained in the spinal animal treated with nialamide after an injection of reserpine and we have recently shown that this effect with reserpine is not obtained in the chronic spinal animal.

We have now found that Deseril (UML) or bromolysergic acid diethylamide (BOL 148) produced a release of the descending control of inhibition from the FRA in the decerebrate cat, but there was no release of excitation or of the dorsal root potentials. Furthermore, the amine oxidase inhibitor nialamide increased the decerebrate control of inhibition from the FRA. These observations support the theory that 5-hydroxytryptamine may be a transmitter liberated from a descending pathway which controls inhibition from FRA. They also indicate that more than one transmitter may be released from the descending pathways which control the flexor reflex. Unfortunately, similar experiments with an adrenergic blocking agent (phenoxybenzamine) have so far failed to show that noradrenaline is released from descending pathways.

A number of neurones were found to be highly sensitive to noradrenaline and to a lesser extent to 5-hydroxytryptamine applied electrophoretically to single neurones (Engberg and Ryall, unpublished observations). These neurones had no spontaneous activity and lay fairly deep (more than 2 mm from the cord dorsum) in the lumbar cord. It was difficult to activate them synaptically and some were inhibited from the FRA. However, when firing was induced by the electrophoretic application of an amino acid, the simultaneous ejection of noradrenaline or of 5-hydroxytryptamine from the micropipette caused a marked reduction or a cessation of firing within a few seconds. Mere diffusion of noradrenaline from the micropipette, which was obtained when the retaining current was reduced, was usually sufficient to produce a marked decrease in the firing rate. It was clearly shown, by appropriate tests, that these effects were not due to current changes. Furthermore, there were many neurones which were spontaneously active and easily activated synaptically but which were unaffected by the amines.

Since neither noradrenaline nor 5-hydroxytryptamine have been observed to excite spinal neurones, it is suggested that the physiological action of these substances is an inhibitory effect. Following the depression, a rebound excitation was sometimes observed. However, a similar rebound excitation sometimes followed the depression produced by current on some cells and this effect cannot be considered primarily as an excitatory effect.

Although some of these neurones which were depressed by noradrenaline or 5-hydroxytryptamine may well be concerned with flexor reflex effects, it is likely to be extremely difficult to show that they are indeed the neurones upon which the descending pathways operate.

ON THE FUNCTION OF THE NIGRO-NEOSTRIATAL DOPAMINE PATHWAY

NILS-ERIK ANDÉN

THE existence and course of the dopamine (DA) pathway to the neostriatum have been established: the cell bodies are localized in the substantia nigra, mainly in its pars compacta, the axons are ascending in the crus cerebri and in the internal capsule and the abundant axon terminals containing very high amounts of DA are present in the neostriatum, i.e. the caudate nucleus and the putamen (Andén, Carlsson, Dahlström, Fuxe, Hillarp and Larsson, 1964; Andén, Dahlström, Fuxe and Larsson, 1965). In the rat brain we have found that about 70 per cent of the DA occur in the corpus striatum. The levels of NA (0·1 μg/g) and 5-HT (0·3 μg/g) in the corpus striatum are very low compared to that of DA (about 5 μg/g). The nigro-neostriatal DA pathway appears to be completely uncrossed. Therefore we have performed biochemical and functional studies after unilateral lesions, the intact side serving as a control (Andén, Dahlström, Fuxe and Larsson, 1965b). After unilateral electrolytic lesions in the substantia nigra or in the internal capsule the forebrain DA of the same side can be reduced by at least 70–80 per cent compared to the intact side ten days or more after the operation. After reserpine treatment the lesion effect on the DA neurons can be estimated biochemically by the loss of the striatal L-DOPA decarboxylase activity, determined *in vitro* (Andén, Magnusson and Rosengren, 1964). Lesions in the internal capsule produce a reduction by at least 75 per cent of this enzyme activity in the ipsilateral corpus striatum. Many of the lesions have produced changes only in the corpus striatum but in some cases a small disappearance of NA and DOPA decarboxylase activity from other forebrain structures have also been observed. There is, however, a clearcut correlation between the striatal changes and the functional effects described below.

Little is usually seen in the operated animals without drug influence. A rat with a lesion of the DA pathway in the left internal capsule has a tendency to flex the tail, especially the root, and the head towards the left or operated side and to have the left extremities adducted below the trunk while the right ones are abducted. When the animals are excited they may rotate to the operated side.

If reserpine (10 mg/kg i.v.) is given after inhibition of the monoamine oxidase (nialamide 100 mg/kg i.p., 2 hr before reserpine) the monoamines in the storage granules will reach the receptors in a free and active form

(Carlsson, Rosengren, Bertler and Nilsson, 1957). A rat with the DA pathway unilaterally damaged in the internal capsule and treated in this way displays a typical syndrome. Within 10 min after the reserpine injection the rat starts to rotate to the same side as the lesion, and does so for hours. The head and tail are flexed to the operated or left side even during the few moments without moving and the extremities of this side are adducted below the trunk. A similar but somewhat weaker syndrome is produced by monoamine oxidase inhibitors alone in larger doses and after a longer time (nialamide 500 mg/kg i.p., 6 hr, 23°C; pargyline 375 mg/kg i.p., 4 hr, 29°C).

If reserpine is given alone quite another syndrome will appear in a rat with the nigro-neostriatal DA pathway damaged in the left internal capsule. The tail and head are flexed to the right, i.e. to the side opposite to the lesion. The back is curved with the concavity to the right or unoperated side. The extremities of the operated side are abducted. The extremities on the side contralateral to the lesion are adducted below the trunk. Sometimes it is possible to feel a distinct rigidity in the extremities on the unoperated side and to see a rapid tremor in them. The onset of the asymmetric syndrome after reserpine is temporarily correlated to the disappearance of DA but not to that of NA. The first signs appear after one hour when 30 per cent of the normal forebrain DA is left. The full syndrome is developed after 2 hr when no DA can be detected while still 20–30 per cent of the forebrain NA is found.

The same syndrome as after reserpine treatment is also produced by haloperidol (2 mg/kg i.p.) and chlorpromazine (10 mg/kg i.p.). The adrenergic α receptor blocker phenoxybenzamine (20 mg/kg i.p.) also causes the reserpine syndrome but it seems to be much weaker than the other three drugs.

In a few experiments we have electrophysiologically studied the excitability of the motoneurons to the tail and the hind limbs (Andén, Larsson and Steg). Without drugs the α motoneurons on the operated side have a higher excitability than those on the unoperated side. After nialamide plus reserpine the α motoneuron excitability is reduced mainly on the unoperated side. Reserpine alone (or haloperidol) produces an increase of the α motoneuron excitability on the unoperated side.

The effect of reserpine after monoamine oxidase inhibition can be evoked only from the neostriatum on the unoperated or right side as there is no DA in store on the opposite side. The reduced α motoneuron excitability on the unoperated side suggests that the descending pathways to the spinal cord, at least in rats, are uncrossed. The effect on the α motoneuron excitability after the excess of free DA produced by nialamide plus reserpine is in agreement with the earlier finding by Dr. Steg and Dr. Roos after injection of the DA precursor DOPA. On account of a higher α motoneuron activity on the operated or left side after nialamide plus reserpine the animals turn and rotate to the operated side.

Very surprisingly, DOPA causes an operated rat to turn and rotate in the opposite direction to that after nialamide plus reserpine. This apparent contradiction may be explained by a denervation supersensitivity to DA in the neostriatum of the operated side. Actually we have observed in the denervated corpus striatum a certain DA formation which possibly has occurred outside the nerve-endings (cf. Andén, Magnusson and Rosengren, 1965). Since 5-HTP does not produce any turning the DOPA effect, in any case, probably indicates that all the asymmetric syndromes are due to changes in the striatal DA and not in the striatal 5-HT.

The fact that reserpine causes the opposite clinical and electrophysiological asymmetry to that after nialamide plus reserpine indicates that also this effect is evoked from the neostriatum of the unoperated or right side. The stronger reserpine effect on the intact than on the denervated neostriatum may be explained if the lesion in the internal capsule, in addition to the degeneration of the nigro-neostriatal DA neurons, also produces damage of an efferent pathway from the corpus striatum or of a nerve tract to the neostriatum antagonistic to that containing DA. In fact we have observed that these lesions cause degeneration of a nigro-neostriatal neuron system with the cell bodies mainly in the pars reticulata of the substantia nigra. These neurons do not contain DA but possibly acetylcholine (Andén, Carlsson, Dahlström, Fuxe, Hillarp and Larsson, 1964). The small effect seen normally may also be due to such damage of a non-DA nerve tract.

The same syndrome as is produced by the presynaptic blockade of the dopaminergic transmission after reserpine is also produced by chlorpromazine, haloperidol and phenoxybenzamine. These drugs are thought to block the dopaminergic transmission postsynaptically. The phenothiazine derivative promethazine (20 mg/kg i.p.) has no neuroleptic action and does not cause any asymmetry. The slight turning after the adrenergic α receptor blocker, phenoxybenzamine, and the no turning after the β receptor blocker, Inderal®, indicates that the DA receptors in the neostriatum are of the α type but that chlorpromazine and haloperidol have a much higher affinity to them than phenoxybenzamine.

A PHYSICO-CHEMICAL MODEL OF HISTAMINE RELEASE FROM MAST CELLS

BÖRJE UVNÄS and INGA LISA THON

RELEASE of histamine from mast cells is usually, if not regularly, accompanied by partial or total degranulation of the cells. Any theory to explain the mechanism of release therefore has to take into account, among other things, the mode of binding of the histamine, the mechanism of degranulation of the mast cell and the relationship between these two phenomena.

MODE OF BINDING OF HISTAMINE

In Stockholm we have made our observations on mast cells isolated from rat abdomen.[7] Some of the properties of these cells and their granules are summarized in Fig. 1.

Calculations show the histamine concentration in the mast cell granules to be very high, corresponding to a molarity of 0·3.[6] There are two main opinions about the mode of storage of granular histamine. According to one, histamine occurs in the granules in a "free" diffusible form,[2] the granule being surrounded by a semipermeable membrane. As soon as this membrane is damaged, histamine leaks out. According to the other opinion, the histamine is electrostatically bound to the granular heparin,[3,1] histamine release requiring a break of the electrostatic binding.

Rat mast cells offer an excellent opportunity to test the validity of the two hypotheses. The protein–heparin structure which forms the main constituent of the mast cell granules, is poorly soluble in water. If histamine were electrostatically linked to the heparin, mast cell granules would retain histamine even after lysis of mast cells in an ion-free medium, e.g. distilled water. The amount of extragranular histamine should depend on the dissociation of the protein-heparin-histamine complex.

Mast cells containing ^{35}S-heparin were subjected to lysis by suspension in distilled water (Fig. 2). On centrifugation the lysed material could be separated into three fractions. At $90 \times g$ a sediment was obtained, which contained clusters of granules adhering to cell debris. A second sediment at $2700 \times g$ contained free granules. Calculations of the ^{35}S activity of the fractions indicated that the heavy sediment contained 60 per cent and the lighter sediment 35 per cent of the heparin. The activity of the supernatant after $2700 \times g$ showed that only about 5 per cent had gone into solution.

The partition of histamine showed a different pattern. 50 per cent was recovered from the sediment $90 \times g$, 25 per cent from the sediment $2700 \times g$ and 25 per cent from the supernatant $2700 \times g$.

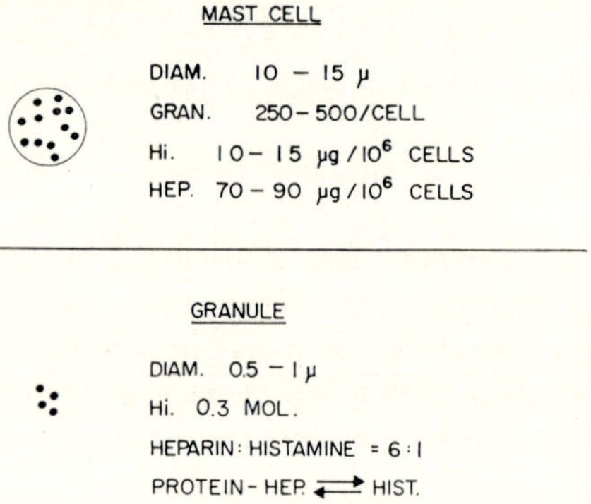

FIG. 1. Contents of mast cells and mast cell granules.

The fact that 75 per cent of the histamine remained in the granules in spite of lysis of the mast cells in water strongly argues in favour of an electrostatic binding of histamine. The size of the supernatant fraction should then be due to the dissociation of histamine from the protein–heparin–histamine complex. The firmness of the heparin–histamine binding is evident from the high concentration gradient between granular and extragranular histamine, the concentration of histamine in the suspended granules being 10,000–100,000 times higher than in the suspension fluid. The fact that all histamine was released from the granules on addition of sodium chloride will be commented on below.

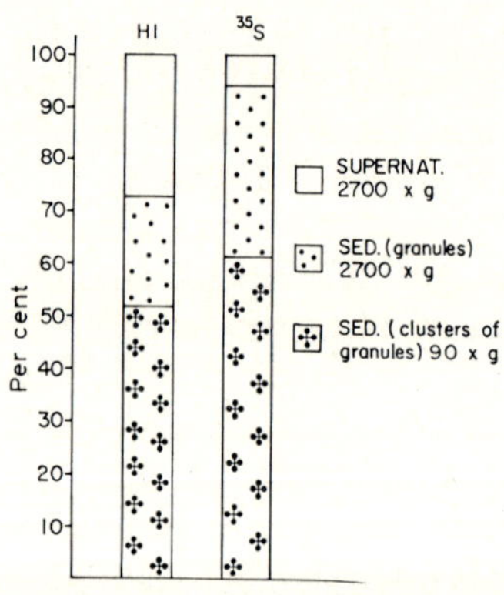

FIG. 2. Release of histamine (Hi) and ^{35}S-heparin (^{35}S) from mast cells lysed in aq. dest. *Note:* Histamine in sediment $90 \times g$ and $2700 \times g$ was released by NaCl.

If histamine were electrostatically linked to granular heparin, addition of extragranular heparin would disturb the equilibrium to the advantage of the extragranular fraction. Mast cells were disrupted by repeated freezing in isotonic sucrose and the total granular matter ($2700 \times g$) was resuspended in isotonic sucrose. On addition of heparin to the suspension fluid, increased amounts of the granular histamine was recovered in the suspension fluid (Fig. 3). Interestingly enough, 50 per cent of the granular histamine was

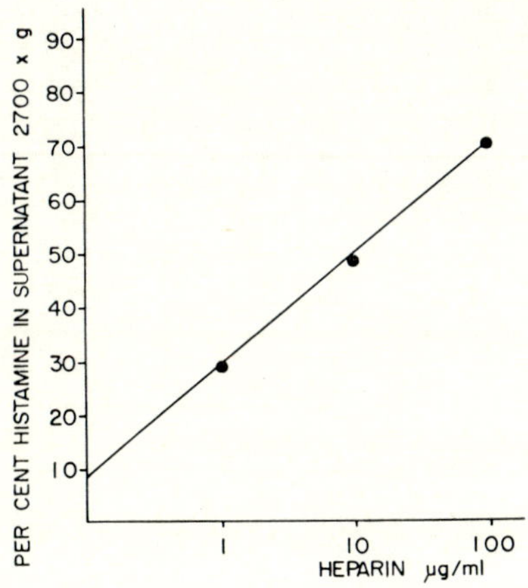

FIG. 3. Influence of extragranular heparin on partition of mast cell histamine between granules and suspension fluid. 2×10^5 cells per ml. frozen in isotonic sucrose.

observed to have left the granules when the amount of the added extragranular heparin equalled the amount of the granular heparin calculated to be present in the suspension.

Granular matter from mast cells was resuspended in water and exposed to sodium chloride (Fig. 4). A successively increasing release of histamine occurred from a concentration of 10^{-3} M and upwards. At isotonicity all histamine was released. Also, when isotonicity was upheld by suitable mixing of sucrose and sodium chloride a similar release curve was obtained. Cations like K^+ and Ca^{++} also released histamine from suspended granular matter. When suspended in isotonic solutions of biogenic composition such as Tyrode, physiological NaCl solution, serum, etc., granular matter released all its histamine (Table 1).

A quantitative study of the histamine-releasing potency of sodium is illustrated in Fig. 5. Granules from water-lysed mast cells were suspended in ^{24}Na-containing sodium phosphate buffer pH 7. The release of histamine was found to increase parallel with the concentration of the buffer and was total around a sodium concentration of 10^{-2} M. Of special interest was the observation of a constant ratio—2 : 1—between the granular uptake of sodium and the release of histamine throughout the whole release curve; whether or not this ratio of 2:1 indicates a two-point ionic attachment of histamine to the heparin molecule remains to be clarified.

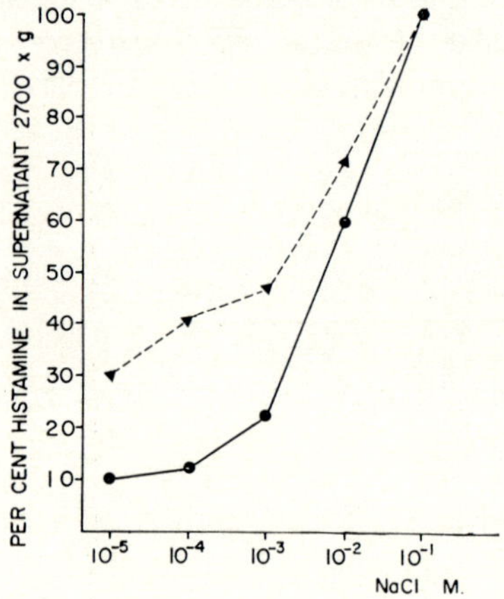

FIG. 4. Release of histamine from granular matter of water-lysed mast cells on addition of NaCl to suspension fluid. 0°C 5min. 6×10^5 cells in 2 ml. ●——●, 12 ml. ▶——▶.

The release of histamine from granular matter occurred rather instantaneously as observed from the contraction of the guinea pig ileum. No significant difference was observed in the latency of the contractions elicited on addition of granular bound histamine and of a corresponding does of pure histamine dihydrochloride. Also, more elaborate experiments with perfusion of granular material showed the cation induced histamine release at room temperature or at 37°C to be a matter of a few seconds.

TABLE 1

Incubation medium	*Histamine, % in supernat. 2700 × g*
Fresh rat serum	100
Isoton Tyrode	100
Isoton NaCl	100
Isoton K-phosphate	100
Aqua dest.	8

Release of histamine from water lysed mast cells, 6×10^5/ml, on suspension of sediment 2700 × *g* in various cation-containing solutions, 5 min at 0°C.

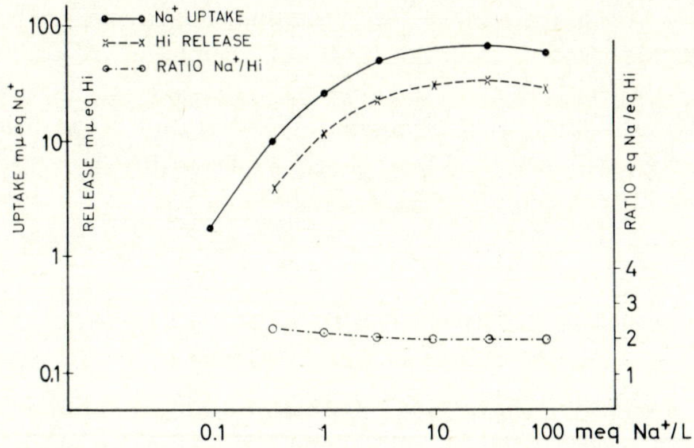

FIG. 5. Sodium uptake and histamine release in mast cell granules suspended in sodium phosphate buffer pH 7, 0°C. *Note* ratio 2:1 between uptake of sodium and release of histamine (Unväs, Åborg and Novotný. To be published).

To summarize: The mast cell histamine is linked to the granular matter in such a way that it is instantaneously released on exposure to cations of physiological concentration.

MECHANISM OF DEGRANULATION

Mast cells respond with degranulation and release of histamine to numerous stimuli, physical and chemical.[5] The chemicals form a very heterogeneous group. No common denominator has been found in their mechanisms of action. The chemical releasers may be divided into two main groups; those which induce an "active discharge" of histamine and those which elicit a "non-active" release of histamine.

To the first group causing an "active discharge" belong the polymer releasing agents such as antigens, principles in lower animals such as the burning jellyfish, and synthetic compounds such as the polymer amine compound 48/80. The mast cell response to polymer releasers is dependent on factors indicating the process to be the result of a rather intricate mechanism.[4] The response is pH dependent with an optimal pH range around 7. Degranulation and histamine release are not induced at 0°C. Heating of the mast cell to 45°C for a few minutes makes the cell irreversibly non-responsive to releasers. Substances which block oxidative and glycolytic metabolism and several other enzyme inhibitors annul the action of polymers. Under the circumstances mentioned both degranulation and histamine release is blocked. As a whole, we have never seen histamine release without a concomitant degranulation in our laboratory.

The polymer releasers do not seem to seriously interfere with the vital functions of the mast cells. The mast cells discharge only part of their granules and histamine, and *in vivo* abdominal mast cells refill within two weeks. The observation that the mast cell membrane is not functionally deranged by the degranulation also emerges from the fact that the partially degranulated cells retain their remaining granules and histamine in spite of repeated washings in isotonic saline. Such washed cells again react normally to polymer releasers (unpublished observations).

RELATIONSHIP BETWEEN DEGRANULATION AND HISTAMINE RELEASE

Since no observations have been presented to show that polymer releasing agents penetrate into the mast cell, thereby reaching the granular sites of histamine, a reasonable explanation of their mode of action would be that they trigger a degranulation mechanism and that the histamine release is a secondary phenomenon, due to an extracellular cation exchange.

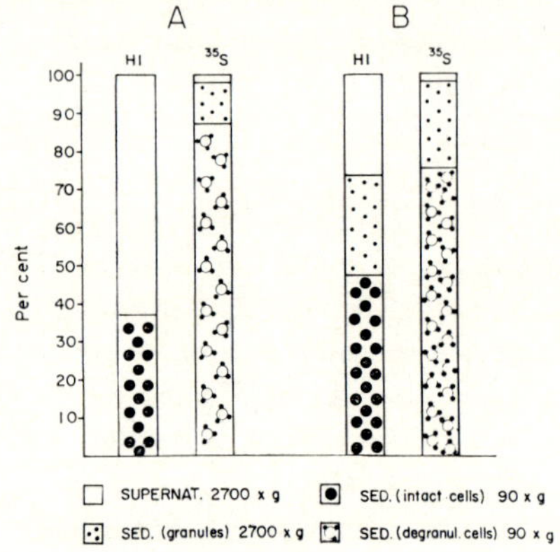

FIG. 6. Release of histamine (Hi) and ^{35}S-heparin (^{35}S) from mast cells exposed to compound 48/80 1 μg/ml. in isotonic saline pH 7 (A) and in isotonic sucrose pH 7 (B). *Note:* Extrusion of histamine-carrying granules into ion-poor medium (isotonic sucrose).

In order to show that polymers cause the discharge of histamine-carrying granules, compound 48/80 was given to mast cells suspended in isotonic sucrose buffered to pH 7 by adding KOH to 10^{-4} M, a cation concentration too low to cause any appreciable release of granular histamine. As a comparison 48/80 was given to mast cells suspended in isotonic salt solution, pH 7 (Fig. 6). 48/80, 1 μg/ml caused a degranulation of most of the mast cells in the suspension. Centrifugation $90 \times g$ yielded a sediment containing mast cells with adhering granules; $2700 \times g$ yielded a free granular fraction and the supernatant contained a few per cent dissolved heparin. The partition of heparin between the three fractions was similar whether the mast cells were suspended in isotonic salt solution or in isotonic sucrose solution. In contrast the partition of histamine differed between the two experiments. In salt solution 60 per cent of the histamine was recovered in the supernatant $2700 \times g$. The remaining 40 per cent fraction went down with the sediment $90 \times g$ containing degranulated mast cells. No histamine was found in the "free" granular fraction (sediment $2700 \times g$). The histamine remaining in the mast cell containing sediment was released when the cells were lysed with distilled water and the cell debris was exposed to isotonic sodium chloride. In sucrose (Fig. 5) only 25 per cent of the histamine was found in the supernatant $2700 \times g$. The rest was found in the "free granular" fraction and in the sediment $90 \times g$ containing cells with adhering granules. On addition of sodium chloride 25 per cent of this histamine went into solution. Evidently it originated from the free and adhering granules. Most of the adhering granules could be removed from the mast cells, by repeated washing in salt solutions at pH 8 without any further loss of histamine from the cell material. The washed cell could be seen to contain granules and when lysed in water and then exposed to isotonic sodium chloride, the remaining histamine was released.

The occurrence of "free" histamine after degranulation with compound 48/80 might be ascribed to an ionic exchange between the 48/80 and the extruded granules. However, the concentrations of this amine which suffice to cause a maximal degranulation and concomitant release of histamine are unable to release any histamine from histamine-bearing granular matter (Fig. 7). The appearance of "free" histamine after degranulation with 48/80 probably reflects the dissociation of histamine from the heparin–histamine complex in the discharged granules (compare Fig. 1).

The demonstration that compound 48/80 is able to cause a discharge from mast cells of histamine-carrying granules favours the view that degranulation is the primary process in the response of mast cells to polymer releasers. Histamine release is a secondary phenomenon due to ion exchange in the granular protein–heparin complex between extracellular cations and histamine.

Reported data have not revealed any principle differences between different polymer agents as to their mode of action on mast cells. It is therefore reasonable to assume that anaphylactic release of histamine from mast cells, as well as the release due to other biogenic polymers show the same principle course—an initial degranulation and a secondary extracellular histamine release (Fig. 8).

As to the mechanism of the numerous monomer chemical agents, only a few studies have been reported. Most of these substances can be assumed to have a lytic or otherwise damaging action on the mast cell membrane. The consequent osmotic disturbances of the cell will ultimately lead to intra- or extracellular exposure of the histamine-carrying granules to cations. The result will be an immediate release of histamine (Fig. 8).

A cation exchange mechanism could theoretically be assumed to hold for the release of many electrically charged biogenic substances (biogenic

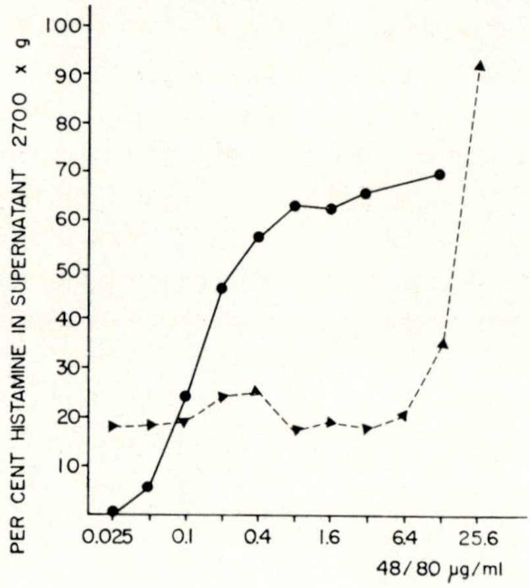

FIG. 7. Release by compound 48/80 of histamine from intact mast cells (A) and from granular matter of water-lysed cells (B).

amines, hormones, etc.) stored in particles and linked to acid polymers (heparin, ATP, etc.). Although attractive, the general applicability of such a mechanism remains to be investigated.

The mechanism of "active" degranulation is unknown and can only be speculated on. Several reports agree that energy-yielding processes are essential for the responsiveness of the mast cells to polymer releasers. On the other hand, no increase of the oxygen consumption has been observed to

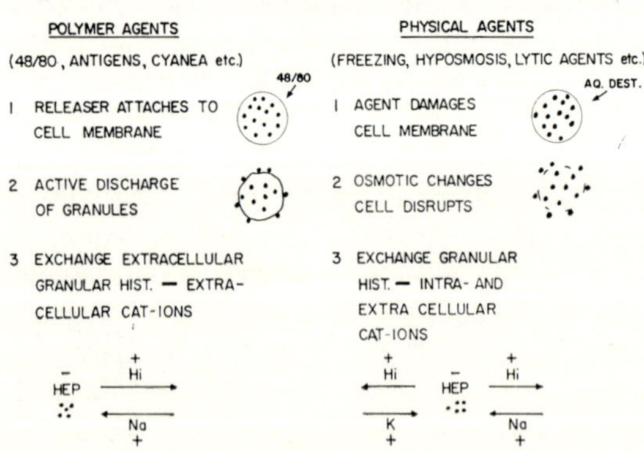

FIG. 8. Proposed mechanisms of release of histamine from mast cells caused by polymers (left) and physical agents (right).

occur *during* the degranulation. If this is correct it seems as if the energy required is utilized to build up and maintain the excitability of the mast cells to the releasers. This excitability deteriorates if the cell is starved by blocking the aerobic and anaerobic metabolism. The nature of the mechanism which transports the granules to the surface of the mast cell is obscure. One very speculative thought is to ascribe the granular transport to a cellular water pump mechanism triggered by polymer releasers.

ACKNOWLEDGMENTS

The investigations described in this communication have been sponsored by the Swedish Medical Research Council (Grant Y 141), by the US Public Health Service (Grant AM 04063-05) and by Karolinska Institutet, Stockholm.

REFERENCES

1. Benditt, E. P. and D. Lagunoff (1964) *Progress in Allergy* **8**, 195–223.
2. Garcia-Arocha, H. (1961) *Canad. J. Biochem. Physiol.* **39**, 395–402.
3. McIntire, F. C. (1956) *Ciba Symposium, London*, p. 416.
4. Uvnäs, B. (1962) *Proc. Int. Union Physiol. Sc.* **1**, 106–14. XXIInd Int. Congr. Leiden.
5. Uvnäs, B. (1964) *Ann. N.Y. Acad. Sci.* **116**, 880–90.
6. Uvnäs, B. (1965) *Arch. Exp. Path. Pharmakol.* **250**, 137–49.
7. Uvnäs, B. and I. L. Thon (1959) *Exp. Cell Res.* **18**, 512–20.

DISCUSSION

Goth: First, what is the effect of proteins on the histamine release from granules by ions? The reason I am asking is that a few years ago Hagen injected granular suspensions intravenously into dogs and got no effect on blood pressure unless he ruptured the granules first.

Secondly, what is the anticoagulant activity of the heparin–protein complex in the granules?

Uvnäs: 1. As far as I know proteins do not protect the granules from release by cations. Hagen's granules were isolated from sucrose homogenates. Long exposure to sucrose —isotonic or above—reduces the releasing action of cations probably by causing physical changes of the granules.

2. I have no experience of my own about the anticoagulant activity of the granular heparin–protein complex.

Lagunoff: When mast cell granules are isolated by differential centrifugation in 0·34 M sucrose and then washed in water, the granules do not in our hands physically disintegrate as you suggest and in fact much of the histamine remains in the granule. Perhaps this difference in results stems from other factors than effects of sucrose on the granules, since your results for salt effects with granules prepared in distilled water are generally consistent with our findings for granules prepared in sucrose.

Unväs: The tendency of the granules to deteriorate (not disintegrate) after suspension in sucrose depends on the sucrose concentration. The deteriorating influence of sucrose on the releasing action of cations increases with exposure time and sucrose concentration.

Shore: When you say that granules are less sensitive to cations after the incubation in sucrose, do you mean that in your experiments you add salt to the sucrose suspension or do you spin down the granules and resuspend in distilled water or in ion medium?

UVNÄS: You can resuspend the granules in sucrose or water with or without salt. Provided that the sucrose solution is of isotonicity or above and the exposure time sufficiently long the results are principally the same, i.e. the release by cations is impaired.

GREEN: It should not be presumed that when mast cells are isolated after administering radioactive sulphate to a rat that all the radioactive sulphur is confined to heparin. In our experiments, the label in mast cells is also associated with cerebroside sulphate, taurine, cysteric acid, and probably cysteinsulfinic acid (Green, J. P. and Robinson, J. D. J. *Biol. Chem.* **235**, 1621, 1960; Green, J. P., Day, M. and Robinson, J. D. *Biochem. Pharmacol.* These also have been shown to be capable of binding histamine and other amines (Green, J. P., Robinson, J. D. and Day, M., *J. Pharmacol. Exper. Therap.* **131**, 12, 1961; and *Biochem. Pharmacol.*).

UVNÄS: Of course radioactive sulphur will be incorporated in various sulphate-containing mast cell components but from a quantitative point of view the extraheparin sulphates will play a very minor role, I presume, and will not invalidate our conclusions.

MECHANISMS OF HISTAMINE RELEASE BY POLYMERIC COMPOUNDS

ANDRES GOTH

Department of Pharmacology, The University of Texas
Southwestern Medical School, Dallas, Texas, U.S.A.

HISTAMINE release by large molecular polymers is characterized by certain unique features. Various species have their own characteristic histamine releasing polymers. Thus, dextran and ovomucoid are effective in the rat, polyvinylpyrrolidone and the tweens in the dog, and horse serum in the cat. Although this species specificity could suggest an immunological explanation, the manifestations of the reaction to polymeric compounds are quite different from those of anaphylaxis. Histamine release by these polymers differs also from the reaction of various species to compound 48/80 and related amines in that tolerance to the large-molecule compounds in a given species does not extend to the compound 48/80 type releaser. Finally, a remarkable feature of the action of large-molecule compounds which are active in the rat is the ability of elevated blood sugar levels (Goth, Nash, Nagler and Holman, 1957; Adamkiewitz and Langlois, 1957) and 2-deoxyglucose (Goth, 1959) to block the reaction.

Studies on the basic mode of action of compounds such as dextran and ovomucoid were difficult in view of the demonstration (Lagunoff and Benditt, 1960) that isolated peritoneal mast cells of the rat do not respond to these large molecular agents with degranulation and histamine release, despite the fact that pieces of rat skin, which contain many mast cells, release histamine upon the addition of dextran (Halpern, 1956).

The purpose of this paper is to show that dextran or ovomucoid will release histamine from suspended rat peritoneal mast cells provided that certain tissue factors are added to the cell suspension. Furthermore, evidence will be presented for the existence of endogenous compounds in brain and skin which are capable of acting on mast cells in a manner analogous to dextran or ovomucoid.

MATERIAL AND METHODS

Peritoneal mast cell suspensions were obtained in the following manner: 15 to 20 ml of a glucose-free metabolic solution was injected by the intraperitoneal route into Sprague–Dawley or Holtzmann rats. The metabolic solutions were either Tyrode's solution or one buffered with phosphate (pH 7) and used previously for this purpose (Moran, Uvnäs and Westerholm, 1962), but used here free of glucose. All solutions also contained heparin (5 μg/ml). The fluid was withdrawn in five minutes under ether anesthesia. Staining with toluidine blue showed numerous mast cells in the fluid. In some experiments, the fluid was centrifuged and the cells were resuspended in fresh metabolic solutions. Best results were obtained when only a short time elapsed between the withdrawal of the fluid and the addition of the various releasing agents. For this reason, in many experiments, the original fluid withdrawn from the peritoneal cavity was used as the cell suspension. For the same reason no attempt was made at separating mast cells from other cells by gradient centrifugation.

Most experiments were performed at room temperature (20–24°C). The peritoneal cell suspension was pipetted into 15 ml centrifuge tubes in the amount of 2.1 ml. The final volume of the incubation mixture was brought up to 3 ml by means of various additions. Most experiments were performed in duplicate tubes.

Twenty minutes following the various additions the tubes were centrifuged and histamine was determined in the supernatant by a fluorimetric method (Shore, Burkhalter and Cohn, 1959). For the determination of total histamine the samples were homogenized in a glass homogenizer prior to the addition of perchloric acid. Results were expressed as the quantity and percentage of histamine released in 20 min, although most of the release occurred within the first few minutes.

In the experiments in which the morphological appearance of the mast cells was investigated, the cells were deposited on Millipore filters and were stained with toluidine blue (Lagunoff and Benditt, 1960).

Dextran (Clinical Grade) having an average molecular weight of 81,000 was purchased from Mann Research Laboratories, New York. Ovomucoid was obtained from Worthington Chemical Co., Freehold, New Jersey, and 2-deoxyglucose from Aldrich Chemical Co., Inc., Milwaukee, Wisconsin. Rabbit brain acetone powder was purchased from Nutritional Biochemicals Corporation, Cleveland, Ohio. Brain thromboplastin from Difco Laboratories, Detroit, had essentially the same activity as rabbit brain acetone powder. The supernatant of brain acetone powder was prepared in the following manner: 150 mg of the powder was suspended in 4 ml of normal saline and was allowed to stand at 45°C for 10 min. After centrifugation, the supernatant yielded approximately 16 mg of solids per ml supernatant, of which 9 mg should be NaCl contained in the saline used for suspension.

The ether extract of brain acetone powder was prepared in the following manner: 150 mg of brain acetone powder was shaken with 20 ml of ether. After centrifugation the ether was evaporated. The residue was resuspended in 4 ml of saline and was homogenized by means of a stirring rod and manual shaking.

RESULTS

The effect of dextran on histamine release from peritoneal cells is shown in Fig. 1, which represents a typical experiment. Dextran alone even at a concentration of 6 mg/ml had no effect, since the amount of histamine appearing in the supernatant was the same as in control tubes. Increasing

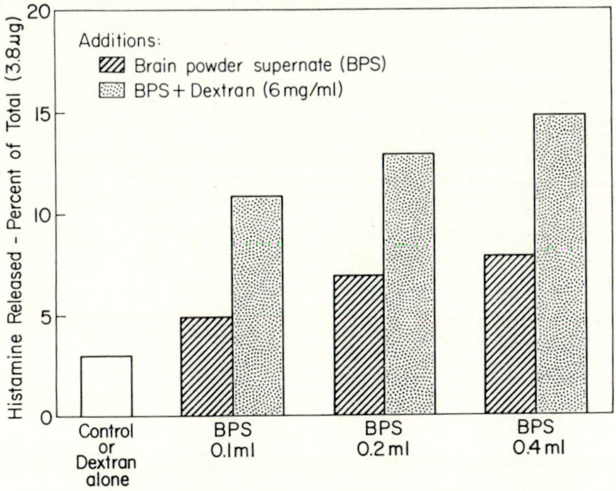

FIG. 1. Effect of dextran alone and in the presence of brain powder supernatant on histamine release from peritoneal mast cells of the rat.

amounts of brain acetone powder supernatant alone caused increasing amounts of histamine to be released. The addition of dextran to cells pretreated with brain powder supernatant caused a considerable increase in the amount of histamine released. It was found important in these experiments to add the brain powder supernatant just before the dextran. In fact, the two solutions could be mixed and added together, a procedure which tended to increase somewhat the amount of histamine released.

Table 1 shows a similar experiment. Results were qualitatively the same as the ones shown in Fig. 1. In some experiments ovomucoid was substituted for dextran with very similar results. On the other hand, polivinylpyrrolidone had no effect. Preliminary experiments indicate that pieces of rat skin extracted with acetone and homogenized in saline give results similar to those with brain powder. Liver acetone powder was much less active.

Figure 2 depicts a morphological counterpart of the previous observations. Figure 2A shows the appearance of control mast cells on a Millipore filter stained with toluidine blue. In Fig. 2B the cells are shown following pretreatment with dextran (6 mg/ml) alone. The mast cells do not appear significantly different from the control cells. Figure 2C shows the effect of brain powder supernatant and Fig. 2D the action of combined brain powder

TABLE 1. EFFECT OF BRAIN THROMBOPLASTIN AND DEXTRAN ON HISTAMINE RELEASE FROM RAT PERITONEAL MAST CELLS

HISTAMINE RELEASE FROM RAT PERITONEAL CELLS *

ADDITIONS	HISTAMINE RELEASED	
	MICROGRAMS	PERCENT OF TOTAL
None	0.13	4
Brain Thromboplastin	0.44	14
Dextran 6 mg/ml	0.12	4
Brain Thromboplastin plus Dextran 6 mg/ml	1.20	39

* 2.1 ml of cell suspension in Tyrode's. Final volume brought up to 3 ml with various additions and Tyrode's.

supernatant and dextran. The cells are somewhat degranulated in these two preparations and many have been disrupted by the combined treatment.

It was of interest to study the effect of 2-deoxyglucose in these experiments, since it was shown previously (Goth, 1959) that this glucose analogue blocked

TABLE 2. EFFECT OF 2-DEOXYGLUCOSE ON HISTAMINE RELEASE FROM RAT PERITONEAL CELLS BY BRAIN THROMBOPLASTIN AND DEXTRAN

HISTAMINE RELEASE FROM RAT PERITONEAL CELLS

ADDITIONS	HISTAMINE RELEASED	
	MICROGRAMS	PERCENT OF TOTAL
None	0.15	5
2-deoxyglucose	0.25	8
Brain Thromboplastin	0.34	11
Brain Thromboplastin plus Dextran (6 mg/ml)	0.76	26
2-deoxyglucose plus Brain Thromboplastin plus Dextran (6 mg/ml)	0.16	5

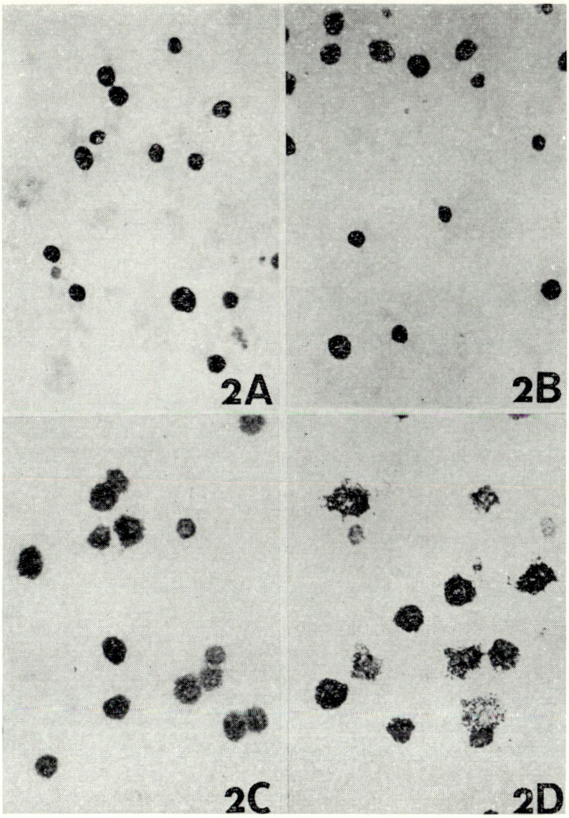

FIG. 2. Mast cells incubated on Millipore filters, fixed and stained with toluidine blue. 2A: Incubation in Tyrode's solution. 2B: Incubation in dextran, 6 mg/ml. 2C: Incubation in the presence of brain powder supernatant. 2D: Incubation in combined brain powder supernatant and dextran, 6 mg/ml. For explanation see text.

the effects of dextran *in vivo* in the rat. Table 2 shows that 2-deoxyglucose when added in a concentration of 1 mg per ml blocked the combined effects of the brain substance and dextran. The slight difference between the 2-deoxyglucose and control supernatants as regards histamine release was probably not significant since many other experiments failed to show any histamine release by 2-deoxyglucose alone.

Figure 3 shows the effect of an ether extract of brain acetone powder supernatant on histamine release by varying concentrations of dextran. It may be seen that the ether extract alone, in contrast with the brain powder supernatant shown previously, causes no histamine release. There is, on the

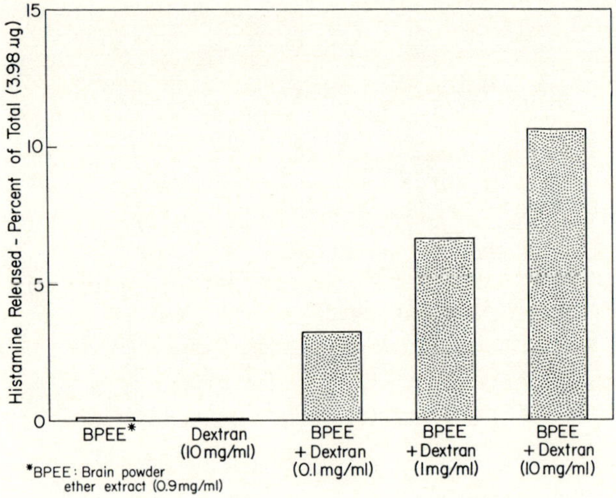

FIG. 3. Effect of dextran with and without an ether extract of brain powder supernatant on histamine release from peritoneal mast cells of the rat.

other hand, a factor in the ether extract which appears to be essential for dextran to exert its effect, since increasing concentrations of the polymer caused increasing histamine release in its presence.

DISCUSSION

The findings of this work suggest that the large-molecule histamine releasers, dextran or ovomucoid require an ether soluble tissue factor in order to exert an effect on rat peritoneal mast cells. In addition, it appears from these studies that there are endogenous compounds in various tissues which are capable of causing mast cell degranulation and histamine release *in vitro*.

It is remarkable that dextran or ovomucoid can cause histamine release when injected into rats, or when added to pieces of rat skin but they have no effect on suspended peritoneal mast cells unless a tissue cofactor is present. There are several possible explanations for this paradox: (*a*) The polymers may combine with a factor in the body before they exert their effect, or they may release endogenous mediators which act on the mast cells. This possibility is not attractive in view of several experiments in which the serum of rats injected with large doses of dextran was also ineffective on rat mast cells *in vitro*. (*b*) Perhaps tissue mast cells differ in some significant way from suspended cells as regards their response to polymeric compounds. It is possible for example, that tissue mast cells contain on their surface a substance which is essential for an interaction with dextran and which is eluted during the process of suspension of the cells in the medium. Our results generally favor this interpretation, although the possibility of some radical difference between peritoneal and other tissue mast cells cannot be excluded.

An interesting finding in this work is the presence of endogenous compounds in brain, skin and to a lesser extent in other organs, which can degranulate mast cells and can release histamine from these cells. The fact that 2-deoxyglucose blocks the action of these endogenous compounds as well as that of dextran and ovomucoid suggests an hypothesis concerning the injurious nature of polymeric compounds.

Mast cells in the rat may have specific receptors for endogenous compounds which perhaps play a role in some cellular function, most likely related to inflammation. Large-molecule compounds which can interact with the receptors will elicit inflammation in that species. The main evidence for this hypothesis is based on the ability of 2-deoxyglucose to block not only the action of the polymers but also that of the endogenous mast cell degranulating substance. It should be stressed at the same time, that 2-deoxyglucose is not a non-specific blocking agent for rat mast cells, since it has no effect on histamine release from such cells by Russell Viper Venom, lung thromboplastin (Goth and Knoohuizen, 1962) or compound 48/80 (unpublished experiments). This hypothesis must remain speculative until more is learned about the chemical nature of the endogenous mast cell degranulating substance whose action is blocked by 2-deoxyglucose.

Although the best known action of 2-deoxyglucose is on glucose transport (Kipnis and Cori, 1959) it is impossible at present to link the action of the large-molecule compounds on histamine release with an interaction with glucose transport. It is conceivable, however, that what is referred to in this paper as a dextran receptor is in fact the same as the glucose carrier envisaged by workers in the field of membrane transport.

SUMMARY

The effect of large-molecule polymeric compounds, such as dextran and ovomucoid on suspended peritoneal cells of the rat was investigated. These polymers do not cause mast cell degranulation or histamine release from suspended cells *in vitro* unless the suspending medium is fortified with certain tissue factors which are present in brain acetone powder and also in skin. These tissue powders also contain compounds which act on mast cells of the rat in a manner similar to dextran and ovomucoid, as evidenced by the fact that mast cell injury caused by all of these agents is blocked by 2-deoxyglucose. It is likely that the polymeric compounds are active in mast cells in a given species because they resemble some endogenous compound for which the cells of that species have a special affinity or receptor.

REFERENCES

ADAMKIEWICZ, V. W. and Y. LANGLOIS (1957) Sensitization by insulin to the dextran "anaphylactoid" reaction, *Canad. J. Biochem.* **35**, 251–6.

GOTH, A. (1959) Inhibition of anaphylactoid edema in the rat by 2-deoxyglucose, *Amer. J. Physiol.* **197**, 1056–8.

GOTH, A. and M. KNOOHUIZEN (1962) Tissue thromboplastins and histamine release from mast cells, *Life Sci.* **1**, 459–65.

GOTH, A., W. L. NASH, M. NAGLER and J. HOLMAN (1957) Inhibition of histamine release in experimental diabetes, *Amer. J. Physiol.* **191**, 25–8.

HALPERN, B. N. (1956) *Ciba Foundation Symposium on Histamine.* Boston, Little, Brown and Company, pp. 108.

KIPNIS, D. M. and C. F. CORI (1959) Studies of tissue permeability 1. The penetration and phosphorylation of 2-deoxyglucose in the rat diaphragm, *J. Biol. Chem.* **234**, 171–7.

LAGUNOFF, D. and E. P. BENDITT (1960) Mast cell degranulation and histamine release observed in a new *in vitro* system, *J. Exp. Med.* **112**, 571–80.

MORAN, N. C., B. UVNÄS and B. WESTERHOLM (1962) Release of 5-hydroxytryptamine and histamine from rat mast cells, *Acta Physiol. Scand.* **56**, 26–41.

SHORE, P. A., A. BURKHALTER and V. H. COHN (1959) A method for the fluorimetric assay of histamine in tissues, *J. Pharmacol.* **127**, 182–6.

DISCUSSION

GREEN: It is good that Dr. Goth has mentioned the species and organ differences that are seen in histamine release. Some of these differences have been collected in a review (Green, J. P. *Advances in Pharmacology* **1**, 1962). These differences suggest that we exercise restraint in our search for generalities in the mechanisms of release.

CHAKRAVARTY: How is 2-deoxyglucose blocking histamine release by dextran? If you postulate a receptor site for the activation of the release mechanism by dextran, does 2-deoxyglucose compete with this receptor, or is its effect due to its antimetabolic activity? I have used 2-deoxyglucose to block the release of histamine from sensitized guinea pig lung *in vitro*, and this inhibitory effect seems to depend on the metabolic antagonism of 2-deoxyglucose.

GOTH: Well, it is obvious that 2-deoxyglucose is a great metabolic inhibitor and it blocks glucose transport and we have considered that. The trouble is that the dextran reaction can be blocked by high concentrations of glucose or certain other sugars, but not certain different ones. There seems to be a competitive relationship here. I have published work before indicating that what 2-deoxyglucose does is a sort of displacing the dose-response curve to the right with these polymers. In anaphylaxis we have tried 2-deoxyglucose many times in whole animal and it has very little effect if any, and *in vitro* we cannot block the 48/80 type of histamine release with 2-deoxyglucose in concentrations of 1 mg/ml. So, while 2-deoxyglucose may have some general inhibitory effect on histamine release, it has a far greater effect on histamine release induced by dextran or ovomucoid in the rat.

ADDENDUM

It has been reported (Da Silva, W. D. and Lemos Fernandez, A. D., *Experientia*, **21**, 96, 1965) that dextran can cause histamine release from peritoneal mast cells of the rat without the addition of the lipid cofactor discussed in this paper. The authors attribute their different results to the fact that their experiments were carried out at 37°C. Preliminary work in our laboratory suggests that the differing results may be attributed to genetic differences. Sprague-Dawley or Holtzmann rats were used in our experiments while Wistar rats were employed by Da Silva and Lemos Fernandez.

ANAPHYLACTIC RELEASE OF HISTAMINE

J. L. MONGAR
Department of Pharmacology, University College London

THE anaphylactic release of histamine has been studied using chopped guinea pig lung tissue and isolated rat peritoneal mast cells. The process involves several stages:

1. Sensitization of mast cells by antibody γ-globulin.
2. Activation of an enzyme system following union of antigen with cell-fixed antibody.
3. Extrusion of granules from mast cell.
4. Histamine release fιom granules.

Stages 3 and 4 are common to histamine release by other agents and will not be discussed in this communication. In the rat stage 2 also appears to be triggered off by certain histamine liberators but in the guinea pig this stage is involved only in the anaphylactic reaction.

Stage 1. Sensitization

The mechanism of this process has been studied by *in vitro* passive sensitization. The sensitization achieved under various conditions was determined by challenging with the specific antigen under optimal conditions so that stages 2–4 of the reaction went to completion. The extent to which stage 1 had proceeded was assessed either by:

1. Counting the number of mast cells that have lost their granules (Humphrey and Mota, 1959).
2. Measuring the amount of histamine that has been released (Mongar and Schild, 1960; Brocklehurst, Humphrey and Perry, 1961).
3. Measuring some effect of the released histamine such as smooth muscle contraction as in the Schultz–Dale reaction (Halpern *et al.*, 1959; Nielsen *et al.*, 1959), or increase in capillary permeability as in the studies on passive cutaneous anaphylaxis where the amount of dye that escapes from the damaged capillaries is measured (Ovary, 1958; Biozzi *et al.*, 1959).

These methods have given results which agree satisfactorily. The histamine release method is probably the most useful since it is quantitative and gives a graded response over a wide range of sensitization.

METHOD

In this procedure the lungs of guinea pigs are perfused with Tyrode solution and then chopped into rods of $\frac{1}{2}$ to 1 mm thickness which are then incubated with antibody, usually rabbit antiserum, diluted with Tyrode solution. After removal of the antibody, the antigen is added to the solution and the amount of histamine diffusing out of the tissue in a standard time, say 15 min, is determined by biological assay. The quantity of histamine thus released can be regarded as an index of sensitization. In a strongly sensitized preparation this may amount to about 50 per cent of the histamine content of the tissue.

RESULTS

The amount of sensitization obtained in an *in vitro* experiment depends on several factors which can be varied independently: concentration of antibody, time and temperature of incubation.

Sensitization is not a simple adsorption of antibody on to tissue cells. Brocklehurst, Humphrey and Perry (1961) have shown in chopped guinea pig lung tissue that antibody which does not sensitize is adsorbed as readily as that which does. They estimated directly the amount of labelled protein which stays on the tissue after thorough washing. The special nature of the "fixation" necessary for sensitization is, however, demonstrated most clearly from studies in which temperature is varied. In the guinea pig ileum the rate of uptake of antibody is not greatly different at 20° and 37°C but the amount of histamine released by antigen at 37° is greatly increased as the temperature of the adsorption process is raised (Feigen, Nielsen and Terres, 1962). In guinea pig lung adsorption is only about twice as great at 37°C as at 0°C, but the rate of sensitization is 15 times as great (Mongar and Schild, 1960).

Figure 1 shows the results of experiments in which lung tissue was incubated with antibody at different temperatures for a fixed time. Incubation below 20°C produced little sensitization, between 30° and 40°C sensitization increased steeply reaching a maximum at about 40°C. Above 40°C there was a diminished release of histamine. This decrease may have been due to damage to the anaphylactic mechanism rather than to diminished fixation of antibody. The temperature coefficient was determined in the range 30–35°C. The Q_{10} was found to be about 5. This value is high: it corresponds to an activation energy of about 30,000 cal/mol and is very similar to the value recently found by Boursnell (1964) for the energy of exchange of antibody adsorbed on to specific cellular antigens.

Very small amounts of antibody are needed to sensitize tissues. In the guinea pig the uptake has been estimated as less than 1 μg/g. The concentration of antibody needed in the incubation medium to obtain sensitization is

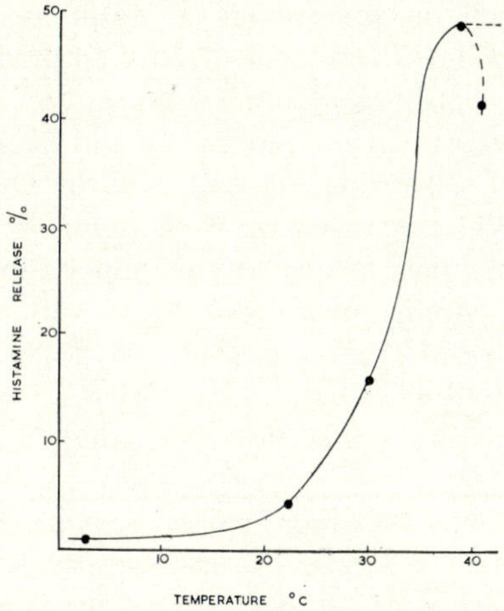

FIG. 1. Passive sensitization of chopped guinea pig lung by 2 per cent rabbit anti-ovalbumin serum incubated for 4 hr at various temperatures followed by challenge with antigen at 38°C.

similarly small. Halpern *et al.* (1959) found that 0·1 μg per ml will sensitize isolated guinea pig ileum if it is left in contact for 8 hr.

Mongar and Winne (1965) have found a similar value with guinea pig lung: 10^{-5} g/ml antiserum containing about 0·1 μg/ml antibody is active; a small effect is produced in 3 hr (Fig. 2). A maximum effect is obtained with 100 to 1000 times this concentration. With higher concentrations an inhibition is observed with increase of time.

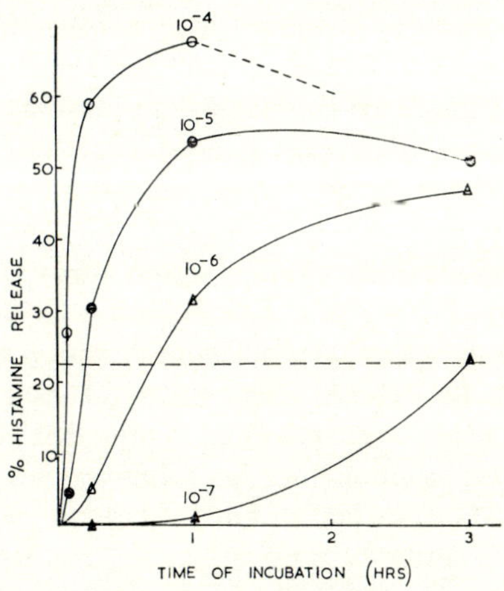

FIG. 2. Rate of passive sensitization of chopped guinea pig lung by varying concentrations (10^{-7} to 10^{-4} g/ml.) of rabbit antibodies to BSA. Dilutions of rabbit serum containing 9 mg/ml. precipitating antibody were used.

Sensitization depends on the species of antibody used. Only certain species produce γ-globulins which form effective antibodies: these are rabbit, guinea pig and man. Antibodies from the horse, cattle, goat, rat and fowl do not sensitize. Moreover in the guinea pig and in man only part of the antibody is capable of sensitizing. Recent work by Ovary, Benacerraf and Bloch (1963) and by White, Jenkins and Wilkinson (1963) has shown that the 7S antibodies of guinea pigs belong to two populations of molecules with quite distinct electrophoretic mobilities—the γ_1 with a high mobility can sensitize but the slower moving γ_2 cannot. No such distinction has been observed with rabbit antibodies but in man antibodies of the 7S fraction will sensitize guinea pig tissues whilst the γ_1-macroglobulins will not (Ovary *et al.*, 1960).

Attempts have been made to inhibit passive sensitization and these studies throw some light on the mechanism of sensitization.

The marked dependence of sensitization on temperature suggested that perhaps an enzymic process was involved or perhaps an active uptake of antibody into the cell. However, Mongar and Schild (1960) found that a number of metabolic inhibitors were inactive. Similarly the exposure to abnormal ionic environments produced no inhibition of sensitization. Unlike the anaphylactic reaction passive sensitization does not require calcium nor any other physiological ion; it will proceed quite well in 0·3 M sucrose. Moreover, Liacopoulos (1962) has shown that in the guinea pig ileum sensitization is actually more readily obtained by removing all ions or by the addition of 0·8 M urea. Thus metabolic inhibitors and abnormal ionic environments do not inhibit.

Passive sensitization can, however, be inhibited by non-antibody γ-globulins. This was first demonstrated with passive cutaneous anaphylaxis (PCA). Ovary and Bier (1953) showed that more antibody was required for sensitization if it was dissolved in serum than if it was dissolved in Tyrode solution. This inhibitory effect was later shown to be due entirely to the γ-globulin content of the serum. Like sensitization itself, inhibition of sensitization is specific for the γ-globulin of certain species nor is there any inhibition with α- or β-globulins or with serum albumin.

Mongar and Schild (1960) studied the inhibition by γ-globulins of passive sensitization of chopped guinea pig lung. Figure 3 shows the effect of human gamma globulin (HGG). The ratio of non-antibody : antibody γ-globulin is about 50 : 1. The reaction is inhibited for nearly 6 hr. The effect of homologous γ-globulin as an inhibitor has also been studied: 10 per cent normal rabbit serum inhibited 2 per cent anti-ovalbumin rabbit serum (non-antibody : antibody ratio about 10 : 1). In the controls nearly maximum sensitization was reached after 2 hr. In the presence of normal rabbit serum sensitization was much slower; no histamine release occurred after 1 hour incubation and after 2 hr it was still only 20 per cent of the

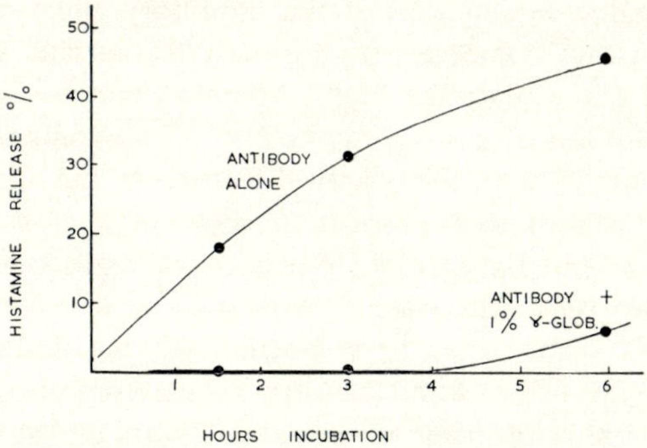

FIG. 3. Inhibition of passive sensitization. Rabbit antibody and human non-antibody γ-globulin were incubated with chopped guinea pig lung at 38°C.

control. A similar inhibition was produced in experiments in which the antibody and normal γ-globulin were both from the guinea pig. Bovine γ-globulin did not inhibit even at a concentration of 10 mg/ml.

Mongar and Schild have also investigated whether an established sensitization can be reversed by γ-globulin. The lung tissue was first sensitized by incubation with antibody for 2 hr. Figure 4 shows that mere washing in Tyrode solution produced no reversal of passive sensitization. Indeed, sensitization increased rather than decreased after washing in Tyrode solution. A reversal could, however, be produced by means of γ-globulins. In 3 hr the reversal by 10 mg/ml HGG was almost complete.

Reversal is only possible with passive sensitization that is of recent origin; passive sensitizations of long standing, or active sensitizations, are difficult or impossible to reverse.

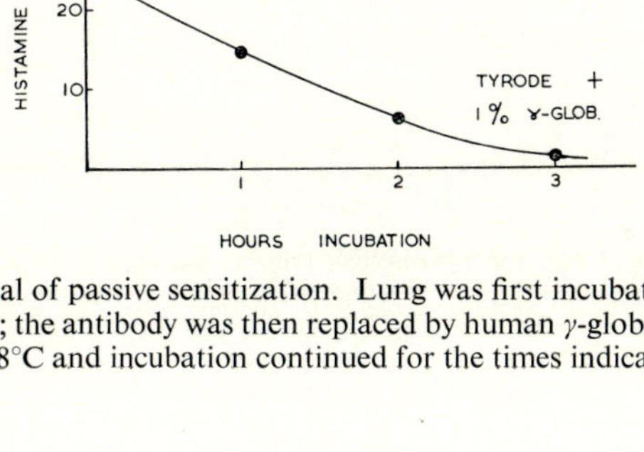

FIG. 4. Reversal of passive sensitization. Lung was first incubated for 2 hr with rabbit antibody; the antibody was then replaced by human γ-globulin (10 mg/ml.) at 38°C and incubation continued for the times indicated.

The most probable explanation of the inhibitory effect of non-antibody γ-globulins is that they compete with antibody for specific attachment sites on the cell. This can explain the highly specific character of the inhibitory effect. The assumption of a simple competitive antagonism is nevertheless insufficient to account for all the observed features of the inhibitory effect of γ-globulins, e.g. it cannot account for the reversal of an established sensitization by γ-globulin since the rate of desorption of antibodies should be the same whether the tissue was washed with Tyrode or γ-globulin solution. Nor is the simple theory easy to reconcile with the finding that passive sensitization becomes firmer with time. It was therefore postulated that each antibody molecule attaches itself to several receptors on the cell:

$$A+R \rightleftharpoons AR$$
$$AR+R \rightleftharpoons ARR$$
$$ARR+R \rightleftharpoons ARRR\ldots$$

These reactions would tend in time to shift towards multi-attachment and in this way become less easily reversible.

Dr. D. Winne of Tübingen has recently been working with me on this problem. He has worked out the implications for the shape of the concentration-effect curves of one- and two-point attachments for sensitization with antibody alone and in the presence of an inhibitor.

Assuming that mass action equations can be applied (which means working under equilibrium conditions), Fig. 5 shows the concentration-effect curves for sensitization in the absence of inhibitor.

Curve 1—only one point attachment is possible—this gives a symmetrical S-shaped log concentration-response curve.

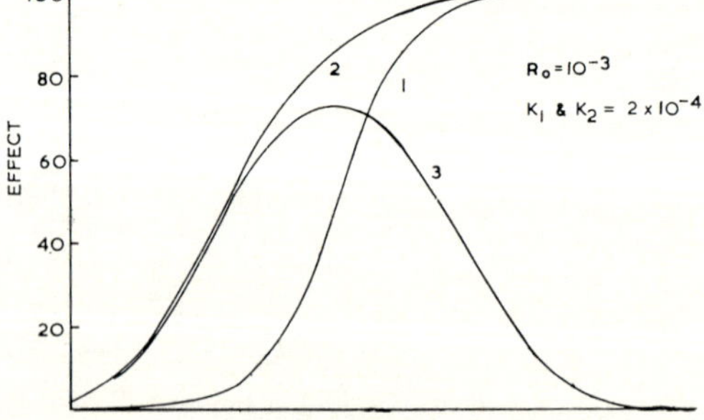

FIG. 5. Calculated concentration-effect curves: curve 1, only one-point attachment possible; curve 2, two-point attachment possible but one-point attachment sufficient for sensitization; curve 3, only two-point attachment effective for sensitization.

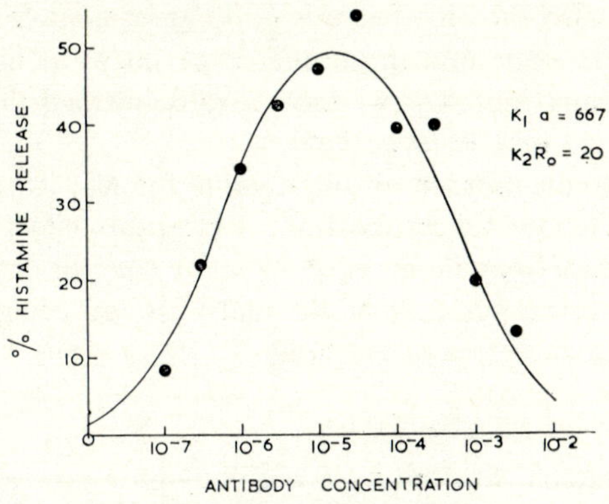

FIG. 6. Comparison of calculated concentration-effect curve with experimental results (mean of five experiments).

Curve 2—two point attachment is possible but one point attachment is sufficient for sensitization—this gives an unsymmetrical S-shaped curve.

Curve 3—only two point attachment is effective for sensitization—this gives a bell-shaped curve.

Figure 6 shows the mean results from five experiments and the calculated curve for two-point attachment. The fit is good.

Similar mass action relations are also involved when considering the effect of inhibition by non-antibody γ-globulin.

Figure 7 shows the calculated curves for increasing concentrations for non-antibody γ-globulin, each representing a 10-fold increase in inhibitor concentration. The experimental points, when the inhibitor was normal

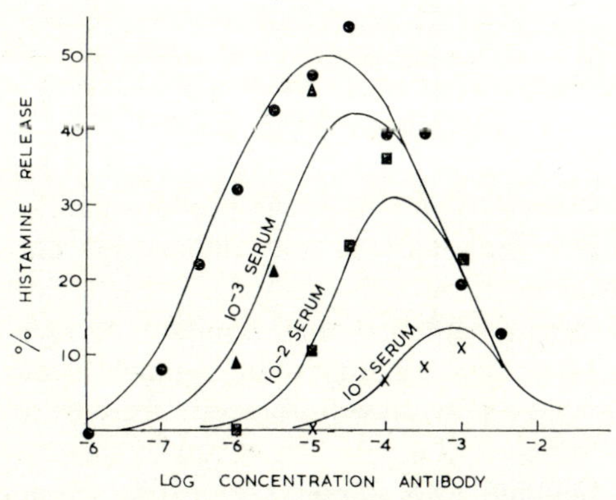

FIG. 7. Comparison of calculated concentration-effect curves for increasing concentrations of non-antibody γ-globulin with experimental results: ●, no inhibitor; ▲, 10^{-3} normal serum; ■, 10^{-2} normal serum; ×, 10^{-1} normal serum, present with antibody.

rabbit serum, are also shown. There is good agreement concerning position of curves and their slope and in another experiment at high concentration of both antibody and inhibitor we have also confirmed that the curves are bell-shaped (Mongar and Winne, 1965).

It is possible to calculate an absolute value for K_1: it is about 10^7 l/mol. We have also figures for the proportion of receptors effectively occupied by antibody to produce maximum effect—i.e. at the top of the bell-shaped curves. Some 60 to 80 per cent of the receptors are occupied, the rest are either singly attached or free of antibody.

Stage 2. Activation of Enzyme System

The union of antigen with cell-fixed antibody initiates a sequence of events leading to the degranulation of mast cells and histamine release. Most of our knowledge of these events has come from the use of inhibitors. The most active inhibitors are those reagents which block sulphydryl groups: iodoacetate, *p*-chloromercuribenzoate and N-ethylmaleimide. Concentrations of

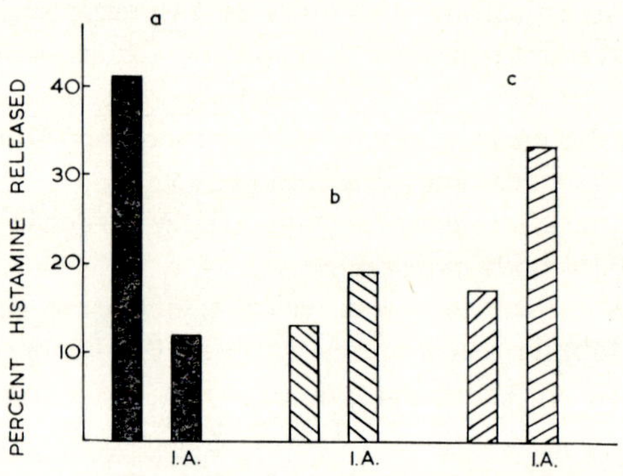

FIG. 8. Effect of 1 mM iodoacetate (IA) on histamine release from chopped, sensitized guinea pig lung by (a) antigen (10^{-3} egg albumin); (b) compound 48/80 (10^{-3}); (c) octylamine (2×10^{-4}).

0·1 mM are effective and 1·0 mM solutions produce almost complete inhibition. The action of organic bases is not inhibited but on the contrary it is potentiated (Fig. 8).

It has recently been found that thiol compounds such as cysteine and thioglycollate are also active inhibitors but at much higher concentrations; 10 mM solutions actually potentiate histamine release by antigen but 50 mM isotonic solutions inhibit it (Fig. 9). The effects of the thiol compounds are fully reversible. Sulphite also inhibits reversibly presumably through a reduction of S-S bridges (Fig. 10) (Edman, Mongar and Schild, 1964).

Pretreatment of sensitized tissue with protein denaturants such as alcohol, urea, guanidine and heat produces an irreversible anaphylactic inhibition.

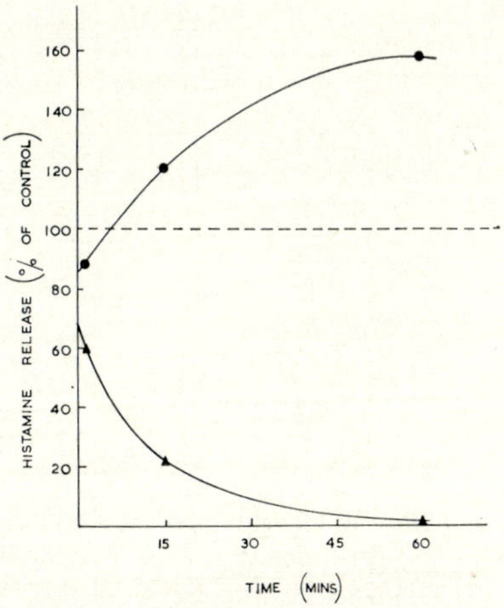

FIG. 9. Effect of duration of preliminary incubation with thioglycollate on histamine release by antigen. Thioglycollate added to Tyrode and left in solution throughout. Upper curve, increasing potentiation by 10 mM; lower curve, increasing depression by 50 mM. Single experiment, duplicate samples.

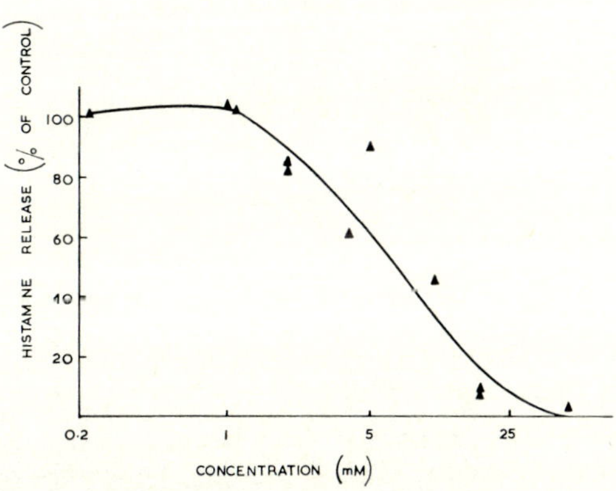

FIG. 10. Inhibition of histamine release by sulphite. Sensitized lung was incubated for 15 min with various concentrations on sodium sulphite in Tyrode solution followed by addition of antigen to the solution and a further 15 min incubation.

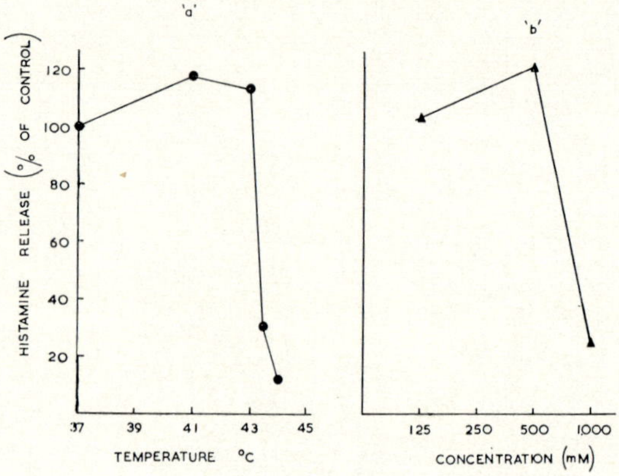

FIG. 11. Comparison of effect of raised temperature and alcohol: (a) histamine release after exposure of sensitized lung to various temperatures for 5 min. before addition of antigen at 37°C; (b) exposure of sensitized lung for 15 min. to various concentrations of ethanol in Tyrode followed by rinsing with Tyrode and addition of antigen 15 min later.

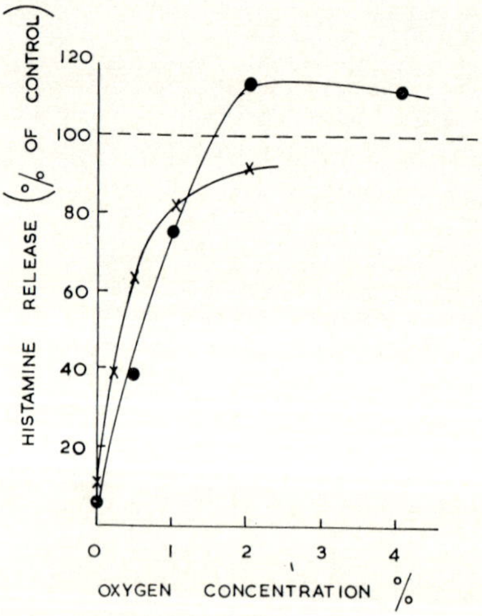

FIG. 12. Effect of histamine release of concentration of oxygen in a stream of nitrogen bubbled through the incubation fluid. Two separate experiments: slow rate of bubbling (●), rapid rate of bubbling (×).

In slightly lower concentrations alcohol and urea produce an activation of the anaphylactic mechanism which may be due to an unmasking of SH groups in the course of incipient protein denaturation (Fig. 11). A similar mechanism may account for the initial potentiation by thiol compounds.

The anaphylactic reaction does not function under anaerobic conditions but oxygen is probably not a participant in the sequence of reactions leading to histamine release (Mongar and Perera, 1965). It is necessary to reduce oxygen to very low concentrations (below 0·5 per cent) to produce appreciable inhibition (Fig. 12). The inhibition by oxygen lack can be rapidly reversed by reintroducing oxygen. Probably one effect of profound oxygen lack is to reduce S-S bonds, and in this way to produce inhibition by a mechanism which is essentially similar to that of S-S reducing compounds (Table 1).

TABLE 1. COMBINED EFFECTS OF CYSTEINE AND OXYGEN LACK

Cysteine concn. (*mM*)	*Oxygen concn.* (%)	*Histamine release* (% *of control*)		
a	*b*	*a*	*b*	*a*+*b*
10	1	123	60	44
20	1	95	60	57
40	1	49	60	18
30	2	68	54	32
10	4	103	88	58
20	4	74	88	40

Phenol inhibits histamine release by antigen; its effects are reversible and this has made it possible to analyse its mode of action. It does not interfere with the antigen–antibody reaction for it does not prevent desensitization. In the presence of phenol there is no histamine release nor does any release occur after removal of the phenol, i.e. the reaction which was started in the presence of the inhibitor fails to resume its course when the inhibitor is no longer present. The antigen–antibody reaction has somehow become uncoupled from the histamine release reaction. The uncoupling is probably due to the rapid inactivation of the enzyme system activated by the antigen–antibody reaction so that after removal of the phenol no active enzyme is left to complete the reaction (Mongar and Schild, 1957a).

Further evidence for the occurrence of an enzymic reaction in anaphylaxis is provided by the effects of temperature upon histamine release. The temperature curve is a typical enzyme curve with a maximum within a narrow range around 40–41°C and falling to zero at 15–20°C and 44–45°C (Fig. 13). Exposure of sensitized tissue to 45°C for 5 min produces irreversible inhibition. Further tests have shown that the heat labile factor is not antibody but a constituent of normal tissue, probably an enzyme precursor whose

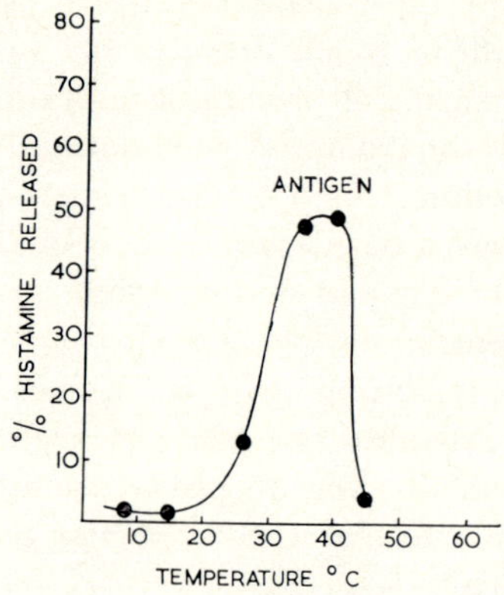

FIG. 13. Effect of temperature on histamine release by Tyrode solution + antigen (10^{-3} egg albumin: mean of two experiments).

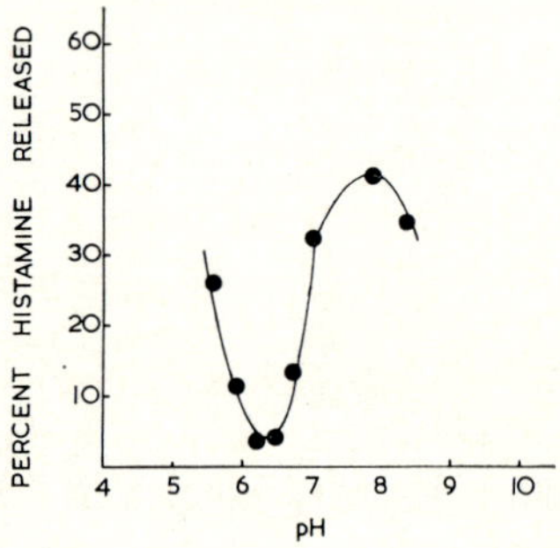

FIG. 14. Effect of pH on histamine release in anaphylaxis (phosphate buffer).

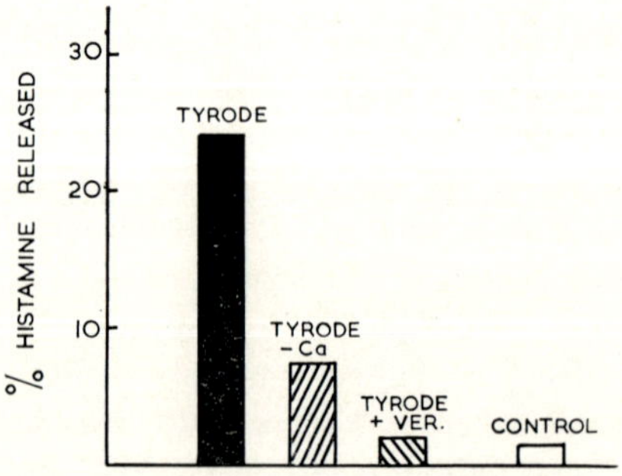

FIG. 15. Effect of removal of calcium from Tyrode solution on histamine release by antigen. The columns show partial inhibition of histamine release when $CaCl_2$ is omitted from the Tyrode solution and complete inhibition when 0·05 per cent versene is added to the Tyrode solution.

denaturation would block any further reaction steps (Mongar and Schild, 1957b).

The anaphylactic reaction functions only in a narrow range of pH with a maximum at pH 7·5; the reaction is almost completely inhibited at pH 6·3 (Fig. 14).

The ionic requirements of the histamine release mechanism in anaphylaxis are specific and limited. Calcium is the only essential ion (Fig. 15). It may be required for the activation of the anaphylactic enzyme in much the same way

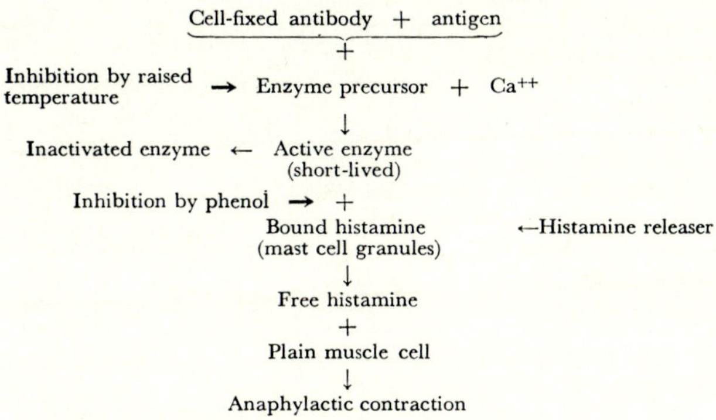

FIG. 16. Scheme illustrating reaction sequence in the anaphylactic reaction.

as calcium is required for the activation of the first component of complement. The inhibiting effect of calcium lack can be overcome by a raised pH and, conversely, the inhibition produced by low pH can be overcome by a raised calcium concentration. A possible interpretation of these effects is that calcium combines with imidazole groups of a protein since imidazole can form complexes with divalent cations which dissociate at about pH 6 (Mongar and Schild, 1958). Figure 16 summarizes the information we have obtained by the use of inhibitors.

Stage 2 of the anaphylactic histamine release process appears to involve a pH-sensitive, heat labile, calcium-requiring, short-lived, energy-requiring, SH enzyme system (Mongar and Schild, 1962).

REFERENCES

BIOZZI, G. B., B. N. HALPERN and R. BINAGHI (1959) The competitive effect of normal serum proteins from various animal species on antibody fixation in passive cutaneous anaphylaxis in the guinea pig, *J. Immunol.* **82**, 215–18.

BOURSNELL, J. C. (1964) Studies with marked antisera. III. Interchange between unmarked antibody and ^{131}I-marked antibody in combination with cellular antigens, *Immunology* **7**, 127–41.

BROCKLEHURST, W. E., J. H. HUMPHREY and W. L. M. PERRY (1961) The *in vitro* uptake of rabbit antibody by chopped guinea pig lung and its relationship to anaphylactic sensitization, *Immunology* **4**, 67–87.

EDMAN, K. A. P., J. L. MONGAR and H. O. SCHILD (1964) The role of SH and S-S groups and oxygen in the anaphylactic reaction of chopped guinea pig lung, *J. Physiol., Lond.* **170**, 124–37.

FEIGEN, G. A., C. B. NIELSEN and G. TERRES (1962) Effect of antibody concentration and temperature upon physical adsorption and histamine release, *J. Immunol.* **89**, 717–29.

HALPERN, B. N., P. LIACOPOULOS, M. LIACOPOULOS-BRIOT, R. BINAGHI and F. VAN NEER (1959) Patterns of *in vitro* sensitization of isolated smooth muscle tissues with antibody, *Immunology* **2**, 351–62.

HUMPHREY, J. H. and I. MOTA (1959) The mechanism of anaphylaxis: specificity of antigen induced mast cell damage in anaphylaxis in the guinea pig, *Immunology* **2**, 31–43.

LIACOPOULOS, P. (1962) L'anaphylaxie passive *in vitro*. III. Modalités de la sensibilisation *in vitro*, *J. Physiol., Paris* **54**, 501–16.

MONGAR, J. L. and B. A. V. PERERA (1965) Oxygen consumption during histamine release by antigen and compound 48/80, *Immunology* **8**, 511–18.

MONGAR, J. L. and H. O. SCHILD (1957a) Inhibition of the anaphylactic reaction, *J. Physiol., Lond.* **135**, 301–19.

MONGAR, J. L. and H. O. SCHILD (1957b) Effect of temperature on the anaphylactic reaction, *J. Physiol., Lond.* **135**, 320–38.

MONGAR, J. L. and H. O. SCHILD (1958) The effect of calcium and pH on the anaphylactic reaction, *J. Physiol., Lond.* **140**, 272–84.

MONGAR, J. L. and H. O. SCHILD (1960) A study of the mechanism of passive sensitization, *J. Physiol., Lond.* **150**, 546–64.

MONGAR, J. L. and H. O. SCHILD (1962) Cellular mechanisms in anaphylaxis, *Physiol. Rev.* **42**, 226–70.

MONGAR, J. L. and D. WINNE (1965) Further studies of the mechanism of passive sensitization, *J. Physiol., Lond.* In press.

NIELSEN, C. B., G. TERRES and G. A. FEIGEN (1959) Adsorption of antibody *in vitro* and magnitude of the Schultz–Dale reaction of guinea pig ileum, *Science* **130**, 41–2.

OVARY, Z. (1958) Immediate reactions in the skin of experimental animals provoked by antibody–antigen interaction, *Progr. Allergy*, *Basel*, New York, S. Karger, vol. **5**, pp. 459–508.

OVARY, Z., B. BENACERRAF and K. J. BLOCH (1963) Properties of guinea pig 7S antibodies. II. Identification of antibodies involved in passive cutaneous and systemic anaphylaxis, *J. Exp. Med.* **117**, 951–64.

OVARY, Z. and O. G. BIER (1953) Action empêchante du sérum normal de lapin sur l'anaphylaxie cutanée passive du cobaye, *Ann. Inst. Pasteur* **84**, 443–5.

OVARY, Z., H. FUDENBERG and H. G. KUNKEL (1960) Anaphylactic reactions in skin of guinea pigs with high and low molecular weight antibodies and γ-globulins, *J. Exp. Med.* **112**, 953–61.

WHITE, R. G., G. C. JENKINS and P. C. WILKINSON (1963) Production of skin sensitizing antibody in the guinea pig, *Int. Arch. Allergy* **22**, 156–65.

DISCUSSION

UVNÄS: When you showed this last picture where the action of octylamine and 48/80 was shown not to be dependent on metabolic processes, I think you should emphasize what Dr. Green and Dr. Goth said about the specificity of the reaction of mast cells in different species because in other tissues 48/80 acts by inducing an active and metabolism depending degranulation mechanism.

MONGAR: Yes, I think I should have pointed out that in the rat mast cell his famine release by 48/80 is inhibited in jnst the same way as the antigen–antibody induced histamine release and it appears that both are triggering the same mechanism, which is quite different from guinea pig tissue.

GREEN: These observations on the effect of calcium may be related to the effects noted by Dr. Douglas and suggest that calcium is not acting directly on the granules.

MONGAR: Yes, I had a short discussion with Dr. Douglas yesterday, trying to arrive at some common mechanism but we did not succeed. I think all we can say about calcium is that it must be present for the enzyme to be activated. If you add antigen in the absence of calcium, you get no effect, but then if you add calcium the activation proceeds.

PLETSCHER: Is there evidence of complement fixation during the antigen–antibody reaction on mast cells?

MONGAR: You do not need what the immunologist calls complement. You can carry this reaction out in the absence of serum. Moreover if you inactivate the tissue by heating to 45°, you do not reactivate it by adding fresh serum, so there is no evidence of complement being directly involved.

BORÉUS: I wonder, Dr. Mongar, if you would comment upon the problem of the variation in anaphylactic histamine release which exists between different tissues in the same individual animal. For instance, why is it that the small intestine of the guinea pig which contains much histamine and a large mast-cell population gives off so little histamine when exposed to antigen when most other tissues release a lot? In work with Dr. Westerholm we have found the same to be true in the rat, both for histamine and serotonin release. One wonders, too, if the mast cells isolated from the peritoneal fluid of the rat are similar to the cells in the intestine. If so, they would be rather non-typical for the rat in the sense that they do not react to antigen as easily as mast cells in most other tissues.

MONGAR: The work I described was done on guinea pig tissues. We have used a number of tissues other than lung, including ileum, and we do not find any basic difference in the mechanism.

THE METABOLISM OF RAT PERITONEAL MAST CELLS

NIRMAL CHAKRAVARTY

Department of Pharmacology, University of Umeå, Sweden

ANAPHYLACTIC histamine release is dependent on the presence of oxygen or glucose. The release is completely blocked by anoxia in a substrate-free medium. The effect of anoxia is reversed by glucose, while in the presence of oxygen glucose does not seem to make any difference in the amount of histamine released. These findings have led to the assumption that energy is required at some stage of the histamine releasing process and respiratory or glycolytic formation of ATP might provide the required energy. This focuses attention on the metabolism of mast cells. The data presented here deal with the respiratory and the glycolytic activity of mast cells obtained from rat's peritoneal cavity. The work was done at the Carlsberg Foundation Biological Institute, Copenhagen, the respiration studies being carried out in collaboration with Dr. Erik Zeuthen.

The mast cells were isolated in human or bovine albumin solution. On the average the isolated cells were 98·7 per cent pure mast cells. The mean diameter of the mast cells was 12·75 μ. Both respiration and glycolysis were studied by the ampulla diver technique (see Zajicek and Zeuthen, 1961) with modifications to permit measurements in a few hundred cells (for details of the method see Chakravarty and Zeuthen, 1965, and Chakravarty, 1965).

Figure 1 shows in a schematic form how the cells were placed in the diver during the experiment. The part occupied by the cell suspension was previously coated with agar. Sodium hydroxide was introduced at the tail end to absorb carbon dioxide for the respiration studies. For the glycolysis experiments a medium containing bicarbonate buffer was introduced to both the ends of the diver. With such an arrangement the size of the sample is determined by direct observation under the microscope as shown in Fig. 2. The photomicrograph was taken with a microscopic enlargement of $\times 96$. This may be obtained with a dissection microscope and the number of mast cells introduced into the diver can thus be counted during preparation.

Table 1 shows the respiration rates of rat peritoneal mast cells with and without substrate. When no substrate was added to the medium the endogenous respiration rate was $0{\cdot}29 \times 10^{-6}$ μl per cell per hour. Glucose caused 60 per cent rise in the respiration rate which now reached $0{\cdot}47 \times 10^{-6}$ μl

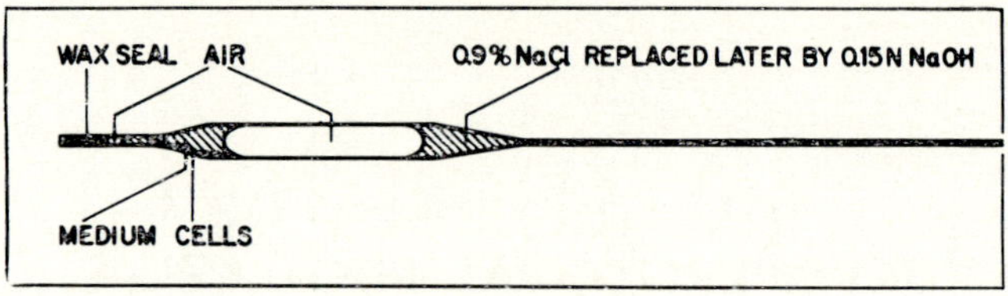

FIG. 1. A diver filled for the determination of the respiration rate of rat peritoneal mast cells.

per cell per hour. The increase by succinate and pyruvate was not striking and isocitrate did not cause any change at all. However, there might be insufficient entry of the substrates into the cells and the possibility remains that these substrates may be utilized at a faster rate. Indeed, it has been shown by Martin, McKinney and Green (1955) that succinate does not influence the respiration of intact polymorphonuclear leukocytes but doubles the respiration rate of the broken cells. Sodium azide (5 mM) caused 80 per cent inhibition of the respiration rate of the mast cells.

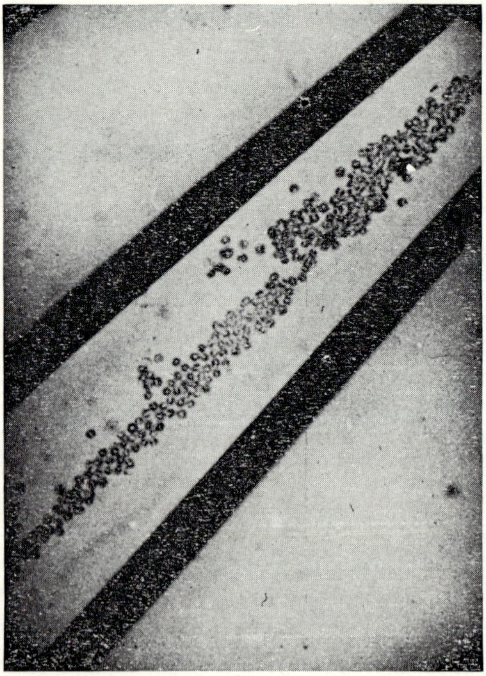

FIG. 2. Photomicrograph of cells in the diver as they are counted. Microscopic enlargement: ×96.

TABLE 1. RESPIRATION OF RAT PERITONEAL MAST CELLS

*Substrate**	*Inhibitor**	*O_2 uptake per cell per hour ($\times 10^{-6}$ μl.)*
—	—	0·29±0·04 (4)
Glucose	—	0·47±0·04 (6)
Succinate†	—	0·43±0·06 (3)
Pyruvate†	—	0·39 (2)
Isocitrate†	—	0·27 (2)
Glucose	Azide†	0·09 (1)

* 5 mM. † Sodium salts used.
Number of observations in parentheses in the last column.

Table 2 shows the anaerobic and aerobic glycolytic activity of rat peritoneal mast cells. The amount of carbon dioxide evolved from the medium containing bicarbonate was measured as shown here, and the corresponding lactic acid values are given in the last column assuming that all carbon

TABLE 2. GLYCOLYSIS IN RAT PERITONEAL MAST CELLS

	Glucose	*Inhibitor*	*CO_2 evol. per cell per hour) ($\times 10^{-6}$ μl.)*	*Lactic acid produced per cell per hour ($\times 10^{-8}$ μM)*
Anaerobic glycolysis	—		1·43±0·17	6·38 (3)
	—	Fluoride, 15–20 mM*	0·61±0·08	2·72 (3)
	5 mM	—	1·70±0·15	7·60 (6)
	5 mM	Iodoacetate, 2 mM*	0·56±0·04	2·50 (3)
Aerobic glycolysis	5 mM	—	0·93±0·08	4·15 (5)

* Sodium salts used.
Number of observations in parentheses in the last column.

dioxide evolved anaerobically is produced by glycolytic lactic acid formation. The glycolysis rate without glucose was $1{\cdot}43 \times 10^{-6}$ μl. CO_2 per cell per hour and that with glucose $1{\cdot}70 \times 10^{-6}$ μl. CO_2 per cell per hour. There is thus no significant difference between the rates of anaerobic glycolysis with and without substrate. Fluoride and iodoacetate caused about 60 per cent inhibition of the anaerobic glycolytic activity. The aerobic glycolysis was measured assuming respiratory quotient to be 1. In a bicarbonate buffer system the change in the gas volume then reflects the non-respiratory carbon dioxide evolution or aerobic glycolysis. The rate of aerobic glycolysis was $0{\cdot}93 \times 10^{-6}$ μl. per cell per hour being thus 55 per cent of the anaerobic value but still much higher than the respiratory oxygen uptake.

If the respiratory and the glycolytic activity in the presence of glucose are expressed in terms of glucose utilization it is found that under aerobic conditions $2{\cdot}43 \times 10^{-8}$ μM glucose is consumed per cell per hour, and of this amount 15 per cent is oxidized and 85 per cent converted to lactic acid. Under anaerobic conditions the rate of its conversion to lactic acid is $3{\cdot}80 \times 10^{-8}$ μM per cell per hour. Total glucose utilization is thus higher in an anaerobic medium than under air, suggesting a mild Pasteur effect.

TABLE 3

Cells	*Species*	*O_2 uptake** ($\times 10^{-6}$ *μl.*) *per cell per hour*	*Aerobic glycolysis** *CO_2 evol.* ($\times 10^{-6}$ *μl.*) *per cell per hour*
Peritoneal mast cells	Rat	0·47	0·93
Neoplastic mast cells (from Fredholm's data)	Mouse	1·93	—
Polymorph. leukocytes (Martin *et al.*)	Human	0·21	1·66

* In presence of 5–10 mM glucose.

Table 3 compares the respiration and glycolysis rates of peritoneal mast cells with the rates for neoplastic mast cells and polymorphonuclear leukocytes. The respiration of neoplastic mast cells is calculated from data provided by Fredholm (1964). The respiration rate of neoplastic mast cells from mouse appears to be four times that of rat peritoneal mast cells. This difference seems to be compatible with the observation made by electron microscopy that fewer mitochondria are noticeable in peritoneal mast cells as compared to neoplastic mast cells (see discussion: Bloom, 1963). It would be of interest to compare the glycolytic activity of peritoneal mast cells with that of neoplastic mast cells, but I have not come across any data on the glycolysis of neoplastic mast cells. The polymorphonuclear leukocyte has a high rate of glycolysis. There is thus a resemblance in the metabolic pattern of the mast cells and the leukocytes, both having a high glycolysis/respiration ratio.

REFERENCES

BLOOM, G. D. (1963) Electron microscopy of neoplastic mast cells, *Annals N.Y. Acad Sci.* **103**, 53.

CHAKRAVARTY, N. (1965) Glycolysis in rat peritoneal mast cells, *J. Cell Biol.* **25**, 123.

CHAKRAVARTY, N. and E. ZEUTHEN (1965) Respiration of rat peritoneal mast cells, *J. Cell Biol.* **25**, 113.

FREDHOLM, B. (1964) Personal communication.

MARTIN, S. P., G. R. MCKINNEY and R. GREEN (1955) The metabolism of human polymorphonuclear leucocytes, *Annals N.Y. Acad. Sci.* **59**, 996.

ZAJICEK, J. and E. ZEUTHEN (1961) *General Cytochemical Methods*, vol. 2, Academic Press, pp. 131.

HISTAMINE SYNTHESIS IN BASOPHILS

G. WINQVIST

Department of Anatomy and Histology
Royal Veterinary College, Stockholm

APART from mast cells there is another important cellular source of histamine in the body, i.e. the basophils of the blood and bone marrow. Amazing amounts of histamine can circulate in the vessels bound to basophils. Thus a human patient with myelogenous leukaemia was reported (Valentine, Lawrence, Pearce and Beck, 1955) to carry almost 50 μg histamine per ml. blood.

In normal guinea-pigs the average amount of histamine in the blood was 0·16 μg/ml and in the bone marrow 12·7 μg/ml. By giving subcutaneous injections of foreign protein to guinea-pigs for 12 to 14 days one will obtain a tenfold increase in the number of basophils as well as in the amount of histamine in blood and bone marrow.

Recently (Aures, Winqvist and Hansson, 1965) we studied the histidine decarboxylase activity of the bone marrow and the buffy coat of the blood in guinea-pigs by measuring $C^{14}O_2$ liberated from histidine $C^{14}OOH$. In guinea-pigs treated with foreign protein the histidine decarboxylase activity of the bone marrow increased ten times as did the number of basophils and the amount of histamine. 6-Methyl-prednisolone depressed all values proportionately. Table 1 also shows that the rate of histamine production of the bone marrow is high, about 5 per cent of the total amount being produced per hour.

TABLE 1. RELATION BETWEEN HISTAMINE CONTENT AND FORMATION TO THE NUMBER OF BASOPHILS IN THE BONE MARROW OF GUINEA-PIG

Treatment	*No. of expts.*	*Histamine µg/ml. bone marrow*	*Histamine-formation* µg/ml. in 2 hr*	*Total basophils (10⁶/ml.)*	*Histamine µg/10⁹ basophils*	*Histamine-formation* µg/10⁹ basophils in 2 hr*
Untreated controls	11	12·7 ±0·1	0·9 ±0·1	17·1 ±1·3	765 ±56	56 ±9
For. prot.	6	120·3 ±34·1	11·9 ±3·3	147·7 ±39·5	815 ±121	79 ±12
For. prot. + 6-methyl-prednisolone	12	74·0 ±10·4	7·4 ±1·1	97·9 ±15·9	756 ±104	81 ±8
6-methyl-prednisolone	4	14·30 ±·7	1·4 ±0·2	15·6 ±1·5	918 ±155	91 ±8

Values are means ±S.E.

* Histidine decarboxylase activity expressed as μg histamine formed per ml. bone marrow in 2 hr.

† Histidine decarboxylase activity expressed as μg histamine formed per 10^9 basophils in 2 hr.

We thus conclude that the basophils not only store histamine but also produce it in considerable amounts.

REFERENCES

AURES, D., G. WINQVIST and E. HANSSON (1965) Histamine formation in the blood and bone marrow of the guinea-pig, *Am. J. Physiol.* **208**, 186–9.

VALENTINE, W. N., J. S. LAWRENCE, M. L. PEARCE and W. S. BECK (1955) The relationship of the basophil to blood histamine in man, *Blood* **10**, 154–9.

UPTAKE AND RELEASE OF H^3-HISTAMINE

B. B. Brodie, M. A. Beaven, F. Erjavec
and H. L. Johnson

Laboratory of Chemical Pharmacology
National Heart Institute
Bethesda, Md., U.S.A.

Since histamine is widely distributed in body tissues and exerts potent biological effects, it is reasonable to assume that it has a physiological role in the tissues where it is located. Current thinking on the possible role of histamine has been colored by the assumption that most of the amine is located in mast cells. However, it is difficult to see how mast cell histamine could regulate physiological function in view of its slow turnover time of several weeks (Schayer, 1959).

Since mast cell releasers, such as compound 48/80, deplete only a fraction of the histamine stores (Mota, Beraldo, Ferri and Junqueira, 1956), the question rises whether non-mast cell stores of histamine might be important in the regulation of physiological processes. This is also suggested by the relative lack of mast cells in the rabbit (Ehrlich, 1877) and the failure of compound 48/80 to release histamine from most tissues in this species (Burkhalter, Cohn and Shore, 1959). Moreover, in certain species a number of mast cells also contain 5-hydroxytryptamine or dopamine, suggesting that the structure of the amine might not be particularly important to these cells (Adams-Ray, Dahlström, Fuxe, and Hillarp, 1964).

The gastric mucosa contains large amounts of histamine, but relatively few mast cells, and this amine is generally believed to mediate the gastric secretion evoked by vagal stimulation. It seems unlikely that this interaction would be restricted to only one exocrine gland, hence it is pertinent that administered histamine, like parasympathetic stimulation, increases the secretion of saliva, tears, sweat, and of the intestinal and pancreatic juice and bile.

The present paper describes a procedure for the selective labeling of non-mast cell histamine. Our results suggest that secretions from exocrine glands are controlled neurohormonally and hormonally through release of non-mast cell histamine.

DO ALL TISSUES CONTAIN NON-MAST CELL STORES OF HISTAMINE?

From examination of the literature one would gather that most of the histamine in the body is located in mast cells. This impression is at best a half-truth, for the bulk of the mast cell histamine is located in a single tissue—the skin, and all tissues contain histamine stores that resist the action of 48/80 (Paton, 1958).

In our experiments on the effects of compound 48/80 in various rat tissues, amine stores were depleted by more than 70 per cent in mesentery, diaphragm, oesophagus and submaxillary gland, hardly affected in gastric mucosa, lung and ileum, and depleted by less than half in duodenum, colon, liver and heart (Table 1). Even in skeletal muscle considerable amounts of

TABLE 1. HISTAMINE CONTENT OF VARIOUS RAT TISSUES BEFORE AND AFTER INJECTION OF COMPOUND 48/80

Organ	*Histamine content* ($\mu g/g$)		
	Total	*Not released*	*% Depletion*
Skin	37	5	87
Mesentery	18	2	89
Diaphragm	17	5	69
Submaxillary gland	21	4	81
Oesophagus	17	5	69
Stomach (glandular)	23	20	13
Ileum	14	13	7
Colon	10	8	20
Liver	1.8	1.4	22
Lung	4.3	5.2	—
Heart	2.3	1.3	45
Skeletal muscle	3.3	1.4	58

Rats received 6mg/kg of 48/80 in 5 divided doses intraperitoneally over a period of 22 hr. The histamine was assayed fluorimetrically (Shore, Burkhalter and Cohn, 1959). Each value is the mean of 5 to 10 animals.

histamine were resistant to the action of 48/80. These results indicate that about 50 per cent of histamine in rat tissues, other than the skin, may not be located in mast cells.

In other species the histamine stores resisted the action of 48/80 to an even greater extent. For example, we found that in the cat most of the histamine was depleted from the skin but almost none from pylorus of stomach, duodenum, ileum, colon, liver, spleen and kidney, and only a minor part from lung, skeletal muscle, corpus of stomach and pancreas.

These results suggest that practically all tissues contain non-mast cell histamine.

SELECTIVE LABELING OF NONMAST CELL HISTAMINE

Efforts to reveal a general function for histamine have been hampered by the difficulty of distinguishing non-mast cell histamine from that in mast cells. Because of the slow turnover of histamine in these cells, we considered the possibility that tracer amounts of labeled amine might be selectively taken

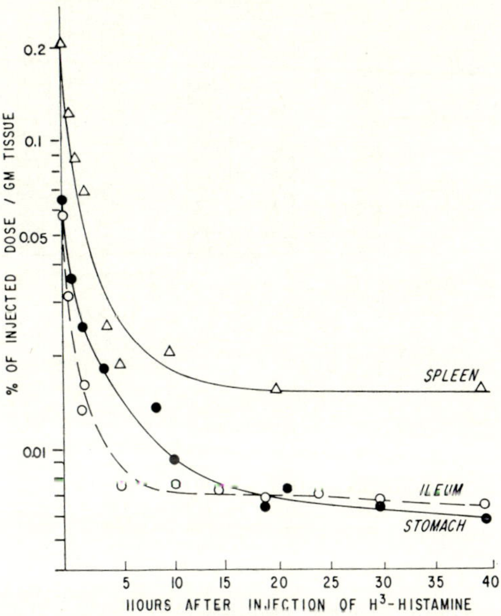

FIG. 1. Concentration of H^3-histamine in rat tissues at various times after the intravenous injection of the labeled amine (250 μc or 27.5 μg/kg). Each point represents the value of tissues pooled from 3 rats. Histamine was extracted from tissue homogenates by the method of Shore, Burkhalter and Cohn (1959) and assayed for H^3-histamine by scintillation counting.

up by histamine stores involved in physiological function. We had the opportunity to test this possibility when H^3-histamine of extremely high specific activity (1–10 curie/mM) became available.

After the intravenous injection of H^3-histamine (5 μg per rat) most of the labeled amine was excreted as metabolic products, but a sizeable portion was taken up by various tissues where its concentration exceeded that in blood. Figure 1 shows that the initial decline in tissue H^3-histamine in spleen, ileum, and gastric mucosa was rapid, then proceeded much more slowly and in a few hours became exponential with an apparent half-life of about 150 hr. The pattern of uptake and retention of H^3-histamine was similar in other body tissues.

In a number of tissues the butanol-extractable radioactivity present in 15 min and 4 hr was identified as histamine by paper chromatography, using the solvent system n-butanol : ethanol : concentrated NH_4OH (8 : 1 : 3). The results showed a single spot (R_f, 0·58), the same as that of authentic

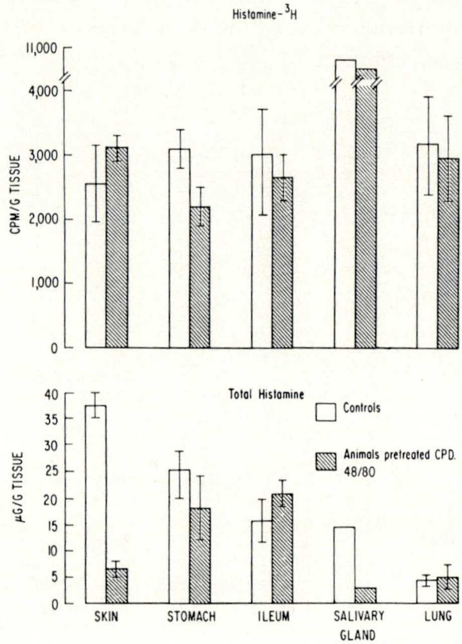

FIG. 2. Effect of pretreatment with compound 48/80 on the tissue levels of labeled and endogenous histamine. Rats were pretreated with a total of 8.5 mg/kg of compound 48/80, given in 5 divided doses over a period of 16 hr. Six hours after the last injection, the animals received H^3-histamine (500 μc/kg) and were killed 1 hr later. Control animals were given H^3-histamine alone. Open bars represent tissue levels of control animals, and striped bars of animals pretreated with 48/80. Each bar represents the mean from 5 animals with standard deviation. In case of salivary glands, tissues from 5 animals were pooled before assay.

H^3-histamine. The material was also identified by isotope dilution after forming the benzene sulfonate derivative of histamine (Schayer, Rothschild and Bizony, 1958).

Studies of rats pretreated with compound 48/80 showed that H^3-histamine was not taken up by mast cells. In these animals the incorporation of H^3-histamine was practically the same as that in normal rats, although compound 48/80 depleted a large portion of endogenous mast cell stores in skin and salivary glands (Fig. 2).

IS H^3-HISTAMINE INCORPORATED INTO ENDOGENOUS STORES OF NON-MAST CELL HISTAMINE?

Dr. Shore has demonstrated that reserpine and insulin lowered the histamine content in rat gastric mucosa by 30 per cent. This effect was associated with the parasympathetic effects of the drugs and was counteracted by atropine or by cutting the vagus (Kim and Shore, 1963). In our studies we showed that after the injection of cholinergic agents such as reserpine (5 mg/kg) and pilocarpine (25 mg/kg), labeled and unlabeled histamine in gastric mucosa and ileum were released to the same extent (25–30 per cent), and that the loss of histamine was prevented by pretreatment with atropine. These results suggested that H^3-histamine had been incorporated into the endogenous amine stores, and that the parasympathetic system is closely linked with these histamine stores.

At this stage of our experiments we discovered an error. We assumed that since the butanol-extractable radioactivity in tissues was largely histamine in 4 hr, this would hold for later times, for it was not possible to identify the label by paper chromatography because of the small number of counts. When the identity of extractable radioactivity in 10 and 20 hr was determined by isotope dilution, we discovered that it was no longer histamine. It was fortunate, however, that the error was not discovered early in our studies. We would have promptly dropped the problem since our minds were not prepared to accept the view that histamine stores could have such a rapid turnover rate. As it was, we went on to show a possible relationship between the parasympathetic system and histamine.

The radioactive material remaining in tissues after 10 hr had interesting properties. It was distributed throughout body water and like water was excreted very slowly (half-life about 150 hr). In fact it *was* water–tritiated water formed by exchange of a small proportion of the tritium from H^3-histamine with the hydrogen of water.

Figure 3 shows the revised levels of tissue H^3-histamine as measured by isotope dilution. The prolonged second phase of the decline is now absent and the half-life of H^3-histamine is 2.5 to 3 hr. These results indicate that non-mast cell stores of histamine are formed continuously and inactivated at a rather rapid rate, even when they are not utilized. The H^3-histamine was located in the cellular fraction of blood with little or none free in plasma.

UPTAKE AND DISTRIBUTION OF H^3-HISTAMINE IN CAT TISSUES

H^3-histamine injected into cats also was taken up by various tissues, the level being particularly high in salivary gland (Fig. 4). H^3-histamine was not released by compound 48/80 although endogenous stores in skin were depleted by 70 per cent. In a number of tissues the radioactivity was identified

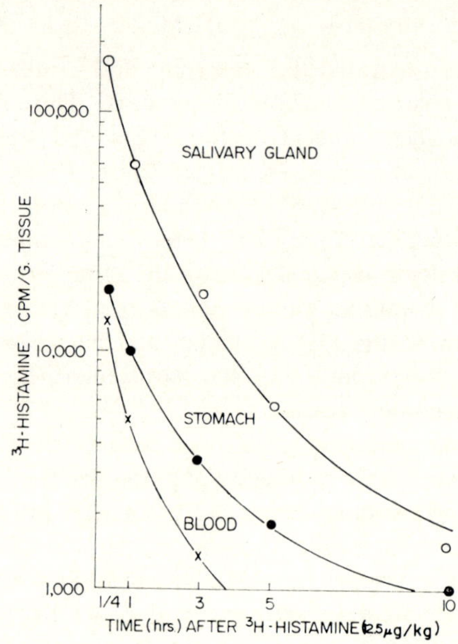

FIG. 3. Levels of H^3-histamine in rat submaxillary gland, stomach and blood at various times after the intravenous injection of the labeled amine (1 mc/kg). Each point represents the value of tissues pooled from 3 rats. The H^3-histamine was determined by the isotope dilution method described by Schayer, Rothschild and Bizony (1958).

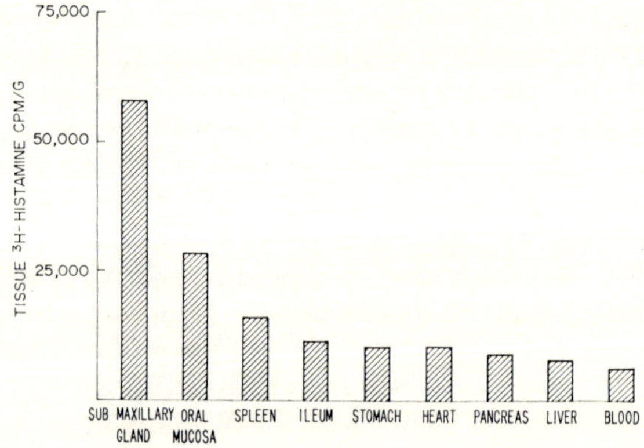

FIG. 4. Levels of H^3-histamine in cat tissues, 3 hr after the intravenous injection of the labeled histamine (100 μc/3 kg cat).

as histamine by paper chromatography and by isotope dilution (Schayer, Rothschild and Bizony, 1958).

The concentration of H^3-histamine in various tissues declined rapidly (half-life about 2 hr). After 24 hr, little or no H^3-histamine remained in tissues.

IS HISTAMINE IN SUBMAXILLARY GLAND OF CAT ASSOCIATED WITH SALIVARY SECRETION?

After giving compound 48/80, almost no histamine was released from the submaxillary gland of the cat indicating that this tissue contained few, if any, mast cells. For this reason we have carried out most of our studies with this tissue.

After the injection of H^3-histamine (5 μg ≡ 100 μc i.v. to cats weighing 2 to 3·5 kg) the submaxillary gland was stimulated electrically (chorda tympani) and saliva collected through a cannula inserted in the duct. Considerable amounts of radioactive material were released into the saliva as histamine metabolites.

These metabolites were not extractable by butanol at any pH. Preliminary results suggest that the material may be the riboside of imidazole acetic acid or of histamine. The nature of the inactivating enzyme is not known, but might be important in controlling the duration of response to parasympathetic stimulation.

The volume of saliva and the amount of radioactivity were determined after stimulation of the chorda tympani at various frequencies (Fig. 5). The salivary flow was proportional to frequency up to 5 stimuli/sec; at about 10 stimuli/sec, the flow was maximal; and at a higher frequency it declined, becoming virtually nil at 25 stimuli/sec. The mobilization of H^3-histamine from the gland followed an almost identical pattern. Up to 5 stimuli/sec, the amount of label released was proportional to frequency; at about 10 stimuli/sec the release was maximal and at a higher frequency it declined and was almost nil at 25 stimuli/sec. The amount of label released into the saliva was closely related to the volume of saliva.

The volume of saliva also paralleled the output of radioactivity after stimulation of the chorda tympani at various voltages. It may be concluded from these results that the amount of labeled histamine mobilized by para-sympathetic stimulation is closely related to the rate of salivary secretion.

After prolonged stimulation of the chorda tympani for several hours, the content of H^3-histamine in the stimulated gland was reduced to about 40 per cent of that of the control gland (Fig. 6). Despite the loss of H^3-histamine, the content of endogenous histamine remained constant, suggesting that the formation of histamine is closely coupled with its utilization.

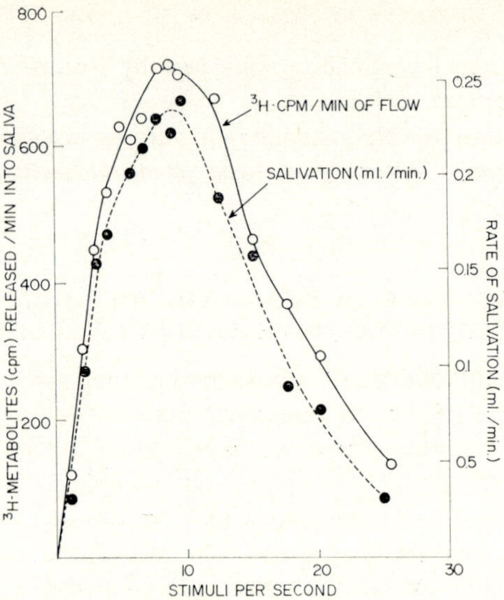

FIG. 5. The rate of salivation and output of H^3-histamine metabolites in saliva a different frequencies of nerve stimulation. H^3-histamine was injected intravenously into the cat 1 hr before the experiment. The saliva was collected through a cannula inserted into the duct of the submaxillary gland. Voltage of stimuli applied to the chorda tympani was supramaximal (10 V) and of 1 msec duration.

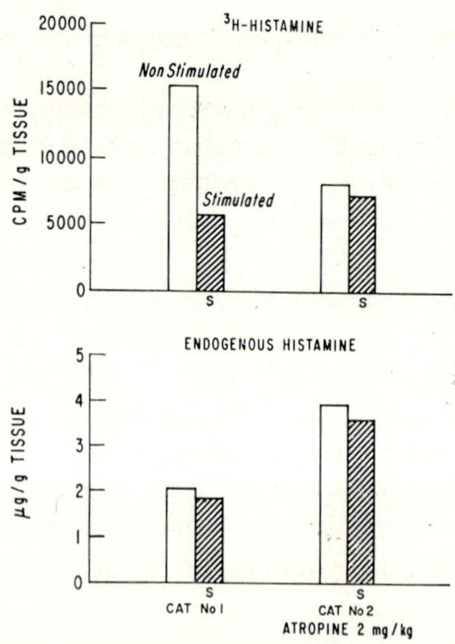

FIG. 6. The levels of histamine (μg/g) and H^3-histamine (cpm/g) in the cat submaxillary gland after maximal nerve stimulation (10 V, 10/sec at 1 msec) for 3 hr (striped bars). The histamine content of the nonstimulated gland is shown by the open bars. The second cat was pretreated with atropine and one gland stimulated for 3 hr (striped bars). The levels in the nonstimulated gland are shown by the open bars. H^3-histamine (100 μc/cat) was injected 30 min. before stimulating the glands.

Pilocarpine produced a rather dramatic effect (Fig. 7). A dose of 1 mg/kg increased the flow of saliva 3 times and released the label 9 times over that produced by nerve stimulation. As the response to the drug diminished, the output of H^3-metabolites and of saliva declined in parallel.

A dose of 50 μg/kg of pilocarpine elicited a maximal salivary response. The effects of this dose of pilocarpine could be repeated a number of times.

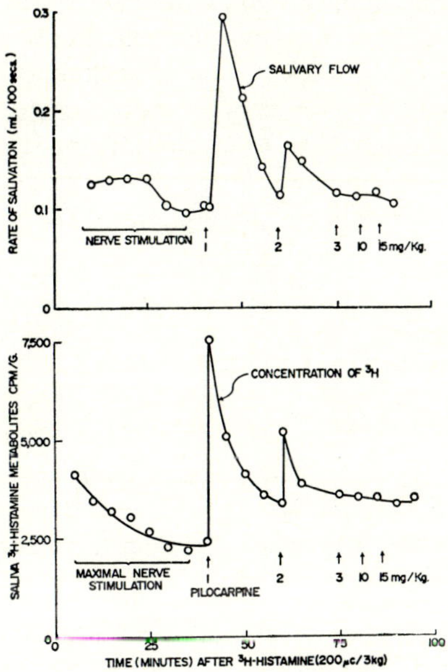

FIG. 7. The effect of successive doses of pilocarpine on salivary flow (upper graph) and on concentration of H^3-histamine metabolites released into saliva (lower graph). H^3-histamine was injected intravenously (100 μc/cat) and salivation elicited by nerve stimulation (10 V, 10/sec at 1 msec). The saliva was collected through the cannulated submaxillary duct. After 40 min, electrical stimulation was stopped and pilocarpine injected intravenously. Changes in flow of saliva were measured and concentration of H^3 determined by scintillation counting.

Pilocarpine increases the output of histamine to a far greater degree than that of saliva, suggesting that the drug might release histamine by a direct action. It may be significant that pilocarpine has an imidazole ring.

Larger doses of pilocarpine increased the duration but not the intensity of the response (Fig. 7). Thus, 1 mg/kg of drug depleted 50 per cent of H^3-histamine from the gland in a few minutes; 20 min later a larger dose of pilocarpine (2 mg/kg) produced much less salivation and release of H^3-histamine. Subsequent doses maintained a constant salivary secretion and release of histamine as though the gland were now working maximally at a reduced rate.

EFFECTS OF ATROPINE ON CHOLINERGIC RESPONSE AND ON TURNOVER OF HISTAMINE

Pretreatment of cats with atropine prevented both the salivation and the loss of H^3-histamine elicited by nerve stimulation or pilocarpine (Fig. 6). In addition, atropine reduced the loss of H^3-histamine from nonstimulated glands (in rats), suggesting that it blocks, in part, the normal turnover of the amine. Preliminary results also indicate that nerve stimulation increases the incorporation of H^3-histamine from H^3-histidine into histamine stores in the salivary gland and that this incorporation is reduced by atropine. These results are further confirmation of a close tie-up between the parasympathetic system and histamine in salivary secretion; in addition they suggest that the formation of histamine is closely related to its utilization.

CLOSE ARTERIAL INJECTION OF H^3-HISTAMINE

To measure the release of histamine into saliva and blood, we studied the effects of cholinergic stimulation on the labeled amine injected directly into the submaxillary gland by close arterial injection. With this preparation it was possible to measure venous and salivary blood flow as well as the amounts of label released into these fluids. In addition, the potential error due to entry of label from blood into saliva is reduced. Finally, the local response to very small doses of pilocarpine could be studied.

After the injection of H^3-histamine (50 ng or 3.5×10^6 cpm) 20 per cent or more of the amine was cleared by salivary gland. The radioactivity in the gland declined exponentially (half-life 40 min) and appeared in the venous blood; at 100 min, when the concentration of label in venous blood approached that in arterial circulation, a considerable concentration of H^3-histamine was still retained by the gland.

On stimulation of the chorda tympani, the concentration of radioactivity in saliva was several times that in arterial blood. This result clearly indicates that parasympathetic stimulation had mobilized H^3-histamine from the gland itself.

The injection of pilocarpine (12 μg) released a large fraction of the histamine from the gland, 50 per cent or more in less than 10 min. Increases in the flow of saliva and blood are accompanied by a large mobilization of radioactivity into saliva and venous blood, though that in saliva was larger. Pretreatment of animals with atropine blocked the salivation but did not prevent the increase in blood flow nor the enhanced release of H^3-histamine into blood. These results suggest that histamine might have a role in parasympathetic stimulation of blood flow through the salivary gland.

POSSIBLE ROLE OF HISTAMINE IN OTHER PARASYMPATHETIC END ORGANS

If histamine mediates parasympathetic control of salivary and gastric secretion, it might also be involved in lacrimation, pancreatic, biliary, and other exocrine secretions. Of particular interest in this regard are the effects of Histalog, a histamine analogue 3-(β-aminoethyl) pyrazole, which might prove a useful tool in histamine studies if it mimics all the actions of non-mast cell histamine. Histalog was found to provoke salivation, bile formation and pancreatic secretion. We do not know whether Histalog acts on all exocrine organs and whether it acts directly or by displacement of non-mast cell histamine.

THE EFFECT OF POLYPEPTIDES ON EXOCRINE GLANDS

The injection of gastrin into rats elicited an increase in gastric secretion and a decline of 25 per cent in the levels of labeled and endogenous histamine in gastric mucosa. No loss of histamine was observed in a number of other tissues. Similarly the intravenous injection of secretin produced a release of H^3-histamine into bile and pancreatic juice.

DISCUSSION

Our results indicate that considerable amounts of histamine are contained by cells of various body tissues, as well as by mast cells. Non-mast cell histamine differs from that in mast cells as follows: (1) It can be selectively labeled by H^3-histamine injected in tracer amounts. Other workers have reported that animal tissues are able to take up labeled histamine from the circulation, but the amine was given in nontracer amounts and no proof was submitted that the uptake was selective (Halpern *et al.*, 1959; Waton, 1964; Snyder *et al.*, 1964). (2) It is not released by compound 48/80. (3) Its rate of formation is singularly high, indicating that it has an important physiological role in the tissue cells where it is located. (4) The content of non-mast cell histamine is particularly high in exocrine glands where it is released by cholinergic stimulation; labeled and unlabeled histamine in rat stomach and cat submaxillary gland is lost in the same proportion, indicating that the H^3-histamine has completely mixed with endogenous stores.

At the normal steady-state, non-mast cell histamine is formed and inactivated continuously, with the metabolites diffusing into the blood stream. Because of its high turnover, nonmast cell histamine probably accounts for the bulk of the amine synthesized and metabolized in the body. Since much of this histamine is formed in exocrine glands, the fate of exogenous histamine might not give a true picture of endogenous amine released from normal sites.

Our experiments support the view that histamine triggers the production of saliva and other exocrine secretions. After cholinergic stimulation, labeled histamine in the submaxillary gland disappears and is found in saliva almost entirely as metabolites. Thus, without having labeled the histamine stores, the correlation between salivary flow and histamine would not be evident.

Pilocarpine may release histamine by two mechanisms: stimulation of the parasympathetic receptors and by a direct action on histamine stores. Our results show that large doses of drug deplete both labeled and unlabeled histamine from the salivary gland which then becomes unresponsive to nerve stimulation. These results suggest that the formation of saliva requires the presence of histamine stores. It is possible that pilocarpine may be a general depleting agent for histamine in exocrine glands.

An interesting aspect of non-mast cell histamine is the close link between utilization and synthesis. Release of histamine speeds up the formation of the amine whereas atropine, by decreasing utilization, slows up the formation. In fact, prior treatment with atropine reduces the incorporation of H^3-histamine from H^3-histidine into endogenous stores.

Histidine decarboxylase has been described as an "adaptive" enzyme, capable of changes in activity (Schayer, 1963). One might also speculate that histamine can inhibit histidine decarboxylase, thereby acting as a brake on its synthesis. When histamine content is lowered, the brake is released and synthesis enhanced. According to this view, the formation of histamine is a direct result of its utilization. This view would also explain why atropine decreases the rate of H^3-histamine turnover and in addition inhibits the conversion of H^3-histidine to H^3-histamine and why histamine stores are not depleted by nerve stimulation.

The effects of gastrin and secretin on other exocrine glands suggest that the actions of histamine are exerted locally in the very cells in which it is released. Thus gastrin-induced production of gastric juice and the secretin-induced secretion of pancreatic juice and bile, are associated with the release of histamine. Thus histamine is not a hormone in the strict sense, since it would not reach the receptors from the circulation. Possibly histamine, like cyclic AMP in sympathetic target cells, is a common mediator for secretion whatever the stimulus.

Of interest are the effects of drugs like Priscoline (which has an imidazoline structure) on non-mast cell histamine. This drug stimulates gastric, salivary, lacrimal, respiratory tract and pancreatic secretions, as well as sweating. In addition it produces excitation of the isolated gut (not blocked by atropine), coronary dilatation and other histamine-like effects. The drug also produces peripheral dilatation. It is tempting to postulate that in producing this effect, it acts through histamine stores that normally regulate blood flow in the small blood vessels. This would be in accord with the considerable amounts of non-mast cell histamine in the skeletal muscle of various mammalian species (about 1.5 to 2 μg/g).

We have also found that the drug Histalog (β-ethylamine pyrazole), claimed to be specific for gastric secretion, also provokes salivation, bile formation and pancreatic secretion.

SUMMARY

Injected H^3-histamine is taken up by histamine stores that resist depletion by compound 48/80. This "non-mast cell" histamine constitutes a large fraction of the body histamine and may play an important role in exocrine secretion since it is released from rat small intestine, gastric mucosa, and salivary glands by cholinergic agents. We also studied the uptake and release of H^3-histamine in cats. On parasympathetic stimulation, H^3-histamine was released from salivary gland and recovered in saliva as metabolites. Prolonged stimulation of the chorda tympani greatly reduced glandular levels of H^3-histamine without lowering endogenous levels, suggesting that synthesis of non-mast cell histamine is coupled to its loss. In addition, infused acetylcholine released H^3-histamine from the gland suggesting that non-mast cell histamine is released by acetylcholine liberated from nerve endings. Pilocarpine also produced a profuse flow of saliva and reduced glandular levels of histamine by 50 per cent. Profuse salivation was elicited by infusing histamine itself and by the histamine analogue, 3(β-ethylamine) pyrazole. Histamine may mediate other exocrine secretions since acetylcholine, pilocarpine and the polypeptide hormones gastrin and secretin released the labeled histamine into gastric juice, bile and pancreatic juice.

REFERENCES

ADAMS-RAY, J., A. DAHLSTRÖM, K. FUXE and N.-Å. HILLARP (1964) Mast cells and monoamines, *Experientia* (Basel) **20**, 80–6.

BURKHALTER, A., V. H. COHN and P. A. SHORE (1959) Studies on histamine in mammalian organs, *Fed. Proc.* **18**, 373.

EHRLICH, P. (1877) Beitrage zur Kenntnis der Anilinfarbungen und ihrer Verwendung in der mikroskopischen Technik, *Archiv. mikr-Anat.* **13**, 236–77.

HALPERN, B. N., T. NEVEU and C. W. M. WILSON (1959) The distribution and fate of radioactive histamine in the rat, *J. Physiol., Lond.* **147**, 437–49.

KIM, K. S. and P. A. SHORE (1963) Mechanism of action of reserpine and insulin on gastric amines and gastric acid secretions, and the effect of monoamine oxidase inhibition, *J. Pharmacol.* **141**, 321–5.

MOTA, I., W. T. BERALDO, A. G. FERRI and L. C. U. JUNQUEIRA (1956) Action of 48/80 on the mast cell population and histamine content of the wall of the gastro-intestinal tract of the rat. In: *Ciba Foundation Symposium, Histamine.* Boston, Little Brown & Co., pp. 47–50.

PATON, W. D. M. (1958) The release of histamine, *Progr. in Allergy*, vol. **5**. Basel, S. Karger, pp. 79–148.
SCHAYER, R. W. (1959) Histamine metabolism in the mammalian organism. In *Mechanism of hyper-sensitivity*. Henry Ford Hospital International Symposium. Boston, Little Brown & Co., pp. 227–33.
SCHAYER, R. W. (1963) Induced synthesis of histamine, microcirculatory regulation and the mechanism of action of the adrenal glucocorticoid hormones, *Progr. in Allergy*, vol. **7**. Basel, S. Karger, pp. 187–212.
SCHAYER, R. W., Z. ROTHSCHILD and P. BIZONY (1958) Increase in histidine decarboxylase activity of rat skin following treatment with compound 48/80, *Amer. J. Physiol.* **196**, 295–8.
SHORE, P. A., A. BURKHALTER and V. H. COHN (1959) A method for the fluorimetric assay of histamine in tissues, *J. Pharmacol.* **127**, 182–6.
SNYDER, S. H., J. AXELROD and H. BAUER (1964) The fate of C^{14}-histamine in animal tissues, *J. Pharmacol.* **144**, 373–9.
WATON, N. G. (1964) Uptake and formation of histamine by cat tissues, *J. Physiol., Lond.* **172**, 475–81.

DISCUSSION

GREEN: In a recent review on histamine and the nervous system (Green, J. P., *Fed. Proc.*, 1964) it was emphasized that the catabolites of histamine (and other amines) may produce physiological effects different from those produced by the amine but nonetheless important. Examples to support this idea have been presented in the aforementioned reference, and at these meetings Dr. Axelrod presented a further example. It is time we stopped using the term "inactivation" products, since we cannot conclude that these catabolites are inactive. We have seen that in the brain, histamine is found primarily in the microsomal fraction, showing a distribution different from that of 5-hydroxytryptamine (Carlini, E. A. and Green, J. P., *Brit. J. Pharmacol.*, 1963, **20**, 264–77; *Biochem. Pharmacol.*, 1963, **12**, 1448–9). These results are different from those obtained by others (Michaelson, J. A. and Dowe, G., *Biochem. Pharmacol.*, 1963, **12**, 949) who used the fluorimetric method for measuring histamine whereas we used a bioassay. It has been demonstrated by us (loc. cit.) and by H. Green and Erikson (*Int. J. Neuropharmacol.*, 1964) and by R. Levine, Sato and Sjoerdsma (personal communication) that the fluorimetric procedure is not specific when applied to brain, a view that is now held by Burkhalter (*Fed. Proc.*, 1964), one of the authors of the method. We have preliminary evidence suggesting that diamines may be a major source of the high values obtained by the fluorimetric method. The finding of Axelrod that intraventricularly injected histamine associates with the large particles of brain is not surprising, for the large particles can bind amines non-specifically (Robinson, J. D., Anderson, J. and Green, J. P., *J. Pharmacol.*, 1965, in press) and this distribution is not necessarily relevant to the distribution of endogenous histamine in brain.

RAŠKOVÁ: Dr. Brodie do you know the old work of Danielopalu concerning relations of histamine and the parasympathetic system?

BRODIE: I am glad that Dr. Rašková has so aptly brought to our attention the views of Danielopalu, a Roumanian who published in the 'thirties and 'forties, mainly in French journals. In addition to his thesis that acetylcholine was a primary agent in shock, he suggested a relationship between acetylcholine and histamine. For example, he pointed out that many symptoms of shock resemble those arising from parasympathetic stimulation and that increased histamine formation might be a result of hyperproduction of acetylcholine. These were deductions of his and I do not believe that they were supported by direct experimental evidence.

His emphasis was on acetylcholine rather than histamine, since he found that antigens release acetylcholine. He postulated that the liberated acetylcholine stimulated exocrine organs, giving rise to the symptoms of shock, and that free acetylcholine increased the production of histamine. According to this view, acetylcholine elicits the main symptoms of shock and histamine, the secondary and more persistent effects. As evidence, he cited that (1) atropine blocks many of the symptoms of shock and eserine exacerbates them; (2) caffeine also alleviates shock—he thinks by interfering with acetylcholine action; (3) hypotension, and the hypermobility of smooth muscle is induced both by acetylcholine and histamine.

AXELROD: Soon after the injection of C^{14}-histamine most of the radioactivity present in tissues is in the metabolites, imidazole acetic acid and its riboside (Snyder, Axelrod and Bauer, *J. Pharmacol.*, 1964; Robinson and Green, *Nature*). Is it possible that the radioactivity released by nerve stimulation or pilocarpine is in imidazole acetic acid and its riboside?

BRODIE: Dr. Axelrod has drawn my attention to one important point that I forgot to mention. Both radioactive histamine and metabolites are present in tissues after injection of H^3-histamine. However, only H^3-histamine is released from these tissues after cholinergic stimulation. For example, radioactive histamine metabolites are not released from cat salivary gland after stimulation of the chorda tympani.

WHITTAKER: In view of Dr. Brodie's account of peripheral non-mast cell histamine, it may be of interest to mention Dr. I. A. Michaelson's work on the subcellular distribution of non-mast cell histamine in brain. He found that total bound brain histamine was bimodally distributed in fractions and guinea-pig brain homogenates, just over half being recovered in the crude nuclear fraction and the remainder in the crude mitochondrial fraction. Experiments with homogenates tagged with mast cells and comparisons between two areas, pituitary, rich in mast cells and hypothalamus, poor in mast cells but rich in non-mast cell histamine, showed that the histamine in the crude nuclear fraction was mast cell histamine, and the histamine in the crude mitochondrial fraction was non-mast cell histamine. On subfractionating this fraction, the non-mast cell histamine was mainly recovered in the pinched-off nerve ending fraction.

AXELROD: I just want to add a comment to Dr. Whittaker's discussion. We have found that when H^3-histamine is introduced into the brain it has a somewhat different fate from that in the periphery (Snyder, Axelrod and Glowinski). First it persists in the brain for long periods of time and it is highly localized in the synaptosomal fraction in the brain.

ACETYLCHOLINE RELEASE FROM MOTOR NERVE ENDINGS

B. Katz and R. Miledi
Department of Biophysics, University College London

Some 12 years ago, I* had the privilege of contributing to a Symposium in Stockholm organized by Professors Zotterman and Schmiterlöw in their new laboratory. The subject of the conference was the mechanism of receptors, rather different from the present discussion which is concerned with the release of neural transmitter substances. A possible reason—or at least a reasonable excuse—for appearing on both occasions, and giving now an account which directly continues the earlier one, is that two different kinds of physiological problems happen to find a meeting place at the neuromuscular junction. The presynaptic release of acetylcholine has an obvious bearing on processes of neural and cellular secretion generally; while the depolarizing action of acetylcholine on the end-plate has an equally direct bearing on the mechanism of chemoreceptors and on the transducer effect at nerve endings in chemical sense organs.

At the previous Symposium, I described a peculiar phenomenon which Dr. Paul Fatt and I had come across, a spontaneous electric activity which is localized at motor end-plates (Fatt and Katz, 1952, 1953). We know now that this kind of subthreshold activity, consisting of a spontaneous random discharge of minute synaptic potentials goes on not only at skeletal nerve-muscle junctions of vertebrate animals, where it has been found in many different species, but also at other synapses where chemical transmission operates, be it cholinergic or not, in invertebrate muscles, in smooth muscle, in autonomic ganglia and in the spinal cord.

Since my previous report, much work has been done to clarify the origin and physiological significance of this phenomenon. The basic findings, however, have remained unaltered. At the motor end-plate, the miniature end-plate potentials are due to a spontaneous secretion of multi-molecular, that is quantal, packets of acetylcholine from random spots of the nerve endings. During nerve stimulation, the rate of this secretory activity is enormously increased, so that a very large number of the same parcels of acetylcholine is delivered at each junction within less than a millisecond. The spontaneous event represents the basic unit of the end-plate response

* The paper was presented by B. Katz.

which is made up of a statistical fusion of an integral number of such sub-cellular unit responses.

The evidence may be briefly recapitulated as follows (see del Castillo and Katz, 1956; Katz, 1958, 1962). The spontaneous potentials, recorded with an intracellular electrode, have the same localization and time course as the end-plate potential (e.p.p.) evoked by a nerve impulse, and both show the same pharmacological reactions as the electric change produced by locally applied acetylcholine. Curare diminishes, anti-cholinesterase drugs increase the size of all three effects. The spontaneous discharges differ from the normal end-plate potential in two respects, in their relatively minute size (approximately 0.5 as against 50 mV, therefore "miniature e.p.p.s"), and in their spontaneous, random appearance.

The origin of the individual potentials can be localized with great precision by probing along the outside of the muscle fibre with a micro-electrode. External miniature potentials are picked up only at focal points very close to the synaptic terminals. In practice, the micro-electrode has to be placed within less than 5–10 μ of a motor nerve ending. In recent experiments, Dr. Miledi and I have mapped out the distribution of such active spots along individual junctions in the frog's sartorius (Katz and Miledi, 1965a). As was verified by histological staining after the experiment, the landmarks presented by the external miniature potentials invariably lie along the non-myelinated endings of the motor nerve and the associated synaptic structures. If one records externally from 2 spots along the same terminal branch, spontaneous potentials are observed at each point, but if the electrodes are more than 15 μ apart, the two series of discharges are quite independent random events and individual potentials recorded at A do not coincide with those registered at B.

The pharmacological tests already mentioned strongly indicate that the miniature potentials are due to acetylcholine which reaches the end-plate receptors of the muscle fibre. The source of the ACh is on the outside of the muscle fibre, for it has been shown by ionophoretic micro-application that ACh is effective only if it is discharged externally, but not if it is released inside the muscle fibre.

That the miniature potentials are caused by release of ACh from the nerve endings, is shown by several findings. The frequency of the discharge can be controlled, rapidly and over a wide range, by passing electric current through the terminal part of the motor axon; the miniature potentials disappear after denervation at a time when the nerve endings degenerate; they are abolished by local botulinum poisoning at a time when the release of ACh stops. In denervated frog muscle, the disappearance of the spontaneous discharge is not permanent, but is followed one or two weeks later by a resumption of activity on a much reduced scale (Birks, Katz and Miledi, 1960). By contrast, in denervated mammalian muscle (rat diaphragm), a resumption of miniature potentials is only very rarely observed (Miledi and Slater, 1963). Electron-microscopic evidence suggests that the re-activation depends upon the development of close contact between the terminal Schwann cells and the end-plate. After nerve section, the terminal Schwann cells proliferate, engulf and subsequently absorb the remnants of the nerve terminals. It looks as if these satellite cells are capable not only of phagocytosis of the degenerating nerve ending, but of taking over, in this process, some of its biochemical and secretory properties.

If one compares the miniature end-plate potential with potentials produced by ionophoretic pulses of acetylcholine, the most obvious difference is that the latter are subject to continuous gradation in size as well as time course, depending on the intensity of the pulse and the proximity of the ACh pipette, while the miniature e.p.p.s are of a discrete size and standard time course. This observation effectively rules out the possibility that the spontaneous potentials could be produced by the impact of single molecules of acetylcholine. The continuous gradation of the applied ACh-potential shows that the effects of single molecules are undetectable by our present methods; a discrete effect of the size of the miniature potential must therefore be due to a large multimolecular packet of acetylcholine synchronously discharging from the terminal. The actual size of this quantal unit is still uncertain. An upper limit can be derived from the quantity of acetylcholine needed to produce an equivalent electrical change, under optimal conditions of close-range ionophoretic application. Dr. Miledi obtained a figure of 10^5 molecules in experiments of this kind (Miledi, 1961). Now it is clear that even under optimal conditions a micropipette cannot be placed as effectively against the receptive site of the membrane as is the axon terminal in the synaptic groove. The 100,000 molecules must therefore be greatly in excess of the actual quantal size, and this may well turn out to be no more than a few thousand molecules.

The rate of secretion of the ACh packets is controlled by the membrane potential of the nerve terminals: depolarization causes the frequency of the miniature e.p.p.s to increase. This effect can be produced by applying steady current to the nerve with the cathode near the myoneural junction, or by raising the external potassium concentration. It has been estimated that over a fairly wide range the average discharge rate increases 10-fold per 15 mV depolarization (Liley, 1956). The brief, but very intense depolarization which occurs during the presynaptic spike raises the frequency of the miniature potentials more than 10^5 times for about 1 msec and so produces the large e.p.p. which initiates the muscle impulse.

Depolarization is necessary, but not sufficient to bring about this enhancement of the rate of ACh-secretion. For depolarization to be effective, calcium ions are required. If the solution is depleted of calcium, or if an excess of magnesium is added, neuromuscular block ensues because the depolarization of the terminals by the impulse no longer raises the frequency of the miniature potentials sufficiently. The importance of the external calcium concentration is well illustrated in Fig. 1 from a recent study by Miledi and me (Katz and Miledi, 1965c). The nerve-muscle preparation was placed in a calcium-free medium to which about 1 mM Mg had been added

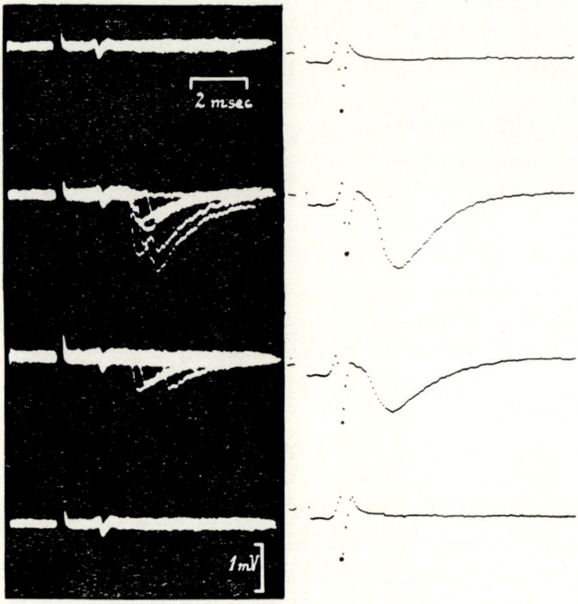

FIG. 1. Effect of calcium on neuro-muscular transmission. Focal external recording from a junctional spot of frog's sartorius. Muscle was immersed in Ca-free solution containing 0·84 mM Mg. An 0·5 M $CaCl_2$ pipette was used for recording; efflux of Ca was controlled electrophoretically. In the four records, starting from below, Ca-efflux was stopped initially by applying sufficient negative voltage to the pipette; the bias was then reduced in two steps, and finally re-applied. On the left, several superimposed traces are shown at each stage; on the right, automatically obtained average of 600 traces at each stage, using a C.A.T. digital computer (averaging in time steps of *ca.* 80 μsec). Negative peaks of averaged nerve spikes have been retouched. Time calibration applies to records on the left only. (From Katz and Miledi, 1965c.)

as a Ca-substitute. Spontaneous miniature e.p.p.s still occur. Also, nerve impulses still arrive in the motor terminals and can be recorded from focal spots with an external micro-electrode. But no ACh is released by the nerve impulse, and consequently the presynaptic spike appears alone, not followed by a postsynaptic response. If one now uses a $CaCl_2$-filled micro-pipette as a recording electrode and by electrophoretic control allows some calcium to diffuse from the pipette, quantal postsynaptic responses appear at once. By adjusting the efflux of calcium, the local release of ACh-packets by the impulse can be controlled and graded at will, from zero up to some 10 or 20 quantal packets per impulse. These changes can be brought about within fractions of a second, and without appreciable alteration in the size of the focal nerve spike. It is clear, therefore, that the calcium action is

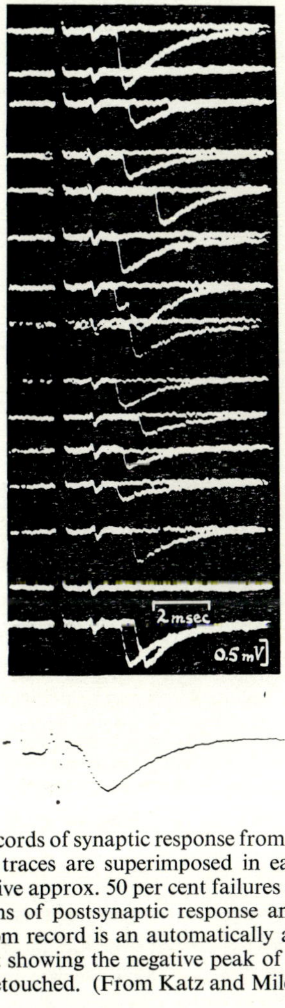

FIG. 2. Extracellular records of synaptic response from a junctional spot. Frog's sartorius, 17·5°C. Two traces are superimposed in each record. Bias on Ca-pipette was adjusted to give approx. 50 per cent failures of postsynaptic response. Note quantal fluctuations of postsynaptic response and variations of synaptic "unit delay". The bottom record is an automatically averaged response to 600 nerve impulses. The dot showing the negative peak of the averaged nerve spike has been retouched. (From Katz and Miledi, 1965b.)

specifically concerned with the electro-secretory coupling process, not—as one might have thought—with the size or spread of the impulse.

Finally, a word about the time relations between impulse and ACh-release. By the method of focal synaptic recording, we restrict our attention to a small portion of a single myoneural junction, effectively confined within a radius of 10 μ or less. By adjusting the release of ACh to a low level in the manner just described, we can determine not only the number of individual units which make up the end-plate response, but the synaptic delay, that is the time interval between the local arrival of the nerve impulse and the start of local postsynaptic current flow. Furthermore, we can measure the temporal dispersion of the release of ACh-packets after the arrival of the impulse. Figures 2 and 3 show that this dispersion is large; miniature potentials evoked by an impulse do not all appear at the same instant, some start as early as 0.5 msec, others as late as 2–3 msec after the impulse has

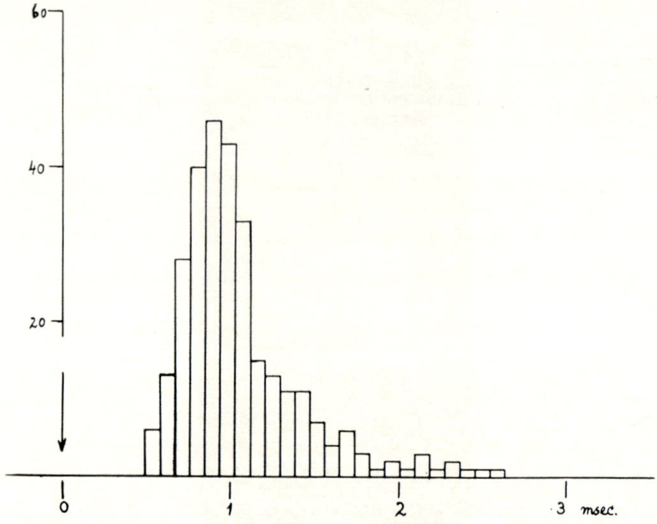

FIG. 3. Distribution of synaptic "unit delays" from experiment illustrated in Fig. 2. Zero time (marked by arrow) corresponds to negative peak of focal nerve spike. (From Katz and Miledi, 1965b.)

reached the synaptic spot. The statistical distribution of the synaptic delays gives us a good measure of the time span during which the transmitter is released.

The minimum latency is indicative of a finite delay between the presynaptic potential change and the moment at which secretion of ACh packets starts. It is true that some fraction of this minimum latency must be taken up by other processes such as diffusion of ACh across the synaptic cleft and first reaction steps at the receptor level. However, there are good reasons for believing that these events occupy much less time than the 0.5 msec in question (Katz and Miledi, 1965b) and that most of this initial delay is inherent in the electro-secretory coupling process.

There is little I want to add to the well-known hypothesis linking the quantal packets of ACh which presumably are pre-formed inside the motor terminals with the synaptic vesicles seen in the electron microscope. I regarded this as an attractive working hypothesis 10 years ago and, to me at any rate, it has retained its attraction. The work of de Robertis (1963) and of Whittaker (1964) and their colleagues has added considerable strength to this view, but to obtain final proof, one ought to look for a technique of histochemically localizing ACh, similar in directness to the remarkable methods which have been developed for the structural localization of the catecholamines.

REFERENCES

BIRKS, R., B. KATZ and R. MILEDI (1960) Physiological and structural changes at the amphibian myoneural junction, in the course of nerve degeneration, *J. Physiol., Lond.* 150, 145–68.

DEL CASTILLO, J. and B. KATZ (1956) Biophysical aspects of neuromuscular transmission, *Progr. Biophys.* **6**, 121–70.

DE ROBERTIS, E., G. RODRIGUEZ DE LORES ARNAIZ, L. SALGANICOFF, A. PELLEGRINO DE IRALDI and L. M. ZIEHER (1963) Isolation of synaptic vesicles and structural organization of the acetylcholine system within brain nerve endings, *J. Neurochem.* **10**, 225–35.

FATT, P. and B. KATZ (1952) Spontaneous subthreshold activity at motor nerve endings, *J. Physiol., Lond.* **117**, 109–28.

FATT, P. and B. KATZ (1953) Chemo-receptor activity at the motor end-plate, *Acta Physiol. Scand.* **29**, 117–25.

KATZ, B. (1958) Microphysiology of the neuromuscular junction, *Johns Hopk. Hosp. Bull.* **102**, 275–312.

KATZ, B. (1962) The transmission of impulses from nerve to muscle, and the subcellular unit of synaptic action, *Proc. Roy. Soc.* B **155**, 455–77.

KATZ, B. and R. MILEDI (1965a) Propagation of electric activity in motor nerve terminals, *Proc. Roy. Soc.* B **161**, 439–68.

KATZ, B. and R. MILEDI (1965b) The measurement of synaptic delay, and the time course of acetylcholine release at the neuromuscular junction, *Proc. Roy. Soc.* B **161**, 469–81.

KATZ, B. and R. MILEDI (1965c) The effect of calcium on acetylcholine release from motor nerve terminals, *Proc. Roy. Soc.* B **161**, 482–9.

LILEY, A. W. (1956) The effects of presynaptic polarization on the spontaneous activity of the mammalian neuromuscular junction, *J. Physiol., Lond.* **134**, 427–43.

MILEDI, R. (1961) From nerve to muscle, *Discovery* **22**, 444–50.

MILEDI, R. and C. R. SLATER (1963) A study of rat nerve-muscle junctions after degeneration of the nerve, *J. Physiol., Lond.* **169**, 23–4P.

DISCUSSION

GIACOBINI: I would like to ask Dr. Katz a "simple" question: how many molecules acetylcholine (ACh) are liberated per second in one single end-plate? The reason for my curiosity is that looking at the recent findings on the localization of AChE in the myoneural junction by electron microscopy (Barnett, 1961; Karnowsky, 1965) pointing to a localization of the enzyme both at the post- and at the pre-junctional membranes and within both primary and secondary clefts and within axonal terminals, it seems that a molecule of ACh has rather many chances to get hydrolyzed. If we look at some quantitative data that we obtained a few years ago (Giacobini, 1959), we can see that the activity of AChE varies in the different parts of the motoneurone and as compared to the end-plate and to the muscle. In fact we found that also the muscle fibres had some AChE activity (Giacobini and Holmstedt, 1960). However, looking at the figures which are reported in activity per unit volume we can see that the motor end-plate has a three times higher activity than the terminal part of the axon and a 7–14 times higher activity than the muscle. Of course this could be expected. However, when we calculated in terms of ACh hydrolyzed (per unit volume) we get figures in the order of magnitude of about $7\text{–}100 \times 10^{-13}$ g ACh/h/μ^3. These are the values in the muscle and in the end-plate respectively. The total amount of ACh hydrolyzed per hour from a single end-plate is in the order of 7×10^{-8} g ACh. If my figures are correct they represent both in the muscle, in the axon and in the motor end-plate a remarkable number of molecules hydrolyzed per hour. But before I am stressing the word "remarkable" I would like to get the figure from Dr. Katz of how many molecules are liberated per second in one single end-plate.

KATZ: It has been reported that in mammalian muscle a nerve impulse releases several million molecules of acetylcholine at each end-plate, within less than one millisecond. This figure is not accurate and still surrounded by some controversy, but one may use it as a tentative estimate of the rate of release during activity. From our own experiments, we can tell how many quantal packets of the transmitter substance are being released (a few hundred per impulse per end-plate), but as I have said, we do not know the number of molecules which make up each packet.

GREEN: We have presented evidence that the sciatic nerve of the rabbit contains a choline-ester different from acetylcholine. Before making this conclusion, we observed the well-known precautions that are necessary to avoid artifacts that attend work on choline esters. The unknown choline-ester differed from acetylcholine in being found mostly in the particulate fraction of nerve, in showing more resistance to antagonism by *d*-tubocurarine on the rectus abdominis muscle of the frog, and in migrating rapidly on paper chromatography (Carlini, E. A. and Green, J. P., *Science* **141**, 901–2, 1963; *Biochem. Pharmacol.* **12**, 1367–76, 1963).

No attempt was made to isolate this substance because at the same time we noted the presence of material in sciatic nerve (and, subsequently, in brain) which released acetylcholine from the guinea-pig ileum. (Green, J. P., Carlini, E. A. and Robinson, J. D., *Nature* **200**, 1108–9, 1963; Green, J. P. and Carlini, E. A., *J. Pharmacol. Exper. Therap.* **143**, 96–106, 1964.) Attempts by us and others to isolate the active material(s), which is a lipid, have been unsuccessful because of the extreme instability of the material. This substance, as can be deduced from the activity of impure extracts, is extremely potent, and it may function physiologically in the release of acetylcholine.

ACETYLCHOLINE RELEASE FROM THE BRAIN

J. F. MITCHELL

Department of Pharmacology, University of Cambridge
Downing Place, Cambridge, England

INTRODUCTION

The complex and relatively inaccessible nature of structures within the cerebral hemispheres has made it impossible to demonstrate the presence of cholinergic synapses directly. In recent years, however, considerable evidence has been accumulating from a variety of sources and techniques which suggests the presence of cholinergic networks within the brain. As well as the release of acetylcholine (ACh) from central nervous structures, this evidence includes the results of iontophoretic studies on single neurones, the staining of nervous pathways and cells for cholinesterase and the distribution of choline acetylase and ACh in the brain. When these factors are considered together, a picture begins to emerge of the role that cholinergic pathways might play in the brain, but the evidence required to reach firm conclusions is, as yet, incomplete.

This paper is concerned mainly with the spontaneous and evoked release of ACh from central structures, but this must be considered in relation to the important results obtained by other techniques where they are relevant to ACh liberating systems. Thus an attempt will be made to present an overall picture of our present concept of cholinergic systems in the brain, but inevitably an account of much of the important work in this field must be omitted.

METHODS

Considerable information about the release of ACh from the surface of the brain has been obtained using the collecting cup technique introduced by MacIntosh and Oborin (1953) and described in detail by Mitchell (1963). The various experimental procedures used (Mitchell, 1963) and the assay and controls required in these types of experiments have also been described (Mitchell, 1963; Szerb, 1963).

A study of ACh release from more localized areas of the cortex and from deeper structures of the brain was made possible by the introduction of the push–pull cannula (Fig. 1) by Gaddum (1961) and by assaying samples on a longitudinal strip of the dorsal muscle of the leech in a microbath (Szerb, 1961).

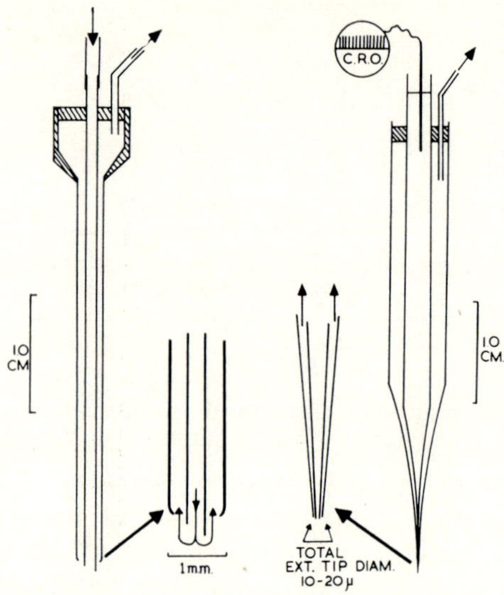

FIG. 1. *Left*. Push–pull cannula with an enlarged diagram of the tip and the direction of flow of the perfusing fluid. *Right*. Concentric barrel glass micropipette. The inner barrel (tip diameter 1–2 μ) is filled with electrolyte solution and is used for recording the electrical activity of cortical units. The outer barrel is filled with artificial cerebrospinal fluid and connected to the sucking device illustrated in Fig. 2.

The release of ACh from the region of single cortical neurones has been studied using concentric glass micropipettes; the original technique has been described briefly (Mitchell, 1964). The central barrel of the micropipette (Fig. 1) is used to record unit electrical activity while fluid is slowly withdrawn up the outer barrel. Several methods have been used to draw small

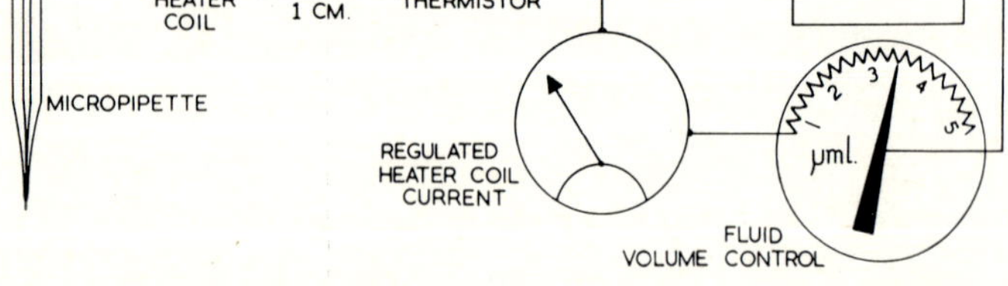

FIG. 2. The complete micropipette sucking system shown in diagrammatic form. Details of the electronic control system have been omitted.

volumes of fluid up the pipette (0·001 ml./min), but the most reliable system so far developed is shown in Fig. 2.

This device, which is being used primarily to withdraw and then eject very small quantities (2–10 μml.) of non-cholinergic transmitter from the region of spontaneously active cortical cells, consists basically of a small thermal chamber (volume 0·3 cm^3) containing a heater coil and a sensitive thermistor. The chamber is filled with fluid and connected to the outer pipette barrel. Changes in the current through the heater coil will produce rapid contraction or expansion of the surrounding fluid and so produce a small pumping or sucking effect at the pipette. The very rapid changes in temperature required and the maintenance of a given temperature for long periods is achieved by placing the thermistor inside the heating coils. This provides a fast responding feedback loop which allows the fluid temperature to change very rapidly but will anticipate the final temperature required and alter the heater current to prevent overshoot. This thermistor also controls, in a proportional manner, the fluid temperature to within $\pm$0·04°C of the required setting. The electronic feedback control for this system has been developed from a circuit described by Krnjević and Mitchell (1961).

SUMMARY OF RESULTS

ACh Release from the Cerebral Cortex

(*a*) *Collecting cup technique*. In the presence of eserine sulphate (10^{-4} g/ml.) ACh was found to be spontaneously released from the post-cruciate gyrus and the parietal, temporal and occipital lobes of anaesthetized sheep, cats and rabbits. The rate of release varied widely from one animal to another and with the depth of anaesthesia, but under conditions of medium anaesthesia was 1·0–5·5 ng ACh/min/cm^2 cortex. This rate of release was obtained after atropine (1 mg/kg i.p. or i.v.) or after atropine had been placed in the cup solutions (Szerb, 1964). Without atropine the spontaneous ACh release was reduced to 20–40 per cent of this level.

The resting release was similar using ether, cyclopropane, sodium pentobarbitone and allobarbitone, but was depressed below the level of detection (<0·3 ng/min/cm^2 cortex) by chloralose (60–80 mg/kg). The depth of anaesthesia affected the resting release of ACh (Fig. 3), the highest outputs being obtained during light anaesthesia or during mild convulsions induced by leptazol (120 mg/kg).

The cortical origin of the released ACh has been established by several methods including undercutting of the cortex, collection of ACh with push–pull cannulae and analysis of venous effluent from the cortex (MacIntosh and Oborin, 1953; Mitchell, 1963).

Direct electrical stimulation of the primary sensory cortex underlying the

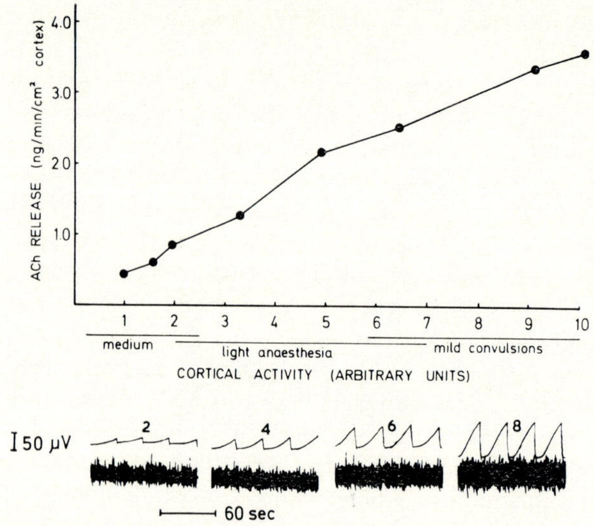

FIG. 3. The release of ACh from the parietal cortex of three sheep anaesthetized with cyclopropane. Each point represents the mean of six samples. The samples were obtained in pairs, one from each side of the cortex, from the three animals. The abscissa represents the degree of electrical activity of the cortex expressed in arbitrary units obtained from measurements of the integrated electroencephalogram records. These records and the electroencephalogram recorded simultaneously, are shown below. The figures above the integrated records refer to the arbitrary units assigned to them.

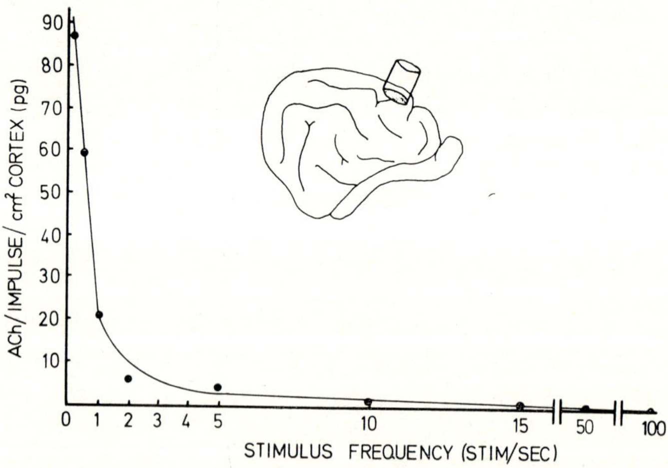

FIG. 4. The amount of ACh released/stimulus/cm^2 cortex from the primary somato–sensory cortex of 9 cats during stimulation of the contralateral forepaw at different frequencies.

collecting cups and transcallosal stimulation both caused an increase in ACh release but this was only 1·5–6·0 per cent of the increase found when a contralateral peripheral nerve was stimulated at low frequencies. The maximum release obtained from the somato-sensory cortex was 93 pg ACh/stimulus/cm^2 cortex when the nerve was stimulated at a frequency of 0·25 stim./sec. (Fig. 4).

Recent experiments by Szerb and Kanai (1965) have shown that stimulation of a peripheral nerve under conditions of light anaesthesia also increases the ACh release from the medial ectosylvian gyrus to a small extent. These authors have also found that stimulation of the tegmental reticular formation produces a marked rise in ACh release from both the ipsilateral post-cruciate gyrus and ectosylvian gyrus. Similar experiments by Phillis (personal communication) confirm that sensory volleys may have a widespread effect on ACh release from the cortex.

Recent experiments on the primary visual receiving areas show that the resting ACh output, which is lower than from the post-cruciate gyrus, may be increased up to eight times by stimulating the retina with light.

(*b*) *Push–pull cannulae.* Results obtained with push–pull cannulae inserted in different areas of the cortex confirmed those obtained with collecting cups, but a more precise localization of the site of ACh release was possible. The resting release from the post-cruciate gyrus under these experimental conditions was 10–40 ng ACh/ml. when the cannula tip was within 3 mm of the surface of the brain. At greater depths, when the cannula was in the white matter, no ACh could be detected ($<1{\cdot}0$ ng/ml.).

It was also found that not all areas of the occipital lobes of cats released ACh but that the ACh releasing structures appeared to be grouped together in patches of about 3 mm^2.

(*c*) *Micropipettes.* In preliminary experiments using this technique it has been possible to collect small volumes of fluid from spontaneously firing cortical cells. With eserine ($1{\cdot}0 \times 10^{-5}$ g/ml.) present in the outer barrel ACh has been collected from the region of these cells by sucking up 0·1 ml. of fluid over a period of about 100 min, the amount of ACh present in the recovered solution varying from 8–30 ng. ACh has only been recovered on six occasions out of twenty attempts with this technique, but always from regions of the cortex at least 1 mm below the surface. (Eight attempts within 1 mm of surface, 12 attempts at depths greater than 1·0 mm.)

ACh Release from the Cerebellar Cortex

Using collecting cups, ACh has been found to be released from the dorsal surface of the cerebellum in the presence of an anticholinesterase. The amounts collected were only 5–20 per cent of those collected simultaneously from the post-cruciate gyrus (Fig. 5), but the rate of release was affected in a similar way by changes in the depth of anaesthesia.

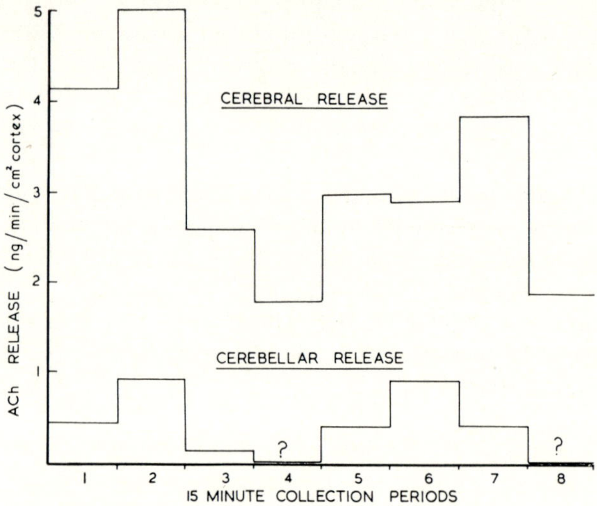

FIG. 5. The ACh released simultaneously from the cerebral and cerebellar cortex of a cat anaesthetized with ether. The depth of anaesthesia was varied during the course of the experiment to show the similar changes in the rate of ACh release from the two sites.

ACh Release from the Caudate Nucleus

Experiments were performed with push–pull cannulae inserted stereotaxically into the medial part of the caudate nucleus. Flow rates of 0·03–0·05 ml./min were used with an anticholinesterase present in the perfusion fluid. The resting rate of release (0·7–3·0 ng ACh/min) was considerably higher than that found in the cortex under similar conditions. Electrical stimulation of the contralateral and ipsilateral forepaws, the post-cruciate gyrus and a large part of the pre-cruciate gyrus produced no change in the resting output, but stimulation (1–10 stimuli/sec) of a small region of the cortex anterior to the primary somatic motor area produced an increase of 7–10 times the resting output (Figs. 6 and 7). Direct stimulation of the caudate nucleus with a wide range of stimulus frequencies evoked increases in ACh output up to five times the resting level (Szerb and Mitchell, 1962).

Similar experiments by McLennan (1964) showed a resting output of only 0·6 ng/min from the caudate nucleus probably due to the less favourable geometry of the tip of the push–pull cannula used. McLennan also failed to confirm that stimulation of the premotor area evoked an increase in ACh release from the nucleus, but it appears that this may be due to stimulation of the wrong area (McLennan, 1964). It was shown, however, that low frequency stimulation of the nucleus ventralis anterior of the thalamus increased ACh output from the caudate nucleus, this being attributed by McLennan to an indirect effect on the caudate nucleus via the frontal regions of the cortex.

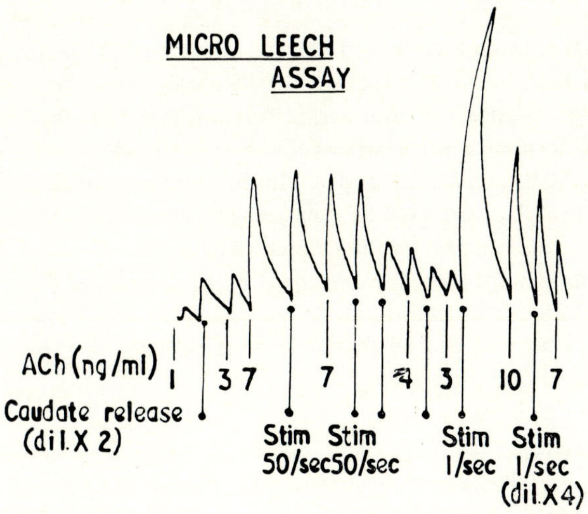

FIG. 6. The assay, against ACh, of four consecutive samples obtained from a push–pull cannula inserted into the medial part of the caudate nucleus. The samples obtained without stimulation contained 2–7 ng ACh/ml. During stimulation of the pre-motor cortex at 50/sec the output was 14 ng/ml. and during stimulation at 1/sec the output rose to about 34 ng/ml. Assays on the dorsal muscle of the leech in a microbath (Szerb and Mitchell, 1962).

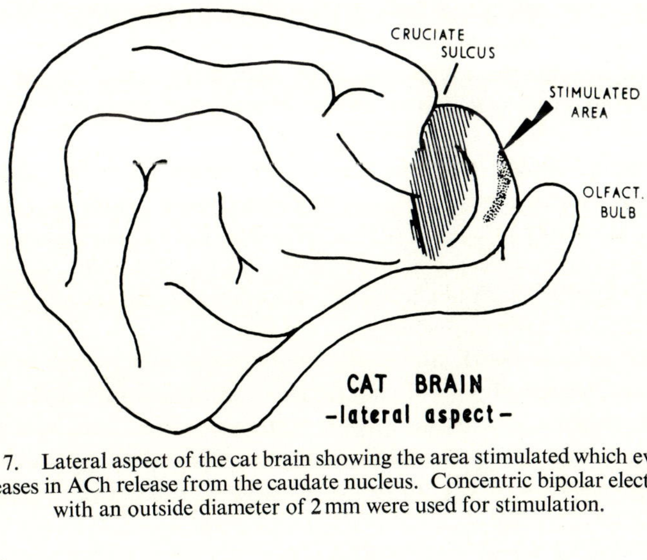

FIG. 7. Lateral aspect of the cat brain showing the area stimulated which evoked increases in ACh release from the caudate nucleus. Concentric bipolar electrodes with an outside diameter of 2 mm were used for stimulation.

DISCUSSION

Although the spontaneous and evoked release of ACh has been demonstrated from many central structures of the brain, this alone does not constitute strong evidence for cholinergic transmission. If, however, it can be supported by evidence for cholinoceptive cells in the same areas as those from which ACh is released and if cholinesterase-staining cells and tracts and choline acetylase are also associated with these sites, then the existence of cholinergic mechanisms becomes more likely.

It is particularly the evidence obtained by Krnjević and Phillis (1963a and b) using iontophoretic techniques to demonstrate the presence of cholinoceptive cells, that will be considered here in relation to the release of ACh from the central nervous system.

Cerebral Cortex

(*a*) *Location of cholinergic synapses.* The release of ACh and the detection of cholinoceptive neurones by iontophoretic techniques both suggest that cholinergic structures may be widely distributed over the surface of the brain. It has not been possible to show significantly different rates of ACh release from different areas of the cortex (post-cruciate gyrus, temporal lobe, occipital lobe) but iontophoretic studies suggest that the primary sensory receiving areas and the pre-cruciate gyrus contain most cholinoceptive cells and that fewer are found in the association areas (Krnjević and Phillis, 1963a). The collecting cup technique would be unlikely to detect these differences and no systematic study of ACh release from different areas has yet been made with more refined methods.

Both the iontophoretic application of ACh and a study of ACh release with micropipettes suggest that the cholinergic structures are not located less than 0·8–1 mm from the brain surface.

Grouping of cholinergic structures in the cortex has been suggested by iontophoresis and by push–pull cannula experiments in the primary visual area. The clustering of cholinoceptive cells detected by iontophoresis is on a very much smaller scale than that detected by push–pull cannula methods, but it does appear that the cholinergic structures are not evenly distributed over the areas of cortex in which they are present.

(*b*) *Effect of anaesthesia.* Both iontophoretic and ACh release experiments have shown that the activity of cholinergic structures in the cortex is vulnerable to anaesthetics. Perhaps the most remarkable effects have been observed using chloralose. This anaesthetic abolishes ACh release from the cortex and also depresses the excitability of cholinoceptive cells when the anaesthetic is injected intravenously (20–50 mg/kg) (Krnjević and Phillis, 1963b). Although chloralose does not appear to block directly the presynaptic release of ACh at Renshaw cells, it has been suggested by Biscoe and Krnjević (1963) that it depresses recurrent inhibition of motoneurones by blocking inhibitory synapses.

It might be reasonable to suppose that in the cortex, chloralose also does not directly block ACh release from nerve endings but acts indirectly by blocking cortical repetitive responses and reverberatory thalamo-cortical activity (Adrian, 1941). This action of chloralose provided the first suggestion that much of the ACh liberated by the cerebral cortex might be associated with these pathways. The presence of a non-specific cholinergic system of this type is supported by other experiments in which the reticular system has been stimulated (Szerb and Kanai, 1965).

The alterations in ACh release seen with changes in depth of anaesthesia are consistent with the idea that during increased synaptic activity more ACh is released from cell structures to be available for collection if an anticholinesterase is present. The experiments of Tobias, Lipton and Lepinat (1946), Richter and Crossland (1949) and Elliott, Swank and Henderson (1950), in which the total ACh content of the brain was shown to be directly related to the depth of anaesthesia, support this view.

(*c*) *Electrical stimulation.* Stimulation of peripheral nerves causes a rise in ACh output from the contralateral primary receiving area of the cortex and to a smaller extent, from at least the contralateral medial ectosylvian area of the cortex; an effect which may be mediated through the reticular formation. Direct stimulation of the mesencephalic reticular formation produces an increase in ACh release from both the ipsilateral primary receiving areas and ectosylvian gyrus. This lends support to the suggestion already made, that sensory volleys activate a specific pathway to the primary receiving area and also a more diffuse ascending cholinergic network to the cortex resulting in the widespread release of ACh which may be concerned with cortical "arousal". This view is corroborated by the work of Shute and Lewis (1963), Lewis, Shute and Silver (1964) and Krnjević and Silver (1962, 1963) who have shown that there is an ascending network of cholinesterase-staining fibres radiating from the reticular formation to wide areas of cortex.

Although these cholinergic networks may be widespread in the brain, evidence from ACh release experiments and from the study of cholinoceptive neurones make it clear that cholinergic synapses account for not more than about 20 per cent of the synapses present in the cortex.

Cerebellar Cortex

ACh collection techniques have established the liberation of small quantities of ACh from the surface of the cerebellum. The existence of cholinoceptive cells has been suggested by McCance and Phillis (1964) in the granular layers and cholinesterase-staining tracts have been shown to ascend through all three cerebellar peduncles to terminate mainly in the granular layers of the cortex (Shute and Lewis, 1965). Choline acetylase (Hebb, 1961) and ACh (MacIntosh, 1941) have also been found to be present in the cerebellum in small amounts. Thus, there is some evidence that cholinergic synapses may be present at least in the granular layers of the cerebellum.

Caudate Nucleus

The high content of ACh (MacIntosh, 1941), choline acetylase (Feldberg and Vogt, 1948; Hebb and Silver, 1956) and cholinesterase (Burgen and Chipman, 1951) in the caudate nucleus has always made this structure a likely site for cholinergic transmission.

Push–pull cannula techniques and the ventricular perfusion experiments of Beleslin, Carmichael and Feldberg (1964) confirm a high spontaneous output of ACh from this nucleus but, as yet, there is little information available on the nervous pathways associated with the release of ACh. Although two areas of the brain, the pre-motor cortex and the nucleus ventralis anterior of thalamus, appear to alter the resting output of ACh when stimulated, it would be premature to draw conclusions from these effects, especially since iontophoretic studies on this region of the brain have not yet been attempted.

General Conclusions

The evidence for cholinergic transmission in the brain has been briefly surveyed with special regard to regions where both ACh is found to be

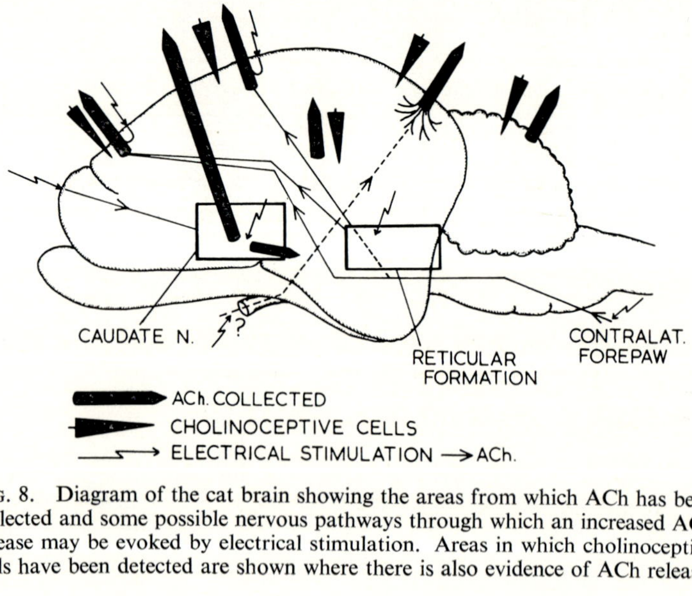

FIG. 8. Diagram of the cat brain showing the areas from which ACh has been collected and some possible nervous pathways through which an increased ACh release may be evoked by electrical stimulation. Areas in which cholinoceptive cells have been detected are shown where there is also evidence of ACh release.

released and where cholinoceptive cells are known to exist. A large amount of supporting evidence for central cholinergic systems has not been dealt with in any detail, especially the evidence concerning the distribution of choline acetylase, cholinesterase and acetylcholine, but when all this work is considered as a whole, the picture becomes more complete and few serious discrepancies arise. Direct evidence for the existence of cholinergic synapses in the brain cannot yet be obtained, but all the available evidence points to the existence of a widespread and non-specific ascending cholinergic network possibly concerned with cortical "arousal" and an associated but more specific system which may be activated by stimulation of particular sensory pathways.

It should be remembered that if a cholinergic pattern of this nature is confirmed, and this will require a considerable amount of further experimental work, it will still account for only a very small proportion of the total nervous pathways and synapses in the brain. The first real clues to the identity and even to the existence of true transmitter agents in central non-cholinergic pathways are still awaited.

SUMMARY

1. Using collecting cup techniques, push–pull cannulae and concentric glass micropipettes, the spontaneous and evoked release of ACh has been studied from areas of the cerebral and cerebellar cortex and from the caudate nucleus.
2. The rate of release of ACh from all regions studied is reduced by increasing the depth of anaesthesia and chloralose is found to abolish ACh release from the cortex.
3. The effect of stimulating peripheral nerves on ACh output from the primary receiving areas and other regions of the cortex is discussed.
4. The results of ACh release experiments are correlated with work demonstrating the presence of cholinoceptive cells in the brain, cholinesterase-staining tracts and cells and the distribution of choline acetylase.
5. It is concluded that, although the presence of cholinergic synapses within the brain cannot be directly demonstrated, there is now considerable evidence suggesting specific cholinergic pathways which may be activated by sensory stimulation and of associated non-specific networks of widely distributed ascending cholinergic fibres in the brain, perhaps concerned with cortical "arousal" responses.

REFERENCES

ADRIAN, E. D. (1941) Afferent discharges to the cerebral cortex from peripheral sense organs, *J. Physiol., Lond.* **100**, 159–91.

BELESLIN, D., E. A. CARMICHAEL and W. FELDBERG (1964) The origin of acetylcholine appearing in the effluent of perfused ventricles of the cat, *J. Physiol., Lond.* **173**, 368–76.

BISCOE, T. J. and K. KRNJEVIĆ (1963) Chloralose and the activity of Renshaw cells, *Exp. Neurol.* **8**, 395–405.

BURGEN, A. S. V. and L. M. CHIPMAN (1951) Cholinesterase and succinic dehydrogenase in the central nervous system of the dog, *J. Physiol., Lond.* **114**, 296–305.

ELLIOTT, K. A. C., R. L. SWANK and N. HENDERSON (1950) Effects of anaesthetics and convulsants on the acetylcholine content of brain, *Amer. J. Physiol.* **162**, 469–74.

FELDBERG, W. and M. VOGT (1948) Acetylcholine synthesis in different regions of the central nervous system, *J. Physiol., Lond.* **107**, 372–81.

GADDUM, J. H. (1961) Push–pull cannulae, *J. Physiol., Lond.* **155**, 1–2P.

HEBB, C. O. (1961) Cholinergic neurones in vertebrates, *Nature, Lond.* **192**, 527–9.

HEBB, C. O. and A. SILVER (1956) Choline acetylase in the central nervous system of man and some other mammals, *J. Physiol., Lond.* **134**, 718–28.

KRNJEVIĆ, K. and J. F. MITCHELL (1961) A simple and reliable device utilizing transistors for the maintenance of a constant body temperature, *J. Physiol., Lond.* **158**, 6–8P.

KRNJEVIĆ, K. and J. W. PHILLIS (1963a) Acetylcholine-sensitive cells in the cerebral cortex, *J. Physiol., Lond.* **166**, 296–327.

KRNJEVIĆ, K. and J. W. PHILLIS (1963b) Pharmacological properties of acetylcholine-sensitive cells in the cerebral cortex, *J. Physiol., Lond.* **166**, 328–50.

KRNJEVIĆ, K. and A. SILVER (1962) Cholinesterase staining in the cerebral cortex, *J. Physiol., Lond.* **165**, 3–4P.

KRNJEVIĆ, K. and A. SILVER (1963) The distribution of "cholinergic" fibres in the cerebral cortex, *J. Physiol., Lond.* **168**, 39–40P.

LEWIS, P. R., C. C. D. SHUTE and A. SILVER (1964) Confirmation from choline acetylase analyses of a massive cholinergic innervation to the hippocampus, *J. Physiol., Lond.* **172**, 9–10P.

MACINTOSH, F. C. (1941) The distribution of acetylcholine in the peripheral and the central nervous system, *J. Physiol., Lond.* **99**, 436–42.

MACINTOSH, F. C. and P. E. OBORIN (1953) Release of acetylcholine from intact cerebral cortex, Abstr. XIX Int. Physiol. Congr., pp. 580–1.

MCCANCE, I. and J. W. PHILLIS (1964) Action of acetylcholine on cells in cat cerebellar cortex, *Experientia* (Basel) **20**, 217–18.

MCLENNAN, H. (1964) The release of acetylcholine and of 3-hydroxytyramine from the caudate nucleus, *J. Physiol., Lond.* **174**, 152–60.

MITCHELL, J. F. (1963) The spontaneous and evoked release of acetylcholine from the cerebral cortex, *J. Physiol., Lond.* **165**, 98–116.

MITCHELL, J. F. (1964) The collection and reapplication of fluids at central synapses, *J. Physiol., Lond.* **171**, 23–4P.

RICHTER, D. and J. CROSSLAND (1949) Variation in acetylcholine content of the brain with physiological state, *Amer. J. Physiol.* **159**, 247–55.

SHUTE, C. C. D. and P. R. LEWIS (1963) Cholinesterase-containing systems of the brain of the rat, *Nature, Lond.* **199**, 1160–4.

SHUTE, C. C. D. and P. R. LEWIS (1965) Cholinesterase-containing pathways of the hind-brain; afferent cerebellar and centrifugal cochlear fibres, *Nature, Lond.* **205**, 242–6.

SZERB, J. C. (1961) The estimation of acetylcholine, using leech muscle in a microbath, *J. Physiol., Lond.* **158**, 8–9P.

SZERB, J. C. (1963) Nature of acetylcholine-like activity released from brain *in vivo*, *Nature, Lond.* **197**, 1016–17.

SZERB, J. C. (1964) The effect of tertiary and quaternary atropine on the cortical acetyl-choline output and on the electroencephalogram in cats, *Canad. J. Physiol. Pharmacol.* **42**, 303–14.

SZERB, J. C. and T. KANAI (1965) Mesencephalic reticular activating system and cortical acetylcholine output, *Nature, Lond.* **205**, 80–2.

SZERB, J. C. and J. F. MITCHELL (1962) The spontaneous and evoked release of acetylcholine from the caudate nucleus, Abstr. XXII Int. Physiol. Congr., p. 819.

TOBIAS, J. M., M. A. LIPTON and A. A. LEPINAT (1946) Effects of anaesthetics and convulsants on brain acetylcholine content, *Proc. Soc. Exp. Biol.* (N.Y.) **61**, 51–4.

DISCUSSION

GIACOBINI: 1. How can you exactly measure the release of ACh after very short periods of time (minutes or seconds).

2. Would you please comment on the mechanism by which ACh is diffusing out? How can it escape cholinesterase?

3. Which is the final evidence that ACh is the transmitter in these cerebral neurones? The iontophoretic experiments show that many compounds besides ACh may have a similar exciting effect.

MITCHELL: 1. The release of acetylcholine shown to occur over short intervals of time (minutes) are calculated from the total release obtained over 15–20 min. Assays were done on the leech muscle preparation in a small organ bath.

2. The cholinesterases are inhibited by eserine placed in the collecting fluid. The acetylcholine has been shown to be released by a process of simple diffusion into the collecting cup from the cortex.

3. There is as yet, no direct evidence that acetylcholine is a central transmitter. The substance released in our experiments is acetylcholine or a very similar choline-ester as indicated by routine alkali treatment, cross-assays and other tests on the collected samples. Szerb has also studied the active substance chromatographically and concluded that we are dealing with acetylcholine.

COSTA: I would like to know the evidence that allows one to relate quantitatively the integrated voltage of the EEG, the rate of neuronal discharge and the amount of acetylcholine released in a cup applied to the cerebral cortex.

MITCHELL: The integrated EEG voltage is not a good guide to neuronal activity, much better would be the recording of unit activity during changes in the depth of anaesthesia. The records of the integrated EEG shown were simply intended to indicate that a change in the depth of anaesthesia was occurring. It is however known, that as anaesthesia is deepened, so spontaneous unit activity in the cortex falls.

TREMORGENIC AGENTS AND BRAIN ACETYLCHOLINE

Bo Holmstedt and Gösta Lundgren

Department of Toxicology, Swedish Medical Research Council,
Karolinska Institutet, Stockholm 60, Sweden

INTRODUCTION

Tremorgenic agents have been used in the past, particularly for the animal screening of anti-Parkinson drugs. Excellent summaries of these procedures are available (Stumpf, 1962; Friedman and Everett, 1964). Although tremor is only one of the symptoms inherent in the clinical picture of Parkinson's disease, the tremorgenic agents have served a useful purpose. However, certain pitfalls exist. Thus, it has recently been pointed out, that some presumed central cholinolytics counteract the tremor by influencing the drug metabolizing enzymes of the liver (Sjöqvist and Gillette, 1964; Leslie and Maxwell, 1964). This concerns principally tremorgenic agents like tremorine, which is converted in the body into its active metabolite oxotremorine (Cho *et al.*, 1961 and 1964). Tolerance to tremorine develops with repeated injections (Decsi *et al.*, 1961). Also, an inhibitor of the liver microsomal enzyme activity like SKF 525 A prevents the tremor due to tremorine (Welch and Kocsis, 1961). By using tremorine and an equipment for quantitative recording of tremor such as the one described in this paper it is thus conceivable to study also the influence of various new drugs on the microsomal enzyme fraction of the liver.

However, the most important role of the tremorgenic agents can well be the light they may be able to throw upon transmission in the central nervous system. In this paper will be described the action of three tremorgenic compounds (Table 1). Their influence on acetylcholine and other brain amines and the relationship of this to intensity and duration of tremor has been studied. Based upon the results the mechanism of action of the compounds is discussed.

TABLE 1

NAME AND FORMULA	ORIGIN	DOSE
ARECOLINE METHYL- 1,2,5,6, TETRAHYDRO- METHYLNICOTINATE 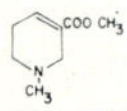	ARECA CATECHU BETEL NUT	5-75 mg/kg
TREMORINE 1,4-DIPYRROLIDINO-2-BUTYNE 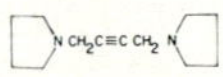	SYNTHETIC	5-20 mg/kg
OXOTREMORINE 1(2-OXOPYRROLIDINO)-4- PYRROLIDINO-2-BUTYNE 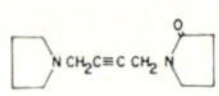	SYNTHETIC OR BY LIVER FROM TREMORINE	02-20 mg/kg

MATERIAL AND METHODS

Recording of Tremor

Principle. A rat is placed on a plastic air cushion contained in an isolation box. The movements of the rat are transferred from the air cushion via the air in a plastic tube to a transducer. The pressure changes are here converted into a low frequency a.c. voltage. This voltage is amplified and made to pass a bandpass filter which only lets through the frequency of the tremor (9–30 c/s). After filtration in the bandpass filter the signal is amplified, rectified and fed into a voltage to a frequency converter which gives positive pulses with a frequency proportional to the inlet voltage (0–1 V gives 0–2000 pulses/sec). The pulses are counted with a scaler, equipped with a printer control and a printer. With the help of an interval timer the number of pulses are recorded on the printer with 1 min intervals.

Recording apparatus (see block diagram Fig. 1).

Box of 2 mm polyethylene
Size 18×200×270 mm. The box is air tight and connected with a polyethylene tubing (inner diameter 8 mm) to a volumetric strain-gauge transducer (Grass PT-5).

Isolation box
Size 280×450×460 mm. The box is equipped with illumination, ventilation and facilities for observing the animal.

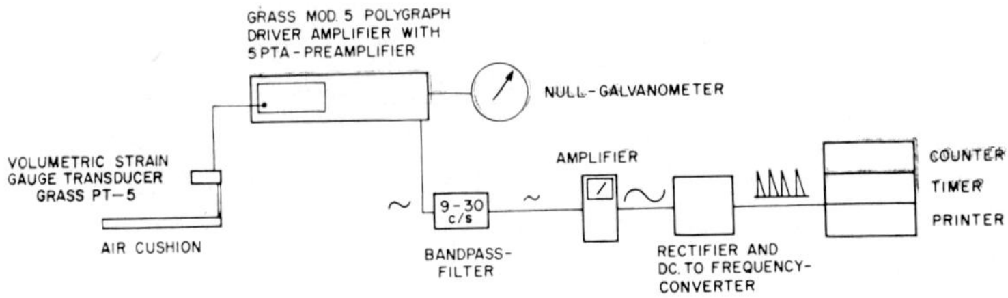

FIG. 1. Block diagram of equipment for recording of tremor. Explanation see text.

Amplifier

The Grass Polygraph d.c. Driver amplifier and low-level d.c. preamplifier equipped with nulling galvanometer to be used upon calibration of the amplifier and nulling of the measuring bridge.

Bandpass filter

The bandpass filter has changeable frequency intervals (6–35, 9–30, 12–22, 14–29 and 17–33 c/s).

Amplifier

Eico VTM 250 (Electronic Instr. Co. Inc. 22-00 Northern Blvd. L.I.C., New York, U.S.A.).

Rectifier and voltage to frequency converter

The equipment has been constructed to give positive pulses with a frequency proportional to the inlet voltage.

Scaler

Philips PW 4230 with supply PW 4210 and PW 4200.

Printer

Victor printer with control.

Timer

Industri AB Reflex. Type T3 with 4 sec latency.

Determination of Brain Acetylcholine

The whole brain from the rat was removed and wiped off and transferred to a 100 ml. wide-necked plastic bottle, containing ice-cold 0·4 N perchloric acid. After mincing with a pair of scissors additional perchloric acid was added to correspond to 3 ml. per g brain. The brain was then homogenized directly in the plastic bottle by means of an Ultra Turrax homogenizer. The homogenate was then left for 30 min in the refrigerator, whereupon it was centrifuged 15 min at $9000 \times g$ at $+2°C$. The supernatant was filtered into a measuring cylinder and the volume noted. The solution was then again transferred to a plastic centrifuge tube and left at −19°C overnight. Upon thawing, 5 N potassium carbonate was added for precipitation of the perchlorate. The addition was stopped at pH 6. To achieve a complete precipitation the test tube was then held at a temperature of +2° for 20 min, whereupon it was centrifuged 20 min at 15000 × *g* and at +2°C. The solution was then filtered into a 15 ml. graded test tube and the contents adjusted to pH 4 with 0·5 N HCl using bromocresol green as indicator. The solution was then stirred under vacuum to remove CO_2. After this procedure, the solution at pH 4 can be kept in the refrigerator without loss of activity for a considerable time or directly utilized for bioassay.

Acetylcholine was determined essentially according to Whittaker (1963) and MacIntosh and Perry (1950). Modifications of the recording technique have, however, been introduced. To prevent potentiation caused by CO_2 ions the amount of KCl in the frog ringer was decreased from 0·016 to 0·008 per cent. The sensitivity for acetylcholine is approximately 0·2 mμmol in a 3 ml. test bath.

Equipment and procedure. The frog rectus was mounted in a 3 ml. test bath equipped with sintered glass as a bottom and bubbled through with air. The upper end of the muscle was connected to a Grass FT-03 transducer without spring. (Grass Instr. Co., Quincy, Mass., U.S.A.) A Grass Polygraph d.c. Driver amplifier + Low-level d.c. preamplifier was used.

For recording, a Hi-Speed recorder HS/1, Control Instr. Ltd., Liverpool, England, was utilized. The transducer and organ bath were mounted on a stand disconnected from the test table to prevent disturbances.

The frog rectus muscle was stretched by a weight (2–4 g). After 30 min the weight was removed and the sensitivity of the amplifier was adjusted to give a suitable baseline for the recording. Frog ringer containing 2 mg eserine salicylate per litre was used for all testings.

Incubation experiments. The rat brain was homogenized in saline (0·4 g/ml.) in a 100 ml. plastic bottle. Eserine salicylate in final concentration 10^{-5} M was added from the beginning.

Oxotremorine, arecoline, atropine, etc., were added to the homogenate which was incubated with stirring in a water bath at +20°C for various times. The incubation was stopped by the addition of perchloric acid to a final molarity of 0·4, whereupon the incubate was placed in the refrigerator during 30 min. After centrifugation during 15 min at 9000 × *g* at +2°C, acetylcholine was determined as above.

Animals Used

White male rats of an inbred strain (Sprague–Dawley) weighing around 300 g were uniformly used for both tremor experiments and analysis of brain amines. All injections were given i.p.

Compounds Used

Tremorine base	Abbott Laboratories, Chicago
Oxotremorine base or oxalate	Synthesized by professor Richard Dahlbom, Kungl. Farmaceutiska institutet, Stockholm
Arecoline hydrobromide Pilocarpine hydrochloride	Commercially available
Atropine sulphate Pentobarbital sodium Atropine methylnitrate Eserine salicylate	Commercially available
F23A = Diethoxydimethylamino ethylthio phosphineoxide F37A = methylisopropoxydimethylamino ethylthio phosphineoxide	Synthesized by Dr. L. Larsson, Research Institute of National Defence, Dept. 1, Sundbyberg 4

RESULTS

Tremor Activity and Dose Response Curves

The frequency of the tremor produced by the chosen compounds was first recorded. Apparently the frequency which can be elicited by drugs in rats is within the same range as that described for Parkinson's disease in man. Jenkner and Ward (1953) give a frequency interval of 3–24 per sec. In mice Ahmed and Taylor (1959) studying drug-induced tremor found 14–18 per sec. Our own results in rats gave a frequency interval of 13–22 per sec and Chalmers and Yim (1962) have recorded 10 per sec in chronic spinal rats.

Quantitative investigations of tremor, including that induced by tremorine, have been made previously (Ahmed and Taylor, 1959; Triolo *et al.*, 1961). A variety of equipment has been described. For quantitative investigations an apparatus is needed which can record both qualitative and quantitative changes in the motility of the animal (Bonta *et al.*, 1960; Cutting *et al.*, 1959; Triolo *et al.*, 1961; Vernier and Unna, 1963).

The apparatus for recording of tremor described under Methods differs from previous equipment in the following way: A frequency filter eliminates all frequencies except those of the tremor and it records tremor activities in one minute intervals or less, so that time curves as well as dose curves can be recorded.

The experiments were carried out in such a way that the tremor recording apparatus was zeroed and the rat placed on the air cushion. The selective filter eliminates most of the animal's spontaneous movements but it is inevitable, that some inherent activity with the characteristics of the tremor

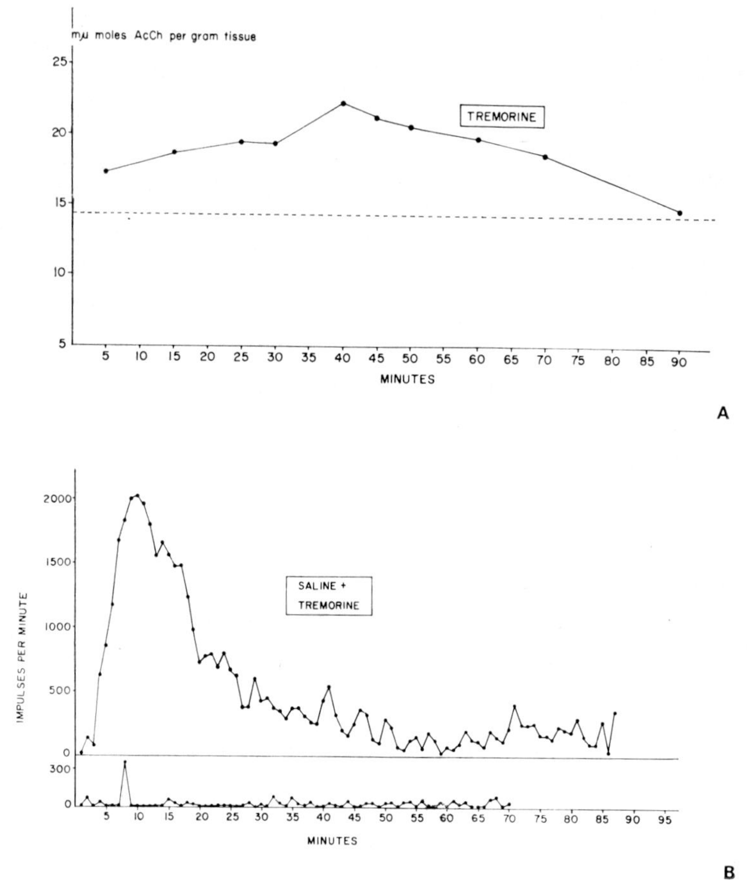

FIG. 2a. A. Time curve for rise in the brain acetylcholine of the rat after the injection of tremorine 20 mg/kg. B. Tremor curve for the corresponding dose.

frequency shows up as a background. This background activity is the same whether an injection of saline or methylatropine is used as pretreatment. The background activity was subtracted from the total tremor activity.

When tremor was recorded quantitatively with the aforementioned equipment it immediately became apparent, that the duration varied for the three agents used. Thus, tremorine (20 mg/kg) produced a tremor effect with a peak activity between 10 and 15 min and a duration of between 70 and 90 min.

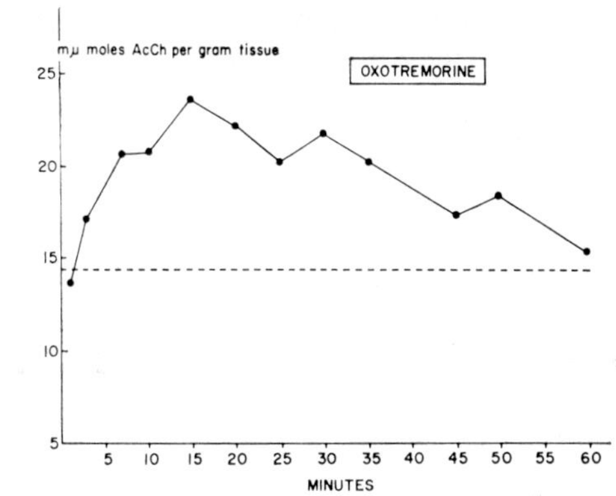

A

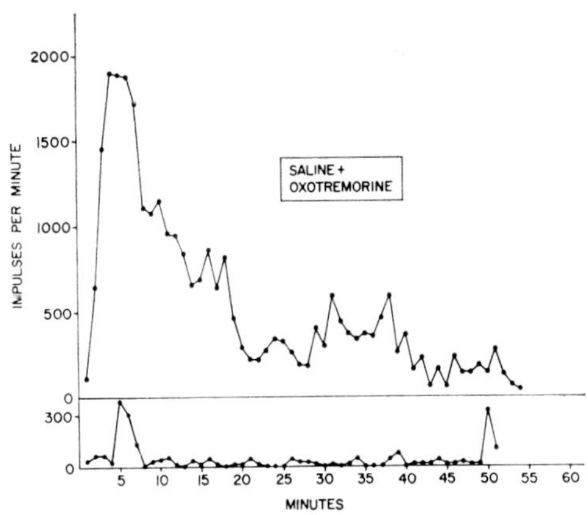

B

FIG. 2b. A. Time curve for rise in the brain acetylcholine of the rat after the injection of oxotremorine 2 mg/kg. B. Tremor curve for the corresponding dose.

The long duration of the tremorine effect is very likely due to its being slowly converted in the body to the active metabolite oxotremorine. (Fig. 2a).

When the latter compound was tested (2 mg/kg), it was found that the peak activity was reached already at 5 min, but the duration of the tremor activity lasted only about 60 min (Fig. 2b).

Arecoline on the other hand had its peak activity at 5 min and a duration scarcely over 15 min (Fig. 2c). When increased doses of arecoline were given to rats, a straight-line log dose-tremor activity curve could be plotted. This demonstrates the usefulness of the equipment for the quantitative recording of tremor (Fig. 3).

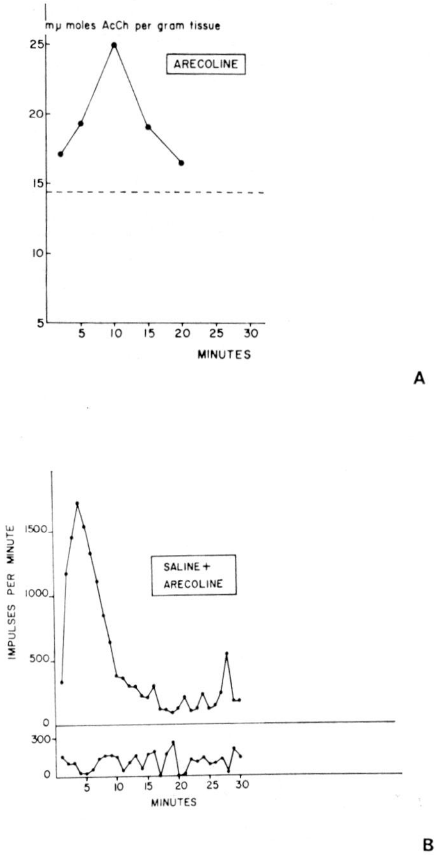

FIG. 2c. A. Time curve for rise in the brain acetylcholine of the rat after the injection of arecoline 25 mg/kg. B. Tremor curve for the corresponding dose.

In the experiments shown in Figs. 2a, 2b, 2c the backgrounds were obtained with the injection of saline. When instead of saline an injection of methylatropine was given, this did not significantly alter the background activity or the tremor curve (Fig. 4). In most experiments methylatropine 5 mg/kg was therefore used as a pretreatment 10 min before the other injections to abolish the peripheral parasympathomimetic symptoms and to obtain a tremor of purely central origin. When on the other hand atropine was given before the tremorgenic agents, no tremor but a typical background activity was obtained (Fig. 4).

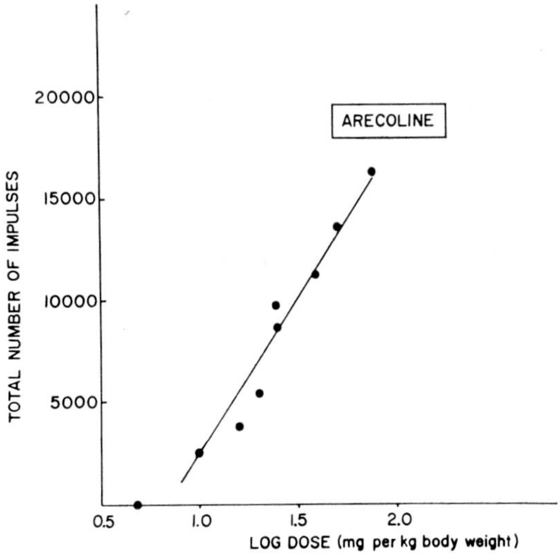

FIG. 3. Recording of tremor after injection of various doses of arecoline. Total number of impulses during tremor period counted. Dose range 5–75 mg/kg.

Influence on Brain Acetylcholine

Acetylcholine was the only brain amine which in our experience could be appreciably influenced by the tremorgenic agents during the period of tremor. The time of tremor activity also roughly corresponds to the rise in brain acetylcholine, as is seen from Figs 2a c. In all three instances the peak activity of the tremor occurs, however, before that of brain acetylcholine. This is most marked with the long acting compound tremorine which has been proven to be metabolized in the body (Fig. 2a). It is less evident with oxo-tremorine and arecoline. On the other hand, the rise in tremor activity is concomitant with the rise in brain acetylcholine.

Dose-response curves with regard to increase in brain acetylcholine were established for all three agents. Arecoline was most potent in this respect increasing brain acetylcholine to about 25 mμmoles per gram tissue (Fig. 5a). A maximal dose response was found for tremorine with 15 milligrams per kg body weight, for oxotremorine with 1·5 mg per kg body weight and for arecoline with 25 mg per kg body weight. Above this dosage no corresponding rise in brain acetylcholine could be achieved. In some instances even a slight decrease occurs at the highest dose level.

Atropine, Methylatropine and Brain Acetylcholine

Since atropine and methylatropine were used in a great number of experiments their effect on brain acetylcholine was studied in some detail. Figure 6 demonstrates that 5 mg per kg body weight of atropine sulphate decreases

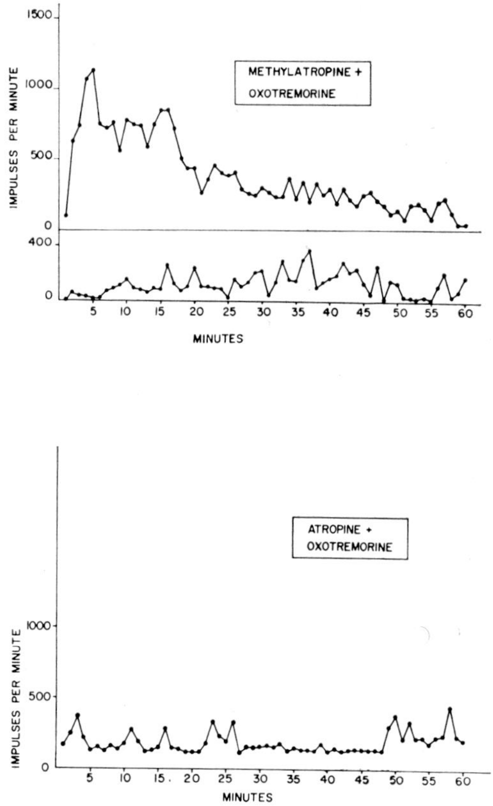

A

B

FIG. 4. Recording of tremor in the rat after the injection of oxotremorine 2 mg/kg. A. Pretreatment with methylatropine 5 mg/kg 60 min before. B. Pretreatment with atropine 100 mg/kg 10 min before.

brain acetylcholine to approximately 10 mμmoles per gram tissue. Increasing the dose further does not result in a progressive decline of brain acetylcholine and in this respect 5 mg of atropine are as effective as 100 mg.

When the highest dose 100 mg/kg was given to an animal and the time curve studied, it was found, that the decline in brain acetylcholine lasted for 40 min and probably longer. In subsequent experiments the highest dose, 100 mg/kg, was used to achieve complete atropinization.

The effect of pretreatment with atropine and methylatropine on the rise in brain acetylcholine produced by the tremorgenic agent oxotremorine is demonstrated in Fig. 7. Methylatropine affected the rise only little, if at all, in the dose given (5 mg/kg).

Atropine, on the other hand, completely prevented the rise caused by the

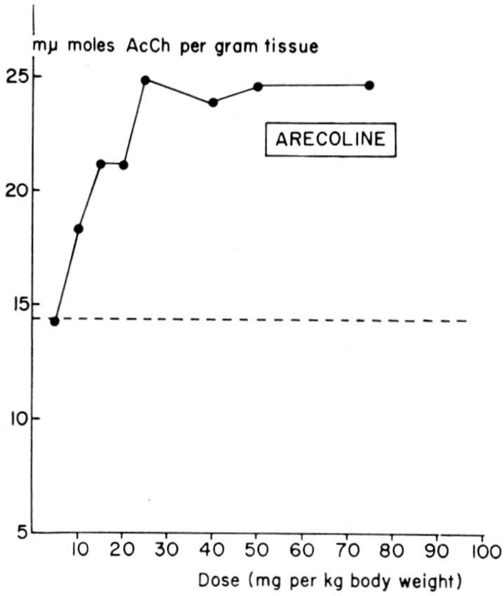

FIG. 5a. Rise in the brain acetylcholine of the rat after the injection of various doses of arecoline. Animals killed 10 min after the injection.

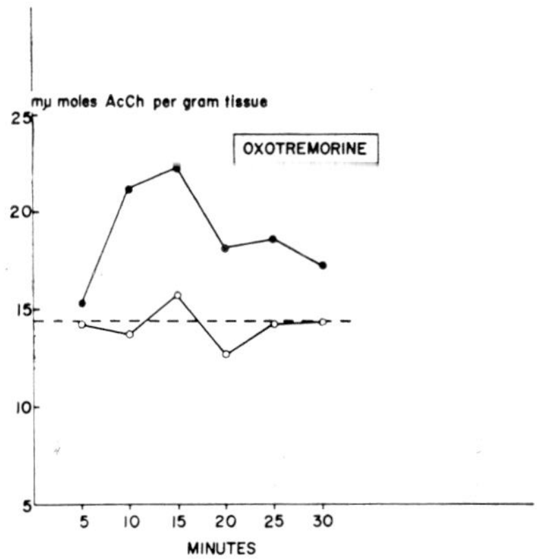

FIG. 5b. Incubation experiment. Effect of oxotremorine 10^{-5} M added to eserinized brain homogenate from rat (●——●). Control without oxotremorine (○——○).

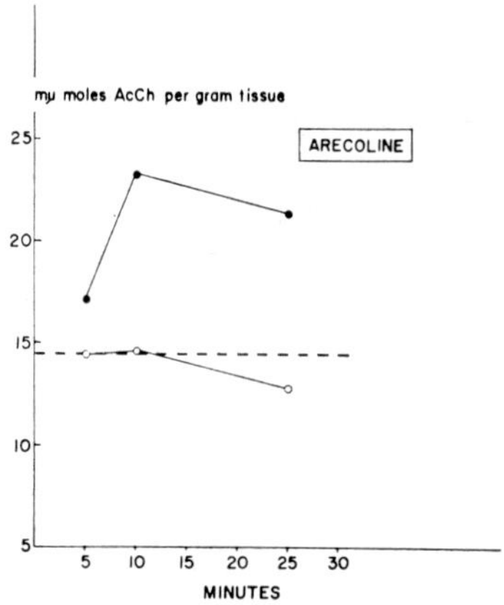

FIG. 5c. Incubation experiment. Effect of arecoline 10^{-4} M added to eserinized brain homogenate from rat (●——●). Control without arecoline (○——○).

tremorgenic agents and even caused a decrease in brain acetylcholine below that of the controls and to the same level as that of atropine given alone. Identical results were obtained with arecoline.

Its non-interference also in this respect corroborated the use of methylatropine throughout the experiments. The details of the analysis of the various combinations of tremorgenic agents with atropine or methylatropine are found in Table 2.

Incubation Experiments

A series of incubation experiments were carried out in which oxotremorine 10^{-5} M and arecoline 10^{-4} M were added to eserinized brain homogenates from rat. When the time curves were plotted, it was found that a considerable rise in brain acetylcholine had occurred also *in vitro*. Both compounds produced a rise above that of the controls to between 20 and 25 mμmoles acetylcholine per gram tissue. Even after 30 min the brain acetylcholine was still above normal (Figs. 5b and 5c).

Atropine in vitro *and* in vivo

Incubation experiments were also carried out with atropine added in various ways. The results are presented in Fig. 8 and Table 3. When atropine was added *in vitro*, prior to the addition of oxotremorine to the brain homogenate, this prevented the rise in acetylcholine which normally occurs with oxotremorine. Added *in vitro*, atropine did not decrease the normal value of acetylcholine in brain (Table 3, exps. 5 and 7). The pretreatment of the animal with atropine prevented the rise normally caused by the addition of oxotremorine to the homogenate. In this case atropine affected the normal value of acetylcholine in brain by causing a decrease (Table 3, exp. 2).

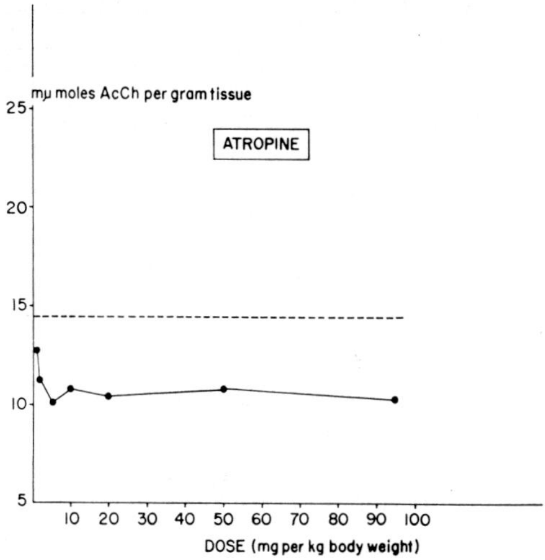

A

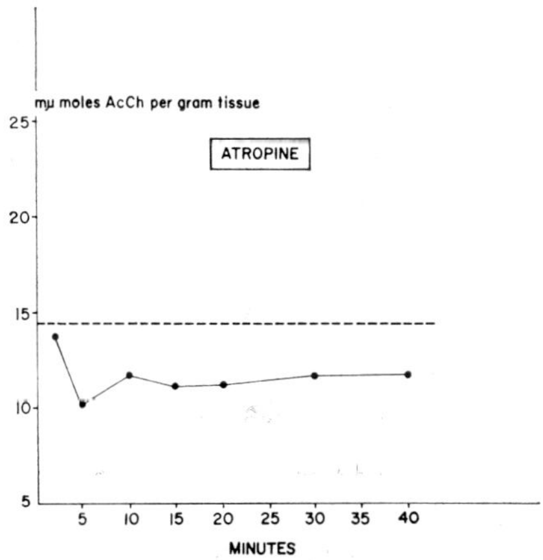

B

FIG. 6. The effect of various doses of atropine on brain acetylcholine of the rat. A. Dose curve. Animals killed after 25 min. B. Time curve after injection of 100 mg atropine per kg.

In related experiments two cholinesterase inhibitors of the organophosphorus type were used, F23A and F37A. These were injected into untreated animals, and animals treated with atropine. The atropine prevented the rise in

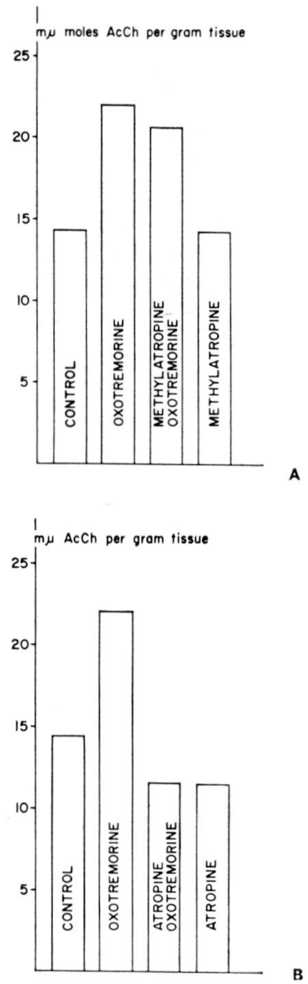

FIG. 7. Effect of pretreatment on the rise in brain acetylcholine caused by oxotremorine. A. Methylatropine 5 mg/kg given 10 min before. Animals killed 30 min after oxotremorine administration. B. Atropine 100 mg/kg given 10 min before. Animals killed 30 min after oxotremorine administration.

brain acetylcholine and even reduced the level below that of the controls to roughly in the region of the atropine alone (Fig. 9A).

In contrast to this, the rise in brain acetylcholine caused by the injection of a barbiturate (nembutal) could not be prevented by pretreatment with atropine (Fig. 9B).

TABLE 2

Exp. no.	Pretreatment	Test subst. after min.	Test substance	Determin. min after last injection	No. of tests	mμmoles AcCh per g tissue M±SD	Significantly different from (P < 0·001)	Not significantly different from (P > 0·1)
1	—	—	saline		56	14·39±1·28		
2	methylatropine 5 mg/kg			40	5	14·27±1·73		1
3	methylatropine 5 mg/kg	10	oxotremorine 2 mg/kg	30	5	20·69±2·00	1, 2	4
4			oxotremorine 2 mg/kg	30	10	22·08±2·26	1, 2	3
5	atropine 100 mg/kg	10	oxotremorine 2 mg/kg	30	9	11·57±1·95	1, 4	
6	atropine 100 mg/kg			40	8	11·56±1·36	1	
7	atropine 25 mg/kg			20	11	10·77±1·44	1, 10	8
8	atropine 100 mg/kg			20	10	11·06±1·11		7
9			arecoline 25 mg/kg	10	4	24·90±1·58	1, 13	
10	atropine 25 mg/kg	10	arecoline 25 mg/kg	10	4	13·35±0·64	7	
11	atropine 100 mg/kg	10	arecoline 25 mg/kg	10	8	13·12±0·85	1	
12	methylatropine 5 mg/kg			20	5	13·45±0·55		7
13	methylatropine 5 mg/kg	10	arecoline 25 mg/kg	10	4	23·73±1·03		9

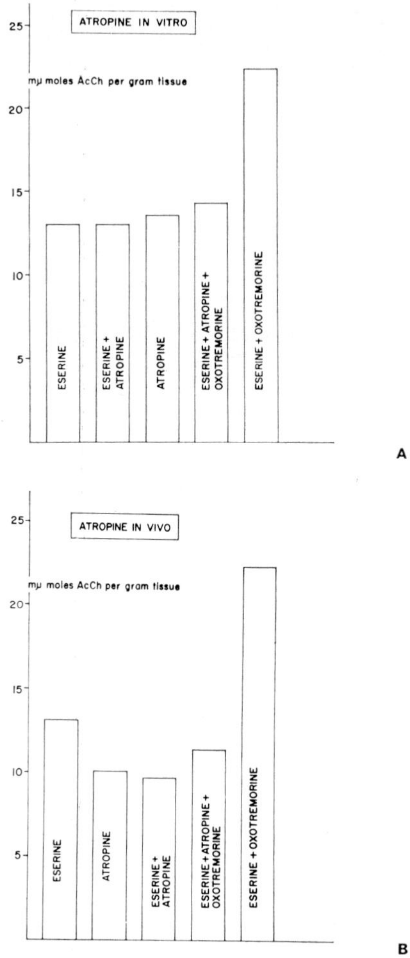

FIG. 8. Incubation of brain homogenate. Ordinate brain acetylcholine. A. Atropine added *in vitro*. B. Animals pretreated with atropine.

Pilocarpine

Pilocarpine hydrochloride in doses as high as 60 mg/kg produced no tremor that could be recorded. The animals were pretreated with methylatropine 5 mg/kg. After approximately 60 min they regained a noticeable salivation indicating that the protective effect of methylatropine was wearing off but that the action of pilocarpine persisted. No rise in brain acetylcholine comparable to that of the tremorgenic agents could be observed with pilocarpine.

TABLE 3 INCUBATION EXPERIMENTS. INCUBATION TIME 15 MIN AT 20° C.

Exp. no.	*Pretreatment* in vivo	*Killed after min.*	*Substances added Molarity in homogenate*	*mμmoles AcCh/g tissue M±SD*	*No. of tests*	*Significantly different from*	*Not significantly different from*
1	atropine 100 mg/kg	10	eserine 10^{-5} M oxotremorine 10^{-5} M	9·65±2·14	36	7 0·01 > P > 0·001 8 P < 0·001 4 0·01 > P > 0·001	2 P > 0·1
2	atropine 100 mg/kg	10	eserine 10^{-5} M	9·43±2·20	35	5 0·01 > P > 0·001 7 0·01 > P > 0·001	1 P > 0·1 3 P > 0·1
3	atropine 100 mg/kg	10		9·47±1·38	12	6 0·05 > P > 0·02 7 0·05 > P > 0·02	2 P > 0·1
4			eserine 10^{-5} M atropine 4×10^{-5} M (0·1 mg/g tissue) oxotremorine 10^{-5} M	11·80±2·89	23	1 0·01 > P > 0·001 8 P < 0·001	7 P > 0·1
5			eserine 10^{-5} M atropine 4×10^{-5} M (0·1 mg/g tissue)	11·35±2·59	30	2 0·01 > P > 0·001	7 P > 0·1 6 P > 0·1
6			atropine 4×10^{-5} M (0·1 mg/g tissue)	11·60±2·59	11	3 0·05 > P > 0·02	7 P > 0·1 5 P > 0·1
7			eserine 10^{-5} M	11·22±2·71	45	1 0·01 > P > 0·001 2 0·01 > P > 0·001 3 0·05 > P > 0·02	5 P > 0·1 4 P > 0·1 6 P > 0·1
8			eserine 10^{-5} M oxotremorine 10^{-5} M	20·95±3·36	13	1 P < 0·001 4 P < 0·001	

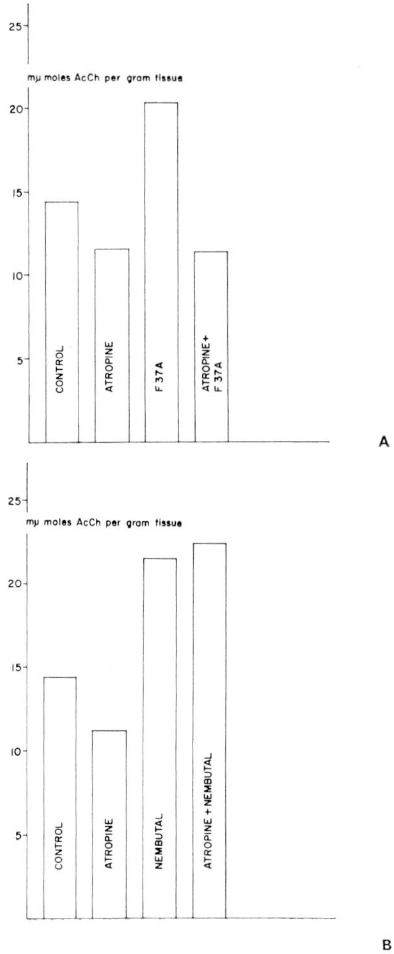

FIG. 9. Pretreatment with atropine. 100 mg/kg. Ordinate brain acetylcholine. A. Injection with cholinesterase inhibitor F37 A. B. Injection of nembutal 50 mg/kg 5 min after atropine 100 mg/kg.

Cholinesterase Inhibitors, Oxotremorine and Atropine

When a cholinesterase inhibitor like F37A is given at increasing doses to rats the resulting enzyme-inhibition is accompanied by an increase in total brain-acetylcholine. At near lethal doses of the organophosphate, brain acetylcholine doubles. Whether the rise is 50 or a 100 per cent, pretreatment with atropine will normalize the brain acetylcholine content. At the lower dosage level the inhibitor can be combined with oxotremorine without lethal results to the animal. An experiment of this kind is demonstrated in Fig. 10.

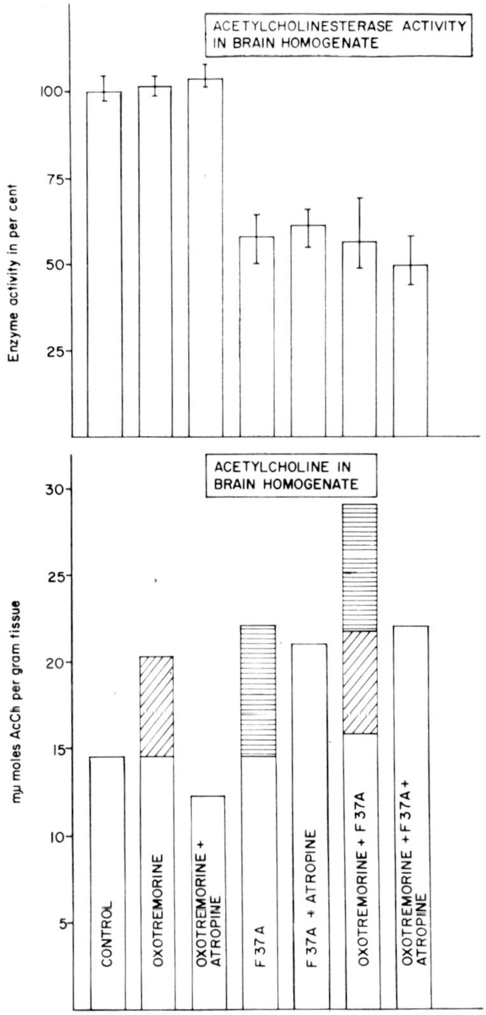

FIG. 10. Acetylcholinesterase activity and acetylcholine level in brain homogenate from rat, treated with different combinations of oxotremorine (2 mg/kg), a cholinesterase inhibitor (F37 A) (100 μg/kg) and atropine (100 mg/kg). Animals sacrificed after various times. (1) Control; (2) Oxotremorine 35′; (3) Oxotremorine 25′+atropine 10′; (4) F37 A 25′; (5) F37 A 15′+atropine 10′; (6) Oxotremorine 10′+F37 A 15′; (7) Oxotremorine 10′+F37 A 15′+atropine 10′.

The upper part of the figure shows that only when F37A has been given is there a decrease in enzyme activity. Oxotremorine and F37A both give a rise in brain acetylcholine which in the former case can be counteracted by atropine given afterwards. When both inhibitor and oxotremorine are given at the same time the result is a doubling of brain acetylcholine. In such animals one component in the rise, in all probability that due to oxotremorine, can be abolished by giving atropine to the animals after the two other drugs.

DISCUSSION

At the time when the experiments described in this paper were undertaken, strangely enough no report had appeared on the effect of the direct acting tremorgenic agents on brain acetylcholine despite the fact that they are known to produce pronounced parasympathomimetic effects. Experiments were therefore initiated to study in detail the effect on brain acetylcholine (Holmstedt, Lundgren and Sundwall, 1963). A rise in brain acetylcholine was discovered and has since been confirmed by other authors with regard to tremorine (Pepeu, 1963).

In our experiments the rise in brain acetylcholine is a constant phenomenon during the tremor caused by tremorine, oxotremorine, and arecoline. As evident from Figs. 2a–c this rise lasts approximately for the duration of the tremor. The peak rise, however, does not coincide with but is preceded by the peak in tremor activity. This is more evident for a compound like tremorine which is known to be comparatively slowly metabolized than for oxotremorine and arecoline. *Atropine given either before or after* the tremorgenic agents abolishes tremor and gives on the whole back normal values of brain acetylcholine except perhaps when it is given before in which case the brain acetylcholine value may be below normal and more like that resulting from atropine given alone. Methylatropine neither influences tremor nor brain acetylcholine level appreciably. On the whole, the rise in brain acetylcholine seems to give a better correlation with tremor as assessed with our equipment than any other brain amine studied.

Previous studies of the effect of tremorgenic agents have included analysis of some brain amines. Only tremorine has been used, however. It has been reported to have a delayed effect on increasing the brain level of histamine. (Ungar and Witten, 1963). This effect occurs only after the active tremor period. Friedman has analyzed catecholamines and serotonin in the brain stem and other brain areas upon injection of tremorine. In the rat, there was a significant decrease of the concentration of norepinephrine in the brain stem which was maximal in 30 min. Serotonin rose to highly significant values in 90 min (Friedman *et al.*, 1963). The author states that tremor activity in the rat begins within 15 min after intraperitoneal administration of tremorine, rises to a maximum within an hour and then disappears slowly over a period of several hours. In our own experiments when tremor was recorded quantitatively versus time (Fig. 2a), it will be seen that tremor activity starts within 5 min, reaches its peak activity between 10 and 15 min and then rapidly declines to reach normal values in about 90 min. On the other hand it would seem as though the rise in brain serotonin recorded by Friedman has a curve which is the reverse of that of the tremor. Concerning the rise of norepinephrine in the brain stem, there seems to be no direct relationship to the tremor curve in Friedman's experiments. We determined norepinephrine, epinephrine and dopamine in rat brain 30 min after the injection of tremorine and obtained an increase of acetylcholine by 30 per cent at the same time as the values for dopamine, adrenaline and noradrenaline were within normal limits. Analogous experiments with oxotremorine also gave 30 min values within normal limits for the catecholamines. The reasons for the discrepancies may possibly be due to the fact that we have used a different technique which allows the analysis of all the brain amines in the whole brain homogenate from the same rat. It may be mentioned that atropine either before or after tremorine decreases the rise in brain acetylcholine to normal without influencing the catecholamines.

The brain level of acetylcholine can be influenced by various kinds of drugs, e.g. cholinesterase inhibitors. There are, however, surprisingly few reports on the effect of cholinesterase inhibitors on the acetylcholine content in tissues in general and on the brain in particular. Previous results have been summarized by Holmstedt (1959) (see also Giarman and Pepeu, 1962). A rise of about 100 per cent can often be obtained. In our own series of experiments irreversible cholinesterase inhibitors of the organophosphorus type give rise to an increase in brain acetylcholine which can be counteracted by atropine given before but not after the cholinesterase inhibitor (Figs. 9 and 10). It may be mentioned here, that this refers to total brain acetylcholine. According to experiments by Michaelis *et al.* (1954) and DuBois *et al.* (1949) atropine given before or after DFP did not change the level of free acetylcholine despite the alleviation of the clinical symptoms.

Richter and Crossland were the first to report in 1949 that barbiturates and related compounds give rise to an increase in brain acetylcholine. The initial discoveries have been frequently corroborated. The effects of sedatives and hypnotics on acetylcholine cannot be attributed to hypoxia, hypercapnia or hypothermia. In contrast with pentobarbital, neither cyclopropane nor nitrous oxide nor xenon in anesthetic concentrations has any effect on brain acetylcholine. No rigid relationship exists between sedation and brain AcCh (Maynert and Buck, 1964).

It is dubious at the moment exactly what kind of acetylcholine is produced in excess by nembutal and related drugs or to what store or compartment in the brain it belongs. Our own experiments confirm those of others, that the rise achieved by the sedatives is insusceptible to atropine given either before or after (Giarman and Pepeu, 1964). One wonders in this context, how brain acetylcholine extracted after treatment with nembutal behaves if subjected to chromatography according to Whittaker and Dowe (1964). One practical lesson from the above-mentioned experiments is that the anesthetic should be carefully considered in animal experiments where acetylcholine is studied. Also Mitchell (1963) recording the spontaneous and evoked release of acetylcholine from the cerebral cortex found that chloralose interfered with the acetylcholine output.

Jenkner and Ward in 1953 reported that there appears to be neurons embedded in the diffuse matrix of the bulbar reticular formation, activation of which produces rhythmic peripheral motor responses analogous to the spontaneous static tremor seen in monkeys and man.

High-frequency electrical stimulation of the medial reticular formation of the brain stem in monkeys produced a rhythmically alternating peripheral movement analogous to tremor. Based on their experiments Jenkner and Ward advanced the hypothesis that the lesions in the rostral tegmentum or lower diencephalon which produce tremor such as that seen in Parkinsonism interrupt projections to these neurons in the medial reticular formation. These de-afferented cells may subsequently become hypersensitive to acetylcholine. Activation by locally liberated acetylcholine then would yield the observed tremor which can be counteracted by suitable anticholinergic drugs.

Mitchell (1963) has reported that stimulation of specific afferent pathways leading to the somatosensory cortex causes an increased release of AcCh from this area. The observations by Kanai and Szerb (1965) also demonstrate that cholinergic pathways are part of at least two distinct cortical systems. In a recent investigation it has been shown that an extensive neuronal system, probably corresponding to the ascending reticular activating system, could be demonstrated in the rat forebrain by a combined surgical and histochemical technique. (For complete references see Lewis and Shute, 1963; Lewis, Shute and Silver, 1964a and b; and Shute and Lewis, 1965.)

Microinjection of oxotremorine into the pontomesencephalic reticular formation in conscious cats causes atonia, cortical desynchronization and synchronization of hippocampal activity (Cho and Jenden, 1964). The effects are antagonized by atropine sulfate administered either locally into the same sites, or systemically.

These findings constitute additional evidence for the existence of cholinergic mechanisms in this area of the reticular formation as well as for the fact, that the receptor sites involved are muscarinic. It appears possible according to Cho and Jenden, that these neurons may be a major site of action of oxotremorine, not only in relation to the effects described, but also in the production of the generalized tremor and disturbances of muscular tone which ensue upon the systemic administration of the drug.

In our own experiments both oxotremorine and other tremor producing agents have been shown to increase the content of total acetylcholine in homogenates of whole brain. Acetylcholine in brain occurs in both "bound" and "free" form and various "compartments" have been described (Birks and MacIntosh, 1961; de Robertis, 1964; and Whittaker, 1965). Proposing a generalization on the mechanism of release of acetylcholine carries with it certain hazards because of the non-homogeneity of synapses, but it is not entirely inconceivable, that the cells in the brain structures mentioned above are particularly sensible to the local piling up of acetylcholine caused by oxotremorine. However, this hypothesis will have to be proved by an analysis of free and bound acetylcholine as well as an analysis of acetylcholine in various regions of the brain after the injection of tremorgenic agents. Such investigations are under way.

Previous work (Berry and Starz, 1956; Giarman and Pepeu, 1962, 1964) has established that doses of atropine and hyoscine and other anticholinergic drugs in the rat produce significant reductions in total level of acetylcholine in the whole brain, while causing either no gross change in behaviour, or occasionally a pattern of alternating excitation and depression (Wikler, 1952).

Bovet and Longo (1956) and Loeb, Magni and Rossi (1959) have implicated a suprathalamic site of action of atropine. In accordance with this the effect of hyoscine on brain AcCh levels is restricted primarily to the cerebral hemispheres and does not occur in subcortical regions (Giarman and Pepeu, 1964). More recent information shows, however, that total acetylcholine in olfactory lobes, parietal cortex and caudate nucleus is significantly reduced as early as 20 min after 25 mg/kg of scopolamine; it approaches the control values after 3 hr. No change takes place in the thalamus and cerebellar vermis. (Beani *et al.*, 1964). The scopolamine tissue concentration and the acetylcholine depletion seem to have different time courses.

Hyoscine has no influence on the synthesis of AcCh by acetone-extracted powder of rat brain as measured during the time of tremor activity (Giarman and Pepeu, 1964; Beani *et al.*, 1964).

Our results with atropine essentially confirm those of previous workers suggesting that these alkaloids saturate central receptors in a manner analogous to their known peripheral actions. In fact, 5 mg/kg of atropine produced a reduction in brain acetylcholine as great as that of 100 mg/kg (Fig. 6). From the time curve it will be seen that there is a very slow recovery, the values for 10 and 40 min being essentially the same.

One could suppose that these cholinergic blocking agents act to increase AcCh discharge in brain; and evidence that this is the case has, as already mentioned, been supplied by Mitchell (1963), who used the eserine-pool technique described by Oborin and MacIntosh (MacIntosh, 1963).

It has been shown in this investigation, that atropine given before a near to lethal dose of an organophosphorus cholinesterase inhibitor normalizes brain acetylcholine and this would lend support to the theory that the acetylcholine released by atropine is in the "free form" susceptible to cholinesterase. Once cholinesterase has been inhibited no decrease in brain acetylcholine can be seen.

Relatively few papers dealing with the detailed action of tremorgenic agents on the central nervous system are available.

The last pharmacological review on arecoline was written by Dixon (1924). It dealt exclusively with its cardiovascular effects. Before the discovery of acetylcholine this drug played an important role in providing a means of eliciting peripheral parasympathomimetic symptoms. Only in very high doses and after atropinization may nicotinic symptoms be elicited (Euler and Domeij, 1945). Since then it has also been used, particularly by workers in Russia and Czechoslovakia, to produce tremor for the screening of e.g. "central cholinolytics" (Anichkov, 1959; Anichkov and Poskalenko, 1959; Zejmal and Votava, 1961). It is one of the relatively few tertiary amines that has a high muscarinic activity (Burgen, 1964).

Arecoline is also an analgesic and the pain threshold is markedly increased already by doses which do not cause tremor (Herz, 1962). It also evokes an activating response of the EEG (Riehl *et al.*, 1962). The results of the findings were interpreted as evidence that the EEG activating response may be evoked by stimulating "muscarine" synapses and influences conditioned behaviour (Pfeiffer and Jenney, 1957).

Tremorine and oxotremorine are also known to exert effects both peripherally and centrally. Since the discovery that oxotremorine was the active metabolite of tremorine its peripheral pharmacology has been extensively investigated. According to Cho *et al.* (1962) this compound is a muscarinic agent equal in potency to acetylcholine, but completely lacking in nicotinic activity. It has been shown to exert its effects solely by stimulation of the post-ganglionic cholinergic receptors of muscarine-like type. It should be noted, however, that peripheral actions of tremorine and oxotremorine, that are presumably not muscarine-like have been reported, such as in the ganglia (DeGroat and Volle, 1963) and at the motor end-plate (Csillic, 1964). For tremorine a combined central and peripheral effect has been suggested with regard to the tremor (Chalmer and Yim, 1962; Bŏskovic *et al.*, 1964).

In a preliminary communication George, Haslett and Jenden (1962) have investigated the central actions of oxotremorine in various species. After premedication with methylatropine tremor, ataxia and spasticity were observed. There were, however, also several other effects which were very striking and indicative of central stimulation. Cats and monkeys became markedly excited and developed a rage-like state. Like arecoline, oxotremorine is also a powerful analgesic.

The general pharmacology of these tertiary amines thus presents obvious similarities, both with regard to peripheral and central action. In the past it has been difficult to reconcile these tertiary amines with the structure of the postsynaptic cholinergic receptor. Upon quaternization arecoline loses much of its muscarinic action (Magazanik *et al.*, 1965). As far back as 1938, Renshaw, Green and Ziff suggested that stable choline derivatives produce their cardiovascular effects by liberating AcCh (Renshaw *et al.*, 1938). This is a theory which has been advanced many times since then, but measurements of the amounts of acetylcholine have been lacking under the experimental conditions.

The finding in the present work that the tremorgenic agents give rise to an increased brain acetylcholine content raises the important question, whether these agents act exclusively through acetylcholine. In this context it is relevant to quote Cho, Haslett and Jenden (1962): "Oxotremorine is a muscarinic agent equal in potency to acetylcholine, but completely lacking in nicotinic activity. It has been shown to exert its effects solely by stimulation of the post-ganglionic cholinergic receptors. The extraordinary potency of oxotremorine as a muscarinic agent is somewhat surprising in view of its chemical dissimilarity to other agents in this class. Its activity is not far short of the most potent muscarinic agents known, yet it appears to violate several of the principles which have been found to describe structure-activity relationships in muscarinic compounds. The structural differences between oxotremorine and other potent muscarine agents are perhaps more striking than the similarities, and suggest that some alternative to the conventional 'lock and key' concept of drugs and receptors should be considered."

The present authors advance the hypothesis that this alternative may be the increased amount of acetylcholine which can be observed with the tremorgenic agents by both *in vivo* and *in vitro* experiments. In this case one would however expect the compounds to show all the characteristics of acetylcholine and not only to elicit muscarinic symptoms. It has been documented that arecoline has nicotinic action as well, although at high dosage levels, but then it takes also large amounts of acetylcholine after atropinization to elicit the nicotinic symptoms. As mentioned earlier, there also have been described for oxotremorine actions at the level of the ganglia, the neuromuscular junction and the spinal cord. Furthermore, with a supposedly indirectly acting compound the classical tool to elicit nicotine-like symptoms, atropinization, may not be as suitable as with a direct-acting one like acetylcholine.

Two explanations for the increase in total brain acetylcholine exist.

A. *Stimulated synthesis.* An increased biosynthesis appears however, very unlikely in view of the fact that the tremorgenic agents do not stimulate the final step of acetylcholine formation, cholinacetylase (Holmstedt *et al.*,

1965); neither could any effect on acetylcoenzyme A in brain be detected by a newly developed method (Schuberth *et al.*, 1965).

B. The tremorgenic agents could *mobilize acetylcholine from an inactive otherwise undetected precursor*; such precursors have previously been postulated to exist.

Since it has been discovered in this investigation, that the duration of tremor is accompanied by a corresponding increase in brain acetylcholine it would seem, as though those structures in the brain reticular formation that have been linked to peripheral tremor are particularly susceptible to the local formation of acetylcholine. This thought has been previously expressed in connection with lesions in that particular region.

An outstanding feature of oxotremorine and arecoline is the fact that the increase in brain acetylcholine they give rise to can be counteracted by atropine given either before or after. Very likely, the acetylcholine formed is so-called free acetylcholine, since in *in vitro* experiments it is necessary to use a cholinesterase inhibitor in order to pick up the increase.

Further *in vitro* experiments with oxotremorine in combination with other drugs and studies of free and bound acetylcholine should provide more information about the mode of action of this important group of compounds and in so doing will also shed more light on central cholinergic transmission.

SUMMARY

1. An apparatus for the recording of tremor in rats and other animals has been constructed. A filter eliminates all frequencies except those of the tremor. The equipment records tremor activities in one minute intervals or less, so that time curves as well as dose curves can be recorded.

2. The tremor produced by tremorine, oxotremorine and arecoline has been studied and simultaneous determinations of brain amines have been carried out. Tremorine tremor lasts 70–90 min, oxotremorine tremor around 60 min and arecoline tremor around 15 min.

3. Acetylcholine was the only brain amine studied which could be appreciably influenced by the tremorgenic agents during the period of tremor. In all three cases a rise in total brain acetylcholine was observed. The values returned to normal at about the same time as the tremor disappeared.

4. Atropine given before or after the tremorgenic agent abolishes the tremor. Methylatropine does not influence the tremor.

5. When oxotremorine (10^{-5} M) or arecoline (10^{-4} M) are added to an eserinized brain homogenate a considerable rise in brain acetylcholine occurs also *in vitro*. Atropine added *in vitro*, prior to the addition of oxotremorine to the brain homogenate, prevents the rise in acetylcholine which normally occurs with oxotremorine. The pretreatment of the animal with atropine

prevents the rise normally caused by the addition of oxotremorine to the homogenate.

6. Atropine alone has a long-lasting lowering effect on brain acetylcholine even at low doses.

7. The rise in brain acetylcholine caused by cholinesterase inhibitors can be counteracted only if atropine is given before the inhibitor.

8. The rise in brain acetylcholine caused by barbiturates is uninfluenced by atropine given either before or after.

9. When both oxotremorine and a cholinesterase inhibitor are given to an animal and atropine subsequently administered, atropine only eliminates the component of the increased brain acetylcholine due to oxotremorine.

10. Pilocarpine although a tertiary amine with cholinergic action does not give tremor and no rise in brain acetylcholine comparable to that of the tremorgenic agents can be observed.

11. The authors advance the following hypothesis: oxotremorine and arecoline do not act *per se* but by mobilizing acetylcholine. Since the tremorgenic agents are not cholinesterase inhibitors they could do this either by a stimulated synthesis or by the unmasking of acetylcholine from an otherwise undetected precursor. From the evidence presented in this paper the latter possibility appears to be the more likely one.

ACKNOWLEDGMENT

This work has been supported by the Swedish Medical Research Council, project 40X–199–01.

REFERENCES

Ahmed, A. and Taylor, N. R. W. (1959) The analysis of drug-induced tremor in mice, *Brit. J. Pharmacol.* **14**, 350.

Anichkov, S. V. (1959) Pharmacology of the central cholinergic synapses. XXI Congreso Internacional de Ciencias Fisiologicas, Buenos Aires, 9–15 de Agosto 1959, pp. 1–5.

Anichkov, S. V. and A. Poskalenko (1959) Einfluss der zentralen Cholinolytica auf die Aktivität der Nebennierenrinde, *Naunyn-Schmiedeberg's Arch. Exp. Path. Pharmak.* **236**, Heft 1.

Beani, L., C. Bianchi and P. Megazzini (1964) Regional changes of acetylcholine and choline acetylase activity in the guinea-pig's brain after scopolamine, *Experientia* (Basel) **20**, 677–8.

Berry, J. F. and E. Starz (1956) Acetylcholine and acetone in brain during acetaldehyde intoxication, *Quart. J. Stud. Alcohol.* **17**, 190–4.

Birks, R. and F. C. MacIntosh (1961) Acetylcholine metabolism of a sympathetic ganglion, *Canad. J. Biochem. Physiol.* **39**, 787–827.

Bonta, I. L. *et al.* (1960) A newly developed motility apparatus and its applicability in two pharmacological designs, *Arch. Int. Pharmacodyn.* **129**, 381–94.

Bošković, B., S. Hadžović, H. Krnjević and P. Stern (1964) Site of production of tremorine effects, *Arch. Int. pharmacodyn.* **149**, 521–6.

Bovet, D. and V. G. Longo (1956) Abstr. Rev. Twentieth Intern. Physiol. Congr., p. 806.

Burgen, A. S. V. (1964) The comparative activity of arecoline and arecoline N-metho salt, *J. Pharm. Pharmacol.* **16**, 638.

CHALMERS, R. K. and G. K. W. YIM (1962) Tremorine tremor in chronic spinal rats, *Proc. of the Soc. for Experim. Biol. and Med.* **109**, 202–5.

CHO, A. K., W. L. HASLETT and D. J. JENDEN (1961) The identification of an active metabolite of tremorine, *Biochem. and Biophys. Res. Communications* **5**, 276–9.

CHO, A. K., W. L. HASLETT and D. J. JENDEN (1962) The peripheral actions of oxotremorine, a metabolite of tremorine, *J. Pharmacol. Exp. Ther.* **138**, 249–57.

CHO, A. K. and D. J. JENDEN (1964) The mechanism of tremorine antagonism, by related compounds, *Internat. J. Neuropharmacol.* **3**, 27–37.

CSILLIK, B. (1964) Tremorine: Its peripheral action on striated muscle, *Science* **146**, 765.

CUTTING, W. *et al.* (1959) Evaluation of compounds affecting the psyche, *Arch. Int. Pharmacodyn.* **121**, 14.

DECSI, L., M. VÁRSZEGI and GY. MÉTHES (1961) Tolerance to tremorine, *Acta Physiol. Academiae Scientiarum Hungaricae* **18**, 353–6.

DEGROAT, W. C. and R. L. VOLLE (1963) Ganglionic actions of oxotremorine, *Life Sci.* **8**, 618–23.

DE ROBERTIS, E. (1964) *Histophysiology of Synapses and Neurosecretion.* Vol. 20 in International Series of Monographs on Pure and Applied Biology. Pergamon Press.

DIXON, W. E. (1927) Areca-Alkaloide. *Heffter-Heubner Handbuch der Experim. Pharmakologie.* Vol. II, pp. 813–16.

DIXON, W. E. and F. RANSOM (1924) Physostigmin. *Heffter-Heubner Handbuch der Experim. Pharmakologie.* Band II, 786–812.

DUBOIS, K. P., J. DOULL, P. R. SALERNO and J. M. COON (1949) Studies on the toxicity and mechanism of action of p-nitrophenyl diethyl thionophosphate (Parathion), *J. Pharmacol. Exp. Ther.* **95**, 79–91.

EULER, U. S. VON and B. DOMEIJ (1945) Nicotine-like actions of arecoline, *Acta Pharmacol.* (Kbh.) **1**, 263.

FRIEDMAN, A. H., R. J. AYLESWORTH and G. FRIEDMAN (1963) Tremorine: Its effect on amines of the central nervous system, *Sci.* **141**, 1188–90.

FRIEDMAN, A. H. and G. M. EVERETT (1964) Pharmacological aspects of Parkinsonism, *Advances in Pharmacology* **3**, 83–127.

GEORGE, R., W. L. HASLETT and D. J. JENDEN (1964) A cholinergic mechanism in the brainstem reticular formation: Induction of paradotical sleep, *Int. J. Neuropharmacol.* **3**, 541–52.

GEORGE, R., W. L. HASLETT and D. J. JENDEN (1962) The central action of a metabolite of tremorine, *Life Sci.* **8**, 361–3.

GIARMAN, N. J. and G. PEPEU (1962) Drug-induced changes in brain acetylcholine, *Brit. J. Pharmacol.* **19**, 226–34.

GIARMAN, N. J. and G. PEPEU (1964) The influence of centrally acting colinolytic drugs on brain acetylcholine levels, *Brit. J. Pharmacol.* **23**, 123–30.

DEGROAT, W. C. and R. L. VOLLE (1963) Ganglionic actions of oxotremorine, *Life Sci.* **8**, 618–23.

HERZ, A. (1962a) Über die Wirkung der optischen Isomere atropinartiger Substanzen auf das Zentralnervensystem, *Naunyn-Schmiedeberg's Arch. Exp. Path. u. Pharmak.* **242**, 508–21.

HERZ, A. (1962b) Wirkungen des Arecolins auf das Zentralnervensystem, *Naunyn-Schmiedeberg's Arch. Exp. Path u. Pharmak.* **242**, 414–29.

HOLMSTEDT, B. (1959) Pharmacology of organophosphorus cholinesterase inhibitors, *Pharm. Rev.* **11**, 567–688.

HOLMSTEDT, B., G. LUNDGREN, J. SCHUBERTH and A. SUNDWALL (1965) Tremorine and oxotremorine effects on acetylcholinesterase and choline acetylase from rat brain. Short communication, *Biochem. Pharmacol.* **14**, 189–91.

HOLMSTEDT, B., G. LUNDGREN and A. SUNDWALL (1963) Tremorine and atropine effects on brain acetylcholine, *Life Sci.* **10**, 731–6.

JENKNER, F. L. and A. WARD, JR. (1953) Bulbar reticular formation and tremor, *Arch. Neurol. Psych.* **70**, 489–502.

KANAI, T. and J. C. SZERB (1965) Mesencephalic reticular activating system and cortical acetylcholine output, *Nature* **205**, 80–1.

LESLIE, G. B. and D. R. MAXWELL (1964) Inhibition of tremorine by a group of non-anti-Parkinson phenothiazine derivatives, *Nature* **202**, 97–8.

LEWIS, P. R. and C. C. D. SHUTE (1963) Tracing presumed cholinergic fibres in rat forebrain, *Proc. Physiol. Soc.* 25 May 1963. *J. Physiol.* **168**, 33–5 P.

LEWIS, P. R., C. C. D. SHUTE and A. SILVER (1964a) *J. Physiol.* **171**, 65 P.

LEWIS, P. R., C. C. D. SHUTE and A. SILVER (1964b) Confirmation from choline acetylase analyses of a massive cholinergic innervation to the hippocampus, *J. Physiol.* **172**, 9–10P.

LOEB, C., F. MAGNI and G. F. ROSSI (1959) Studio degli effetti electroencefalografici dell'atropine nell gatto, *Boll. Soc. ital. Biol. Sper.* **35**, 1168–9.

MACINTOSH, F. C. (1963) Synthesis and storage of acetylcholine in nervous tissue, *Canad. J. Biochem. and Physiol.* **41**, 2555–71.

MACINTOSH, F. C. and W. L. M. PERRY (1950) Biological estimation of acetylcholine, *Meth. Med. Res.* **3**, 78–92.

MAGAZANIK, L. G., N. K. FRUENTOV, E. K. ROSHKOVA, R. S. RYBOLOVLEV and M. YA. MIKHELSON (1965) On the evolution of cholinoreceptive sites of locomotor muscle, *Pharmacology of Cholinergic and Adrenergic Transmission* (Edited by G. B. Koelle, W. W. Douglas and A. Carlsson). Pergamon Press, pp. 113–27.

MAYNERT, E. W. and E. G. BUCK (1964) Effects of CNS depressants on brain acetylcholine, *Federation Proceedings* **150**.

MICHAELIS, M., J. E. FINESINGER, F. DE BALBIAN VERSTER and R. W. ERICKSON (1954) The effect of the intravenous injection of DFP and atropine on the level of free acetylcholine in the cerebral cortex of the rabbit, *J. Pharmacol. Exp. Ther.* **111**, 169–75.

MITCHELL, J. F. (1963) The spontaneous and evoked release of acetylcholine from the cerebral cortex, *J. Physiol.* **165**, 98–226.

PEPEU, G. (1963) Effect of "Tremorine" and some anti-Parkinson's disease drugs on acetylcholine in the rat's brain, *Nature* **200**, 895–7.

PFEIFFER, C. C. and E. H. JENNEY (1957) The inhibition of the conditioned response and the counteraction of Schizophrenia by muscarinic stimulation of the brain, *Ann. N.Y. Acad. Sci.* **66**, 753.

RENSHAW, R. R., D. GREEN and M. ZIFF (1938) A basis for the acetylcholine action of choline derivatives, *J. Pharmacol.* **62** 430–48.

RICHTER, D. and J. CROSSLAND (1949) *Amer. J. Physiol.* **159**, 247.

RIEHL, J. L., J. PAUL-DAVID and K. R. UNNA (1962) Comparison of the effects of arecoline and muscarine on the central nervous system, *Int. J. Neuropharmacol.* **1**, 393–401.

SCHUBERTH, J., J. SOLLENBERG, A. SUNDWALL and B. SÖRBO (1965) Determination of acetylcoenzyme A in brain, *J. Neurochem.* **12**, 451–4.

SHUTE, C. C. D. and P. R. LEWIS (1960) The salivary centre in the rat, *J. Anat.* **94**, 59–73.

SHUTE, C. C. D. and P. R. LEWIS (1963) Cholinesterase-containing systems of the brain of the rat, *Nature* **199**, 1160–4.

SHUTE, C. C. D. and P. R. LEWIS (1965) Cholinesterase-containing pathways of the hind-brain: Afferent cerebellar and centrifugal cochlear fibres, *Nature* **205**, 242–5.

SJÖQVIST, F. and J. R. GILLETTE (1964) Potentiation of oxotremorine by "antitremorine" drugs, *The Pharmacologist* **6**, No. 2.

STUMPF, CH. (1962) Pharmakologische Methoden. Experimentelle Erzeugung von Tremor durch Pharmaka, *Handb. der Experim. Pharmakol. Heffter-Heubner. Ergänzungswerk*, Teil 7, 1–105.

TAKAHASHI, R. and M. H. APRISON (1964) Acetylcholine content of discrete areas of the brain obtained by a near-freezing method, *J. Neurochem* **11**, 887–98.

TRIOLO, A. J. *et al.* (1961) An analysis of tremorine-induced involuntary movement, *Arch. Int. Pharmacodyn.* **134**, 1–2.

UNGAR, G. and J. W. WITTEN (1963) *Fed. Proc.* **22**, 273.

VERNIER, V. G. and K. R. UNNA (1963) The central nervous system effects of drugs in monkeys with surgically-induced tremor, *Arch. Int. Pharmacodyn.* **144**, 1–2.

WELCH, R. M. and J. J. KOCSIS (1961) Activation of tremorine by liver, *Proc. Soc. Experim. Biol. & Med.* **107**, 731–4.

WHITTAKER, V. P. (1963) Identification of acetylcholine and related esters of biological origin, *Heffter-Heubner Handbuch der Experim. Pharmakologie* **15**, 1–39.

WHITTAKER, V. P. (1965) The application of subcellular fractionation techniques to the study of brain function, *Proc. Biophys. Mol. Biol.* **15**, 38–88.

WHITTAKER, V. P. and G. H. C. DOWE (1964) The identification of the acetylcholine-like substance in synaptosomes derived from guinea-pig brain as acetylcholine itself, *Int. J. Neuropharmacol.* **3**, 593–7.

WIKLER, A. (1952) Pharmacologic dissociation of behavior and EEG "sleep patterns" in dogs: morphine, N-allylnormorphine, and atropine, *Proc. Soc. Exp. Biol. Med.* **79**, 261–5.

ZEJMAL, E. V. and Z. VOTAVA (1961) Antagonism against the arecholine tremors of oxyphenonium and its tertiary analogue after intravenous and intracerebral administration in mice, *Experientia* (Basel) **17**, 471–5.

DISCUSSION

GREEN: In the studies on the activity of a lipid extract of sciatic nerve and brain to release acetylcholine from the guinea-pig ileum, we made observations that were best interpreted on the basis that atropine releases acetylcholine (Green, J. P. and Carlini, E. A., *J. Pharmacol. Exper. Therap.* **143**, 96–106, 1964). A search of the literature revealed experiments that could best be interpreted on the basis that atropine releases acetylcholine, and the relevant papers are quoted in this paper.

GENERAL DISCUSSION

UVNÄS: As Professor Katz mentioned, we should now try to ventilate our agreements and disagreements about the various aspects of the storage and release of biogenic amines which we have discussed during these three days. Professor Brown has promised to take over the chair and put some provocative questions before you. Sir Lindor, please.

BROWN: I'm not doing this particularly willingly and I doubt whether I can be really provocative, because I don't know enough about the subject. I only agreed to do this because I am, as you well know, used to public speaking, but most important, I know perfectly well that I look upon the subject rather from the outside and I have no real reputation to lose in it. I look at it as a physiologist and I constantly keep in mind the fact that the Creator put these peculiar structures between nerves and muscles in order that nerve impulses should be transmitted to effector cells. Professor Uvnäs looks rather sceptical about this because he feels that I'm leaving him out since nobody has yet, as far as I know, really been able to attribute a transmitter function to histamine however hard they have tried. That may well be wrong too, but the point is that all these granules and vesicles and structures are not just there for pharmacologists to put false transmitters in and for biochemists to extract enzymes out of. They are there for performing a function about which we know remarkably little. It is rather shame making, I speak now as a catecholaminologist, that we know so very little about the intimate details of the transmission process as compared with what, for instance, Katz can produce in terms of the acetylcholine transmission process, although I must say that nowadays we seem to know much more about the synthesis and disappearance and binding, indeed, of catecholamines than we know about the binding, synthesis, liberation and so on of acetylcholine. Of course Katz has the great advantage of working with a neuromuscular junction that you can see easily, and we, if I may speak fairly generally, work on neuro-smooth muscular junctions which are beastly.

I have got a few questions from Professor Uvnäs, but he is too modest to read them out himself. What I am going to do with these questions is to ask two people only, and I shall try to get a foreign visitor and one of our local hosts to speak. The first topic is about subcellular particles. I think a little more discussion about this would be useful and the difference between a non-synaptic granule and a synaptic granule has not really been brought out. It is perfectly clear that the granules that contain histamine are different from all the others. I think the people who should start talking about that are Dr. Robertson if he is here and Professor von Euler.

ROBERTSON: Now, if you want me to tell you what I think as a morphologist about the definition of a vesicle versus a granule, I think this is clearcut in my mind, and I hope I can make it clear to you. Let me start out with a granule inside the cytoplasm. I think the essential thing about a granule is that it is some kind of organized structure that lies within cytoplasm in the sense that it is on the cytoplasmic side of a unit membrane which may surround it. Another kind of granule that does not fit the definition I am building up, a granule which, in fact, can apparently be in a cell but which is outside the cell, is the kind that you see in the mast cell, in which you have, in essence, a kind of situation with the membrane invaginated. This is a granule and it is included in a membrane but that granule is effectively outside of the cell.

Now, a vesicle may appear to be a granule but it does not fit the definition given above. A vesicle is a membrane-limited structure inside the cytoplasm. There is nothing in it that we can see. The vesicle is homogenous. There is no formed element in it that we see in electron micrographs which will stand up, which can exist independently of its environment. This is not to say there isn't anything there but what I mean is that there is nothing that we can see with our standard technics. There is nothing there that we can isolate as a formed element. Now, vesicles can be of various sizes but there is one important point about the vesicle which I think is worth mentioning, and that is that they do not get smaller than a certain size which is something like, let's say 300 Å. At least I have never seen one much smaller than that. Perhaps they may be a little smaller but they do have a definite limit on the minimal size that they can reach. This probably has some kind of molecular explanation. The vesicle has practically no maximal size, except that I think that once it reaches a size of a few thousand Ångström units we stop calling it a vesicle and start speaking of a cistern. Particularly if it begins to take on folds, pretty soon it becomes a component of the endoplasmic reticulum, a cistern of the endoplasmic reticulum. I think there is a continuous gradation between a vesicle and a cisternal component of the endoplasmic reticulum. But they all have in common all the way this feature of the polarity of the unit membrane that bounds them and they all have in common the fact that there is a unit membrane bounding them. Well, I don't know if I covered that adequately?

BROWN: I think we could have general comments on what Dr. Robertson has said.

LAGUNOFF: I think perhaps the differentiation between a vesicle and a granule can be framed in terms that depend less on the technique used for preserving the structure and I will propose that the main difference between a vesicle and a granule is the character of the principal phase and I would suggest that in a granule the principal phase is solid and in a vesicle the principal phase is liquid. Now obviously, in granules there is water, I think, but the question is whether the solid phase is the principal phase with the water in among it or whether you have solids in a liquid phase as you do in a vesicle. This serves pretty well to distinguish it beyond the technical ability to preserve the substance in one or the other.

ROBERTSON: There is only one difficulty with that, that comes to mind, and that is that a lipid droplet can exist in a cell as a granule which is liquid.

LAGUNOFF: Maybe it has to be changed to aqueous.

ROBERTSON: All right, perhaps so.

WHITTAKER: The spherical appearance of vesicles calls for comment. Artificial lipid bilayers tend to assume a collapsed laminar form. We have to assume that there is something very special which prevents collapse. Dr. Robertson has stressed the homology between the vesicle membrane and the external membrane. But biochemical analysis of isolated membranes shows that there is considerable specialization, and room must be found for this in any theory of unit membranes.

ROBERTSON: Well, let me say immediately with reference to your last comment that I have said nothing, nothing at all, about the particular molecular species which make up this unit membrane pattern. This is obviously different, in different parts of the cell and different kinds of membranes. Well, what I have said is that there is a consistent pattern which runs through all membranes and that there is a chemical difference between the outside and the inside surface which is apparent. But the nature, the exact nature, of the chemical differences I have not specified. Now, with regard to keeping one of these things spherical, the fact is that in cells they are not spherical. If they get beyond a certain size then they do tend to collapse. In fact they are going into layers but when they reach this minimal limiting size, they are spherical, and I would like to suggest that there are special reasons for this.

UNVÄS: I think that Dr. Lagunoff hit the nail on the head. I wonder if there are any principal differences between granules and vesicles. Couldn't it be that all these structures have similar characteristics but it depends on how we treat them or mistreat them, whether they turn into vesicles or granules. If we think that when a histologist fixes the tissue and stains it the content may precipitate and then you have a granule. If you use another technique he may dissolve the content and then you have a vesicle.

CARLSSON: There is one question I should like to ask Dr. Robertson. Physiologically there is a tremendous difference between the different kinds of membranes that you have been drawing here. The granule or vesicle membrane, whatever you may call it, is freely permeable to lots of fairly large molecules, for example catecholamines, sucrose and several other molecules. The same kind of membrane at another place is not at all that permeable and I should like to ask you why this is so.

ROBERTSON: All right. I have been very careful in all I have said about the unit membrane structure, never to refer to the chemical species that make up this pattern. I am talking about a pattern of organization of lipids and proteins, a structural arrangement. If I may use an analogy with a railroad track, I am not attempting to distinguish between a regular track and a narrow gauge track. A train that can run on a regular-sized track, certainly gets stopped when it gets to the narrow gauge region and I think the same kind of analogy applies to membranes, the different regions of membranes are highly specific. But this is not to say that there is not a common structural pattern, which is present through them all.

VON EULER: I would like to start with the granules or vesicles from adrenergic nerves. Their size in our preparations is from 300 Å to about 2000 Å, so in this respect they fit very well within the group which was described by Dr. Robertson. Whether they have an internal structure or not is difficult to say with certainty but it sometimes appears that they have a substructure of some kind. In the adrenergic nerve endings the granules do not seem to be freely swimming particles. In many electron micrographs there is an indication of some connection between them and I would like to bring forth the concept that they are being structurally arranged in some kind of rows or chains inside the varicosities. This is also supported by electron-microscopic pictures of suspensions of granules from adrenergic nerves or from homogenized organs like the rabbit heart. They very often have bipolar attachments like handles or stalks. Sometimes the granules form short rows, and it seems plausible that this pattern also occurs *in vivo*.

On incubation of nerve granules in a medium containing the amine in a low concentration, the transmitter leaves the granules at a fixed rapid rate. On the other hand, if the amine concentration outside is sufficiently high, of the order of 10^{-4} M, the amine content in the granules is unchanged owing to a compensatory uptake. Depleted granules take up amine from the outside. After injection of amines *in vivo* previously depleted organs are refilled to about a normal content. In addition one finds the normal proportion between the amount of amine in the vesicles, which can be sedimented, and the high-speed supernatant, suggesting a rapid equilibration process.

AXELROD: We speak a great deal about the storage in the granules and vesicles and I think we are neglecting an important storage mechanism which is the extragranular storage of noradrenaline. I think all of us agree that there is a large amount of noradrenaline which is outside the granule. Now the question is, is the extragranular noradrenaline free or bound. This is a crucial question we have to answer. I think that most of the extragranular noradrenaline is bound, because if it were free it wouldn't be there, it would be metabolized very rapidly because of the large amounts of monoamine oxidase, which we know is present within the neuron. Now, if it's bound, how is it bound and how is it liberated? I would think that the extragranular noradrenaline will be functionally more important than the intragranular because it's more available. It does not have to come out of the vesicle into the cytoplasm and out of the nerve. We just have to think of experiments to show the nature of this extragranular binding of noradrenaline which we know is present in this particular store.

VON EULER: I would also believe that the greater part of the extragranular transmitter is bound in some way and also that it is of importance for the immediate nerve function. Dr. Udenfriend, when I talked to him in Bethesda last time, facetiously suggested the name garbage cans for the vesicles. The idea is that the transmitter is stored and can be supplied when the available portion, to use Dr. Trendelenburg's expression, is used up. It may be attached to a membrane or loosely bound in some other way and can be released when the membrane is depolarized. Then it would have a chance immediately to cross the synaptic cleft, which has been shown to occur in many cases and then exert its action. The refilling of this store could then be effected from the granules. Resynthesis is then presumably triggered as a result of that and must be very rapid since the total amount of transmitter is almost unchanged, even during prolonged stimulation.

CARLSSON: I quite agree that a kind of feeding from the granules to the membrane is a very likely phenomenon, but my guess would be that the amount of noradrenaline or transmitter that would stick to the membrane would be extremely small and actually I do not think this is what you were talking about earlier because you were talking about a large extragranular fraction and I would like to ask what's your evidence in favour of a large extragranular fraction.

VON EULER: The evidence is that if one prepares the vesicles for instance from nerves or from organs, one gets a bound and a soluble fraction. I admit that the figure we get of about 50 per cent bound transmitter in the rabbit heart is only an approximate value. It could be larger than that, but I don't think it could be smaller. During the treatment there will probably be destruction of some granules. On the other hand, if one uses different methods of preparing the granules the proportions are about the same for instance in the adrenal medulla. Whatever technique we use, we get about 90 per cent in the granules and 10 per cent in the other in the soluble fraction. I think the values indicate a difference in the proportion of the available or loosely bound pool and the more firmly bound pool between different organs.

DOUGLAS: I just wonder how severe Professor Carlsson's criticism is. You wouldn't need very much, surely, bound to the membrane to satisfy the transmission process; and didn't Dr. Whittaker yesterday show us a calculation that all the acetylcholine in a granule could be bound to the outside of the membrane or he could accommodate it all quite nicely if it was distributed around the membrane.

UVNÄS: I think Dr. Carlsson touched a very important problem, a principal problem. What is free and what is bound. When I gave my paper this morning, I told you that when you lyse mast cells in water and then suspend the granules, there will be about 75 per cent of the histamine retained in the granules and 25 per cent will occur in the supernatant. Now you can change this supernatant fraction; for instance if you increase the suspension volume, you will immediately increase the amount of free histamine. If you reduce the volume you can depress the amount of free histamine. If you change the temperature, there will be changes in the amount of free histamine. So I think, at least in a mast cell, you normally do not have any free histamine because the granules swim, if in anything, in a very, very small volume of fluid, but when you suspend the granules in water you will have some histamine free, due to the dissociation of the heparin–histamine binding. And I should not be surprised if it is the same thing with catecholamines you find free in your granular suspensions or in your extracts of tissues.

CARLSSON: We have heard that the evidence in favour of a large, extragranular pool in the nerves is the phenomenon that on homogenization a very large proportion is found in the supernatant. I would suggest that this is an artifact. If we look at the *in vivo* conditions we find that there is a very close correlation between the occurrence of granules and the occurrence of noradrenaline. The granules, as you know, are accumulated in the varicosities of a nerve ending and so is noradrenaline. Furthermore, if the nerve is cut or ligated, then there is a tremendous accumulation of noradrenaline proximally to the cut, and at the same time there is an accumulation of granules. They go hand in hand. Furthermore, we have the effect of reserpine. This again supports the view that there is a correlation between granules and noradrenaline. Yes, I think that is all.

BROWN: I think it is Dr. Kirshner next and then Dr. Axelrod and then Dr. Weiner and then Dr. Green.

VON EULER: May I put in just one word.

BROWN: Yes, please.

VON EULER: Dr. Carlsson seems to think that most of the pool appearing in the supernatant from the rabbit heart is an artifact. I don't agree because I think that one would then find similar proportions in the adrenal medulla. But there, with the same technique, we get a high proportion of granule-bound amines and a low proportion in the supernatant. I could add that in the rabbit kidney, we get up to 70 per cent of the amines in the sediment. Therefore I doubt that the portion, which appears as soluble, that is to say occurs in the supernatant, is an artifact. Precisely how big it is, I would not say.

CARLSSON: But there are reproducible artifacts, are there not?

VON EULER: Sure, but they would hardly be that different.

KIRSHNER: The electron micrographs suggest that the noradrenaline in the adrenergic nerves is not contained in the granules, but in a more or less limp tube, which may have interconnections. I would suggest that in the homogenization of the tissues it is very likely that these structures may be torn and what you are seeing is the results of breaking open the tubule or the larger vesicles, which may be interconnected, and the longer the string of connections the easier it is to break. In connection with the adrenal gland, I think it has already been shown by Hillarp that the amount that one does get in the supernatant and the pellet depends very much upon the degree of homogenization and also on the size of the bore of the pestle. These things can produce these large artifacts.

BROWN: Now the man with the short arm, Dr. Axelrod.

AXELROD: There are several bits of evidence for an extragranular pool. First, the noradrenaline in the superior cervical ganglia and uterus is present in the soluble fraction while that in the heart and most other tissues are present mainly in the granular fraction. Secondly, in the brain reserpine will deplete more noradrenaline from the soluble fraction than the particulate while amphetamine will deplete mainly from the particulate fraction.

VON EULER: Well, I can testify to this because Dr. Gordon Johnson in our department has done similar experiments with reserpine and also found this characteristic change in the proportion between the two.

CARLSSON: Concerning the superior cervical ganglia there is no question I think, that the amines are bound and not free, so if you get them free after homogenization, apparently something has happened. The evidence that the amines are bound in the superior cervical ganglia is very, very convincing. If you follow the recovery following reserpine or α-MMT, you will find that the fluorescence, the specific fluorescence, histochemically, will appear around the nucleus as a very distinct band and that will slowly broaden as the new granules are formed. There is a very nice correlation in that case. The evidence that Dr. Axelrod mentioned was apparently from an experiment on brain where he injected radioactive norepinephrine.

AXELROD: No, it was——

CARLSSON:—but the ratio—it must have been labelled amines.

AXELROD: No, we did both. Labelled and unlabelled. The same result.

CARLSSON: Yes, but the localization in this experiment I wonder about.

GREEN: I do not think any of the evidence that you said to show that it's all granular is incontestable. To talk about reserpine for example, a good parallel is what Dr. Uvnäs found with 48/80. It's true that 48/80 will release histamine from granules but its true mechanism of action is in much lower concentrations than that working on granules. It is working on the whole cell, so it must be acting primarily extragranularly. But what I really wanted to say is this. I suggested an experiment to Julius Axelrod yesterday, because I know that he has not much to do. But he suggested that you do it, Professor Brown. We assume that there is an extragranular store—just for fun. Supposing we would load the adrenergic nerve fibre with cold norepinephrine. It has been shown by Axelrod and others that the stuff will go into granules and that the infused stuff will go into the same granules as the endogenous stuff. Now, after we load those granules, we then begin to put in a lot of radioactive stuff. At one point with the cold norepinephrine we should saturate the granules and then all the radioactive or much of the radioactive stuff, at least a greater amount of radioactive stuff would stay in the clear cytoplasm. Now, this is for Professor Brown to do: stimulate the nerve and then measure in the effluent the specific activity of the noradrenaline. We might be able to tell whether the transmitter is in the vesicles, on the membrane, and is shooting right out of the nerve; this would be mostly cold stuff; or whether it must diffuse out. Maybe very little comes from the granules but the bulk of it comes from the clear cytoplasm.

VON EULER: If we empty this store, as we have done and then inject noradrenaline intravenously, we fill up the whole system in less than one minute.

GREEN: You mean when you are loading it with the radioactive stuff it goes to exchange with the granular stuff.

VON EULER: Exchange may be a slower process.

GREEN: You don't have to do the experiment, Dr. Brown!

BROWN: This is just a sort of experiment you suggest that other people should do. Thank you Dr. Green.

WEINER: Dr. Axelrod mentioned that the free stuff could not exist in the nerve endings because of the monoamine oxidase. I just wonder about this because I think we are facing the fact that the monoamine oxidase is present in the mitochondria and one wonders whether this is a means to protect free norepinephrine. Perhaps the mitochondria changes in some way during nerve stimulation to allow it to be deaminated or something like this. We know very little about kinetics of uptake and changes in mitochondrial permeability in varying conditions.

CARLSSON: The mitochondrial membrane, at least of the liver, is freely permeable to the catecholamines. I am turning now to *in vivo* data and I should like to remind you of the histochemical evidence. If you inject noradrenaline, it's taken up very quickly, not only by the varicosities but along the whole part of the axon and for that matter also by the cell body. Now, if you have the monoamine oxidase intact, the fluorescence disappears very rapidly so you hardly have time to see this. It vanishes almost immediately. If you give α-methyl-noradrenaline instead of adrenaline, it stays. You can easily detect it. If you give a MAO-inhibitor and then noradrenaline, it's taken up and stays for a fairly long time.

von Euler: I have understood that the fluorescence technique could hardly distinguish between granular and extragranular amines in the varicosity.

Carlsson: Well, if it is bound, if we can agree that it is bound, because it is protected from MAO, then we have to say that the binding sites, the hypothetical binding sites outside the granules have exactly the same distribution as the granules. That is a possibility, I do not believe in it.

Brown: Dr. Kopin, yes, it is your turn now.

Kopin: I don't think that it is necessary to say that vesicles are the only means of binding norepinephrine in the sympathetic neurone. It has been argued that the norepinephrine in the soluble fractions is artifactual. It is also possible that vesicles are artifacts. Now, I do not doubt that some vesicles are present in the sympathetic nerve, *in vivo*. They can be seen in the electron microscope and the evidence of Burnstock and Holman regarding quantal release is in favor of some sort of mechanism for bringing together a number of molecules and simultaneously releasing them. Vesicles, of course, make a nice explanation for this. On the other hand, it is perfectly possible that a major portion of the norepinephrine in tissues is bound to the surface of an intraneuronal membrane. During homogenization this membrane could break up and give rise to a soluble or small fractional particle which is not spun down. When a membrane is homogenized, vesicles can form. The size of the vesicles will, in large part, be dependent on the nature of the membrane, as Dr. Robertson has pointed out. His unit membrane has two sides with different structures. These could account for two different types of binding sites for norepinephrine. Because of the rise in specific activity of norepinephrine–^{3}H in the rat heart after release of the catecholamine by tyramine, one is forced to conclude that there are at least two pools of norepinephrine.

Norepinephrine has a number of chemical structures that could serve as binding sites. The catechol group and β-hydroxylamine groups can chelate, etc. The membrane with two sides differing in structure could provide two sites for binding, by different mechanisms.

The relative susceptibility of the two binding sites to depletion by tyramine could explain the rise in specific activity in norepinephrine–^{3}H since it provides two "pools". The fact that the soluble and particulate norepinephrine–^{3}H have equal specific activities before or after tyramine depletion may be readily explained, since they have a common source—the membrane formed into vesicles, and the vesicles or membrane broken up to form a soluble fraction. The formation of vesicles from a membrane at the nerve ending or after sectioning a nerve could be explained by a crumpling up of the membrane, by changes in ionic composition, and "stickiness' of the membrane, etc.

von Euler: I would like to put in one word. The kinds of substance which would be admirably suited for temporarily harbouring the catecholamines are phospholipids, since they can form a reversible kind of binding. Such phospholipids in a membrane are known to occur. I just wanted to bring that up.

Brown: We are getting on to binding, I think. Let's have Professor Uvnäs and Professor Carlsson. Is that the right team? Yes, and Dr. Green and Professor Carlsson will intervene when appropriate.

Uvnäs: Many of the theories which have been presented are to me as difficult to understand as the Indian rope-trick. That I think is due to my ignorance in biochemistry. But I would like to put forward a question. Is it possible that all amines are bound to some kind of polymer substances, for instance some kind of polysaccharides? As you heard, in the mast cells the histamine is bound to heparin.

Now, heparin is bound to protein and is therefore insoluble. But I think Carlsson and Hillarp once suggested that noradrenaline could be bound to ATP. What acetylcholine could be bound to, I do not know. In the mast cell, histamine is released together with heparin. If you stimulate the splanchnic nerves and get release of adrenaline, do you get any ATP out? Do you get any ATP out if you stimulate an adrenergic nerve? There was a paper a couple of years ago by Pamela Holton. She stimulated sensory nerves and produced antidromic vasodilatation, she found ATP coming out at the same time. I wonder if anybody has any experience about the release of possible binding substances at the same time as the release of the transmitter?

DOUGLAS: Well, what we felt was that it would be useful to have some facts rather than a welter of theories and it occurred to us too, that it would be very useful to know something about what substances other than the catecholamines came out of the chromaffin cell. As I discussed in my paper the other day, I think exocytosis is the most reasonable sort of explanation. I think we are forgetting that secretory cells show certain things in common, namely vesicles with membranes. In focussing on all the fancy connections between catecholamines and binding materials we are tending to get too specific here. We should think in terms of general mechanisms, and I suggest that such a general mechanism is exocytosis.

CARLSSON: With regard to this story about the light granules. That such granules exist I think we agree upon. But you said that Hillarp meant that it was from this pool that release occurred on nerve stimulation. Hillarp and I found that if you stimulate the adrenal medulla through the nerves, catecholamines and ATP disappear in equivalent proportions.

DOUGLAS: In his paper Hillarp (*Acta physiol. scand.* **50**, 8, 1960) describes three pools, including the "light" granules, any of which might be the source of amines drawn upon when the cell is called upon to secrete.

CARLSSON: Well, that of course is a possibility but the facts are that ATP and catecholamine disappear along with each other.

DOUGLAS: We are showing that the ATP that disappears from the cell can be recovered, as metabolites in the effluent along with equivalent amounts of catecholamines. This is the new, positive, finding which we have. I am perfectly well aware of the evidence that if you look at the cell which has been stimulated for hours with insulin you do find that it contains less catecholamine and less nucleotide phosphate. I am presently talking of phenomena which can be correlated within seconds. We now know that the catecholamine appearing within, say, 10 sec escapes with equivalent amounts of nucleotide phosphates.

CARLSSON: Then I think there is a nice agreement between these different sets of data so there is no reason to argue. Furthermore, Hillarp found that the protein, the specific intragranular protein, remained in the granules after stimulation.

UVNÄS: I was very happy to receive such a quick answer on my question. If I should assume anything, I should like to agree with this suggestion that the release goes on as exocytosis, because that is in fact just the same thing as happens in the mast cell. There you have the granule content, the whole content being secreted. The only thing is that in mast cells, the carrying structure is insoluble. Here I think it is soluble. Now, you say that one of the arguments against this hypothesis was that you could not find the protein. How should you be able to identify the protein coming out in the blood?

CARLSSON: The protein is still left with the granules within the cell you see, so the exocytosis I think is excluded.

DOUGLAS: That would be a reason for excluding it, but what I argued was that our results are consistent with exocytosis. The difficulty is that if one measures the protein which is coming out—I am not using blood, I am using Locke's solution—the amount is far below that one would expect if the whole granule appeared in the vascular tree—in the blood vessels. Now, this doesn't mean that it is not coming out of the chromaffin cell, because the chromaffin cell, unlike the mast cell, is not swimming in a test tube, it is underneath the endothelium with a basement membrane. One could argue that the stuff is shot out of the cell but the protein doesn't get across the basement membrane, although clearly the nucleotide and the catecholamine do. But the notion of Cramer and others that the whole granule comes out into the blood vessels is not consistent with our data,.

ROBERTSON: I think that as one of the few morphologists here, I really should make one comment. There is another alternative for where the things could get out. Let me say that we have not seen the particular relationship where you have direct evidence of true fusion of membranes. I thought I saw it in Dr. Bloom's micrographs of mast cells. This is the first suggestion of this that I have ever seen, I would like to talk to him about that and it may be that he has got it. We have been looking for it for a long time, but now, if that does not occur, then one is forced to think of a neck which is retained during the life of the cell and by some means or other it is sealed off and opens up at the right time. This might be the mechanism whereby stuff gets out. This is one alternative possibility. But things do not go out, as far as we know, by a mechanism of fusion and then simple migration all the way through, taking the membrane with it. This would reverse the polarity of the membrane in the extruded membrane-limited granule or vesicle and we would expect to see the extruded vesicles.

DOUGLAS: When Dr. Robertson says he sees this, he is operating on general principles. What I mean is that you were not talking of chromaffin cells, you were talking perhaps about the pancreas, extrusion of proteins and so on. Now Dr. de Robertis who has published some micrographs in this field, shows what looks to me very much like granules attached to the membrane.

ROBERTSON: Well, I just have to say about this that I have examined Dr. de Robertis' micrographs on this subject very, very carefully. His diagram shows this very clearly, but his micrographs do not quite convince me.

DOUGLAS: I am glad to have the opinion of a physiologist corroborated by a morphologist.

KIRSHNER: I would just like to ask a question. All the pictures I have seen of electron micrographs of the adrenal medulla and comments that the authors have made indicate that the granules or the vesicles have a single membrane. I just wonder whether anyone has seen a double membrane?

DOUGLAS: I thought that all membranes are double membranes.

BLOOM: I thought I would answer Dr. Robertson's question directly: that with regard to mast cells in the secretion of their granules. We have seen in a number of pictures the same phenomenon repeated and that is that at the periphery of the cell, at the cell membrane, the mast granule is surrounded by its own perigranular membrane, which in these adhered granules that I showed the other day, is separate from the matrix of the granule. In the later stages you will find the phenomenon where the mast granule has turned into a vacuole; bulges out from the periphery of the cell. I too would like to say that these bulges from the periphery may very well be granules that are judged to be entirely free and just adhering to the membrane after being released, because you can find cells with numerous bulges getting the mulberry appearance that Professor Unväs spoke about. Now, the interesting thing is that we have numerous pictures showing that when this bulge has grown sufficiently large the membrane actually disrupts and you can see clearly this phenomenon. Then the granule contents, what was in the granule, and what is also surrounding it, in the way of dissolved substances, is poured out into the surrounding medium.

ROBERTSON: But the crucial point is the intermediate stage. Do you see the intermediate stage of membrane fusion?

BLOOM: Membrane fusion, put it this way: I have looked at numerous pictures where I have not been able to see any gap between the two membranes. I have not been able to see that there is more than one membrane and it does appear a little thicker than the unit membrane.

ROBERTSON: This is such a critical point. What I think one must add here is this. First of all we must see the unit membrane structure resolved, preserved, let's say, in the micrographs. Then we must see the unit membrane surrounding the granule. Then we want to see the intermediate stage in which two distinct unit membranes melt into one.

BROWN: If you tell him too carefully what he has got to see, he will see it.

ROBERTSON: All these membrane structures must be resolved and you must see this crucial junction point.

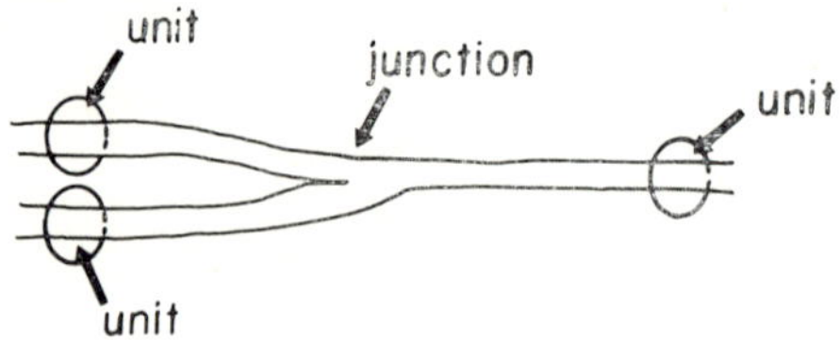

This is what I have been looking for. I thought I had seen it in one of your micrographs but not quite. Perhaps this is because they were fixed with glutaraldehyde and osmium tetroxide methods which routinely don't display the unit membrane structures. You would not be likely to see it in that kind of preparation.

BROWN: One more question. Dr. Stjärne.

STJÄRNE: I would first like to emphasize that a 4 to 1 catecholamine to ATP molar ratio is not obligatory for amine storage. This is well known from the pheochromocytoma, where you can have a molar ratio which is very different from the 4:1 ratio, since the pheochromocytoma is relatively ATP poor. You can have molar ratios of 20 to 1 or above that. Secondly, I want to point out that the catecholamine to ATP ratio of the isolated nerve granules under resting conditions so to speak, is not very far away from 4 to 1 as is also the case in the adrenal medullary granules.

UVNÄS: Would you like to say one word more. It must be a very short one.

DOUGLAS: I think it is very reasonable that material should be coming from vesicles. You see, one of the things that we have not discussed is the question of quantal release. Professor Katz touched on that a little. If the material is packaged you have a very nice explanation for quantal release at the cholinergic nerve ending. Now, the electrophysiological evidence of the sympathetic junction is not all that good as yet, but it looks as though quantal release is occurring there too. Now, release from granules would provide an equally neat explanation there and this is something which I think those who believe in release from non-granular sources should have in mind and should be able to explain.

UVNÄS: Thank you Dr. Douglas. I am afraid we have reached the end of this Symposium and on behalf of the organizing committee I would like to thank you all who have so actively and intelligently taken part in our scientific discussions.

LIST OF PARTICIPANTS

ADAMS-RAY, JACK, M.D., Professor, Dept. of Surgery, Karolinska Sjukhuset, Stockholm 60, Sweden

AHLBORG, ULF, M.D., Dept. of Toxicology, Lab. 60, Karolinska Institutet, Stockholm, 60, Sweden

ALMGREN, OLLE, M.B., Dept. of Pharmacology, University of Göteborg, Göteborg, Sweden

ANDÉN, NILS-ERIK, M.D., Docent, Dept. of Pharmacology, University of Göteborg, Göteborg, Sweden

ANDERSSON, BENGT, D.V.M., Professor, Dept. of Physiology, Kungl. Veterinärhögskolan, Stockholm 50, Sweden

ÅSTRÖM, ARNE, M.D., Lecturer, Dept. of Physiology, Karolinska Institutet, Stockholm 60, Sweden

AXELROD, JULIUS, PH.D., Chief, Section on Pharmacology, Laboratory of Clinical Science, National Institute of Mental Health, NIH, Bethesda 14, Md., U.S.A.

BERGQVIST, ULLA, Mrs., Dept. of Pharmacology, Karolinska Institutet, Stockholm 60, Sweden

BLOOM, GUNNAR, M.D., Docent, Dept. of Cell Research and Genetics, Karolinska Institutet, Stockholm 60, Sweden

BORÉUS, LARS OLOF, M.D., Docent, Dept. of Pediatrics, Karolinska Sjukhuset, Stockholm 60, Sweden

BRODIE, BERNARD B., PH.D., Chief, Laboratory of Chemical Pharmacology, National Heart Institute, NIH, Bethesda 14, Md., U.S.A.

BROWN, LINDOR, Sir, Professor, Laboratory of Physiology, University of Oxford, Oxford, England

CARLSSON, ARVID, M.D., Professor, Dept. of Pharmacology, University of Göteborg, Göteborg, Sweden

CEDERGREN, EBBA, M.D., Professor, Dept. of Anatomy, University of Umeå, Umeå, Sweden

CHAKRAVARTY, NIRMAL, M.D., Docent, Dept. of Pharmacology, University of Umeå, Umeå, Sweden

CORRODI, HANS, M.D., Docent, AB Hässle, Göteborg, Sweden

COSTA, E., M.D., National Heart Institute, NIH, Bethesda 14, Md., U.S.A.

DAHLSTRÖM, ANNICA, M.B., Dept. of Histology, Karolinska Institutet, Stockholm 60, Sweden

DOUGLAS, WILLIAM W., M.D., Dept. of Pharmacology, Albert Einstein College of Medicine, Yeshiva University, New York 61, N.Y., U.S.A.

ELMQVIST, DAN, M.D., Dept. of Pharmacology, University of Lund, Lund, Sweden

ERÄNKÖ, OLAVI, M.D., Professor, Dept. of Anatomy, University of Helsinki, Helsinki, Finland

VON EULER, U. S., M.D., Professor, Dept. of Physiology, Karolinska Institutet, Stockholm 60, Sweden

FALCK, BENGT, M.D., Reader, Dept. of Histology, University of Lund, Lund, Sweden

FLÜCKIGER, E., Professor, Sandoz A.G., Basle, Switzerland

FOLKOW, BJÖRN, M.D., Professor, Dept. of Physiology, University of Göteborg, Göteborg, Sweden

FREDHOLM, BO, M.D., Dept. of Pharmacology, Karolinska Institutet, Stockholm 60, Sweden

FUXE, KJELL, M.D., Docent, Dept. of Histology, Karolinska Institutet, Stockholm 60, Sweden

GARATTINI, SILVIO, M.D., Professor, Istituto di Richerche Farmacologiche "Mario Negri", Milan, Italy

GIACOBINI, EZIO, M.D., Docent, Dept. of Pharmacology, Karolinska Institutet, Stockholm 60, Sweden

GJESSING, LEIV, M.D., Dikemark Sykehus, Asker, Norway

GOTH, ANDRES, M.D., Professor, Chairman, University of Texas, Southwestern Medical School, Dallas 35, Texas, U.S.A.

GOTTFRIES, CARL GERHARD, M.D., Östra Sjukhuset, Malmö, Sweden

GRAHAM, LINDSAY A., M.D., Dept. of Physiology, Karolinska Institutet, Stockholm 60, Sweden

GREEN, JACK PETER, M.D., PH.D., Dept. of Pharmacology, Yale University, New Haven, Conn., U.S.A.

GUNNE, LARS-MAGNUS, M.D., Docent, Dept. of Physiology, Karolinska Institutet, Stockholm 60, Sweden

HAAVALDSEN, R., Mag. Scient., Dept. of Biochemistry, Institute of Occupational Health, Oslo 3, Norway

HABERLAND, G., M.D., Docent, Farbenfabriken Bayer A.G., Bayerwerk, Leverkusen, West Germany

HAEGERMARK, ÖSTEN, M.D., Dept. of Pharmacology, Karolinska Institutet, Stockholm 60, Sweden

HAMBERGER, BERTIL, M.B., Dept. of Histology, Karolinska Institutet, Stockholm 60, Sweden

HANSSON, LENNART, M.D., Dept. of Pharmacology, University of Göteborg, Göteborg, Sweden

HENNING, MATTS, M.B., Dept. of Pharmacology, University of Göteborg, Göteborg, Sweden

HILLARP, N.-Å., M.D., Professor, Dept. of Histology, Karolinska Institutet, Stockholm 60, Sweden

HOLMSTEDT, BO., M.D., Professor, Dept. of Toxicology, Lab. 60, Karolinska Institutet, Stockholm 60, Sweden

HÄGGENDAL, JAN, M.D., Docent, Dept. of Pharmacology, University of Göteborg, Göteborg, Sweden

HÖGBERG, BERTIL, M.D., Research Director, AB Leo, Hälsingborg, Sweden

JOHNSON, DAVID, M.D., Dept. of Physiology, Karolinska Institutet, Stockholm 60, Sweden

KALOW, WERNER, M.D., Professor, Boehringer Sohn, 6507 Ingelheim am Rhein, West Germany

KATZ, B., Professor, Dept. of Biophysics, University College London, London W.C.1, England

KIRSHNER, NORMAN, M.D., Dept. of Biochemistry, Duke University, Durham, North Carolina, U.S.A.

KOPIN, IRWIN J., M.D., Chief, Section on Medicine, Laboratory of Clinical Science, National Institute of Mental Health, NIH, Bethesda 14, Md., U.S.A.

LAGUNOFF, DAVID, M.D., Assistant Professor, Dept. of Pathology, University of Washington, Seattle 5, Wash., U.S.A.

LILJESTRAND, GÖRAN, M.D., Professor, Dept. of Pharmacology, Karolinska Institutet, Stockholm 60, Sweden

LINDGREN, PERCY, M.D., Reader, Dept. of Pharmacology, Karolinska Institutet, Stockholm 60, Sweden

LINDQVIST, MARGIT, Miss, Dept. of Pharmacology, University of Göteborg, Göteborg, Sweden

LISHAJKO, FJODOR, Mr., Dept. of Physiology, Karolinska Institutet, Stockholm 60, Sweden

LUNDBORG, PER, M.B., Dept. of Pharmacology, University of Göteborg, Göteborg, Sweden

MALMFORS, TORBJÖRN, M.D., Docent, Dept. of Histology, Karolinska Institutet, Stockholm 60, Sweden

MARCHISIO, PIER CARLO, M.D., Dept. of Pharmacology, Karolinska Institutet, Stockholm 60, Sweden

MASUOKA, DAVID, M.D., Dept. of Histology, Karolinska Institutet, Stockholm 60, Sweden

MEYERSON, BENGT, M.D., Docent, Akademiska Sjukhuset, Uppsala, Sweden

MITCHELL, J. F., M.D., Dept. of Pharmacology, University of Cambridge, Cambridge, England

MONGAR, J. L., M.D., Reader, Dept. of Pharmacology, University College London, London W.C.1, England

MUSCHOLL, E., M.D., Pharmakologisches Institut der Universität Mainz, Mainz, West Germany

NAESS, KNUT, M.D., Docent, Dept. of Pharm. Therapy, University of Oslo, Oslo, Norway

NGAI, S. H., M.D., Dept. of Pharmacology, Karolinska Institutet, Stockholm 60, Sweden

NORBERG, KARL-AXEL, M.D., Docent, Dept. of Histology, Karolinska Institutet, Stockholm 60, Sweden

NOVOTNY, JOSEF, M.D., Dept. of Pharmacology, Karolinska Institutet, Stockholm 60, Sweden

OBIANWU, HOPÉ, M.D., Dept. of Pharmacology, University of Göteborg, Göteborg, Sweden

OLBE, LARS, M.D., Docent, Dept. of Surgery, Sahlgrenska Sjukhuset, Göteborg, Sweden

OLIVERIO, ALBERTO, M.D., Dept. of Physiology, Karolinska Institutet, Stockholm 60, Sweden

PAUL, KARL-GUSTAV, M.D., Professor, Dept. of Medical Chemistry, University of Umeå, Umeå, Sweden

PEKKARINEN, AIMO, M.D., Professor, Dept. of Pharmacology, Turku University, Turku, Finland

PLETSCHER, ALFRED, M.D., PH.D., Hoffman-La Roche & Co., Basle, Switzerland

POULIOT, MARCEL, M.D., Dept. of Physiology, Karolinska Institutet, Stockholm 60, Sweden

RAŠKOVÁ, HELENA, Professor, Institute of Pharmacology, Charles University, Prague 2, Czechoslovakia

RIDLEY, PETER T., PH.D., Dept. of Pharmacology, Karolinska Institutet, Stockholm 60, Sweden

ROBERTSON, J. DAVID, M.D., PH.D., Dept. of Neuropathology, Harvard Medical School, Boston, Mass., U.S.A.

ROOS, BJÖRN-ERIK, M.D., Docent, Dept. of Pharmacology, University of Göteborg, Göteborg, Sweden

ROSELL, SUNE, M.D., Lecturer, Dept. of Pharmacology, Karolinska Institutet, Stockholm 60, Sweden

ROSS, SVANTE, PH.D., AB Astra, Södertälje, Sweden

RYALL, RON, M.D., Dept. of Physiology, University of Göteborg, Göteborg, Sweden

SCHOU, JENS, M.D., Professor, Dept. of Pharmacology, University of Copenhagen, Copenhagen ø, Denmark

SEDVALL, GÖRAN, M.D., Docent, Dept. of Pharmacology, Karolinska Institutet, Stockholm 60, Sweden

SJÖBERG, BERNT, PH.D., Docent, AB Astra, Södertälje, Sweden

SHORE, PARKHURST, A., PH.D., Professor, Dept. of Pharmacology, University of Texas, Dallas 35, Texas, U.S.A.

SOMERVILLE, ANDREW R., M.D., Imperial Chemical Industries Ltd., Wilmslow, Cheshire, England

STJÄRNE, LENNART, M.D., Docent, Dept. of Physiology, Karolinska Institutet, Stockholm 60, Sweden

STONE, FREDERICK L., PH.D., Director, National Institute of General Medical Sciences, Division of Research Facilities and Resources, North Bethesda Office Center, NIH, Bethesda 14, Md., U.S.A.

SUNDWALL, ANDERS, M.D., Docent, Research Institute of National Defence, Fack, Stockholm 80, Sweden

THAM, RICHARD, M.D., Dept. of Toxicology, Lab. 60, Karolinska Institutet, Stockholm 60, Sweden

THESLEFF, STEPHEN, M.D., Professor, Dept. of Pharmacology, University of Lund, Lund, Sweden

THON, INGA-LISA, Mrs., Dept. of Pharmacology, Karolinska Institutet, Stockholm 60, Sweden

TRENDELENBURG, ULLRICH, M.D., Dept. of Pharmacology, Harvard Medical School, Boston, Mass., U.S.A.

UDENFRIEND, SIDNEY, M.D., Chief, Laboratory of Clinical Biochemistry, National Heart Institute, NIH, Bethesda 14, Md., U.S.A.

UTLEY, JOHN, M.D., Dept. of Pharmacology, University of Göteborg, Göteborg, Sweden

UVNÄS, BÖRJE, M.D., Professor, Dept. of Pharmacology, Karolinska Institutet, Stockholm 60, Sweden

WALDECK, BERTIL, Mr., Dept. of Pharmacology, University of Göteborg, Göteborg, Sweden

WANG, HSUEH-HWA, M.D., Dept. of Physiology, Karolinska Institutet, Stockholm 60, Sweden

WEINER, NORMAN, M.D., Dept. of Pharmacology, Harvard Medical School, Boston, Mass., U.S.A.

WERDINIUS, BENGT, M.D., Dept. of Pharmacology, University of Göteborg, Göteborg, Sweden

WESTERHOLM, BARBRO, M.D., Docent, Dept. of Pharmacology, Karolinska Institutet, Stockholm 60, Sweden

WHITTAKER, V. P., M.D., Biochemistry Dept., Agricultural Research Council, Institute of Animal Physiology, Babraham, Cambridge, England

WINQVIST, GÖSTA, D.V.M., Reader, Kungl. Veterinärhögskolan, Stockholm 50, Sweden

WISTRAND, PER, M.D., Docent, AB Astra, Södertälje, Sweden

WOLD, JENS, PH.D., Dept. of Pharmacology, Karolinska Institutet, Stockholm 60, Sweden

ZOTTERMAN, YNGVE, M.D., Professor, Kungl. Veterinärhögskolan, Stockholm 50, Sweden